T0178129

Lecture Notes in Business Information Processing 353

Series Editors

More information about this series at http://www.springer.com/series/7911

Witold Abramowicz · Rafael Corchuelo (Eds.)

Business Information Systems

22nd International Conference, BIS 2019
Seville, Spain, June 26–28, 2019
Proceedings, Part I

Editors
Witold Abramowicz
Poznań University of Economics and Business
Poznań, Poland

Rafael Corchuelo
ETSI Informática
University of Seville
Seville, Spain

ISSN 1865-1348 ISSN 1865-1356 (electronic)
Lecture Notes in Business Information Processing
ISBN 978-3-030-20484-6 ISBN 978-3-030-20485-3 (eBook)
https://doi.org/10.1007/978-3-030-20485-3

This Springer imprint is published by the registered company Springer Nature Switzerland AG
The registered company address is: Gewerbestrasse 11, 6330 Cham, Switzerland

Preface

During the 22 years of the International Conference on Business Information Systems, it has grown to be a well-renowned event of the scientific community. Every year the conference joins international researchers for scientific discussions on modeling, development, implementation, and application of business information systems based on innovative ideas and computational intelligence methods. The 22nd edition of the BIS conference was jointly organized by the University of Seville, Spain, and Poznań University of Economics and Business, Department of Information Systems, Poland, and was held in Seville, Spain.

The exponential increase in the amount of data that is generated every day and an ever-growing interest in exploiting this data in an intelligent way has led to a situation in which companies need to use big data solutions in a smart way. It is no longer sufficient to focus solely on data storage and data analysis. A more interdisciplinary approach allowing one to extract valuable knowledge from data is required for companies to make profits, to be more competitive, and to survive in the even more dynamic and fast-changing environment. Therefore, the concept of data science has emerged and gained the attention of scientists and business analysts alike.

Data science is the profession of the present and the future, as it seeks to provide meaningful information from processing, analyzing, and interpreting vast amounts of complex and heterogeneous data. It combines different fields of work, such as mathematics, statistics, economics, and information systems and uses various scientific and practical methods, tools, and systems. The key objective is to extract valuable information and infer knowledge from data that then may be used for multiple purposes, starting from decision-making, through product development, up to trend analysis and forecasting. The extracted knowledge allows also for a better understanding of actual phenomena and can be applied to improve business processes. Therefore, enterprises in different domains want to benefit from data science, which entails technological, industrial, and economic advances for our entire society. Following this trend, the focus of the BIS conference has also migrated toward data science.

The BIS 2019 conference fostered the multidisciplinary discussion about data science from both scientific and practical sides, and its impact on current enterprises. Thus, the theme of BIS 2019 was "Data Science for Business Information Systems." Our goal was to inspire researchers to share theoretical and practical knowledge of the different aspects related to data science, and to help them transform their ideas into the innovations of tomorrow.

The first part of the BIS 2018 proceedings is dedicated to Big Data, Data Science, and Artificial Intelligence. This is followed by other research directions that were discussed during the conference, including ICT Project Management, Smart Infrastructures, and Social Media and Web-based Systems. Finally, the proceedings end with Applications, Evaluations, and Experiences of the newest research trends in various domains.

The Program Committee of BIS 2018 consisted of 78 members who carefully evaluated all the submitted papers. Based on their extensive reviews, 67 papers were selected.

We would like to thank everyone who helped to build an active community around the BIS conference. First of all, we want to express our appreciation to the reviewers for taking the time and effort to provide insightful comments. We wish to thank all the keynote speakers who delivered enlightening and interesting speeches. Last but not least, we would like to thank all the authors who submitted their papers as well as all the participants of BIS 2019.

June 2019 Witold Abramowicz

Organization

BIS 2019 was organized by the University of Seville and Poznań University of Economics and Business, Department of Information Systems.

Program Committee

Witold Abramowicz (Co-chair)	Poznań University of Economics and Business, Poland
Rafael Corchuelo (Co-chair)	University of Seville, Spain
Rainer Alt	Leipzig University, Germany
Dimitris Apostolou	University of Piraeus, Greece
Timothy Arndt	Cleveland State University, USA
Sören Auer	TIB Leibniz Information Center Science and Technology and University of Hannover, Germany
Eduard Babkin	LITIS Laboratory, INSA Rouen; TAPRADESS Laboratory, State University – Higher School of Economics (Nizhny Novgorod), Russia
Morad Benyoucef	University of Ottawa, Canada
Matthias Book	University of Iceland, Iceland
Dominik Bork	University of Vienna, Austria
Alfonso Briones	BISITE Research Group, Spain
François Charoy	Université de Lorraine – LORIA – Inria, France
Juan Manuel Corchado Rodríguez	University of Salamanca, Spain
Beata Czarnacka-Chrobot	Warsaw School of Economics, Poland
Christophe Debruyne	Trinity College Dublin, Ireland
Renata De Paris	Pontifical Catholic University of Rio Grande do Sul, Brazil
Yuemin Ding	Tianjin University of Technology, China
Suzanne Embury	The University of Manchester, UK
Jose Emilio Labra Gayo	Universidad de Oviedo, Spain
Werner Esswein	Technische Universität Dresden, Germany
Charahzed Labba	Université de Lorraine, France
Agata Filipowska	Poznań University of Economics and Business, Poland
Ugo Fiore	Federico II University, Italy
Adrian Florea	'Lucian Blaga' University of Sibiu, Romania
Johann-Christoph Freytag	Humboldt Universität Berlin, Germany
Naoki Fukuta	Shizuoka University, Japan
Claudio Geyer	UFRGS, Brazil
Jaap Gordijn	Vrije Universiteit Amsterdam, The Netherlands

Volker Gruhn	Universität Duisburg-Essen, Germany
Francesco Guerra	Università di Modena e Reggio Emilia, Italy
Hele-Mai Haav	Institute of Cybernetics at Tallinn University of Technology, Estonia
Axel Hahn	University of Oldenburg, Germany
Constantin Houy	Institute for Information Systems at DFKI (IWi), Germany
Christian Huemer	Vienna University of Technology, Austria
Björn Johansson	Lund University, Sweden
Monika Kaczmarek	University of Duisburg Essen, Germany
Pawel Kalczynski	California State University, Fullerton, USA
Kalinka Kaloyanova	University of Sofia, Bulgaria
Marite Kirikova	Riga Technical University, Latvia
Gary Klein	University of Colorado Boulder, USA
Ralf Klischewski	German University in Cairo, Germany
Ralf Knackstedt	University of Hildesheim, Germany
Agnes Koschmider	Karlsruhe Institute of Technology, Germany
Marek Kowalkiewicz	Queensland University of Technology, Australia
Elżbieta Lewańska	Poznań University of Economics and Business, Poland
Dalia Kriksciuniene	Vilnius University
Jun-Lin Lin	Yuan Ze University, Taiwan
Peter Lockemann	Universität Karlsruhe, Germany
Fabrizio Maria Maggi	Institute of Computer Science, University of Tartu, Estonia
Andrea Marrella	Sapienza University of Rome, Italy
Raimundas Matulevicius	University of Tartu, Estonia
Andreas Oberweis	Karlsruhe Institute of Technology, Germany
Toacy Oliveira	COPPE/UFRJ, Brazil
Eric Paquet	National Research Council, Canada
Adrian Paschke	Freie Universität Berlin, Germany
Marcelo Pimenta	Universidade Federal do Rio Grande do Sul, Brazil
Biswajeet Pradhan	University of Technology, Sydney, Australia
Birgit Proell	Johannes Kepler University Linz, Austria
Luise Pufahl	Hasso Plattner Institute, University of Potsdam, Germany
Elke Pulvermueller	Institute of Computer Science, University of Osnabrück, Germany
António Rito Silva	Universidade de Lisboa, Portugal
Duncan Ruiz	Pontificia Universidade Católica do Rio Grande do Sul, Brazil
Virgilijus Sakalauskas	Vilnius University, Lithuania
Sherif Sakr	The University of New South Wales, Australia
Demetrios Sampson	Curtin University, Australia
Elmar Sinz	University of Bamberg, Germany
Alexander Smirnov	SPIIRAS, Russia
Stefan Smolnik	University of Hagen, Germany

Milena Stróżyna	Poznań University of Economics and Business, Poland
York Sure-Vetter	Karlsruhe Institute of Technology, Germany
Herve Verjus	Universite Savoie Mont Blanc – LISTIC, France
Krzysztof Węcel	Poznań University of Economics and Business, Poland
Hans Weigand	Tilburg University, The Netherlands
Benjamin Weinert	University of Oldenburg, Germany
Mathias Weske	University of Potsdam, Germany
Anna Wingkvist	Linnaeus University, Sweden
Julie Yu-Chih Liu	Yuan Ze University, Taiwan

Organizing Committee

Milena Stróżyna (Chair)	Poznań University of Economics and Business, Poland
Barbara Gołębiewska	Poznań University of Economics and Business, Poland
Inmaculada Hernández	University of Seville, Spain
Patricia Jiménez	University of Seville, Spain
Piotr Kałużny	Poznań University of Economics and Business, Poland
Elżbieta Lewańska	Poznań University of Economics and Business, Poland
Włodzimierz Lewoniewski	Poznań University of Economics and Business, Poland

Additional Reviewers

Simone Agostinelli
Asif Akram
Christian Anschuetz
Kimon Batoulis
Ilze Birzniece
Victoria Döller
Katharina Ebner
Filip Fatz
Lauren S. Ferro
Umberto Fiaccadori
Dominik Filipiak
Szczepan Górtowski
Lars Heling
Olivia Hornung
Amin Jalali
Maciej Jonczyk
Piotr Kałużny
Adam Konarski
Izabella Krzemińska
Vimal Kunnummel
Meriem Laifa
Pepe Lopis
Ronald Lumper
Pavel Malyzhenkov
Sławomir Mazurowski
Odd Steen
Piotr Stolarski
Ewelina Szczekocka
Marcin Szmydt
Jakub Szulc
Patrick Wiener

Contents – Part I

Big Data and Data Science

Artificial Intelligence

ICT Project Management

Smart Infrastructure

Contents – Part II

Social Media and Web-Based Systems

Applications, Evaluations and Experiences

Big Data and Data Science

Decision-Support for Selecting Big Data Reference Architectures

Matthias Volk(✉), Sascha Bosse(✉), Dennis Bischoff(✉), and Klaus Turowski(✉)

Otto-von-Guericke-University Magdeburg,
Universitaetsplatz 2, 39106 Magdeburg, Germany
{matthias.volk, sascha.bosse, dennis.bischoff, klaus.turowski}@ovgu.de

Abstract. In recent years, big data systems are getting increasingly complex and require a deep domain specific knowledge. Although a multitude of reference architectures exist, it remains challenging to identify the most suitable approach for a specific use case scenario. To overcome this problem and to provide a decision support, the design science research methodology is used. By an initial literature review process and the application of the software architecture comparison analysis method, currently established big data reference architectures are identified and compared to each other. Finally, an Analytic Hierarchy Process as the main artefact is proposed, demonstrated and evaluated on a real world use-case.

Keywords: Big data · Reference architectures · Analytic hierarchy processing · Multi criteria decision making

1 Introduction

In recent years, the topic big data has gained increasing interest and reached a productive status in business and research. Today organizations are conducting big data projects for various reasons, such as fraud detection, predictive maintenance, and optimizations [1]. With the growing interest among users, also the number of new technologies being developed and distributed is constantly rising. Big data itself can be described as a collection "of extensive datasets primarily in the characteristics of volume, variety, velocity, and/or variability that require a scalable architecture for efficient storage, manipulation, and analysis" [2]. Due to their particularities and requirements, organizations are today not only confronted with the question of suitability of big data technologies in their information technology (IT) projects, but also the adequate composition of specific tools and technologies [3]. This combination should follow a comprehensive and scalable system architecture, that can be generally described as "fundamental concepts or properties of a system in its environment embodied in its elements, relationships, and in the principles of its design and evolution" [4]. In order to reduce the complexity accompanied by the introduction of these, best-practices in terms of architectural considerations can be exploited. Reference architectures are designated for these kind of recommendations, supporting the

W. Abramowicz and R. Corchuelo (Eds.): BIS 2019, LNBIP 353, pp. 3–17, 2019.
https://doi.org/10.1007/978-3-030-20485-3_1

engineering procedure in terms of an extensive and complex combination of tools and technologies for a targeted purpose. By applying them, the number of technological decisions and their constellations can be reduced. This leads to a facilitation of the technology decision process and the general support of decision makers, such as enterprise architects. Since the introduction of the first big data reference architecture by Nathan Marz in 2011, called Lambda architecture [5], a large number of innovative approaches for specific big data problem types have been created. However, different constraints, components, and application domains lead to a certain precariousness in the decision-making process and, thus, reinforces the problem of a suitable selection of big data reference architectures. Although a multitude of various approaches for decision-support exists, such as Decision Trees, the Analytical Hierarchy Process (AHP) or Techniques for Order of Preference by Similarity to Ideal Solution (TOPSIS), these have never been applied to this particular problem. Only a few contributions, such as [6] took methods like the AHP in big data environments into consideration. Probably, this could be traced back to the lack of existing big data reference architectures comparisons, that are also available only in a limited way [7]. Hence, this papers aims to provide a decision support in big data related-projects, when it comes to the selection of a reference architecture. Apart from this, organizations will be supported when identifying the appropriate domain experts and needed skills. To achieve the desired solution the following research question will be answered: "How can a decision support in big data environments be provided for the selection of the most suitable reference architecture?". In doing so, the following sub-questions were derived: *RQ1*-Which big data reference architectures are relevant today? *RQ2*-How can big data architectures be examined, compared and selected with regard to their applicability? By answering these, a decision process for the selection of suitable reference architectures for specific use cases will be facilitated, which is exemplary constructed using the well-known AHP. In order to approach this problem, the design science research methodology (DSRM) by Hevner et al. [8] and the recommended workflow by Peffer et al. [9] were applied. The latter divides the execution of DSRM into six consecutive steps. Depending on the research entry point of the DSRM, the recommended workflow starts with the description and motivation of the current problem situation in order to formulate the main objectives of a desired solution. This initial step was already realized within this section by highlighting the necessity of a suitable artifact that supports the decision making process of big data reference architectures. However, before the actual design and development takes place, it is first required to obtain both an overview of the current state of the art as well as the theoretical foundations. This will be provided in the second section. In this context, the approach of a literature review by Webster and Watson [10] was used for the collection of relevant material and for processing RQ1. Due to the highlighted lack of comparisons, all identified architectures are contrasted afterwards in the fourth section of this work using the software architecture comparison analysis method (SACAM) [11] in order to provide answers to RQ2. Finally, all results will be brought together in the fourth section, in which the artifact, an AHP for big data reference architectures, will be constructed, demonstrated, evaluated in a real-world use case, and critically discussed. The work ends with a conclusion and an outlook to further research activities. Together with the respective content of this paper, all steps are depicted in Fig. 1.

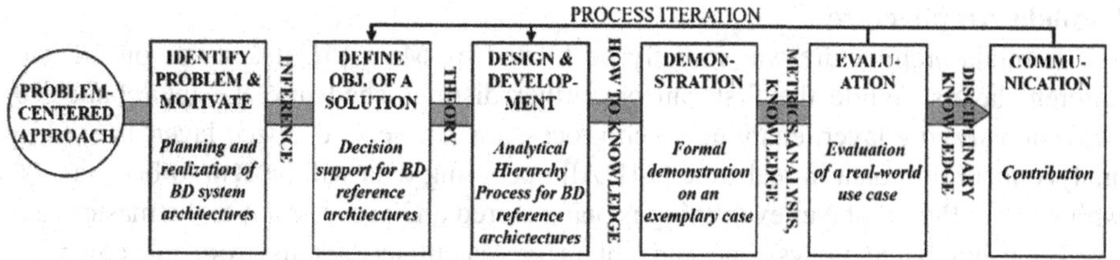

Fig. 1. Conducted workflow of the design science research methodology [9]

2 State of the Art

Today, many big data reference architectures exist. However, there are many differences determining their suitability, such as the intended use, the complexity, or the general system design. Until now, only limited effort was put into the comparison and selection of reference architectures, for instance, when a new approach is introduced [7, 12]. To get a better understanding about the currently existing reference architectures and possible selection process for those, a literature review according to the methodology by Webster and Watson was conducted [10].

2.1 Literature Review

The literature review was carried out in three consecutive steps. In the first step, the databases ScienceDirect and SpringerLink were initially queried using topic-specific keywords like "big data" or "reference architectures" and afterwards filtered, based on titles and abstracts. The second step included an examination of the references from all remaining contributions, in order to conduct forward & backward search with no restriction to the index database [10, 13]. For refinement purposes, three inclusion criteria were formulated to determine the most frequently discussed reference architectures that are described in Table 1. The analysis of the results reveals that existing big data reference architectures differ strongly to one another, e.g. in their complexity, application area and depth of their description. While many of the new approaches are little researched and, thus, have not been included in the analysis, also other architectures exist, fulfilling at least one of the formulated criteria. In the following, all of them are briefly described. For a more detailed description, the referred origins should be consulted.

Table 1. Inclusion criteria of a reference architecture.

Criteria
The work that presents the reference architecture has been cited at least ten times, of which at least five referencing works should be published between 2016 to 2018. This is needed to ensure that the identified approach still has a certain relevancy
The work was published after 2017 and cites at least two papers, which are qualified according the first criteria. In doing so, entirely new approaches that did not received lots of attention in the scientific community so far will be recognized
The architecture was addressed in at least three contributions, covering implementation details

Lambda Architecture

The Lambda architecture was initially postulated by Marz [5]. It consists out of two different layers. While the first can be further distinguished into the interconnected *batch* and *serving* layer, enabling batch processing, the second, *speed* layer, is used to analyze incoming data in real-time [14]. All upcoming data will be sent to both layers, while within the first layer everything is being stored and processed into the master data set. By using batch views, one can obtain pre-calculated results from the complete dataset to allow specific queries. Due to the very time-consuming process, new incoming data cannot be processed. Thus, the speed layer will analyze the new data in real-time. The results can be later aggregated with the results of the predefined batch views [15].

Bolster Architecture

The Bolster architecture represents an extension of the Lambda architecture to allow the use of semantic data. In the respective work [7], the authors compared features of different architectures and came to the result that none of them fulfills all of their desired requirements. Thus, the Bolster architecture, as a new all-encompassing approach, was developed. In addition to the layers of the Lambda architecture, the bolster architecture adds another layer. The *semantic* layer is responsible for storing and providing metadata. This metadata can be used in the batch layer and also in the speed layer [7].

Kappa Architecture

The Kappa architecture follows the idea to reduce the overhead maintaining two different systems for one task, as it appears in the case of the Lambda architecture. Hence, it demands to have only the stream processing system [16]. Although multiple advantages occur with this approach, the solution may cause huge cost. This is especially the case if the volume is very high and lots of resource regarding network capacity and computational power will be required [17]. However, approaches exist for which compression techniques can be used for the cost of additional computational power [18].

Solid Architecture

The Solid architecture presents another approach introducing semantic data into a big data architecture. Within Solid, the Resource Description Framework (RDF) is used to deal with semantic data [19]. This architecture is comprised of a service tier, a content tier and a merge tier. The service tier describes the interface that communicates with components outside of the architecture. Therefore, it gets the required data from the content tier. This consist out of the *online*, *data* and *index* layer. Each of these serve a specific purpose: while the latter provides only access to the data layer, it stores all data except the new one. All new data will be stored by the online layer. The merge tier afterwards handles the integration of all data packages to the respective data layer [20].

The Big Data Research Reference Architecture (BDRRA)

This architecture was developed by comparing existing architectures of large companies and extracting their key features. The components of each of the use cases were mapped to crucial steps of big data projects, finally forming a reference architecture. With the exception of the data storage that is building the foundation, each of the functional components are separated. In particular, the reference architecture consists out of basic components such as data sources, data extraction, data processing, data analysis, interfacing and visualization, data storage as well as job and model specification [21].

Big Data Architecture Framework (BDAF)
In comparison to the previous architectures, the BDAF rather represents a theoretical work listing components that can be used for a big data architecture. Hence, the BDAF has no limitations. It focuses on five different components: data models, big data management, big data analytics and tools, big data infrastructure as well as big data security. Each of these components has dependencies to the other components [22].

2.2 Decision Making Frameworks

Since no selection process is recommended in the literature, also possibilities for decision support frameworks need to be discussed. The complexity of the respective architectures indicates that a multitude of factors has to be considered and special techniques are required to identify a suitable approach. Multi-criteria decision-making (MCDM) techniques observe various aspects of alternatives to be decided. Basically, these kind of problems are composed out of five essential components, namely the "goal, decision maker's preferences, alternatives, criteria's and outcomes respectively" [23]. This involves comparing the individual components using various mathematical methods and procedures. In addition to general approaches that represent stepwise decision processes and are commonly referred to as decision trees [24], also numerous complex techniques. In many cases, these take conflicting objectives and personal preferences into account, to illustrate the inner decision making process as comprehensive as possible. Famous approaches that found acceptance in both business and science are for instance TOPSIS, VIKOR, PROMETHE, ELECTRE and the AHP [23]. Those are also frequently applied in approaches, such as Commercial-Of-The-Shelf (COTS) decision making, at which related products for a system can be identified [25]. One approach that directly implements the AHP, in the targeted domain of big data, identifies a suitable analytics platform based on various criteria and available tools [6].

3 Comparison of Big Data Reference Architectures

Before the actual construction of the artifact and, therefore, the third step of the conducted DSRM [9] is taking place, a comparison must be made first to determine the prevailing differences between the found out architectures. Additionally, it is intended to identify criteria which may influence the decision making process. Although no specific methods for the comparison of big data architectures exist, today there exist a multitude of different methods for evaluating and comparing architectures in terms of information and communication systems. These differ mainly in their structure, the purpose, and the information that are required. SACAM only requires a set of comparison criteria and the documentations of the architectures as input parameter [11]. While the first will be formulized during the procedure, the latter were already determined during literature review. Although the application of the method is primarily oriented towards software [26] and not big data reference architectures, however, it is also focused on the right combination of tools and technologies in the sense of patterns application [27]. Predestined for the comparison of multiple architectures, the method was used, for instance, to compare two architectures to avoid frauds in telephone banking [28]. Consequently, SACAM appears to be a suitable way to contrast the found out approaches. The procedure itself, as depicted in Fig. 2, will be further explained

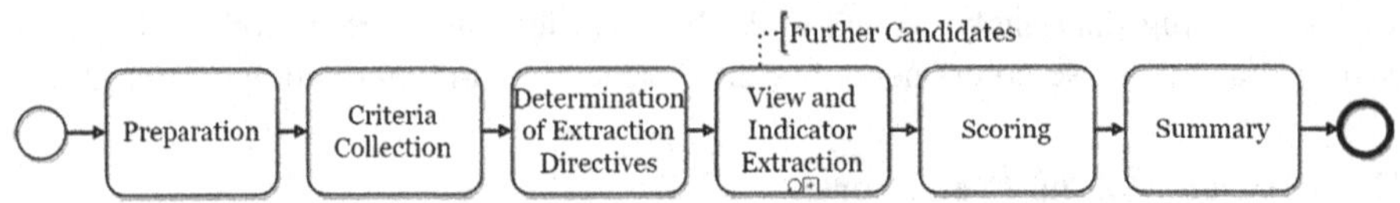

Fig. 2. The SACAM procedure as an BPMN Model according to [11].

The first step serves as a preparation. In here, the targeted architectures, their intended use as well the required documentations are identified. The latter was already realized by the initially performed literature review. After that, the criteria collection is taking place, at which typically exemplary scenarios will be formulated "for capturing quality attributes and refining them into quality attribute scenarios" [11]. Due to the previously provided definition of big data, especially the targeted data characteristics and the scalability appear to be suitable requirements and, therefore, quality attributes. According to those, the required scenarios were formulated, predominantly deduced from the results of the literature review, either in an explicit (condensed) or implicit (concrete references) form. The third step declares the determination of extraction directives, which are required to extract the needed information from the documentations and, thus, assess the applicability of the architectures. This was realized by the observation of the scenarios and the deduction of the main tactics, which were predominantly focusing on concepts such as the processing capabilities. A complete overview of the documented and derived scenarios, tactics, and quality attributes is depicted in Table 2.

After this step, the view and indicator extraction is performed. All available information of the targeted architectures are extracted and compared to the derived tactics from the formulated scenarios, to review the capabilities of its quality attributes fulfillment. The evidence for each of the comparisons is scored afterwards. Both steps are performed similar to the workflow. The same applies for the recommended scoring itself, for which a range from 0 to 10 was chosen, "where 0 means that no support is provided, and 10 completely" [11]. Whenever a reference architecture is fully capable to realize such a tactic, a 10 can be assigned. The complete results of the application of SACAM, as well as the specific component of each big data reference architecture capable to solve this scenario, are listed in Table 3. In case of limitations, such as the use of the Kappa architecture within the first scenario that is not suitable for huge amounts of data, a five is given. The same applies for scenario two and the Solid architecture, due to the lack of scalability in terms of stream processing. The gradation between the upper and lower boundary results out of a comparison between the architectures.

Table 2. Developed scenarios, derived tactics and targeted quality attributes.

No.	Description of the scenario	Tactics	V's
S1	Multi massive analysis of previously made online purchases in E-Commerce. All used data will be saved in a conventional database table formats and analyzed at periodic intervals to give product recommendations	A system that is able to store huge amounts of data and analyze them using batch processing. All obtained results need to be further persisted	Volume
S2	Real-time analysis of viewed products in E-Commerce by (potential) customers. The results shall give immediately recommendations for products that might be interesting	A system that is able to stream process data. All obtained results need to be further persisted	Velocity
S3	Multi massive analysis of historic and real-time data in E-Commerce, to provide tailor-made customer recommendations on the base of preceding and recent behavior	A system that is able to batch and stream process data, either in a combined or separated procedure. The results needs be persisted for further use	Volume, Velocity
S4	Analysis of an on online-performed questionnaire, including multiple questions types and also free text answer options. The latter is context based and requires further analysis [29]	A system that is able to batch process huge amounts of differently structured data using various analyses, such as semantical or statistical methods. Changes may occur over time to the persisted data	Variety, Variability
S5	Development of a real-time network traffic measurement system. The monitored data will provide basic statistics and deliver warnings in case of specific events [30]	A system that is able to analyze huge amounts of structured data by using a stream processing approach	Volume, Velocity
S6	Observation of multiple machine sensors to prevent and react on failures. The required information will be gathered from single sensors, their combination and the underlying metadata [31]	A system that is able to analyze huge amounts of incoming data and metadata by using a batch and stream processing approach. The dataflow may change over time	Volume Velocity, Variability
S7	Giving predictions for stock price changes, by using Twitter feeds. Related data from these feeds will be analyzed with deep learning approaches [32]	A system that is able to analyze huge amounts of differently structured data and metadata by using a batch and stream processing approach. The dataflow may change over time	Volume, Velocity, Variety, Variability

In scenario four, it has to been noted that information may change over time. Apart from that, this scenario is similar to the first one. The Lambda, Solid and Kappa architecture are not intended to perform any of these changes on the master data set. However, the Lambda and Solid architecture can be easily extended, whereas more effort needs to be invested when using the Kappa architecture. Thus, a rating of 7 was given to the latter and the both other approaches received an 8. Comparing all of the given scores, one can notice that some architectures are more dominant than the others. Especially the BDRRA and BDAF received the best scores. This results predominantly out of the generalized structure by which most of the existing requirements and constraints can be covered. However, this generalized point of view may lead to inefficiencies in comparison to specific approaches. Thus, there is a clear distinction between very general and specific approaches. Furthermore, the BDAF architecture strongly dominates all other alternatives. The same applies for the BDRRA, but the specificity is still higher compared to the BDAF and explicit metadata management is not supported here, thus, the BDAF won't be further used.

Table 3. Results of the application of SACAM.

No.	Lambda	Bolster	Kappa	Solid	BDRRA	BDAF
S1	B	B	LSP	DL	ED, DA	C
	10	10	5	10	10	10
S2	RT	RT	SP	OL	SD, SA	C
	10	10	10	5	10	10
S3	B, RT	B, RT	LSP	DL	ED, DA, SD, SA	C
	8	8	6	7	10	10
S4	B, RT	B, RT, DA	LSP	OL, DL	ED, SD, SA	C
	8	10	7	8	10	10
S5	RT	RT	SP	OL, DL	SA	C
	10	10	10	4	10	10
S6	–	RT, S	–	OL, DL	–	C
	0	10	0	4	0	10
S7	–	C	–	OL, DL	–	C
	0	8	0	3	0	10

B: Batch; **BP**: Batch Processing; **C**: Complete; **DA**: Deep Analysis; **DA**: Data Analyst; **DB**: Database; **DL**: Data Layer; **ED**: Enterprise Data; **LSP**: Limited Stream Processing; **MD**: Metadata; **RT**: Real-Time; **S**: Semantic; **SA**: Stream Analysis; **SD**: Stream Data; **SP**: Stream Processing; **OL**: Online Layer.

4 AHP Model for Big Data Reference Architectures

On the basis of those results, it becomes clear that a variety of different criteria play a major role during the selection of the most suitable architecture. After further examinations about the capability of suitable methods mentioned in Sect. 2.1, the AHP has proved to be a promising solution. Compared to other approaches, this is especially due to the reason that the AHP is a widely accepted technique of decision makers and was already applied in contributions within a similar domain [6]. The application itself is initiated by pairwise comparisons of different criteria, with regard to individual preferences. Each of the criteria will be compared to all others, resulting in a total of $(n^2 - n)/2$ comparisons, in which n is the total number of criteria of a single procedure. The outcomes of the comparisons range from 1–9. While a value of 1 means that both criteria are equally important, a value of 9 reflects an "extreme importance" [33] for the criterion to be compared. Intermediate ratings are intended to highlight slight preferences, for instance, 5 means that the criterion to be compared has a higher importance than the other has. The calculatory equivalent, necessary for further computation, is expressed by the reciprocal value. All gradation values of the ratings can be found in [33]. After the initial comparisons, all results will be stored in an $n \times n$ identity matrix for each criterion, called comparison matrix (1). In doing so, the diagonal is represented by ones, all values above the main diagonal represent the corresponding ratings of the comparison and in turn all values below the reciprocal value. After all comparisons have been made, a normalized matrix is generated. The average of each row of this normalized matrix indicates the absolute priority of the targeted criteria, forming all together the weighting vector W. The higher an entry, the higher is the importance of the criteria (2). In order to avoid an undeliberated decision, the consistency of the made comparisons can be further assessed. First, the initial comparison matrix needs to be multiplied with W to obtain the weighted sum vector Ws. Then the reciprocal is formed of each element of W and, afterwards, the scalar product of both vectors is calculated $\left(Ws \circ \{W\}^{-1}\right)$. The number of compared elements then divides the resulting consistency vector, to obtain λ_{Max} as the eigenvalue (3). This value is required to calculate the consistency index (CI). Finally, with the calculation of the Consistency Ratio (CR), a comparison of the CI with a randomized consistency index (RI) takes place, to measure the consistency of the given scores (5). If CR is less than or equal to 0.10, all of the given ratings are consistent, otherwise individual pairwise comparisons should be reviewed again. When all values of the criteria have been determined and W calculated, the alternatives will be inspected in a similar way. Each of them will be compared and rated on the basis of one specific criterion. Once again, a W will be calculated for each criterion (6). This continues until each of them has been checked in terms of the available alternatives. Further, all W vectors will be brought together into one matrix, highlighting the connection between the alternatives and criteria. Finally, this newly created matrix will be multiplied with the initially calculated vector W from the criteria comparison matrix. The resulting vector describes the suitability of each alternative, based on the made comparison (7).

An exemplary realization is depicted in Fig. 3, highlighting that the alternative A_1 appears to be the best solution for this particular case. For a better understanding, each of the previously described steps is linked to the figure via the corresponding number within the brackets.

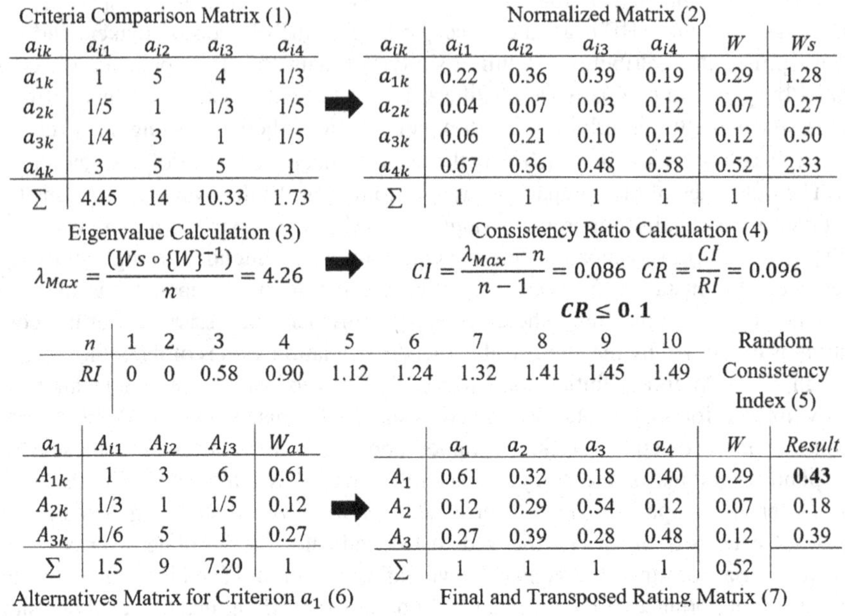

Criteria Comparison Matrix (1)

a_{ik}	a_{i1}	a_{i2}	a_{i3}	a_{i4}
a_{1k}	1	5	4	1/3
a_{2k}	1/5	1	1/3	1/5
a_{3k}	1/4	3	1	1/5
a_{4k}	3	5	5	1
Σ	4.45	14	10.33	1.73

Normalized Matrix (2)

a_{ik}	a_{i1}	a_{i2}	a_{i3}	a_{i4}	*W*	*Ws*
a_{1k}	0.22	0.36	0.39	0.19	0.29	1.28
a_{2k}	0.04	0.07	0.03	0.12	0.07	0.27
a_{3k}	0.06	0.21	0.10	0.12	0.12	0.50
a_{4k}	0.67	0.36	0.48	0.58	0.52	2.33
Σ	1	1	1	1	1	

Eigenvalue Calculation (3)

$$\lambda_{Max} = \frac{(Ws \circ \{W\}^{-1})}{n} = 4.26$$

Consistency Ratio Calculation (4)

$$CI = \frac{\lambda_{Max} - n}{n - 1} = 0.086 \quad CR = \frac{CI}{RI} = 0.096$$

$$\boldsymbol{CR \leq 0.1}$$

Random Consistency Index (5)

n	1	2	3	4	5	6	7	8	9	10
RI	0	0	0.58	0.90	1.12	1.24	1.32	1.41	1.45	1.49

Alternatives Matrix for Criterion a_1 (6)

a_1	A_{i1}	A_{i2}	A_{i3}	W_{a1}
A_{1k}	1	3	6	0.61
A_{2k}	1/3	1	1/5	0.12
A_{3k}	1/6	5	1	0.27
Σ	1.5	9	7.20	1

Final and Transposed Rating Matrix (7)

	a_1	a_2	a_3	a_4	*W*	*Result*
A_1	0.61	0.32	0.18	0.40	0.29	**0.43**
A_2	0.12	0.29	0.54	0.12	0.07	0.18
A_3	0.27	0.39	0.28	0.48	0.12	0.39
Σ	1	1	1	1	0.52	

Fig. 3. An exemplary realization cation of the AHP

4.1 Design and Development

In order to facilitate such a multi-criteria decision support, it is first needed to find suitable criteria at the beginning, which can be later assessed by the user [33]. While comparing the different reference architectures, it became apparent that many of the data characteristics are strongly interconnected to each other (cf. Table 2). In most cases, the basic intention of the different reference architectures can be attributed to a certain combination of data characteristics. For this reason, all of the characteristics should be observed in a combined way to enable decision makers a comprehensive prioritization according to their undertakings. A suitable solution for this was already found out during the comparison of the determined reference architectures. With the identification of quality attribute scenarios and the subsequent extraction of tactics, that allow a comparison of the approaches, suitable criteria were determined (cf. Table 2) [11]. By evaluating those tactics in terms of the relevancy for the planned undertaking, decision makers are not bounded to rely on abstract views, functions and any forms of metrics. Otherwise this would be necessary for the isolated consideration of single data characteristics. Thus, an AHP distributed in the three layers named *objective*, *criteria* and *alternatives* was developed, as it is depicted in Fig. 4. In addition to the mutual relationships between the levels of the AHP, there are also other conspicuous features. The application of the artifact itself takes place as demonstrated in Fig. 3 by the pairwise-comparison of the tactics described in Table 2. First of all, the comparison matrix and the weighting vector *W* need to be determined. Then *W* will be multiplied with the matrix depicted in Table 4.

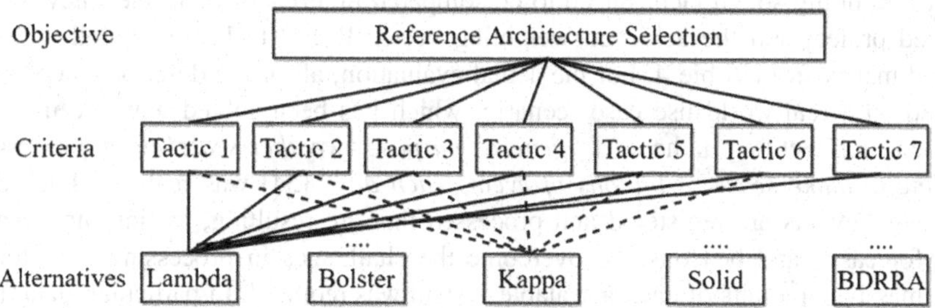

Fig. 4. A conceptual overview of an AHP for big data references architectures.

The resulting vector then reflects the viability for the undertaking. In contrast to the standard procedure, *W* was not calculated for each comparison of the alternatives to one respective criterion, instead the previously determined results by SACAM were used. These were obtained by a similar procedure, comparing the architectures in terms of the developed quality attribute scenarios and the derived tactics. To overcome the different scaling's of the rating, the final resulting matrix was normalized and transposed, as in the case of the standard AHP procedure. Within the table, the first row of each of the architectures describe a matrix containing the BDRRA, as one possible alternative, whereas the second row does not take the general approach into consideration. The final result itself will mostly prioritize the BDRRA, due to its dominancy. However this does not present always the desired decision support, most of all if general approaches are not preferred, due to their missing implementation details. If a specific approach is to be preferred over a general, the BDRRA can be omitted in advance. Nevertheless, at the beginning extensive research always needs to be carried out to allow for a rational and reasonable decision when it comes to the comparison of the tactics. In any case, it is up to the decision maker to interpret the results after the successful application.

Table 4. Normalized and transposed SACAM rating matrix with and without BDRRA

	T1	T2	T3	T4	T5	T6	T7
Lambda	0.22	0.22	0.21	0.19	0.23	0	0
	0.29	0.29	0.28	0.24	0.29	0	0
Bolster	0.22	0.22	0.21	0.23	0.23	0.71	0.73
	0.29	0.29	0.28	0.30	0.29	0.71	0.73
Kappa	0.11	0.22	0.15	0.16	0.23	0	0
	0.14	0.29	0.21	0.21	0.29	0	0
Solid	0.22	0.11	0.18	0.19	0.09	0.29	0.27
	0.29	0.14	0.24	0.24	0.12	0.29	0.27
BDRRA	0.22	0.22	0.26	0.23	0.23	0	0

4.2 Evaluation of the Artifact

After the completed design and development phase, the application of the artifact needs to be demonstrated and evaluated (cf. Fig. 1). Essentially, only a few steps are necessary to apply the artifact, as outlined in Fig. 3. In this sense, a pairwise comparison only has to be made concerning the criteria layer and not additionally for the alternatives. In doing so, all tactics need to be compared in terms of their relevancy for the targeted project and the achieved weighting vector W has to be multiplied with the desired matrix from Table 4. For the actual evaluation, all of the described steps were applied on a real-world use case scenario, which has been solved with the use of a Lambda architecture. In here, a "*general purpose, weather-scale event processing pipeline to make sense of billions of events each day*" [34] was realized. Each day, billions of events are registered and processed that are resulting, for instance, out of logs, forecasts, and beacons. To overcome the challenges in processing and storing these massive amounts of data, a scalable system was required. In particular, data from different sources needed to be analyzed in batch as well as in streaming mode and all results are to be stored for long term access [34]. Thus, a Lambda architecture and different tools were used for the realization. By extracting the required information and comparing the tactics, regarding this specific use case, a comparison matrix was realized and the described procedure was conducted. All results are depicted in Table 5. The main focus in this project was on the combined use of batch and streaming processing. Furthermore the results should be persisted, accessed for long term, and not changed over time. Thus, the third tactic was favored very strongly to the others while the opposite applies for tactic number four. Since no further information about the metadata handling was available, it was tried to assess these as unbiased as possible. For each of the tactics that offered only one processing approach, a moderate importance to tactic three was chosen. Compared to one another, these tactics were equal important and, thus, received the value one. After that, all calculation steps were performed. As one can notice, a CR of 0.069 was achieved and therefore all made

Table 5. Results of the evaluation

T_i	T_1	T_2	T_3	T_4	T_5	T_6	T_7	W	Ws
T_1	1	1	1/7	3	1/3	1/4	1/3	0.6	0.41
T_2	1	1	1/7	3	1/3	1/4	1/3	0.06	0.41
T_3	7	7	1	7	4	8	7	0.48	3.79
T_4	1/3	1/3	1/7	1	1/3	1/3	1/3	0.04	0.27
T_5	3	3	1/4	3	1	3	2	0.16	1.28
T_6	4	4	1/8	3	1/3	1	1	0.12	0.91
T_7	3	3	1/7	1/2	1/2	1	1	0.09	0.73
$\sum$	2.95	26	7.95	7.45	23.00	9.58	7.67	1	–

λ_{Max}: 7.54; CI: 0.09 RI: 1.32; CR: 0.069 $\leq$ 0.1
(1) With BDRRA - Bolster: 0.32; Lambda: 0.17; BDRRA: 0.19; Solid: 0.18; Kappa: 0.13
(2) Without BDRRA - Bolster: 0.37; Solid: 0.23; Lambda: 0.22; Kappa: 0.18

comparisons were consistent. Derived from these results, one can conclude that the *Bolster* architecture appears to be a promising solution, whereas also *BDRRA, Solid, Lambda,* and *Kappa* achieved good results. This applies for both types of the used alternative matrix, with and without the *BDRRA*.

Only the distribution of the results has slightly changed. One of the main reasons for this result might come from the indifferent position towards the metadata handling. However, considering the various data source and data requests, in this particular case, an additional use of the metadata layer might be beneficial. One of the main advantages is the increased level of automation, for instance, if new data sources are getting added. Furthermore, data analyst, data management, and data steward tasks can be better distinguished, which is important in terms of the high interconnection between stakeholders in this particular case [7]. Without these additional considerations, the Lambda architecture is also capable for the desired purpose and has, therefore, been proved to be valid. In addition to the general architectural recommendation, the obtained result can be further used to derive required expert knowledge about related tools and the technical integration of the targeted domain. To enhance this outcome, the artifact should be evaluated in large-scale in future research. In context of this, an extension and integration into a decision support system appears to be suitable. This would ensure that not only decision support regarding the architecture, but also possible tools and technologies could be provided. The artifact itself can be completely adapted as well as extended in terms of alternatives and comparison criteria. This includes new tactics as well as references architectures. In doing so, the tactics could be refined to standard application scenarios, at which different information are presented using an even more complex description. From a practitioner's perspective, also project-related requirements could be recognized, including relevant big data aspects, such as the data characteristics and performed operations. Regarding the extension, however, the rank reversal problem should be considered when alternatives are changed. This may lead to a restructuring of various multi criteria decision making approaches, such as the AHP [35]. Another limitation presents the number of additional criteria and alternatives, which may occur through new tactics, data characteristics, reference architectures or even entirely new objects, such as technical implementation details. In this sense, the introduction of further levels (sub-criteria) appears useful in order to reduce the complexity [33]. While in this work only the AHP in its pure form was used, also other decision making frameworks might be considered and compared in future research.

5 Conclusion

In this paper, an AHP for big data architectures was developed to provide a decision support when it comes to the right choice of currently existing approaches. Based on a thorough literature review process, six relevant big data reference architecture were identified. After an initial description of each of those, a qualitative scenario-based comparison using SACAM was executed. The results were used afterwards for the design and development of the main artifact, the AHP. Finally, the solution was evaluated and discussed using a real world use case scenario. Although the choice of tactics has proofed to be as suitable criteria, in the future, extensions are planned to be able to withstand the constant dynamics of this domain. Beyond the sole extension of new tactics and architectures, also an integration within a comprehensive decision support system for the realization of big data projects will be pursued that additionally considers project-related requirements and other MCDM techniques. In doing so, the additional comparison to an COTS approach appears to be desirable [25]. Same applies for the examination of suitable tools, as it was realized by the application of an AHP in [6].

References

1. Dresner Advisory Services, LLC: Big Data Analytics Market Study. https://www.microstrategy.com/us/resources/library/reports/dresner-big-data-analytics-market-study. Accessed 5 Dec 2018
2. NIST Big Data Interoperability Framework. Volume 1, Definitions. National Institute of Standards and Technology
3. Volk, M., Pohl, M., Turowski, K.: Classifying big data technologies - an ontology-based approach. In: 24th Americas Conference on Information 2018. Association for Information Systems, New Orleans, LA, USA (2018)
4. ISO/IEC/IEEE 42010:2011: Systems and software engineering — Architecture description. Geneva, CH (2011)
5. Marz, N.: How to beat the CAP theorem. http://nathanmarz.com/blog/how-to-beat-the-cap-theorem.html. Accessed 5 Dec 2018
6. Lněnička, M.: AHP model for the big data analytics platform selection. Acta Informatica Pragensia **4**, 108–121 (2015)
7. Nadal, S., et al.: A software reference architecture for semantic-aware Big Data systems. Inf. Softw. Technol. **90**, 75–92 (2017)
8. Hevner, A.R., March, S.T., Park, J., Ram, S.: Design science in information systems research. MIS Q. **28**, 75–105 (2004)
9. Peffers, K., Tuunanen, T., Rothenberger, M.A., Chatterjee, S.: A design science research methodology for information systems research. J. Manage. Inf. Syst. **24**, 45–77 (2007)
10. Webster, J., Watson, R.T.: Guest Editorial: Analyzing the past to prepare for the future: writing a literature review. MIS Q. **26**, xiii–xxii (2002)
11. Stoermer, C., Bachmann, F., Verhoef, C.: SACAM: The Software Architecture Comparison Analysis Method. Pittsburgh, Pennsylvania (2003)
12. Azarmi, B.: Scalable Big Data Architecture. A Practitioner's Guide to Choosing Relevant Big Data Architecture. Apress, Berkeley (2016)
13. Haunschild, R., Hug, S.E., Brändle, M.P., Bornmann, L.: The number of linked references of publications in Microsoft Academic in comparison with the Web of Science. Scientometrics **114**, 367–370 (2018)
14. Kiran, M., Murphy, P., Monga, I., Dugan, J., Baveja, S.S.: Lambda architecture for cost-effective batch and speed big data processing. In: 2015 IEEE International Conference on Big Data (Big Data), pp. 2785–2792. IEEE (2015)
15. Marz, N., Warren, J.: Big Data. Principles and Best Practices of Scalable Real-Time Data Systems. Manning, Shelter Island (2015)
16. Kreps, J.: Questioning the Lambda Architecture. The Lambda Architecture has its merits, but alternatives are worth exploring. https://www.oreilly.com/ideas/questioning-the-lambda-architecture. Accessed 5 Dec 2018

17. Zschörnig, T., Wehlitz, R., Franczyk, B.: A personal analytics platform for the Internet of Things - implementing Kappa Architecture with microservice-based stream processing. In: Proceedings of the 19th International Conference on Enterprise Information Systems, pp. 733–738. SCITEPRESS (2017)
18. Wingerath, W., Gessert, F., Friedrich, S., Ritter, N.: Real-time stream processing for Big Data. it Inf. Technol. **58**(4), 186–194 (2016)
19. Martínez-Prieto, M.A., Cuesta, C.E., Arias, M., Fernández, J.D.: The solid architecture for real-time management of big semantic data. Future Gener. Comput. Syst. **47**, 62–79 (2015)
20. Cuesta, C.E., Martínez-Prieto, M.A., Fernández, J.D.: Towards an architecture for managing big semantic data in real-time. In: Drira, K. (ed.) ECSA 2013. LNCS, vol. 7957, pp. 45–53. Springer, Heidelberg (2013). https://doi.org/10.1007/978-3-642-39031-9_5
21. Pääkkönen, P., Pakkala, D.: Reference architecture and classification of technologies, products and services for big data systems. Big Data Res. **2**, 166–186 (2015)
22. Demchenko, Y., de Laat, C., Membrey, P.: Defining architecture components of the Big Data Ecosystem. In: 2014 International Conference on Collaboration Technologies and Systems (CTS), pp. 104–112. IEEE (2014)
23. Kumar, A., et al.: A review of multi criteria decision making (MCDM) towards sustainable renewable energy development. Renew. Sustain. Energy Rev. **69**, 596–609 (2017)
24. Quinlan, J.R.: Induction of decision trees. Mach. Learn. **1**, 81–106 (1986)
25. Alves, C., Finkelstein, A.: Challenges in COTS decision-making. In: Tortora, G. (ed.) Proceedings of the 14th International Conference on Software Engineering and Knowledge Engineering, p. 789. ACM, New York (2002)
26. Alebrahim, A.: Bridging the Gap between Requirements Engineering and Software Architecture. Springer Fachmedien Wiesbaden, Wiesbaden (2017). https://doi.org/10.1007/978-3-658-17694-5
27. Cloutier, R., Muller, G., Verma, D., Nilchiani, R., Hole, E., Bone, M.: The concept of reference architectures. Syst. Eng. **13**(1), 14–27 (2009)
28. Anwer, H., Fatima, K.N., Saqib, N.A.: Fraud-sentient security architecture for improved telephone banking experience. In: IEEE 7th Annual Information 2016, pp. 1–7 (2016)
29. Schönebeck, B., Skottke, E.-M.: „Big Data" und Kundenzufriedenheit: Befragungen versus Social Media? In: Gansser, O., Krol, B. (eds.) Moderne Methoden der Marktforschung. F, pp. 229–245. Springer, Wiesbaden (2017). https://doi.org/10.1007/978-3-658-09745-5_13
30. Huici, F., Di Pietro, A., Trammell, B., Gomez Hidalgo, J.M., Martinez Ruiz, D., d'Heureuse, N.: Blockmon: a high-performance composable network traffic measurement system. In: Eggert, L. (ed.) Proceedings of the ACM SIGCOMM 2012 Conference, p. 79. ACM Press, New York (2012)
31. Daily, J., Peterson, J.: Predictive maintenance: how big data analysis can improve maintenance. In: Richter, K., Walther, J. (eds.) Supply Chain Integration Challenges in Commercial Aerospace, pp. 267–278. Springer, Cham (2017). https://doi.org/10.1007/978-3-319-46155-7_18
32. Sohangir, S., Wang, D., Pomeranets, A., Khoshgoftaar, T.M.: Big data. Deep learning for financial sentiment analysis. J. Big Data **5**, 210 (2018)
33. Saaty, T.L.: Decision making with the analytic hierarchy process. Int. J. Serv. Sci. **1**, 83–98 (2008)
34. Lambda at Weather Scale – Databricks. https://databricks.com/session/lambda-at-weather-scale. Accessed 4 Sept 2018
35. Wang, Y.-M., Luo, Y.: On rank reversal in decision analysis. Math. Comput. Model. **49**, 1221–1229 (2009)

In-Memory Big Graph: A Future Research Agenda

Deepali Jain, Ripon Patgiri(✉), and Sabuzima Nayak

National Institute of Technology Silchar, Silchar, India
jaindeepali010@gmail.com, ripon@cse.nits.ac.in, sabuzimanayak@gmail.com

Abstract. With the growth of the inter-connectivity of the world, Big Graph has become a popular emerging technology. For instance, social media (Facebook, Twitter). Prominent examples of Big Graph include social networks, biological network, graph mining, big knowledge graph, big web graphs and scholarly citation networks. A Big Graph consists of millions of nodes and trillion of edges. Big Graphs are growing exponentially and requires large computing machinery. Big Graph is posing many issues such as storage, scalability, processing and many more. This paper gives a brief overview of in-memory Big Graph Systems and some key challenges. Also, sheds some light on future research agendas of in-memory systems.

Keywords: Big Graph · Big Data · In-memory Big Graph · Large graph · Semi-structured data · Social networks

1 Introduction

Today, every domain ranging from social networks to web graphs implements Big Graph. There are diverse graph data that are growing rapidly. Big Graph has found its application in many domains, in particular, computer networks [19], social networks [23,39], mobile call networks [40], and biological networks [13]. A prominent example of Big Graph in the field of computer network is Mobile Opportunistic Networks (MONs) [19]. It is a challenging task to understand and characterize the properties of time-varying graph, for instance, MONs. Big Graph has major applications in social networking sites, for example, Facebook friends [39], and Twitter tweets [23]. In Facebook, there are millions of nodes which represent people and billion of edges which represent the relationships between these people. In Twitter, "who is following whom" is represented by Big Graphs. In mobile call networks, Wang et al. [40] uses Big Graph to understand the similarity of two individual relationships over mobile phones and in the social network. In Bioinformatic, Big Graph is used to represent DNA and other protein molecular structure. Moreover, the graph theory is used for analysis and calculation of molecular topology [13].

Handling Big Graph is a complex task. Moreover, there are many challenges associated with large graphs as they require huge computation. Therefore, there

W. Abramowicz and R. Corchuelo (Eds.): BIS 2019, LNBIP 353, pp. 18–29, 2019.
https://doi.org/10.1007/978-3-030-20485-3_2

is a great need for parallel Big Graph systems. Most of the Big Graph frameworks are implemented based on HDD. A few in-memory Big Graph frameworks are available. Because, RAM is volatile storage media, small in size and costly. However, the cost of the hardware is dropping sharply. Therefore, in the future, RAM will be given prime focus in designing a Big Graph framework. Currently, Flash/SSD-based Big Graph framework is developing. Flash/SSD based frameworks are faster than HDD. Naturally, Flash/SSD based Big Graph frameworks are slower than in-memory Big Graph frameworks. Thus, in coming future, more RAM-based Big Graph frameworks will be designed for storing Big Graph. In-memory Big Graph is a key research to be focused on. A few research questions (RQ) on in-memory Big Graph are outlined as follows: (a) **RQ1**: Can In-memory Big Graph able to process more than trillions of nodes or edges? (b) **RQ2**: Can In-memory Big Graph able to handle the Big Graph size beyond terabytes? (c) **RQ3**: Is there any alternative to HDD, SSD or Flash memory (NAND)? (d) **RQ4**: What are the real-time Big Graph processing engines available without using HDD, SSD, or Flash?

The research questions **RQ1**, **RQ2**, **RQ3**, and **RQ4** motivate us to examine the insight on massively scalable in-memory Big Graph processing engine. Besides, implementing in-memory Big Graph Database is a prominent research challenge. Thus, in this paper, we present a deep insight on scalable in-memory Big Graph processing engine as well as a database for future research.

2 Big Graph

A Big Graph comprises of billions of vertices and hundreds of billion of edges. Big Graph is applied in diverse areas [40], namely, biological networks, social networks, information networks and technological network. Big Graph is unstructured and irregular that makes the graph processing more complex. In real world cases, Big graph is dynamic, i.e., there are some temporal graphs which changes with time [29]. Particularly, new nodes are inserted and deleted frequently. Handling the frequent changes in edges and vertices are truly a research challenge.

2.1 In-Memory Big Graph

Big Graph represents a huge volume of data. This huge sized data are usually stored in secondary memory. However, storing graphs in RAM improves the performance of graph processing and analysis. But, in-memory Big Graph system requires a huge amount of resources [34]. Numerous Big Graph processing systems are designed based on in-memory Big Graph with the backing of secondary storage. For example, PowerGraph [15], GraphX [16], and Pregelix [6]. The ability of continuous holding data in RAM in a fault-tolerant manner makes in-memory big graph systems more suitable for many data analytic applications. For instance, Spark [44] uses the elastic persistence model to keep the dataset in memory, or disk or both. However, state-of-the-art Big Graphs do not provide intrinsic in-memory Big Graph without using secondary storage.

3 In-Memory Big Graph Framework

As the data keeps on growing, there should be a system which can efficiently work with the incremental data and has a large memory to hold the data. As the data is growing rapidly, processing of large graph becomes the key barrier. There is also scalability issue associated with in-memory Big Graph systems. So, there is a great need for Big Graph processing systems that can overcome the issues. There are many existing in-memory Big Graph systems like Power Graph [15], GraphX [16]. These systems handle the issues like storage, scalability, fault tolerance, communication costs, workload. In this section some in-memory Big Graph engines are discussed. Table 1 illustrates the evaluation of in-memory Big Graph on the basis of various parameters. Moreover, Table 2 exposes the various sizes of nodes and edges with data sources.

The power-law degree distribution graphs are challenging task to partition. Because, it causes work imbalance. Work imbalance leads to communication and storage issue. Hence, PowerGraph [15] uses Gather-Apply Scatter (GAS) model.

Table 1. Evaluation of existing framework

Name	In-memory	Hybrid	Scalability	Fault tolerance issue	Communication overhead	Framework
PowerGraph [15]	✓	×	✓	×	×	Vertex-centric
GraphX [16]	✓	×	✓	×	×	RDD [16]
Ringo [31]	✓	×	×	×	×	SNAP [22]
Trinity [33]	✓	×	✓	✓	×	Distributed graph engine
Pregelix [6]	✓	×	✓	×	×	Distributed graph system
GraphBig [24]	✓	×	✓	×	-	Vertex-centric
GraphMP [45]	×	✓	✓	×	×	VSW [35]
GraphH [34]	✓	×	✓	×	×	Vertex-centric

Table 2. Evaluation of existing framework. M = Million, B = Billion

Name	Nodes	Edges	Data source
PowerGraph [15]	40M	1.5B	Twitter [20]
GraphX [16]	4.8M	69M	twitter-2010 [4], uk-2007-05 [3]
Ringo [31]	4.8M,42M	69M,1.5B	Twitter [20], LiveJournal [1]
Trinity [33]	1B	13B	Social Graph, Web Graph
Pregelix [6]	6.9M	6 B	BTC [10], Webmap [37]
GraphBig [24]	1.9M	2.8M	CA Road Network [22]
GraphMP [45]	1.1B		Twitter Graph
GraphH [34]	788M	47.6B	Twitter Graph

It uses vertices for computation over edges. In this way, PowerGraph maintains the 'think like a vertex' [38] philosophy. And, it helps in processing trillion of nodes. It exploits parallelism, to achieve less communication and storage costs. PowerGraph supports both asynchronous and synchronous execution. It provides fault-tolerance by vertex replication and data-dependency method.

There is another system, called GraphX [16] which is built on Spark. GraphX supports in-memory system by using Spark storage abstraction, called Resilient Distributed Dataset (RDD) which is essential for iterative graph algorithms. RDD helps in handling trillions of nodes. It also has enough in-memory replication to reduce the re-computation in case of any failure. GraphX retains low-cost fault tolerance by using distributed dataflow networks. GraphX provides ease of analyzing unstructured and tabular data.

Ringo [31] is an in-memory interactive graph analytic system that provides graph manipulation and analysis. Working data set is stored in RAM, and non-working data set are stored in HDD. The prime objective is to provide faster execution rather than scalability. Also, Ringo needs a dynamic graph representation. For efficient graph representation, Ringo uses a Compressed Sparse Row format [17]. Ringo builds on the Stanford Network Analysis Platform (SNAP) [21]. Ringo is easily adaptable due to the integrated processing of graphs and tables. It uses an easy-to-use Python interface and execution on a single machine. However, scalability is a major concern in Ringo. In addition, Ringo is unable to support more than trillions of edges or higher sized Big Graph.

Trinity [33] is a distributed graph engine build over a memory cloud. Memory cloud is a globally addressable, distributed key-value store over a cluster of machines. Data sets can be accessed quickly through distributed in-memory storage. It also supports online query processing as well as offline analysis large graphs. The basic philosophy behind designing the Trinity is (a) high-speed network is readily available today, and (b) DRAM prices will go down in the long run. Trinity is an all-in-memory system, thus, the graph data are loaded in RAM before computation. Trinity has its own language called Trinity specification language (TSL) that minimizes the gap between graph model and data storage. Trinity tries to avoid memory gaps between large numbers of key-value pairs by implementing circular memory management mechanism. It uses heartbeat messages to proactively detect machine failures. Trinity uses Random access data of distributed RAM storage, therefore, it can support trillions of nodes in future. There are many advantages of Trinity, specifically, (1) object-oriented data manipulation of data in the memory cloud, (2) data integration, and (3) TSL facilitates system extension.

Preglix [6] is an open source distributed graph processing system. It supports bulk-synchronous vertex-oriented programming model for analysis of large scale graphs. Pregelix is an iterative dataflow design, and it can effectively handle both in-memory and out-of-core workloads. Pregelix uses Hyracks [5] engine for execution purpose. It is a general-purpose shared-nothing dataflow engine. Pregelix employs both B-Tree and LSM (log-structured merge-tree) B-Tree index structures. These index structures are used to store partitions of vertices on worker machines and these are imported from the Hyracks storage library. The level of fault tolerance in Pregelix is same as other Pregel-like systems. Pregelix system supports larger datasets, and also strengthen multi-user workloads. Pregelix explores more flexible scheduling mechanisms. It gives various data redistribution (allowed by Hyracks) techniques for the optimization of given Pregel algorithm's computation time. It is the only open source system that supports multi-user workloads, has out-of-core support, and allows runtime flexibility. Accommodation of nodes in memory is based on the available memory.

GraphBig [24] is a benchmark suite inspired by IBM System G project. It is a toolkit for computing industrial graphs used by many commercial clients. It is used for performing graph computations and data sources. GraphBig utilizes a dynamic, vertex-centric data representation, which can be oftenly seen in real-world graph systems. GraphBig uses compact format of CSR (Compressed Sparse Row) to save memory space and simplify the graph build complexity. The memory of graph computing shows high cache miss rates on CPUs and also high branch/memory divergence on GPUs.

GraphMP is a semi-external-memory Big Graph processing system. In SEM [45] all vertices of the graph are stored in RAM and edges are accessed from the disk. GraphMP uses vertex-centric sliding window (VSW) computation model. It initially separates the vertices into disjoint intervals. Each interval has a shard. The shard contains the edges that have destination vertices within the interval. During computation, GraphMP slides a window on every vertex and the edges are processed shard by shard. The shard is loaded into RAM for processing. At the end of the program the updates are written to the disk. GraphMP uses a Bloom Filter for selective scheduling to avoid inactive shards. A shard cache mechanism is implemented for complete usage of the memory compressed. GraphMP does not store the edges in memory to handle Big Graph efficiently with limited memory. However, it requires more memory to store all vertices. In addition, it does not use logical locks to improve the performance. It is unable to support trillions of nodes since it is a single machine semi-external memory graph processing system.

GraphH [34] is a memory-disk hybrid approach which maximizes the amount of in-memory data. Initially, the Big Graph is partitioned using two stages. In first stage, the Big Graph is divided into a set of tiles. Each set of tiles uses a compact data structure to store the assigned edges. In the second stage, GraphH assigns the tiles uniformly to computational servers. These servers run the vertex-centric programs. Each vertex maintains a replica of all servers during computation. GraphH implements GAB (Gather-Apply-Broadcast) Computation Model for updating the vertex. Along the in-edges, the data are gathered from local memory to compute the accumulator. GraphH implements Edge Cache Mechanism to reduce the disk access overhead. It is efficient in small cluster or single commodity server. However, it can support trillions of nodes due to GAB model implementation.

4 Key Issues

Power-Law Graph: Power-law graph can be defined as the graph with vertex degree distribution follows a power-law function. It creates many difficulties in the analysis and processing of Big Graph. For example, imbalanced workload in the Big Graph processing systems.

Graph Partitioning: Big Graph uses graph parallel processing technology. The processing requires partitioning of Big Graph into subgraphs. However, the real world graph is highly skewed and have a power-law degree distribution. Hence, Big Graph needs to efficiently partition the graph.

Distributed Graph Placement: Big Graph is stored in cluster of systems. Hence, issues of distributed system are also applicable to the storage and processing of Big Graph. In-memory Big Graph system suffers from scalability issues because RAM is costly and small sized. In addition, in-memory Big Graph must ensure consistency. The data are replicated in several nodes to retain the data even if there is a system fault. In-memory Big Graph systems also has hotspot issue due to the skewed nature of Big Graph.

Cost: In-memory Big Graph processing systems store the graph in RAM. Hence, it requires good quality computing infrastructure to handle such huge size graph. For example, GraphX requires 16 TB memory to process 10 billion edges [16]. During computation the systems require to store the whole graph and also the network-transmitted messages in RAM [34].

Incompetent to Execute Inexact Algorithm: Big Graphs are incapable to execute the inexact algorithms due to graph matching. Most of the inexact algorithms take a very long computational time for processing [8].

5 Key Challenges

Dynamic Graph: Analysis of dynamic graph is an arduous process in which the graph structure changes frequently. In this type of graph, vertices and edges are inserted and deleted frequently [42]. As the dynamic graphs keep on changing, data management and graph analytic takes the responsibility for the sequence of large graph snapshots as well as for streaming data.

Graph-Based Data Integration and Knowledge Graphs: For the analysis purpose, Big Graph data is extracted from original data sources. However, data extraction from the data source is full of obstacles. One key challenge is knowledge graph [12,27]. The knowledge graph provides a huge volume of interrelated information regarding real-world entities. Moreover, key issues with the knowledge graph is integration of low quality, highly diverse and large volume of data.

Graph Data Allocation and Partitioning: Effectiveness of graph processing highly depends on efficient data partitioning of Big Graph. Along with partitioning, load balancing is required for efficient utilization of nodes. Hence, the challenge is to find a graph partitioning technique that balances the distribution of vertices and their edges such that each subgraph have minimum and same number of vertices and vertex cut. But, graph partitioning problem is NP-hard [7].

Interactive Graph Analytic: Interactive graphs with proper visualization is highly desirable for exploration and analysis of graph data. But sometimes this visualization becomes a major challenge to analyze. For example, $k - SNAP$ [38] generates summarized graphs having k vertices. The parameter change k in $k - SNAP$ activates an OLAP-like roll-up and drill-down within a dimension hierarchy [9]. However, due to its dependency on pre-determined parameter, this approach is not fully interactive.

6 Future Research Agenda

An 'in-memory' system requires backing of the secondary storage for consistency due to volatility of RAM. The in-memory system stores data in RAM as well as in HDD/SSD. However, the data of intrinsic 'in-memory' system is stored entirely in RAM. HDD/SSD is used to recover data upon failure of a machine. A few hybrid system stores working datasets in RAM and non-working dataset in secondary storage. Read/write cost is high in secondary storage. Nevertheless, hybrid system becomes more scalable. Hence, there is a trade-off between performance and scalability.

Today, everyone is connected globally through internet. Hence, data is growing exponentially. But, the size of RAM is fixed. Thus, the challenge starts with maintaining large scale data in RAM. Similarly, Big Graph size is also growing. For instance, Twitter and Facebook. Big Graph analytic requires a real-time processing engine which demands in-memory Big Graph system. It is a grand challenge to design a pure in-memory Big Graph database and also a future research agenda. Intrinsic in-memory Big Graph database can be implemented through Dr. Hadoop framework [11]. In this paper, future research agenda is categorized into two categories, namely, data-intensive Big Graph, and compute-intensive Big Graph.

6.1 Data-Intensive Big Graph

Dr. Hadoop: A Future Scope for Big Graph. Dr. Hadoop is a framework of purely in-memory systems [11]. However, Dr. Hadoop backups the data in secondary storage periodically [26]. Even though, Dr. Hadoop is developed for massive scalability of metadata, it can be adapted in various purposes [30]. Dr. Hadoop stores all data in RAM and replicates in other two neighbor RAM, say, left and right node. Figure 1 demonstrates the replication of data in RAM in Dr. Hadoop framework. Dr. Hadoop forms 3-node cluster at the very beginning [30]. Each node must have left and right node for replication of data from RAM.

Any node can leave or join the Dr. Hadoop cluster. Figure 1 illustrates the insertion of a new node in Dr. Hadoop. Dr. Hadoop implements circular doubly linked list where a node failure breaks the ring. But, Dr. Hadoop is unaffected by the failure of any one node at a given time [30]. There are left node and right

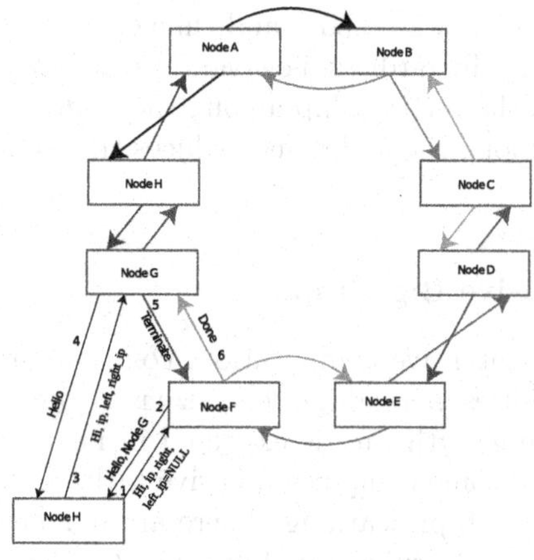

Fig. 1. Insertion of a node in Dr. Hadoop.

node to serve the data. Moreover, Dr. Hadoop can tolerate many non-contiguous node failure at a given time. However, Dr. Hadoop is unable to tolerate consecutive three-node failure at a given time. Three contiguous node failure causes loss of data of a node. Even, Dr. Hadoop can tolerate consecutive two-node failure at a given time. Big Graph can be stored in the RAM and also replicated to two neighboring nodes. The backup is stored in HDD/SSD. The key merits of Dr. Hadoop are- (a) purely in-memory database system, (b) incremental scalability, (c) fine-grained fault-tolerance, (d) efficient load-balancing, and (e) requires least administration. Incorporating Dr. Hadoop with Big Graph can help in overcoming the issues like scalability, fault tolerance, communication overhead associated with existing in-memory system. Hence, implementing Big Graph in Dr. Hadoop framework is future research agendas.

Bloom Filter. Bloom Filter [2] is a probabilistic data structure for approximate query. Bloom Filter requires a tiny on-chip memory to store information of a large set of data. A few modern Big Graph engine deploys Bloom Filter to reduce the on-chip memory requirement. For instance, ABySS [18]. The DNA assembly requires very large size of RAM. Therefore, DNA assembler deploys Bloom Filter for faster processing with low sized RAM. In biological graph such as a de Bruijn graph, Bloom Filter is commonly used to increase it efficiency, for instance, deBGR [28]. [32] uses cascading Bloom Filters to store the nodes of the graph. It reduces construction time of the graph. Gollapudi et al. [14] proposed a Bloom Filter based HITS-like ranking algorithm. Bloom Filter helps in reducing the query time. Similarly, Najork et al. [25] proposed a Bloom Filter based query reduction technique to increase the performance of SALSA. Bloom Filter is used to approximate the neighborhood graph. Bloom Filter has great potential for its implementation in the Big Graph. Bloom Filter is a simple and dumb data structure. However, these two are the parameters for its efficiency.

Its simple data structure makes it space and time efficient. It's dumbness makes its applicable in any field. Regardless, new variants of Bloom Filter are required which are able to store the relationship among the nodes. Such variants help in checking the relation among the nodes and reduces the complexity of Big Graph processing.

6.2 Compute-Intensive Big Graph

In contrast to memory-intensive computation, compute-intensive tasks require more processing capabilities. Nowadays, data size is growing exponentially. However, the decline in the growth rate is expected in near future. Therefore, the future Big Graph will become compute-intensive task. A few research work has been carried out on Big Graph learning. There are numerous machine learning algorithms to evaluate the learning capability on Big Graph data. Hierarchical Anchor Graph Regularization (HAGR) [41] for instance. Deep learning is another example of extreme learning which requires a huge computation capability [36]. Also, machine learning is deployed in spatio-temporal networks [43].

7 Conclusion

Big Graphs are interdependent among each subgraphs. Whole Big Graph cannot be stored in RAM. However, storing whole Big Graph in RAM extremely boosts up the performance. There is future research scope in building in-memory Big Graph database without using HDD/SSD. Big Graph helps in representing the relationship between entities in Big Data. Furthermore, in-memory Big Graph boosts up the system performance. Because, RAM is about 100× faster than SSD/Flash-based Big Graph representation. Moreover, in-memory Big Graphs are nearly 1000× faster than HDD-based representations. In-memory Big Graphs are capable of storing trillions of nodes and edges by using other techniques such as Bloom Filter. Bloom Filter helps in eliminating the duplication in Big Graph. In addition, in-memory Big Graphs have to use scalable framework to increase its storage capacity beyond terabytes. For instance, Dr. Hadoop has infinite scalability and many merits including fault-tolerant, and load balancing. Many in-memory Big Graph techniques are proposed which are discussed in the paper. However, one big challenge in in-memory Big Graph is RAM. RAM is very costly and using a large size is impractical in current scenario. But high performance can be achieved through in-memory representations of Big Graph. The trade-off between cost and performance creates difference among HDD-based, SSD/Flashed-based and in-memory based representation of Big Graph. The choice depends on the priority of the applications. Most of the applications require high performance. Hence, in-memory Big Graph is near future, since RAM cost is dropping.

References

1. Backstrom, L., Huttenlocher, D., Kleinberg, J., Lan, X.: Group formation in large social networks: membership, growth, and evolution. In: Proceedings of the 12th ACM SIGKDD International Conference on Knowledge Discovery and Data Mining, pp. 44–54. ACM (2006)
2. Bloom, B.H.: Space/time trade-offs in hash coding with allowable errors. Commun. ACM **13**(7), 422–426 (1970)
3. Boldi, P., Rosa, M., Santini, M., Vigna, S.: Layered label propagation: a multiresolution coordinate-free ordering for compressing social networks. In: Proceedings of the 20th International Conference on World Wide Web, pp. 587–596. ACM (2011)
4. Boldi, P., Vigna, S.: The webgraph framework I: compression techniques. In: Proceedings of the 13th International Conference on World Wide Web, pp. 595–602. ACM (2004)
5. Borkar, V., Carey, M., Grover, R., Onose, N., Vernica, R.: Hyracks: a flexible and extensible foundation for data-intensive computing. In: Proceedings of the 2011 IEEE 27th International Conference on Data Engineering, ICDE 2011, pp. 1151–1162. IEEE Computer Society (2011)
6. Bu, Y., Borkar, V., Jia, J., Carey, M.J., Condie, T.: Pregelix: Big(ger) graph analytics on a dataflow engine. Proc. VLDB Endow. **8**(2), 161–172 (2014). https://doi.org/10.14778/2735471.2735477
7. Buluç, A., Meyerhenke, H., Safro, I., Sanders, P., Schulz, C.: Recent advances in graph partitioning. In: Kliemann, L., Sanders, P. (eds.) Algorithm Engineering. LNCS, vol. 9220, pp. 117–158. Springer, Cham (2016). https://doi.org/10.1007/978-3-319-49487-6_4
8. Carletti, V., Foggia, P., Greco, A., Saggese, A., Vento, M.: Comparing performance of graph matching algorithms on huge graphs. Pattern Recognit. Lett. (2018)
9. Chen, C., Yan, X., Zhu, F., Han, J., Philip, S.Y.: Graph OLAP: towards online analytical processing on graphs. In: Eighth IEEE International Conference on Data Mining, ICDM 2008, pp. 103–112. IEEE (2008)
10. Cheng, J., Ke, Y., Chu, S., Cheng, C.: Efficient processing of distance queries in large graphs: a vertex cover approach. In: Proceedings of the 2012 ACM SIGMOD International Conference on Management of Data, pp. 457–468. ACM (2012)
11. Dev, D., Patgiri, R.: Dr. Hadoop: an infinite scalable metadata management for Hadoop–How the baby elephant becomes immortal. Front. Inf. Technol. Electron. Eng. **17**(1), 15–31 (2016). https://doi.org/10.1631/FITEE.1500015
12. Dong, X., et al.: Knowledge vault: a web-scale approach to probabilistic knowledge fusion. In: Proceedings of the 20th ACM SIGKDD International Conference on Knowledge Discovery and Data Mining, KDD 2014, pp. 601–610. ACM (2014)
13. Gao, W., Wu, H., Siddiqui, M.K., Baig, A.Q.: Study of biological networks using graph theory. Saudi J. Biol. Sci. **25**, 1212–1219 (2017)
14. Gollapudi, S., Najork, M., Panigrahy, R.: Using bloom filters to speed up HITS-like ranking algorithms. In: Bonato, A., Chung, F.R.K. (eds.) WAW 2007. LNCS, vol. 4863, pp. 195–201. Springer, Heidelberg (2007). https://doi.org/10.1007/978-3-540-77004-6_16
15. Gonzalez, J.E., Low, Y., Gu, H., Bickson, D., Guestrin, C.: PowerGraph: distributed graph-parallel computation on natural graphs. In: Proceedings of the 10th USENIX Conference on Operating Systems Design and Implementation, OSDI 2012, pp. 17–30. USENIX Association (2012)

16. Gonzalez, J.E., Xin, R.S., Dave, A., Crankshaw, D., Franklin, M.J., Stoica, I.: Graphx: graph processing in a distributed dataflow framework. In: OSDI, vol. 14, pp. 599–613 (2014)
17. Gregor, D., Willcock, J., Lumsdaine, A.: Compressed sparse row graph. https://www.boost.org/doc/libs/1_57_0/libs/graph/doc/compressed_sparse_row.html. Accessed 21 June 2018
18. Jackman, S.D., et al.: Abyss 2.0: resource-efficient assembly of large genomes using a Bloom filter. Genome Res. **27**, 768–777 (2017). https://doi.org/10.1101/gr.214346.116
19. Kui, X., Samanta, A., Zhu, X., Li, Y., Zhang, S., Hui, P.: Energy-aware temporal reachability graphs for time-varying mobile opportunistic networks. IEEE Trans. Veh. Technol. **67**, 9831–9844 (2018). https://doi.org/10.1109/TVT.2018.2854832
20. Kwak, H., Lee, C., Park, H., Moon, S.: What is Twitter, a social network or a news media? In: Proceedings of the 19th International Conference on World Wide Web, pp. 591–600. ACM (2010)
21. Leskovec, J.: Stanford network analysis project. http://snap.stanford.edu/. Accessed 22 June 2018
22. Leskovec, J., Perez, Y., Sosic, R.: Snap datasets. http://snap.stanford.edu/ringo/. Accessed 20 June 2018
23. Myers, S.A., Sharma, A., Gupta, P., Lin, J.: Information network or social network?: the structure of the Twitter follow graph. In: Proceedings of the 23rd International Conference on World Wide Web, pp. 493–498. ACM (2014)
24. Nai, L., Xia, Y., Tanase, I.G., Kim, H., Lin, C.Y.: GraphBIG: understanding graph computing in the context of industrial solutions. In: SC15: International Conference for High Performance Computing, Networking, Storage and Analysis, pp. 1–12 (2015). https://doi.org/10.1145/2807591.2807626
25. Najork, M., Gollapudi, S., Panigrahy, R.: Less is more: sampling the neighborhood graph makes salsa better and faster. In: Proceedings of the Second ACM International Conference on Web Search and Data Mining, pp. 242–251. ACM (2009)
26. Nayak, S., Patgiri, R.: Dr. Hadoop: in search of a needle in a Haystack. In: Fahrnberger, G., Gopinathan, S., Parida, L. (eds.) ICDCIT 2019. LNCS, vol. 11319, pp. 99–107. Springer, Cham (2019). https://doi.org/10.1007/978-3-030-05366-6_8
27. Nickel, M., Murphy, K., Tresp, V., Gabrilovich, E.: A review of relational machine learning for knowledge graphs. Proc. IEEE **104**(1), 11–33 (2016)
28. Pandey, P., Bender, M.A., Johnson, R., et al.: deBGR: an efficient and near-exact representation of the weighted de Bruijn graph. Bioinformatics **33**(14), i133–i141 (2017)
29. Paranjape, A., Benson, A.R., Leskovec, J.: Motifs in temporal networks. In: Proceedings of the Tenth ACM International Conference on Web Search and Data Mining, pp. 601–610. ACM (2017)
30. Patgiri, R., Nayak, S., Dev, D., Borgohain, S.K.: Dr. Hadoop cures in-memory data replication system. In: 6th International Conference on Advanced Computing, Networking, and Informatics, 04–06 June 2018 (2018)
31. Perez, Y., et al.: Ringo: interactive graph analytics on big-memory machines. In: Proceedings of the 2015 ACM SIGMOD International Conference on Management of Data, SIGMOD 2015, pp. 1105–1110. ACM (2015). https://doi.org/10.1145/2723372.2735369
32. Salikhov, K., Sacomoto, G., Kucherov, G.: Using cascading bloom filters to improve the memory usage for de Brujin graphs. Algorithms Mol. Biol. **9**(1), 2 (2014)

33. Shao, B., Wang, H., Li, Y.: Trinity: a distributed graph engine on a memory cloud. In: Proceedings of the 2013 ACM SIGMOD International Conference on Management of Data, SIGMOD 2013, pp. 505–516. ACM (2013). https://doi.org/10.1145/2463676.2467799
34. Sun, P., Wen, Y., Duong, T.N.B., Xiao, X.: GraphH: high performance big graph analytics in small clusters. In: 2017 IEEE International Conference on Cluster Computing (CLUSTER), pp. 256–266. IEEE (2017)
35. Sun, P., Wen, Y., Duong, T.N.B., Xiao, X.: GraphMP: an efficient semi-external-memory big graph processing system on a single machine. In: 2017 IEEE 23rd International Conference on Parallel and Distributed Systems (ICPADS), pp. 276–283. IEEE (2017)
36. Sun, Y., Li, B., Yuan, Y., Bi, X., Zhao, X., Wang, G.: Big graph classification frameworks based on extreme learning machine. Neurocomputing **330**, 317–327 (2019). https://doi.org/10.1016/j.neucom.2018.11.035
37. Tabaja, A.: Yahoo!webscope program. https://webscope.sandbox.yahoo.com/. Accessed 20 June 2018
38. Tian, Y., Balmin, A., Corsten, S.A., Tatikonda, S., McPherson, J.: From "think like a vertex" to "think like a graph". Proc. VLDB Endow. **7**(3), 193–204 (2013). https://doi.org/10.14778/2732232.2732238
39. Ugander, J., Karrer, B., Backstrom, L., Marlow, C.: The anatomy of the facebook social graph. arXiv preprint arXiv:1111.4503 (2011)
40. Wang, D., Pedreschi, D., Song, C., Giannotti, F., Barabasi, A.L.: Human mobility, social ties, and link prediction. In: Proceedings of the 17th ACM SIGKDD International Conference on Knowledge Discovery and Data Mining, pp. 1100–1108. ACM (2011)
41. Wang, M., Fu, W., Hao, S., Liu, H., Wu, X.: Learning on big graph: label inference and regularization with anchor hierarchy. IEEE Trans. Knowl. Data Eng. **29**(5), 1101–1114 (2017). https://doi.org/10.1109/TKDE.2017.2654445
42. Yan, D., Bu, Y., Tian, Y., Deshpande, A., Cheng, J.: Big graph analytics systems. In: Proceedings of the 2016 International Conference on Management of Data, pp. 2241–2243. ACM (2016)
43. Yu, B., Yin, H., Zhu, Z.: Spatio-temporal graph convolutional networks: a deep learning framework for traffic forecasting. In: Proceedings of the Twenty-Seventh International Joint Conference on Artificial Intelligence (IJCAI-18), pp. 3634–3640 (2017)
44. Zaharia, M., et al.: Resilient distributed datasets: a fault-tolerant abstraction for in-memory cluster computing. In: Proceedings of the 9th USENIX Conference on Networked Systems Design and Implementation, p. 2. USENIX Association (2012)
45. Zheng, D., Mhembere, D., Lyzinski, V., Vogelstein, J.T., Priebe, C.E., Burns, R.: Semi-external memory sparse matrix multiplication for billion-node graphs. IEEE Trans. Parallel Distrib. Syst. **28**(5), 1470–1483 (2017)

Modelling Legal Documents for Their Exploitation as Open Data

John Garofalakis[1], Konstantinos Plessas[1], Athanasios Plessas[1](✉), and Panoraia Spiliopoulou[2]

[1] Department of Computer Engineering and Informatics, University of Patras, Panepistimioupoli, 26504 Patras, Greece
{garofala, kplessas, plessas}@ceid.upatras.gr

[2] School of Law, National and Kapodistrian University of Athens, Akadimias 45, 10672 Athens, Greece
panoraia@gmail.com

Abstract. As our society becomes more and more complex, legal documents are produced at an increasingly fast pace, generating datasets that show many of the characteristics that define Big Data. On the other hand, as the trend of Open Data has spread widely in the government sector nowadays, publication of legal documents in the form of Open Data is expected to yield important benefits. In this paper, we propose the modelling of Greek legal texts based on the Akoma Ntoso document model, which is a necessary step for their representation as Open Data and we describe use cases that show how these massive legal open datasets could be further exploited.

Keywords: Akoma Ntoso · Legal text modelling · Legal Open Data · Legal big data · Greek legislation

1 Introduction

In all contemporary democratic States, the relationships between the members of society and the actors of social, political and economic activities are regulated by acts passed by legislative bodies such as the Parliament or the Congress. However, each country's legal framework is not formed only by sources of primary legislation, but also from texts known as secondary or delegated legislation (e.g. ministerial decisions, cabinet acts, acts of legislative content etc.). A variety of several other documents related to national and international laws and the details of their implementation exist: legal opinions, court decisions, decisions of independent authorities, decisions of public bodies, administrative circulars, international conventions and transnational agreements are just some examples.

Since our society becomes more and more complex and as the number of freely available online legal sources increases, the volume of these heterogeneous documents that are related to law is constantly rising, forming datasets that show many of the characteristics that define Big Data [1]. Even though the number of documents in these legal datasets is not such to consider them as Big Data under today's sense, especially compared to data collected from sources such as social networks, it is still impossible to

W. Abramowicz and R. Corchuelo (Eds.): BIS 2019, LNBIP 353, pp. 30–44, 2019.
https://doi.org/10.1007/978-3-030-20485-3_3

manually analyze them, and automatic processes are needed to undertake such tasks [2]. Mazzega and Lajaunie [1] note that there exist no definite measures even for the most well-known dimensions of Big Data (volume, variety, velocity) and argue that Big Data approaches are required to analyze legal text corpuses, agreeing that their volume makes manual processing prohibitive.

Initiatives such as the "*free access to law movement*" [3] will result in governments providing access to even more legal information resources in the future. On the other hand, public administrations are expected to apply the Open Data model to the legal documents publishing process, in order to take advantage of the benefits that Open Government Data can bring. Expected benefits include, among others, promotion of transparency and democratic accountability, strengthening of public engagement, improved policy making, stimulation of innovation, competitiveness and economic development [4]. Nevertheless, the ever-expanding datasets of legal information that are currently made available mostly contain unstructured or poorly structured documents [5]. The use of formats that are not machine-processable for data representation is recognized as an important barrier for Open Data exploitation [6], while it is also a practice that hinders interoperability. As shown in the related work section, the described situation has attracted attention from the research community and several projects trying to tackle the problem were undertaken. These efforts are mainly based on the standards that emerged for the representation of legal documents in structured Open Data formats, namely Akoma Ntoso and CEN Metalex [7].

In this paper, we present the modelling of Greek legal documents for their exploitation as Open Data, as a first step to overcome the barrier of unstructured legal texts. Our modelling is based on the Akoma Ntoso document model. Our work is part of a project that aims to automatically transform available datasets of Greek legal documents into Legal Open Data. Moreover, we describe some use case scenarios that show how the resulted datasets could be further exploited.

2 Related Work

Several research efforts focus on modelling legal texts. Researchers either present and discuss the localization of models such as CEN Metalex and Akoma Ntoso or undertake the modelling task in the framework of projects aiming at the representation of legal documents as (Linked) Open Data.

A work belonging to the first category is that of Gen et al. [8], where the authors describe how the Akoma Ntoso standard can be adapted in the case of Japanese legislation. The researchers show how they tackle the problem of structural ambiguity, which is caused by the flexibility of the standard and propose a method of mapping rules for the conversion of texts originally marked up in the Japanese statutory schema to the Akoma Ntoso schema. Another approach of modelling legislation is described in [9], however the presented choices (e.g. metadata management, naming convention etc.) have already well-established foundations in existing standards such as Akoma Ntoso and CEN Metalex. Another work on modelling legal acts following the CEN Metalex specification can be found in [10].

In [11], Marković et al. after reviewing several mechanisms for the effective identification and description of judgments, concluded that Akoma Ntoso meets the requirements for the representation of Serbian court decisions and showed how these documents can be modelled according to the Akoma Ntoso document model.

Other research efforts focus on modelling legislative documents at a higher abstraction level, using ontologies. As an example, one could mention the work of Ceci and Palmirani on an ontology framework for modelling judgments [12] or the work of Oksanen et al. [13], where ontological models (e.g. FRBR, ELI etc.) are extended and combined for the representation of Finnish law and court decisions as Linked Open Data using RDF.

A project that follows a similar approach to ours for the transformation of legislation and case law into Linked Open Data is EUCases [14]. In the framework of this project, the Akoma Ntoso standard was adopted in order to model the collected legislative and judicial documents.

Interestingly, several projects on opening legal documents were undertaken during the last 2–3 years in Greece. Koniaris et al. [15] were the first to present a modelling approach of Greek legal resources (laws, presidential decrees and regulatory acts of the council of ministers) that was based on the Akoma Ntoso standard. Moreover, Angelidis et al. [16] describe the metadata and structure of Greek legal documents (laws, presidential decrees, regulatory acts of the council of ministers, ministerial decisions) and present their modelling using the Nomothesia ontology. The same research team proposed a similar approach [17] for modelling decisions of Greek public administration bodies according to the Diavgeia ontology.

3 The Project

This research work is part of a wider 15-month (June 2018 – September 2019) project entitled "Automated Analysis and Processing of Legal Texts for their Transformation into Legal Open Data" [18], which is funded by the European Social Fund and Greek national funds under the call "Support for Researchers, with Emphasis on Young Researchers" of the Operational Program "Human Resources Development, Education and Lifelong Learning". The project aims to exploit the strict structure of legal documents and take advantage of natural language processing techniques in order to automatically transform available unstructured Greek legal texts into Legal Open Data.

As described in [18], the project sets several research questions, however the work presented in this paper addresses the following:

- Q1: How is it possible to formally model the different types of Greek legal documents (in terms of structure, content and metadata)?
- Q2: Which open standards should be adopted and how should we extend them in order to adapt to the needs of the Greek legal system?
- Q3: In which way can we combine available Legal Open Data and exploit them in new services and applications?

In the framework of this project, we will investigate the possibility to analyze and transform into Open Data the following types of Greek legal texts:

- Primary and secondary legislation, including Laws, Presidential Decrees, Ministerial Decisions and Acts of Legislative Content.
- Court Decisions published from the Supreme Courts (the Supreme Civil and Criminal Court of Greece and the Hellenic Council of State).
- Bills submitted to the Hellenic Parliament and their attached reports.
- Legal opinions of the Legal Council of the Hellenic State.
- Decisions of Independent Authorities.
- Circulars published by public administration bodies.

In the next section we discuss the modelling of court decisions, legal opinions and administrative circulars. We do not refer to documents of Greek primary legislation, since their modelling for Akoma Ntoso representation is covered in [15].

4 Modelling of Legal Texts

4.1 The Akoma Ntoso Standard

Akoma Ntoso, recently accepted as an OASIS standard, provides an XML schema for the machine-readable representation of parliamentary, legislative and judiciary documents and a naming convention for their unique identification, based on the FRBR model [19]. Akoma Ntoso supports both structure and metadata modelling. Moreover, the standard allows for separation of the different layers of legal documents: text, structure, metadata, ontology, legal rules. It implements the first three levels and provides hooks to external ontologies and legal knowledge modelling. Apart from structural and metadata elements, the schema provides also semantic elements, which can be used to capture the legal meaning of parts of the text (e.g. recognize legal references, the name of a judge, dates etc.)

4.2 Modelling Court Decisions

In Greece, there are publicly available (in HTML format) the anonymized decisions of two Supreme Courts: Areios Pagos, which is the Supreme Civil and Criminal Court and the Council of State, which is the Supreme Administrative Court. In both cases, the decisions have similar structure and it is possible to identify and extract the elements needed to build the metadata block of the Akoma Ntoso representation, the structural parts and several semantic elements also (e.g. name of judges and lawyers, the litigant parties etc.).

The metadata block of the Akoma Ntoso model contains a block for uniquely identifying the document (*<identification>* element) according to the FRBR model, as well as other optional blocks such as the publication (contains information about the publication of the document), the lifecycle (lists the events that modify the document), the classification (keyword classification), the workflow (lists the procedural steps necessary for the delivery of the decision), the analysis (contains the result of the decision and the qualification of the case law citations) and the references (models references to other documents or ontology classes). According to the FRBR model, every Akoma Ntoso document can be considered in one of the following levels: Work level (the abstract concept of the legal resource), Expression level (a specific version of the Work) and Manifestation level (any electronic or physical format of the Expression). Figure 1 shows the identification block of the metadata part of a court decision (decision A5515/2012 of the Council of State).

```
<identification source="#openLawsGR">
    <FRBRWork>
        <FRBRthis value="/akn/gr/judgement/COS/2012/A5515/!main"/>
        <FRBRuri value="/akn/gr/judgement/COS/2012/A5515/"/>
        <FRBRalias value="ECLI:EL:COS:2012:1231A5515.03E3256" name="ECLI"/>
        <FRBRdate date="2012-12-31" name=""/>
        <FRBRauthor href="#councilOfState"/>
        <FRBRcountry value="gr"/>
    </FRBRWork>
    <FRBRExpression>
        <FRBRthis value="/akn/gr/judgement/COS/2012/A5515/ell@/!main"/>
        <FRBRuri value="/akn/gr/judgement/COS/2012/A5515/ell@"/>
        <FRBRdate date="2012-12-31" name=""/>
        <FRBRauthor href="#councilOfState"/>
        <FRBRlanguage language="ell"/>
    </FRBRExpression>
    <FRBRManifestation>
        <FRBRthis value="/akn/gr/judgement/COS/2012/A5515/ell@/!main.xml"/>
        <FRBRuri value="/akn/gr/judgement/COS/2012/A5515/ell@.xml"/>
        <FRBRdate date="2018-10-15" name="XMLConversion"/>
        <FRBRauthor href="#openLawsGR"/>
    </FRBRManifestation>
</identification>
```

Fig. 1. The identification block of the metadata part of a court decision

IRIs (Internationalized Resource Identifiers) follow the Akoma Ntoso naming convention. The Work level identifier consists in general of the following parts, separated by slashes: the akn prefix, the ISO 3166 country code, the type of the document, the emanating actor, the date or year and the number or title of the Work. The identifier of the Expression is an extension of the Work identifier containing the ISO 639-2 alpha-3 three-letter code for the language of the document, followed by the @ character. The identifier of the Manifestation is an extension of the Expression identifier, followed by a dot and a three or four-letter string signifying the file format (e.g. xml). IRIs may contain several other optional parts for disambiguation purposes.

Akoma Ntoso provides a specific document type for the representation of court decisions: the Judgment document type. Figure 2 shows the XSD diagram for the *<judgment>* element.

At the beginning of the decisions of the two Supreme courts there is text referring to the number of the decision and the name of the issuing court, the date of the court hearing, the composition of the court, the litigant parties and their lawyers. This part corresponds to the *<header>* element. The rest of the decision, apart from a phrase about the location and date of the court conference and the signatures of the judges that are assigned to the *<conclusions>* element, belongs to the *<judgmentBody>* element, which represents the main body of the decision. We assign the next paragraphs, which contain information about the previous decisions that are appealed and about the trial

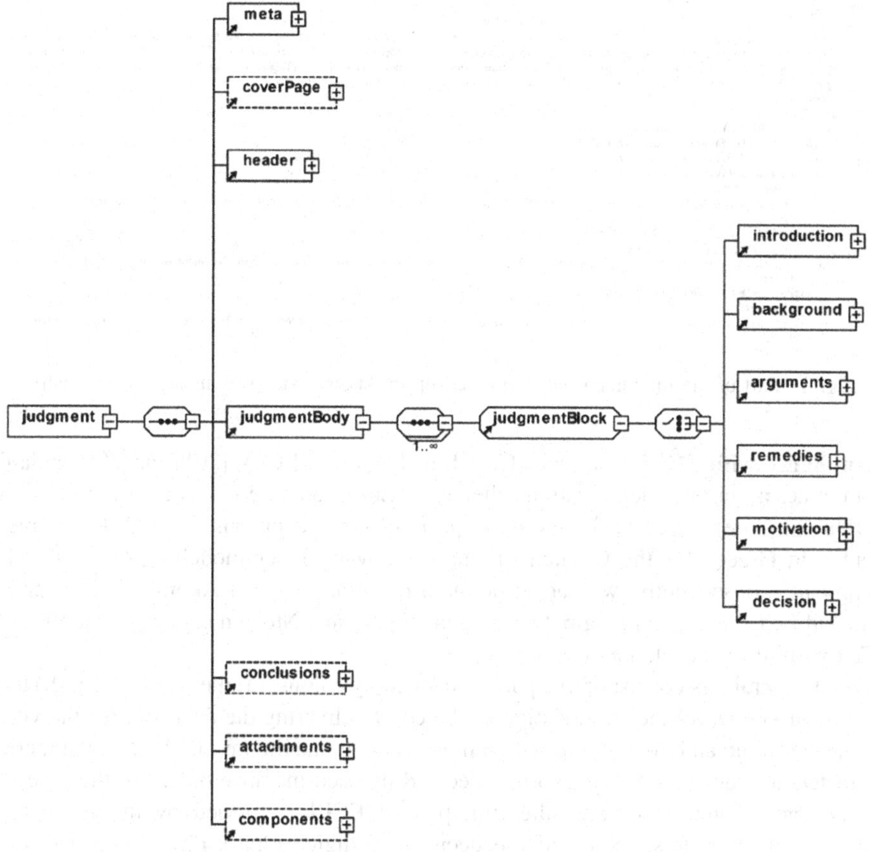

Fig. 2. The Akoma Ntoso Judgment document type

procedure to the *<introduction>* element. A standard phrase (e.g. "*the court after studying the relevant documents, considered the law*") is followed by a listing of points that are explaining how the judges reached the decisions. This part is assigned to the *<motivation>* element, since it contains their argumentation. We dismiss the *<background>* element, since the facts were analyzed in the decisions of lower courts that are appealed and this analysis is not included in the decisions of the Supreme Courts. Some references to the facts may be found within the motivation list, however it is impossible to separate them from the argumentation. The final part of the judgment body, usually beginning with the phrase "*for these reasons*", contains the decision of the court about the case and is included within the *<decision>* element. Several semantic elements of the Akoma Ntoso can be used to provide semantic information about concepts found in the decision text: *<judge>*, *<party>*, *<lawyer>*, *<docDate>*, *<docProponent>*, *<role>*, *<ref>* etc. Figure 3 shows part of a decision of the Council of State (translated to English) marked up in Akoma Ntoso.

```
<judgmentBody>
    <introduction>
        <p>In this application, the applicant bank seeks to annul: 1) the decision No. 1 / 15-11-2002 of the Secondary Board of Appeal of Forest
        <p>The hearing began with the report of the rapporteur, <role refersTo="#privy.councillor">Counselor</role> <judge refersTo="#th.aravani
        <p>The court then heard the assignee of the applicant bank, who also spoke verbally and claimed the application and the representative c
        <p>Following the public hearing, the court held a conference in a courtroom and</p>
    </introduction>
    <motivation>
        <p>Having studied the relevant documents</p>
        <p>Considered the facts according to the Law</p>
        <blockList eId="motivation__list_1">
            <item eId="motivation__list_1__item_1">
                <num>1</num>
                <p>Since the lawful fee (389537, 754017/2003 parcel notes) was paid for the execution of the requested application</p>
            </item>
            <item eId="motivation__list_1__item_2">
                <num>2</num>
                <p>Since this application, as supplemented by an application for additional grounds, seeks the annulment of a) 8213 / 19.11.1998
            </item>
            <item eId="motivation__list_1__item_3">
                <num>3</num>
                <p>Since the first and second acts are inadmissible because they are legally subject to an intra-Community procedure, which was
            </item>
```

Fig. 3. Part of the judgement body marked up in Akoma Ntoso (translated to English)

Compliance with ECLI. European Case Law Identifier (ECLI) [20] is an EU standard that defines a uniform identification scheme for European case law and a minimum set of (Dublin Core) metadata for the description of court decisions. As ECLI is implemented in Greece for the Council of State, we wanted our modelling to be ECLI-compliant. Consequently, we had to define a mapping between Akoma Ntoso identifiers and ECLI and a mapping between available Akoma Ntoso metadata elements and ECLI required metadata elements.

ECLI identifiers consist of five parts separated by colons: (a) the word ECLI, (b) the EU country code, (c) the abbreviation of the court delivering the decision, (d) the year of the decision and (e) an ordinal number with a maximum of 25 alphanumeric characters and dots following a format decided by each member state. For the case of the Hellenic Council of State, the fifth part of ECLI is formed by the following sequence of characters: month of the decision (2-digit format), day of the decision (2-digit format), number of the decision, a dot (.) and finally a string consisted of the year (2-digit format) of the notice of appeal to the court and its number. Figure 4 shows the mappings needed to form ECLI from the Akoma Ntoso elements. The number and year of the notice of appeal is available as a metadata element in the website of the Counsil of State. The element *<FRBRalias>* is a metadata element that can be used to denote other names of the document at the Work level, therefore it is appropriate to handle the value of ECLI.

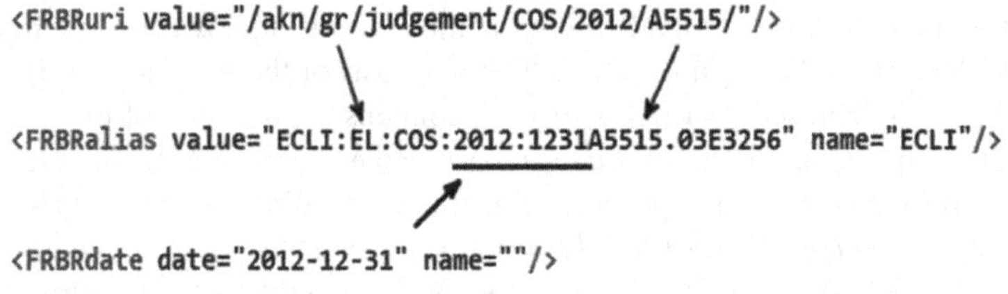

Fig. 4. Mapping between Akoma Ntoso and ECLI identifiers

Table 1 shows the mapping between the required metadata elements of ECLI and the respective metadata elements of Akoma Ntoso.

Table 1. Mapping between ECLI metadata elements and Akoma Ntoso metadata elements

ECLI metadata element	Akoma Ntoso metadata element or default value
dcterms:identifier	The URL from which the text of the decision can be retrieved
dcterms:isVersionOf	The ECLI (<FRBRwork> → <FRBRalias>)
dcterms:creator	<FRBRWork> → <FRBRauthor>
dcterms:coverage	<FRBRWork> → <FRBRcountry>
dcterms:date	<FRBRWork> → <FRBRdate>
dcterms:language	<FRBRExpression> → <FRBRlanguage>
dcterms:publisher	<FRBRWork> → <FRBRauthor>
dcterms:accessRights	"public"
dcterms:type	"judgment"

4.3 Modelling Legal Opinions

In Greece, legal opinions are issued from the Legal Council of State when public administration bodies are making formal questions regarding the application of the law. While legal opinions could be modelled using more generic document types of the Akoma Ntoso standard (e.g. the Statement type can be used to represent formal expressions of opinion or will), we decided to adopt again the Judgment type, since the structure of legal opinions resembles that of court decisions. In most cases, legal opinions are analyzed in the following parts: At the beginning, there is some basic information regarding the opinion (opinion's number, session, composition of the Council, summary of the question etc.). This part is assigned to the *<header>* element, which is followed by the main body of the legal opinion (*<judgmentBody>*). The first part of the main body is the detailed background of the case that prompted the body to submit the question, which is assigned to the *<background>* element, since it corresponds to the description of the facts. The next section usually cites the applicable provisions and the section that follows contains their interpretation relating to the question. Both sections are part of the *<motivation>* block, since they contain the arguments of the Council that led their members to express the opinion. The final part of the legal opinion's body cites the opinion that the members of the Council express regarding the question and it is assigned to the *<decision>* element. The document ends with the signatures (*<conclusions>* block). Figure 5 shows the above described modelling for an indicative part of a legal opinion (translated to English).

The metadata section is formed in a similar manner as in the case of court decisions. Additionally, for each legal opinion published on the website of the Legal Council of State there is a set of keywords that describe the opinion. Akoma Ntoso provides the *<classification>* metadata element for handling keywords. In Fig. 6, we show how we use this metadata block.

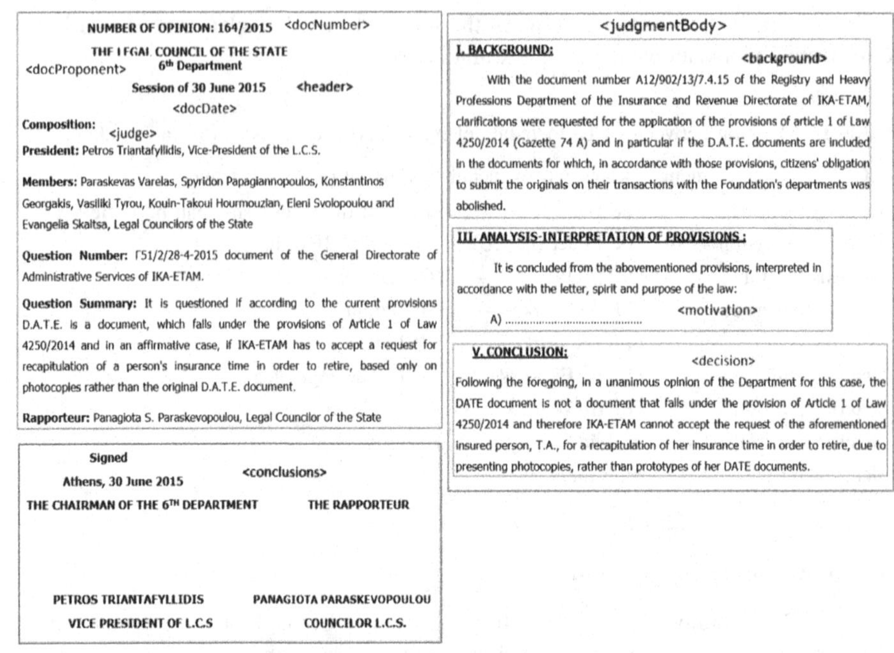

NUMBER OF OPINION: 164/2015 <docNumber>

THE LEGAL COUNCIL OF THE STATE
<docProponent> 6th Department

Session of 30 June 2015 <header>

<docDate>

Composition: <judge>

President: Petros Triantafyllidis, Vice-President of the L.C.S.

Members: Paraskevas Varelas, Spyridon Papagiannopoulos, Konstantinos Georgakis, Vasiliki Tyrou, Kouin-Takoui Hourmouzian, Eleni Svolopoulou and Evangelia Skaitsa, Legal Councilors of the State

Question Number: Γ51/2/28-4-2015 document of the General Directorate of Administrative Services of IKA-ETAM.

Question Summary: It is questioned if according to the current provisions D.A.T.E. is a document, which falls under the provisions of Article 1 of Law 4250/2014 and in an affirmative case, if IKA-ETAM has to accept a request for recapitulation of a person's insurance time in order to retire, based only on photocopies rather than the original D.A.T.E. document.

Rapporteur: Panagiota S. Paraskevopoulou, Legal Councilor of the State

Signed

Athens, 30 June 2015 <conclusions>

THE CHAIRMAN OF THE 6TH DEPARTMENT THE RAPPORTEUR

PETROS TRIANTAFYLLIDIS PANAGIOTA PARASKEVOPOULOU

VICE PRESIDENT OF L.C.S COUNCILOR L.C.S.

<judgmentBody>

I. BACKGROUND: <background>

With the document number A12/902/13/7.4.15 of the Registry and Heavy Professions Department of the Insurance and Revenue Directorate of IKA-ETAM, clarifications were requested for the application of the provisions of article 1 of Law 4250/2014 (Gazette 74 A) and in particular if the D.A.T.E. documents are included in the documents for which, in accordance with those provisions, citizens' obligation to submit the originals on their transactions with the Foundation's departments was abolished.

III. ANALYSIS-INTERPRETATION OF PROVISIONS:

It is concluded from the abovementioned provisions, interpreted in accordance with the letter, spirit and purpose of the law:

A) .... <motivation>

V. CONCLUSION: <decision>

Following the foregoing, in a unanimous opinion of the Department for this case, the DATE document is not a document that falls under the provision of Article 1 of Law 4250/2014 and therefore IKA-ETAM cannot accept the request of the aforementioned insured person, T.A., for a recapitulation of her insurance time in order to retire, due to presenting photocopies, rather than prototypes of her DATE documents.

Fig. 5. Modelling of a legal opinion (translated to English)

```
<classification source="#openLawsGR">
    <keyword eId="keyword_1" value="COMPANIES" showAs="COMPANIES" dictionary="none"/>
    <keyword eId="keyword_2" value="COURT DECISION" showAs="COURT DECISION" dictionary="none"/>
    <keyword eId="keyword_3" value="FINE" showAs="FINE" dictionary="none"/>
    <keyword eId="keyword_4" value="LABOUR INSPECTION" showAs="LABOUR INSPECTION" dictionary="none"/>
</classification>
```

Fig. 6. Metadata section containing keywords describing the opinion (translated to English)

4.4 Modelling Administrative Circulars

Circulars are documents released from public bodies, which clarify how legal provisions should get applied and define the relevant processes or actions. Administrative bodies in Greece publish their circulars on their websites and on the Diavgeia website, which is the national portal where public administration's documents are uploaded. The structure and content of circulars differ from organization to organization, while even circulars of the same organization may show significant differences. For this reason, we adopted the Statement document type (denoting documents with no or limited legal effect) to represent administrative circulars, which provides a more general structure (shown in Fig. 7). The Doc type could also be used, as it provides the same structure as the Statement type.

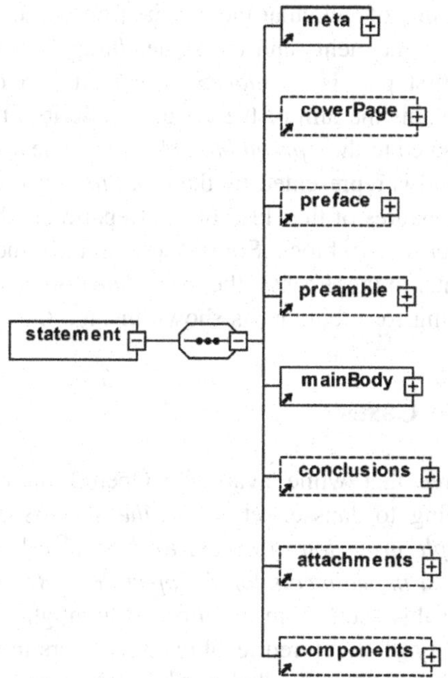

Fig. 7. The Akoma Ntoso Statement document type

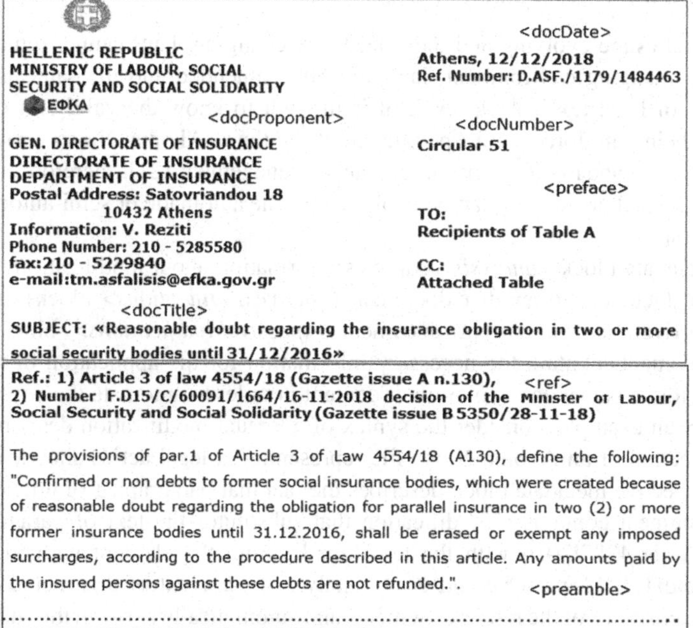

<docDate>

HELLENIC REPUBLIC
MINISTRY OF LABOUR, SOCIAL SECURITY AND SOCIAL SOLIDARITY
ΕΦΚΑ

<docProponent>

GEN. DIRECTORATE OF INSURANCE
DIRECTORATE OF INSURANCE
DEPARTMENT OF INSURANCE
Postal Address: Satovriandou 18
10432 Athens
Information: V. Reziti
Phone Number: 210 - 5285580
fax:210 - 5229840
e-mail:tm.asfalisis@efka.gov.gr

Athens, 12/12/2018
Ref. Number: D.ASF./1179/1484463

<docNumber>

Circular 51

<preface>

TO:
Recipients of Table A

CC:
Attached Table

<docTitle>

SUBJECT: «Reasonable doubt regarding the insurance obligation in two or more social security bodies until 31/12/2016»

Ref.: 1) Article 3 of law 4554/18 (Gazette issue A n.130), <ref>
2) Number F.D15/C/60091/1664/16-11-2018 decision of the Minister of Labour, Social Security and Social Solidarity (Gazette issue B 5350/28-11-18)

The provisions of par.1 of Article 3 of Law 4554/18 (A130), define the following: "Confirmed or non debts to former social insurance bodies, which were created because of reasonable doubt regarding the obligation for parallel insurance in two (2) or more former insurance bodies until 31.12.2016, shall be erased or exempt any imposed surcharges, according to the procedure described in this article. Any amounts paid by the insured persons against these debts are not refunded.".

<preamble>

....

Fig. 8. Part of a circular (translated to English) modelled in Akoma Ntoso

Usually, at the beginning of a circular there is the frontispiece with the details of the issuer, the date, the list of recipients and the subject/title. The Akoma Ntoso element used to represent this first part is *<preface>*. The next part of the document lists relevant documents, explains the aim of the circular, describes the legal basis and the recitals. This part is assigned to the *<preamble>* block. The next section holds the main content of the circular and is represented by the *<mainBody>* element. After the main content, there are the signatures of the Head of the Department that issued the circular, represented by the *<conclusions>* block. Sometimes, circulars include annexes or other documents as attachments. In these cases, the *<attachments>* block is used. Part of the above described modelling for a circular is shown in Fig. 8.

5 Exploitation Use Cases

Several researchers agree that while available, Open Data remain in most cases underexploited. According to Janssen et al. "*...the success of open data systems requires more than simple provision of access to data...*" [4]. Zuiderwijk et al. concluded in [6] that "*...most impediments for the open data process concern the actual use of open data...*". In this section, in an effort to highlight the exploitation possibilities that the availability of structured legal datasets offers in the real-world setting, we discuss how Legal Open Data modelled in Akoma Ntoso could be further used, by presenting three relevant use-cases.

5.1 Automatic Consolidation of Legislation

As our societies are evolving and their needs are changing, laws cannot remain static. Due to the changing environment, laws are often amended and the same is true for other types of legal texts. As a result, it is difficult to know the valid text of a legal document being in force at a specific point in time. Akoma Ntoso provides the appropriate mechanisms for legislative change management [21], therefore a marked-up dataset of legal documents can be exploited for the automatic or semi-automatic text consolidation.

The metadata block *<analysis>* contains information about modifications made by the current document to another document (*<activeModifications>* block) and modifications arrived at the current document (*<passiveModifications>* block). These metadata elements contain the necessary information for the application of modifications, therefore a parser can exploit their content in order to automate the consolidation process. As an example, consider the syntax of a textual modification described in the metadata section of an Akoma Ntoso file representing a legal act as shown in Fig. 9.

The respective metadata block describes the fact that paragraph 1 of article 1 of the current document contains a modification that substitutes the text of paragraph 1 of article 5 of act 4387/2016 with the text found within the element with id equal to 'mod_1__qtext_1'. A parser could easily analyze the content of this metadata block, identify the positions of the referred elements and automatically perform the substitution.

```
<analysis source="#openLawsGR">
    <activeModifications>
        <textualMod type="substitution" eId="amod1">
            <source href="#art_1__par_1"/>
            <destination href="akn/gr/act/2016/4387/main#art_5__par_1"/>
            <new href="#mod_1__qtext_1"/>
        </textualMod>
    </activeModifications>
</analysis>
```

Fig. 9. Example of metadata block describing a textual modification (substitution)

5.2 Legal Network Analysis

Legal documents are highly interconnected, each one usually containing several citations to other documents. Detecting and extracting these legal citations allows for legal data network analysis, a process that can reveal the complexity of legislation in specific domains, enhance information retrieval, allow for visualization of the legal corpus [22] and provide insights on the influence of legal documents and the evolution of legal doctrine [23]. References in Akoma Ntoso are marked up with the *<ref>* tag and the *href* attribute contains the IRI string that identifies the referenced document. Consequently, extraction of citations is possible by parsing the XML file and taking advantage of XPath or XQuery expressions. For example, the XPath expression *//ref/@href* returns the values of href attributes of all references in a specific XML file. If these references strictly follow the Akoma Ntoso naming convention, the implementation of a resolver that parses the IRI and points to the referenced document is straightforward. When a dataset containing various types of legal texts is available, it is possible to construct an enhanced legal graph that captures more accurately the dynamics of the legal system.

5.3 Legal Requirements Modelling and Regulation Compliance

Extracting semantic representations of legal requirements from legal documents plays an important role for achieving or checking regulatory compliance. LegalRuleML [24] is an XML language able to support the modelling and representation of legal norms and rules, which aims to fill the gap between description of legal text and modelling of legal rules. LegalRuleML can be used in combination with Akoma Ntoso to formally describe legal norms and facilitate legal reasoning. As an example, one can consider the work of Palmirani and Governatori [25], where a framework that combines Akoma Ntoso, LegalRuleML, the PrOnto ontology and BPMN (Business Process Model and Notation) for checking compliance with GDPR is presented. In a similar approach [26], the authors show a framework for the visual representation of processes defined in legal texts, which uses documents marked up in Akoma Ntoso with ontologies and visual languages (e.g. BPMN or UML). The proposed model can be useful for understanding legal requirements and re-engineering processes.

6 Conclusions

In this paper, we presented an approach for modelling legal documents using the Akoma Ntoso document model. Although focusing on documents produced within the Greek legal ecosystem, our work is not limited to the case of Greece and the approach can be adapted to support the special features of legal documents produced in other countries. This flexibility is a significant design characteristic of the Akoma Ntoso standard [19].

We believe that our work makes several contributions to the legal research and business communities. Legal data modelling is the first and important step for the automatic transformation of available unstructured legal documents into structured Open Data. As already mentioned, availability of data in unstructured formats consists a barrier for Open Data exploitation. The adoption of manual approaches for the completion of such tasks is not a practical option in terms of required cost and time, mainly due to the volume of available legal datasets. Anderson [27] has already stressed the important role that data modelling holds in the case of Big Data software engineering.

Moreover, our work is built on the Akoma Ntoso document model, a standard that is designed to facilitate interoperability. As shown in the related work section, several researchers adopt custom approaches and schemata, a choice that hinders interoperability. In addition to this, as far as we know, our work is the first research effort to present the Akoma Ntoso modelling for legal opinions and administrative circulars, important documents for the implementation of the law, since most publications focus on laws and court decisions. Another contribution of our work is metadata modelling for Greek legal documents, as Alexopoulos et al. report a low maturity level of Greek Open Government Data sources in terms of metadata management [28].

Finally, the described use-cases for the exploitation of legal documents represented in the Akoma Ntoso format show how new systems and services could be designed to take advantage of available Legal Open Data, assisting to tackle the problem of underexploited Open Data.

Our plans for future work are guided by the goals of our project. The next step in our research is the investigation of the appropriate natural language processing techniques for the development of a parser that will automatically convert legal texts to their Akoma Ntoso representation, following the presented modelling. We have already built a prototype that successfully performs this task for the metadata and the structural parts of court decisions, using the ANTLR parser generator and we intend to extend it for other types of legal documents.

Acknowledgement. The project “Automated Analysis and Processing of Legal Texts for their Transformation into Legal Open Data” is implemented through the Operational Program “Human Resources Development, Education and Lifelong Learning” and is co-financed by the European Union (European Social Fund) and Greek national funds.

References

1. Mazzega, P., Lajaunie, C.: Legal texts as big data: a study case on health issues. In: 3rd International Seminar on Health, Human Rights and Intellectual Property Rights (2017)
2. Wass, C.: Openlaws.eu – building your personal legal network. J. Open Access Law **5**(1) (2017)
3. Greenleaf, G., Mowbray, A., Chung, P.: The meaning of free access to legal information: a twenty year evolution. J. Open Access Law **1**(1) (2013)
4. Janssen, M., Charalabidis, Y., Zuiderwijk, A.: Benefits, adoption barriers and myths of open data and open government. Inf. syst. Manage. **29**(4), 258–268 (2012). https://doi.org/10.1080/10580530.2012.716740
5. Agnoloni, T., Francesconi, E., Sagri, M.T., Tiscornia, D.: Linked data in the legal domain. In: Proceedings of ITAIS 2011, AIS, Rome (2011)
6. Zuiderwijk, A., Janssen, M., Choenni, S., Meijer, R., Alibaks, R.S.: Socio-technical impediments of open data. Electron. J. e-Government **10**(2), 156–172 (2012)
7. Casanovas, P., Palmirani, M., Peroni, S., van Engers, T., Vitali, F.: Semantic web for the legal domain: the next step. Semant. Web **7**(3), 213–227 (2016). https://doi.org/10.3233/SW-160224
8. Gen, K., Akira, N., Makoto, M., Yasuhiro, O., Tomohiro, O., Katsuhiko, T.: Applying the Akoma Ntoso XML schema to Japanese legislation. J. Law Inf. Sci. **24**(2), 49–70 (2016)
9. Hallo, M., Luján-Mora, S., Mate, A.: Data model for storage and retrieval of legislative documents in digital libraries using linked data. In: 7th International Conference on Education and New Learning Technologies, Barcelona, pp. 7423–7430. IATED (2015)
10. Gostojic, S., Konjovic, Z., Milosavljevic, B.: Modeling MetaLex/CEN compliant legal acts. In: IEEE 8th International Symposium on Intelligent Systems and Informatics, Subotica, pp. 285–290. IEEE (2010). https://doi.org/10.1109/sisy.2010.5647452
11. Marković, M., Gostojić, S., Konjović, Z., Laanpere, M.: Machine-readable identification and representation of judgments in Serbian judiciary. Novi Sad J. Math. **44**(1), 165–182 (2014)
12. Ceci, M., Palmirani, M.: Ontology framework for judgment modelling. In: Palmirani, M., Pagallo, U., Casanovas, P., Sartor, G. (eds.) AICOL 2011. LNCS (LNAI), vol. 7639, pp. 116–130. Springer, Heidelberg (2012). https://doi.org/10.1007/978-3-642-35731-2_8
13. Oksanen, A., Tuominen, J., Mäkelä, E., Tamper, M., Hietanen, A., Hyvönen, E.: Law and justice as a linked open data service. Draft Paper
14. Boella, G., et al.: Linking legal open data: breaking the accessibility and language barrier in European legislation and case law. In: Proceedings of the 15th International Conference on Artificial Intelligence and Law, pp. 171–175. ACM (2015). https://doi.org/10.1145/2746090.2746106
15. Koniaris, M., Papastefanatos, G., Vassiliou, Y.: Towards automatic structuring and semantic indexing of legal documents. In: Proceedings of the 20th Pan-Hellenic Conference on Informatics. ACM (2016). https://doi.org/10.1145/3003733.3003801
16. Angelidis, I., Chalkidis, I., Nikolaou, C., Soursos, P., Koubarakis, M.: Nomothesia: a linked data platform for Greek legislation. In: MIREL 2018 Workshop (accepted paper)
17. Beris, T., Koubarakis, M.: Modeling and preserving Greek government decisions using semantic web technologies and permissionless blockchains. In: Gangemi, A., et al. (eds.) ESWC 2018. LNCS, vol. 10843, pp. 81–96. Springer, Cham (2018). https://doi.org/10.1007/978-3-319-93417-4_6
18. Garofalakis, J., Plessas, K., Plessas, A., Spiliopoulou, P.: A project for the transformation of Greek legal documents into legal open data. In: 22nd Pan-Hellenic Conference on Informatics. ACM (2018). https://doi.org/10.1145/3291533.3291548

19. Palmirani, M., Vitali, F.: Akoma-Ntoso for legal documents. In: Sartor, G., Palmirani, M., Francesconi, E., Biasiotti, M. (eds.) Legislative XML for the Semantic Web. LGTS, vol. 4, pp. 75–100. Springer, Dordrecht (2011). https://doi.org/10.1007/978-94-007-1887-6_6
20. Opijnen, M.: European case law identifier: indispensable asset for legal information retrieval. In: Biasiotti, M.A., Faro, S. (eds.) From Information to Knowledge – Online Access to Legal Information: Methodologies, Trends and Perspectives. Frontiers in Artificial Intelligence and Applications, pp. 91–103. IOS Press, Amsterdam (2011)
21. Palmirani, M.: Legislative change management with Akoma-Ntoso. In: Sartor, G., Palmirani, M., Francesconi, E., Biasiotti, M. (eds.) Legislative XML for the Semantic Web. LGTS, vol. 4, pp. 101–130. Springer, Dordrecht (2011). https://doi.org/10.1007/978-94-007-1887-6_7
22. Koniaris, M., Anagnostopoulos, I., Vassiliou, Y.: Network analysis in the legal domain: a complex model for European Union legal sources. J. Complex Netw. **6**(2), 243–268 (2017). https://doi.org/10.1093/comnet/cnx029
23. Branting, L.: Data-centric and logic-based models for automated legal problem solving. Artif. Intell. Law **25**(1), 5–27 (2017). https://doi.org/10.1007/s10506-017-9193-x
24. Athan, T., Governatori, G., Palmirani, M., Paschke, A., Wyner, A.: LegalRuleML: design principles and foundations. In: Faber, W., Paschke, A. (eds.) Reasoning Web 2015. LNCS, vol. 9203, pp. 151–188. Springer, Cham (2015). https://doi.org/10.1007/978-3-319-21768-0_6
25. Palmirani, M., Governatori, G.: Modelling legal knowledge for GDPR compliance checking. In: The 31st International Conference on Legal Knowledge and Information Systems, pp. 101–110. IOS Press (2018)
26. Ciaghi, A., Villafiorita, A.: Improving public administrations via law modeling and BPR. In: Popescu-Zeletin, R., Rai, I.A., Jonas, K., Villafiorita, A. (eds.) AFRICOMM 2010. LNICST, vol. 64, pp. 69–78. Springer, Heidelberg (2011). https://doi.org/10.1007/978-3-642-23828-4_7
27. Anderson, K.M.: Embrace the challenges: software engineering in a big data world. In: IEEE/ACM 1st International Workshop on Big Data Software Engineering, pp. 19–25. IEEE (2015). https://doi.org/10.1109/bigdse.2015.12
28. Alexopoulos, C., Loukis, E., Mouzakitis, S., Petychakis, M., Charalabidis, Y.: Analysing the characteristics of open government data sources in Greece. J. Knowl. Econ. **9**(3), 721–753 (2018). https://doi.org/10.1007/s13132-015-0298-8

Time Series Forecasting by Recommendation: An Empirical Analysis on Amazon Marketplace

Álvaro Gómez-Losada(✉) and Néstor Duch-Brown

Joint Research Centre, European Commission, Seville, Spain
{alvaro.gomez-losada,nestor.duch-brown}@ec.europa.eu

Abstract. This study proposes a forecasting methodology for univariate time series (TS) using a Recommender System (RS). The RS is built from a given TS as only input data and following an item-based Collaborative Filtering approach. A set of top-N values is recommended for this TS which represent the forecasts. The idea is to emulate RS elements (the users, items and ratings triple) from the TS. Two TS obtained from Italy's Amazon webpage were used to evaluate this methodology and very promising performance results were obtained, even the difficult environment chosen to conduct forecasting (short length and unevenly spaced TS). This performance is dependent on the similarity measure used and suffers from the same problems that other RSs (e.g., cold-start). However, this approach does not require high computational power to perform and its intuitive conception allows for being deployed with any programming language.

Keywords: Collaborative Filtering · Time series · Forecasting · Data science

1 Introduction

Broadly speaking, autocorrelation is the comparison of a time series (TS) with itself at a different time. Autocorrelation measures the linear relationship between lagged values of a TS and is central to numerous forecasting models that incorporate autoregression. In this study, this idea is borrowed and incorporated to a recommender system (RS) with a forecasting purpose.

RSs apply knowledge discovery techniques to the problem of helping users to find interesting items. Among the wide taxonomy of recommendation methods, Collaborative Filtering (CF) [1] is the most popular approach for RSs designs [2]. CF is based on a intuitive paradigm by which items are recommended to an user looking at the preferences of the people this user trusts. *User-based* or *item-based* [3] recommendations are two common approaches for performing CF. The first evaluates the interest of a user for an item using the ratings for this item by other users, called neighbours, that have similar rating patterns. On the other hand,

W. Abramowicz and R. Corchuelo (Eds.): BIS 2019, LNBIP 353, pp. 45–54, 2019.
https://doi.org/10.1007/978-3-030-20485-3_4

item-based CF considers that two items are similar if several users of the system have rated these items in a similar fashion [4]. In both cases, the first task when building a CF process is to represent the user-items interactions in the form of a rating matrix. The idea is that given a rating data by many users for many items it can be predicted a user's rating for an item not known to the user, or identify a set of N items that user will like the most (top-N recommendation problem). The latter is the approach followed in this study.

The goal of this study is to introduce a point forecasting methodology for univariate TS using a item-based CF framework. In particular, to study the behaviour of this methodology in short lenght and unvenly spaced TS. On one side, the distinct values of a TS are considered the space of users in the RS. On the other, items are represented by the distinct values of a lagged version of this TS. Ratings are obtained by studying the frequencies of co-occurrences of values from both TS. Basically, the forecast is produced after averaging the top-N set of recommended items (distinct values of the shifted TS) to a particular user (a given value in the original TS).

2 Related Work

TS forecasting has been incorporated into recommendation processes in several works to improve the users' experience in e-commerce sites. However, to the best of the authors' knowledge, the use of a RS framework as a tool for producing forecasts in TS is new in literature.

2.1 Item-Based CF Approach

This section describes the basic and notation of an standard item-based CF RS, with a focus on the approach followed in this study. Mostly, CF techniques use a database as input data in the form of a user-item matrix $\mathbf{R}$ of ratings (preferences). In a typical item-based scenario, there is a set of m users $\mathcal{U} = \{u_1, u_2, ..., u_m\}$, a set of n items $\mathcal{I} = \{i_1, i_2, ..., i_n\}$, and a user-item matrix $\mathbf{R} = (r_{jk}) \in \mathbb{R}^{m \times n}$, with r_{jk} representing the rating of the user u_j $(1 \leq j \leq m)$ for the item i_k $(1 \leq k \leq n)$. In $\mathbf{R}$, each row represents an user u_j, and each column represents an item i_k. Some filtering criteria may be applied by removing from $\mathbf{R}$ those r_{jk} entries below a predefined $b \in \mathbb{N}^+$ threshold. One of the novelties of this study is the creation of an $\mathbf{R}$ matrix from a TS, which is explained in the next Sect. 3. Basically, $\mathbf{R}$ is created after using the TS under study and its lagged copy, and considering them as the space of users and items, respectively. The rating values (r_{jk}) are obtained by cross-tabulating both series. Instead of studying their relation with an autocorrelation approach, the frequency of co-occurrence of values is considered. In order to clarify how the transformation of a TS forecasting problem is adapted in this study using a RS, Table 1 identifies some assumed equivalences.

The model-building step begins with determining a similarity between pair of items from $\mathbf{R}$. Similarities are stored in a new matrix $\mathbf{S} = (s_{ij}) \in \mathbb{R}^{n \times n}$, where

Table 1. Some equivalences used in this study to build the forecasting recommender system (RS) from a given time series (TS).

Concept	Symbol	RS equivalence
Set of distinct values in TS	$\mathcal{U}$	Set of users with cardinality m
Set of distinct values in TS shifted	$\mathcal{I}$	Set of items with cardinality n
Two distinct values in the TS	u_j, u_l	A pair of users
Two distinct values of the shifted TS	i_i, i_j	A pair of items
Number of times a distinct value in the TS and its shifted version co-occurs	r_{jk}	Rating of user u_j on item i_k
TS value to which perform a forecasting	u_a	Active user to which recommend an item

s_{ij} represent a similarity between items i and j ($1 \leq i, j \leq n$). s_{ij} is obtained after computing a similarity measure on those users who have rated i and j items. Sometimes, to compute the similarity is set a minimum number of customers that have selected the (i, j) pair. This quantity will be referred as the $c \in \mathbb{N}_{>2}$ threshold. Traditionally, among the most commonly similarity measures used are the Pearson correlation, cosine, constraint Pearson correlation and mean squared differences [5]. In this study it will be used the Cosine and Pearson correlation measures, but also, the Otsuka-Ochiai coefficient, which is borrowed from Geosciences [8] and used by leading online retailers like Amazon [9].

Some notation follows at this stage of the modelling. The vector of ratings provided for item i is denoted by $\mathbf{r}_i$ and $\bar{r}_i$ is the average value of these ratings. The set of users who has rated the item i is denoted by $\mathcal{U}_i$, the item j by $\mathcal{U}_j$, and the set of users who have rated both by $\mathcal{U}_{ij}$.

Forecasting. The aim of item-based algorithm is to create recommendations for a user, called the active user $u_a \in \mathcal{U}$, by looking into the set of items this user has rated, $I_{u_a} \in \mathcal{I}$. For each item $i \in I_{u_a}$, just the k items wich are more similar are retained in a set $\mathcal{S}(i)$. Then, considering the rating that u_a has made also on items in $\mathcal{S}(i)$, a weighted prediction measure can be applied. This approach returns a series of estimated rating for items different from those in I_{u_a} that can be scored. Just the top ranked items are included in the list of N items to be recommended to u_a (top-N recommended list). The k and N values has to be decided by the experimenter. In this study, each active user (u_a) was randomly selected from $\mathcal{U}$ following a cross-validation scheme, which is explained next. To every u_a, a set of recommendable items is presented (the forecast). The space of items to recommend is represented by the distinct values of the shifted TS. Since the aim is to provide a point forecast, the numerical values included in the top-N recommended list are averaged.

Evaluation of the Recommendation. The basic structure for offline evaluation of RS is based on the train-test setup common in machine learning [5,6]. A usual approach is to split users in two groups, the training and test sets of users ($\mathcal{U}_{train} \cup \mathcal{U}_{test} = \mathcal{U}$). Each user in $\mathcal{U}_{test}$ is considered to be an active user u_a. Item ratings of users in the test set are split into two parts, the *query set* and the *target set*. Once the RS is built on $\mathcal{U}_{train}$, the RS is provided with the query set as user history and the recommendation produced is validated against the target set. It is assumed that if a RS performs well in predicting the withheld items (target set), it will also perform well in finding good recommendations for unknown items [7]. Typical metrics for evaluation accuracy in RS are root mean square error (RMSE), mean absolute error (MAE) or other indirect functions to estimate the novelty of the recommendation (e.g., serendipity). The MAE quality measure was used in this study.

3 Creation of the Rating Matrix

This section describes the steps to transform a given TS (**TS**) in a matrix of user-item ratings (**R**). This should be considered the main contribution of this study. The remaining steps do not differs greatly from a traditional item-based approach beyond the necessary adaptations to the TS forecasting context.

Since the aim it is to emulate **R**, the first step is to generate a space of users and items from **TS**. Thus, **TS** values are rounded to the nearest integer which will be called $\mathbf{TS_0}$. The distinct (unique) values of $\mathbf{TS_0}$ are assumed to be $\mathcal{U}$.

The second step is setting up $\mathcal{I}$. For that, $\mathbf{TS_0}$ is shifted forward in time $h \in \mathbb{N}_{>0}$ time stamps. The new TS is called $\mathbf{TS_1}$. The value of h depends on the forecasting horizon intended. Now, the distinct values of $\mathbf{TS_1}$ set $\mathcal{I}$. Once $\mathcal{U}$ and $\mathcal{I}$ have been set, they are studied as a bivariate distribution of discrete random variables. The last step is to compute the joint (absolute) frequency distribution of $\mathcal{U}$ and $\mathcal{I}$. Thus, it is assumed that $n_{jk} \equiv r_{jk}$, where n_{jk} representes the frequency (co-occurrence) of the u_j value of $\mathbf{TS_0}$ and i_k value of $\mathbf{TS_1}$. These steps are summarized below.

Algorithm. Rating matrix creation from a Time Series

Input:
  Univariate time series **TS**
Parameters:
  h : intended forecasting horizon
  b : chosen threshold value for removing less representative frequencies
Output:
  Rating matrix **R**

Procedure:
  Round to the nearest integer **TS** values $\rightarrow$ $\mathbf{TS_0}$

Shift $\mathbf{TS_0}$ values h time stamps forward in time $\rightarrow \mathbf{TS_1}$
Remove first h values from $\mathbf{TS_0}$ and last h values from $\mathbf{TS_1}$

Obtain $\mathbf{TS_0}$ distinct values $\rightarrow \mathcal{U}$
Obtain $\mathbf{TS_1}$ distinct values $\rightarrow \mathcal{I}$
Cross-tabulate $\mathcal{U}$ and $\mathcal{I} \rightarrow \mathbf{R}$
Remove r_{jk} entries $\leq b$
Return $\mathbf{R}$

Some Considerations. The followed approach experiences the same processing problems that a conventional item-based approach, namely, sparsity in $\mathbf{R}$ and *cold start* situations (user and items with low or inexistent numbers of ratings). In large RS from e-commerce retailers, usually $n \ll m$. However, under this approach $n \simeq m$.

4 Experimental Evaluation

This section describes the sequential steps followed in this study for creating a item-based CF RS with the purpose of TS forecasting.

4.1 Data Sets

Two short length TS were used to evaluate this methodology. These TS were obtained after scraping the best-selling products from the Italy's Amazon webpage between 5th April, 2018 and 14th December, 2018. The scraping process setting was sequential. It consisted of obtaining the price from the first to the last item of each category, and from the first category to the latest. The first TS (TS-1) was created after averaging the evolution of best-selling products' prices included in the *Amazon devices* category. The second TS (TS-2) represents the price change of the most dynamic best-selling product from this marketplace site. Italy's Amazon webpage has experienced changes during the eight month period of the crawling process due to commercial reasons. The main change is related to the number of best-selling products being showed for each category (e.g., 20, 50 or 100). This causes the period of the scraping cycles is not similar and the TS derived from this data (TS-1 and TS-2) are not equally spaced at time intervals. In this study, the data obtained in the scraping process will be considered as a sequence of time events. Therefore, it is worth to note that the values of TS-1 were obtained after averaging a different number of products (20, 50, or 100), according to the number of products shown by Amazon at different times. Their main characteristics are shown in Table 2.

4.2 Creation of the Rating Matrices

A rating matrix $\mathbf{R}$ was created for TS-1 and TS-2 according to the algorithm described in Sect. 3. As mentioned before, the different duration of scraping

Table 2. TS characteristics used in the methodology testing (P: percentile; min: minimum value, max: maximum value; in €).

Abbreviation	Length	min	P50	max	P75-P25	Distinct values (m)
TS-1	2169	13	84	142	14	82
TS-2	1224	7	17	36	8	22

cycles causes unevenly spaced observations in the TS. Also, it represents an additional difficulty when setting a constant forecasting horizon (h) as described in the algorithm. Therefore, in this study, it will be necessary to assume that the forecasting horizon coincides with the duration of the scraping cycle, independently of its length in time. In practice, this means that the TS representing the items ($\mathbf{TS_1}$) was obtained after lagging one position forward in time with respect the original TS representing the users ($\mathbf{TS_0}$). After empirical observation, h approximately takes values 1 h, 2 h or 4 h depending on Amazon shows the 20, 50 or 100 best-selling products for each category, respectively. The lack of proportionality between the duration of scraping cycles and the number of best-selling product shown is explained by technical reasons in the crawling process (structure of the Amazon's webpage for each product affecting the depth of the scraping process). Those ratings with a value $b \leq 3$ were removed from the corresponding $\mathbf{R}$.

4.3 Similarity Matrix Computation

Three symmetric functions were used in this study to calculate different $\mathbf{S}$ for TS-1 and TS-2. The Pearson correlation (1) and cosine (2) similarity functions are standards in the RS field. The third one, the Otsuka-Ochiai coefficient, incorporates a geometric mean in the denominator:

$$s_1(i,j) = \frac{\sum_{u \in \mathcal{U}_{i,j}} (r_{u,i} - \bar{r}_i)(r_{u,j} - \bar{r}_j)}{\sqrt{\sum_{u \in \mathcal{U}_{i,j}} (r_{u,i} - \bar{r}_i)^2} \sqrt{\sum_{u \in \mathcal{U}_{i,j}} (r_{u,j} - \bar{r}_j)^2}} \tag{1}$$

$$s_2(i,j) = \frac{\mathbf{r}_i \bullet \mathbf{r}_j}{\|\mathbf{r}_i\|_2 \, \|\mathbf{r}_j\|_2} = \frac{\sum_{u \in \mathcal{U}_{i,j}} r_{u,i} \bullet r_{u,j}}{\sqrt{\sum_{u \in \mathcal{U}_{i,j}} r_{u,i}^2} \sqrt{\sum_{u \in \mathcal{U}_{i,j}} r_{u,j}^2}} \tag{2}$$

$$s_3(i,j) = \frac{|\,\mathcal{U}_{ij}\,|}{\sqrt{|\,\mathcal{U}_i\,|\,|\,\mathcal{U}_j\,|}} \tag{3}$$

where $\bullet$, $\|\cdot\|_2$ and $|\cdot|$ denote the dot-product, l_2 norm, and cardinality of the set, respectively. These similarity measures were calculated when the minimum number of user rating a given items was $c \geq 3$, and set the value of $k = 3$.

4.4 Generation of the Top N-Recommendation

This step begins by looking at the set of items the active user u_a has rated, I_{u_a}. In particular, the interest is to predict ratings for those items $j \notin I_{u_a}$ rated by user u_a. Then estimated rating ($\hat{r}_{u_a,j}$) for a given item $j \notin I_{u_a}$ was calculated according to (4), where $\mathcal{S}(j)$ denotes the items rated by the user u_a most similar to item j:

$$\hat{r}_{u_a,j} = \frac{1}{\sum_{i \in \mathcal{S}(j)} s(i,j)} \sum_{i \in \mathcal{S}(j)} s(i,j)\; r_{u_a,i} \tag{4}$$

The value of $\hat{r}_{(u,j)}$ can be considered a score that is calculated for each item not in I_{u_a}. Finally, the highest scored items are included in the top-N recommended list.

4.5 Evaluation

The set of users ($\mathcal{U}$) was splitted following a 80:20 proportion for obtaining the training and test sets of users ($\mathcal{U}_{train}$ and $\mathcal{U}_{test}$), respectively. Every user in $\mathcal{U}_{test}$ (20% of distinct values in $\mathbf{TS}_0$) was considered an active user (u_a) to whom recommend a set of items. From the history of each u_a, again the proportion 80:20 proportion was used to obtain the *query* and *target* sets, respectively. The produced recommendation was validated against the target set of each u_a.

Evaluation Metric. The MAE value was obtained according to (5):

$$MAE = \frac{\sum_{i=1}^{n} \mid e_i \mid}{n} \tag{5}$$

where n represent the number of active users to whom a recommendation has been suggested. The N value to generate a top-N recommendation was set dynamically according to the length of items in the *target* set for each u_a. Thus, the value of e is the difference between the average value of the top-N recommended items and the average value of the items in the target set.

4.6 Performance Results

The MAE results for assessing the quality of the forecasting proposal is shown in Table 3, for the two analysed TS and three similarity measures studied.

It can be seen that Otsuka-Ochiai similarity (s_3) yields a better performance on both TS studied. It is necessary to remind the characteristics of the TS used for testing this methodology (TS-1 and TS-2). They are characterized by unevenly spaced observations in the TS, but also, the variable forecasting horizon provided by the data acquisition environment. This represents a very complex environment for performing forecasting. In the case of forecasting the average

Table 3. MAE performance on both TS and the different similarity measures (s_1, s_2 and s_3: Pearson correlation, cosine and Otsuka-Ochiai similarities, respectively), in €.

	TS-1			TS-2		
	s_1	s_2	s_3	s_1	s_2	s_3
MAE	6.2	6.5	5.5	6.0	3.2	3.0

price for a given category, or the price for a given product, the forecasting results could be considered an estimation of the trend of such prices (uptrend or downtrend) more than aiming to forecast an exact value of such prices. Other experiences have been accomplished to test this methodology with ozone (O_3) TS (results now shown). These experiences with conventional TS (evenly spaced observations and constant forecasting horizon) has yield a very good results in term of forecasting. Therefore, TS-1 and TS-2 should be considered to represent an extreme environment in which perform forecasting. One of the advantage of this approach is that does not require high computational power to operate, due to the analysis of the distinct rounded values of a given TS, which is independent of its length. Besides, the proposed approach is very intuitive and allows for further developments to be included and being deployed using any programming language.

4.7 Computational Implementation

The computational implementation was accomplished using *ad hoc* designed Python functions except those specified in Table 4. The first two purposes correspond to the creation of the Rating matrix (**R**), and the remaining ones to the similarity matrix computation (**S**). In particular, the last two ones are used in the computation of the cosine similarity measure used in this work.

Table 4. Python functions for specific tasks.

Purpose	Function	Package
TS shifting	`shift`	pandas
Cross-tabulation	`crosstab`	pandas
To iterate over pairs of items	`itertools`	itertools
Dot product	`dot`	numpy
l_2 norm	`norm`	numpy.linalg

5 Conclusions

This study aims to produce a point forecast for a TS adopting a RS approach, in particular an item-based CF. To that end, basic elements in a RS (the users, items and ratings triple) was emulated using a TS as only input data. An autocorrelation-based algorithm is introduced to create a recommendation matrix from a given TS. This methodology was tested using two TS obtained from Italy's Amazon webpage. Performance results are promising even the analyzed TS represent a very difficult setting in which to conduct forecasting. This is due to the TS are unevenly spaced and the forecasting horizon is not constant. Application of classical forecasting approaches (e.g., autoregression models) to this type of TS is not possible mainly due to the irregular time stamps in which observations are obtained. Thus, the introduced algorithm should be considered a contribution to the forecasting practice in both short length and unevenly spaced TS. Complementary, computational time required to obtain forecasting estimates is short due to such estimates are obtained considering distinct values of the TS are not all the values forming the TS. Further developments include to consider contextual information when building the RS and transforming the sequence of events to a TS with evenly spaced data.

Disclaimer. The views expressed are purely those of the authors and may not in any circumstances be regarded as stating an official position of the European Commission.

References

1. Goldberg, D., Nichols, D., Oki, B.M., Terry, D.: Using collaborative filtering to weave an information tapestry. Commun. ACM Spec. Issue Inf. Filter. **35**, 61–70 (1992). https://doi.org/10.1145/138859.138867
2. Sharma, R., Gopalani, D., Meena, Y.: Collaborative filtering-based recommender system: approaches and research challenges. In: 3rd International Conference on Computational Intelligence & Communication Technology, pp. 1–6. IEEE Press (2017). https://doi.org/10.1109/CIACT.2017.7977363
3. Sarwar, B., Karypis, G., Konstan, J., Reidl, J.: Item-based collaborative filtering recommendation algorithms. In: Proceedings of the 10th International Conference on World Wide Web, Hong Kong, 2001, pp. 285–295. ACM, New York (2001). https://doi.org/10.1145/371920.372071
4. Desrosiers, C., Karypis, G.: A comprehensive survey of neighborhood-based recommendation methods. In: Ricci, F., Rokach, L., Shapira, B., Kantor, P.B. (eds.) Recommender Systems Handbook, pp. 107–144. Springer, Boston, MA (2011). https://doi.org/10.1007/978-0-387-85820-3_4
5. Bobadilla, J., Ortega, F., Hernando, A., GutiéRrez, A.: Recommender Systems Survey. Knowl. Based Syst. **46**, 109–132 (2013). https://doi.org/10.1016/j.knosys.2013.03.012
6. Ekstrand, M.D., Riedl, J.T., Konstan, J.A.: Collaborative filtering recommender systems. Found. Trends Hum. Comput. Interact. **4**(2), 81–173 (2010). https://doi.org/10.1561/1100000009
7. Haslher, M., Vereet, B.: recommenderlab: A Framework for Developing and Testing Recommendation Algorithms (2018). https://CRAN.R-project.org/package=recommenderlab
8. Ochiai, A.: Zoogeographical studies on the soleoid fishes found in Japan and its neighhouring regions-II. Bull. Japan. Soc. Sci. Fish **22**, 526–530 (1957). https://doi.org/10.2331/suisan.22.526

9. Jacobi, J.A., Benson, E.A., Linden, G.D.: Personalized recommendations of items represented within a database. US Patent US7113917B2 (to Amazon Technologies Inc.) (2006). https://patents.google.com/patent/US7113917B2/en
10. Breese, J.S, Heckerman, D., Kadie, C.: Empirical analysis of predictive algorithms for collaborative filtering. In: Proceedings of the 14th Conference on Uncertainty in Artificial Intelligence, pp. 43–52. Morgan Kaufmann Publishers (1998)

A Comparative Application of Multi-criteria Decision Making in Ontology Ranking

Jean Vincent Fonou-Dombeu(✉)

School of Mathematics, Statistics and Computer Science, University of KwaZulu-Natal, King Edward Avenue, Scottsville, Pietermaritzburg 3209, South Africa
fonoudombeuj@ukzn.ac.za

Abstract. The number of available ontologies on the web has increased tremendously in recent years. The choice of suitable ontologies for reuse is a decision-making problem. However, there has been little use of decision-making on ontologies to date. This study applies three Multi-Criteria Decision Making (MCDM) algorithms in ontology ranking. A number of ontologies/alternatives and the complexity metrics or attributes of these ontologies are used in the decision. The experiments are carried out with 70 ontologies and the performance of the algorithms are analysed and compared. The results show that all the algorithms have successfully ranked the input ontologies based on their degree/level of complexity. Furthermore, the results portray a strong correlation between the ranking results of the three MCDM algorithms, thereby, providing more insights on the performance of MCDM algorithms in ontology ranking.

Keywords: Algorithms · Biomedical ontologies · Complexity metrics · Multi-criteria Decision Making · Ontology ranking · Ontology reuse

1 Introduction

Ontology building has been an active area of interest in Semantic Web development in the past years. This is witnessed by the ever increasing number of ontologies made available to the public on the web today. Three main reasons may explain this proliferation of ontologies on the web, namely, the advent of linked data, ontology libraries and the increase interest in semantic-based applications in various domains. The main benefits of making existing ontologies available on the internet to the public is that they can be reused in other applications and/or in research. Ontology reuse is an active subject of research in ontology engineering today [1,2]. In fact, reusing existing ontologies in new applications, may save the users or ontology engineers the time and cost needed to develop new ontologies *de novo*. More importantly, one of the benefits of building ontology in a domain is to provide a common and shared representation of knowledge in that particular domain.

W. Abramowicz and R. Corchuelo (Eds.): BIS 2019, LNBIP 353, pp. 55–69, 2019.
https://doi.org/10.1007/978-3-030-20485-3_5

Therefore, the reuse of ontology may increase the interoperability of applications. In fact, different applications that use the same ontology process the same syntactic and semantic representation of knowledge; this provides the suitable interface for the interoperability of these applications. Ontology reuse consists of using existing ontologies to build new ones in the process of developing semantic-based applications. This entails integrating and/or merging parts of or the entire existing ontologies to constitute new ontologies [3].

With the rise of the number of available ontologies on the web today, the users or ontology engineers are presented with many ontologies to choose from. The task of choosing the suitable ontologies for reuse is a decision making problem [4,5]. Several criteria may be applied to choose ontologies for reuse. The users or ontology engineers may utilize keywords search or structured queries in an automatic or semi-automatic process to find suitable ontologies that match their desire terms [6]. They may find suitable ontologies from ontologies libraries [7,8]. In fact, ontologies libraries host in dedicated locations on the web, the ontologies that model or describe the same domain such as biomedical, agriculture, e-government, etc. Therefore, users or ontology engineers may simply download a number of ontologies they want to reuse from existing libraries. Another criterion that may be utilized to choose suitable ontologies for reuse is the level of complexity of the candidate ontologies. Here, a number of metrics that measure the design complexity of ontologies [9,11] may be applied to rank a list of ontologies describing the same domain based on how complex they are. This information may assist in the choice of suitable ontologies for reuse. In fact, the scale of the application at hand may require users or ontology engineers to want less, medium or highly complex ontologies.

The abovementioned criteria that may assist in discriminating a set of ontologies for reuse amongst existing ontologies can form the basis for the application of Multi-Criteria Decision Making (MCDM) [12,13] in ontology ranking. To this end, authors have applied decision making in ontology ranking in [4,5]; however, the ranking criteria utilized in these studies were based on the conceptual features of ontologies and not on their design complexity as in this study. Furthermore, the three MCDM algorithms experimented in this study are different from the ones implemented in related studies [4,5]; this provides a more wider perspective for assessing the performance of MCDM algorithms in ontology ranking.

The rest of the paper is structured as follows. Section 2 discusses related studies. The materials and methods used in the study are explained in Sect. 3. Section 4 presents the experiments and results of the study and a conclusion ends the paper in the last section.

2 Related Work

Ontologies ranking has been a subject of interest to many researchers in the past years [6,14–20]. A number of studies have developed various criteria to weight and rank ontologies according to their relevance to user's query terms. These studies have proposed different ranking methods including AKTiveRank [6],

ARRO [17], OntologyRank [14], OS_Rank [18], content-based [19], Content-OR [20] and DWRank [16]. The ranking criteria utilized in these methods range from the centrality, class match, density and betweenness [6], semantic relations and hierarchy structure of classes [17], semantic relationships between the classes [14], class name, ontology structure and semantic relations [18] to the centrality and authority of concepts [16]. Some of these ranking methods are the improvement or combination of others; for instance, OS_Rank [18] is an improved version of ARRO [17], whereas, the Content-OR [20] is a combination of the content-based [19] and OntologyRank [14] methods. In particular, the content-based [19] method does not use a set of criteria to weight the relevance of ontologies to the user's query terms as in the other methods; instead, a corpus of terms is built from the user's query and used to rank ontologies by matching their classes to the corpus.

Most of the previous studies above attempted to rank ontologies based on the weighting or measurement of their conceptual features again user's query terms. They all have not considered the use of MCDM in the ranking process. The first attempt to apply MCDM algorithms to rank ontologies was in [5]. The authors selected the versions I&III of the Elimination and Choice Translating the REality (ELECTRE) family of MCDM methods, and applied them on the set of criteria defined in the AKTiveRank method [6] to rank ontologies. The criteria adopted from the AKTiveRank method [6] include the class match, density, semantic similarity and betweenness measures. These criteria were applied to select the candidate ontologies or alternatives based on how best they matched the user's query terms. Thereafter, a process called outranking was applied on pairs of ontologies/alternatives to rank them based on the concepts of concordance, discordance and fuzziness [5].

Another study in [4] applied MCDM in ontology ranking. The authors used the Analytic Hierarchy Process (AHP) MCDM method to select and rank ontologies. Like in [5], a number of metrics adopted from the OntologyRank method [14] are used to evaluate the candidate ontologies based on their coverage of user's keyword terms. The candidate ontologies or alternatives are further evaluated and selected with a set of criteria including domain coverage, size, consistency, cohesion and language expressivity. In the end, a pairwise comparison matrix is computed to weight and rank the selected ontologies.

It can be noted from the above literature review that only a few studies have focused on the application of MCDM in ontologies ranking to date. Furthermore, the ranking criteria used in [4,5] are based on the conceptual features of ontologies adopted from the early ontology ranking methods such as OntologyRank [14] and AKTiveRank [6] methods. However, the authors in [11] have demonstrated that the metrics that measure the design complexity of ontologies are useful indicators for selecting the appropriate ontologies for reuse amongst a set of candidate ontologies. Therefore, the ranking criteria in this study are constituted of various metrics that measure the design complexity of ontologies.

Furthermore, the three MCDM algorithms experimented in this study are different from the ones implemented in [4,5]; this provides a more wider perspective for assessing the performance of MCDM algorithms in ontology ranking. The materials and methods used in the study are presented in the next section.

3 Materials and Methods

3.1 Complexity Metrics of Ontologies

An ontology is inherently a complex artefact. In fact, ontology may be built by collaboration between users and domain experts or through the reuse of existing ontologies. The subjective nature of the development process (modeling decisions, choice of ontologies, datasets, standards or components for reuse, etc.) introduces a level of complexity in the resulting ontology. Furthermore, the formalization of ontology in standard languages such as RDF and OWL introduces another level of complexity inherent to the syntactic and structural features of the languages. Moreover, the scale of the ontology itself may augment its complexity, i.e. the complexity of ontologies may increase from small, medium to large scale ontologies. Therefore, evaluating the quality of ontology based on its design complexity is an active topic of research in ontology engineering. To this end, authors have developed various metrics that measure the design complexity of ontologies [9–11]. These metrics characterize the graph structure of ontology and are utilized to measure its quality during the design, application and maintenance [21,22]. In this study eight popular of these metrics are used as criteria in the MCDM algorithms to rank a set of ontologies or alternatives. These metrics include:

1. **Depth of inheritance** (*DIT*): It is the average number of subclasses per class in the ontology. An ontology with a low DIT may indicate a specific kind of information, whereas, a high DIT may represent a wide variety of information in the ontology. The DIT of an ontology is defined as in Eq. (1):

$$DIT = \frac{\sum_{c_i \in C} NS_{C_i}}{|C|} \tag{1}$$

where C is the set of classes in the ontology and NS_{C_i} the number of subclasses of the class C_i belonging to C.

2. **The average number of paths per concept** (*ANP*): it indicates the average connectivity degree of a concept to the root concept in the ontology inheritance hierarchy. A higher ANP indicates the existence of a high number of inheritance relationships in the ontology; it also shows that there is a high number of interconnections between classes in the ontology. This metric is defined as in Eq. (2):

$$ANP = \frac{\sum_{i=1}^{m} p_i}{|C|} \tag{2}$$

where p_i is the number of paths of a given concept. The value ANP for any ontology must be greater or equal to 1; a $ANP = 1$ indicates that an ontology inheritance hierarchy is a tree.

3. **Tree Impurity** (*TIP*): This metric is used to measure how far an ontology inheritance hierarchy deviates from a tree, and it is given in Eq. (3):

$$TIP = |P^{'}| - |C^{'}| + 1 \tag{3}$$

where $P^{'}$ and $C^{'}$ represent the set of edges that are subclasses and nodes (named and anonymous) in the inheritance hierarchy, respectively [10]. The rational of the TIP is that a well-structured ontology is composed of classes organized through inheritance relationships. A TIP = 0 means that the inheritance hierarchy is a tree. The greater the TIP, the more the ontology inheritance hierarchy deviates from the tree and the greater its complexity is.
4. **Size of vocabulary** (*SOV*): This metric defines the total number of named classes, properties and instances in the ontology; it is defined in Eq. (4):

$$SOV = |P| + |C| + |I| \tag{4}$$

where C, P and I are the sets of classes, properties and instances in the ontology, respectively. A higher SOV implies that the ontology is big in size and would require a lot of time and effort to build it.
5. **The average path length of the ontology** (*APL*): This metric indicates the average number of concepts in a path. A path p between two nodes c_0 and c_n in a sub-graph G' of an ontology is represented as a sequence of unrepeated nodes connected by edges from c_0 to c_n; the length p_t^l of this path is the number of edges on the path. It is defined in Eq. (5):

$$APL = \frac{\sum_{i=1}^{m}\sum_{k=1}^{p_i} p_{i,k}^{l}}{\sum_{i=1}^{m} p_i} \tag{5}$$

were $p_{i,k}^l$ and p_i are the length of the kth path and number of paths of the ith concept, respectively. The APL is the ratio of the sum of the path lengths $pl_{i,k}$ of each of the m concepts in the ontology over the sum of the number of paths p_i of concepts.
6. **Entropy of ontology graph** (*EOG*): This metric is the application of the logarithm function to a probability distribution over the ontology graph to provide a numerical value that can be used as an indicator of the graph complexity [10]. It is defined in Eq. (6):

$$EOG = \sum_{i=1}^{n} p(i) log_2(p(i)) \tag{6}$$

where $p(i)$ is the probability for a concept to have i relations. The minimum value of EOG corresponds to $EOG = 0$, it is obtained when concepts have the same distribution of relations in the ontology, that is, all the nodes of the ontology sub-graphs have the same number of edges. Therefore, an ontology with a smaller EOG can be considered as less complex in terms of relations distribution.

7. **Relationship Richness** (RR): This metric provides an indication of the distribution of relations in an ontology. It is defined in Eq. (7):

$$RR = \frac{|R|}{|SR| + |R|} \tag{7}$$

where $|R|$ and $|SR|$ represent the number of relations between classes and the number of subclass relations, respectively. A RR value close to one indicates that most of the relations between concepts in the ontology are not subclass relations, while a RR close to zero specifies that the subclass relations are predominant amongst the concepts of the ontology.

8. **Class Richness** (CR): the value of this metric provides an indication of the distribution of individuals across the ontology classes [22]. It is defined in Eq. (8):

$$CR = \frac{|C^{'}|}{|C|} \tag{8}$$

where $C^{'}$ and C are the number of classes that have instances and the total number of classes in the ontology, respectively. The value of the CR is a percentage that indicates the amount of instantiation of classes in the ontology. A CR close to one indicates that most of the ontology classes have instances.

The decision making algorithms implemented in this study are presented next.

3.2 Multi-criteria Decision Making Algorithms

Multi-criteria decision-making (MCDM) techniques are classified into two main groups, namely, multi-objectives decision-making (MODM) and multi-attributes decision-making (MADM). The MODM techniques address decision problems on a continuous space, i.e. problems where the alternatives are in the order of infinity. Such decision problems require complex mathematical modelling such as optimization [13] to infer the appropriate solution(s). On the other hand, MADM algorithms operate on a discrete space where the number of alternatives is known [13,23]. This study is concerned about choosing the appropriate ontologies amongst a set of known ontologies that constitute the alternatives. Therefore, the focus is on the MADM techniques in this study.

Many studies have developed various MADM techniques. The most popular include the Weighted Sum Method (WSM), Weighted Product Method (WPM), Analytic Hierarchy Process (AHP), Technique for Order Preference by Similarity to Ideal Solution (TOPSIS), Weighted Linear Combination Ranking Technique (WLCRT) and Elimination and Choice Expressing Reality (ELECTRE) [23,24]. This study implements and compares the performance of three of these techniques, namely, WSM, WPM and TOPSIS in ontology ranking. These three MADM techniques were chosen due to their popularity; in fact, recent studies reported their use in the study of sustainable renewable energy development [25] and housing affordability [26].

Despite the differences in their structures and steps, MADM methods share some common features [26] including:

- **The alternatives** - These are the available choices. The decision making process must guide in the choice of the best solution(s) by providing a ranking of these choices based on certain attributes/criteria. This study uses a list of ontologies as alternatives. The set of M alternatives (A) of a MADM problem is given in Eq. (9).

$$A = \{A_1, A_2, A_3, \ldots, A_{M-1}, A_M\} \tag{9}$$

where $A_i, 1 \leq i \leq M$ is the ith alternative.
- **Attributes or criteria** - They are the measures that are used to analysed the alternatives. They are allocated weights called criteria weights. Equation (10) represents the set of N criteria of a MADM problem.

$$C = \{C_1, C_2, C_3, \ldots, C_{N-1}, C_N\} \tag{10}$$

where $C_j, 1 \leq j \leq N$ is the jth attribute/criterion. Eight metrics for measuring the design complexity of ontologies are computed and used as attributes/criteria in this study.
- **Criteria Weights** - They measure the level of importance of each criterion. They enable to classify the attributes/criteria in terms of their importance to one another. The N criteria weights associated with the attributes/criteria of a MADM problem constitute the vector in Eq. (11).

$$W = (w_1, w_2, w_3, \ldots, w_{N-1}, w_N) \tag{11}$$

where $w_k, 1 \leq k \leq N$ is the kth criteria weight.
- **Decision matrix** - This is the mathematical formulation of a MADM problem with M alternatives and N attributes/criteria as defined in Eqs. (9) and (10). Each element of the decision matrix measures the performance of an alternative against the associated decision criterion. A decision matrix D of MxN elements is represented in Eq. (12).

$$D = \begin{array}{c} \\ A_1 \\ A_2 \\ A_3 \\ \vdots \\ \vdots \\ A_M \end{array} \begin{array}{c} \begin{array}{ccccc} C_1 & C_2 & C_3 \ldots & & C_N \end{array} \\ \begin{pmatrix} d_{11} & d_{12} & d_{13} \ldots & & d_{1N} \\ d_{21} & d_{22} & d_{23} \ldots & & d_{2N} \\ d_{31} & d_{32} & d_{33} \ldots & & d_{3N} \\ \vdots & \vdots & \vdots & \ldots & \vdots \\ \vdots & \vdots & \vdots & \ddots & \vdots \\ d_{M1} & d_{M2} & d_{M3} \ldots & & d_{MN} \end{pmatrix} \end{array} \tag{12}$$

where d_{ij} represents the performance of the alternation Ai, after it has been evaluated against the attributes/criteria $C_j, 1 \leq i \leq M$ and $1 \leq j \leq N$. The following Subsections explain the three MADM algorithms implemented in this study including WSM, WPM [23] and TOPSIS [27].

Weighted Sum Model (WSM). The simplicity of the WSM make it one of the popular MADM methods. The underlying principle of WSM is to calculate a score for each alternative A_i as the sum of the products of its performances d_{ij} in the decision matrix and associated weights w_k in Eq. (11). Equation (13) represents the formula for computing the score of an alternative A_i in WSM.

$$Score(A_i) = \sum_{j}^{N} d_{ij} \cdot w_j, 1 \leq i \leq M \text{ and } 1 \leq j \leq N \tag{13}$$

where M and N are the number of alternatives and attributes/criteria of the decision making problem, respectively.

Weighted Product Model (WPM). The WPM is similar to the WSM. The difference between these two MADM methods lies in the fact that instead of adding up the products of performances to the criteria weights, the score of an alternative A_i is obtained by multiplying the exponential of each of its performances to their respective weights. A score of an alternative A_i in WPM is given in Eq. (14).

$$Score(A_i) = \prod_{j}^{N} d_{ij}^{w_j}, 1 \leq i \leq M \text{ and } 1 \leq j \leq N \tag{14}$$

Technique for Order Preference by Similarity to Ideal Solution (TOPSIS). TOPSIS is one of the most widely adopted MADM. It consist of finding the best solution from all the alternatives. In TOPSIS, decision is made based on the shortest distance from the positive ideal solution (PIS) and the farthest distance from the negative ideal solution (NIS). The preference order of alternatives is ranked according to how closed they are according to the PIS and NIS distance measures [27].

The TOPSIS algorithm consists of the following steps:

1. **Normalize the Decision Matrix** - The decision matrix D in Eq. (12) is normalized into a new matrix R with Eq. (15).

$$r_{ij} = \frac{d_{ij}}{\sqrt{\sum_{i}^{M} d_{ij}^2}}, 1 \leq i \leq M \text{ and } 1 \leq j \leq N \tag{15}$$

2. **Construct the Weighted Normalized Matrix** - A weighted normalized matrix V is built from R as in Eq. (16) by multiplying each column of R by the weight of the associated criterion.

$$V_{ij} = r_{ij} \cdot w_j, 1 \leq i \leq M \text{ and } 1 \leq j \leq N \tag{16}$$

3. **Build the Positive and Negative Ideal Solutions** - This entails creating two sets A^+ and A^- from the weighted normalized matrix V. The A^+ includes the maximum values in each column of V which are positive ideal solutions, whereas, A^- is the set of negative ideal solutions formed of the minimum values in each column of V. A^+ and A^- are calculated as in Eqs. (17) and (18), respectively.

$$A^+ = \{V_1^+, V_2^+, \ldots, A_N^+\} = \{max\ V_{ij}, 1 \leq j \leq N\} \tag{17}$$

$$A^- = \{V_1^-, V_2^-, \ldots, A_N^-\} = \{min\ V_{ij}, 1 \leq j \leq N\} \tag{18}$$

4. **Calculate the Distances from Positive and Negative Ideal Solutions** - Each alternative is considered in turn and its distances from the position and negative ideal solutions d^+ and d^- are calculated. This is done with the weighted normalized matrix V and the sets of positive and negative ideal solutions as in Eqs. (19) and (20), respectively.

$$d_i^+ = \sqrt{\sum_{j=1}^{N} (V_{ij} - V_j^+)^2}, 1 \leq i \leq M \ and \ 1 \leq j \leq N \tag{19}$$

$$d_i^- = \sqrt{\sum_{j=1}^{N} (V_{ij} - V_j^-)^2}, 1 \leq i \leq M \ and \ 1 \leq j \leq N \tag{20}$$

where $V_{ij} \in V$, $V_j^+ \in A^+$ and $V_j^- \in A^-$.

5. **Calculate the Relative Closeness of Alternatives to Ideal Solutions** - The relative closeness C_i of each alternative A_i is computed based on its distances from the positive and negative ideal solutions as in Eq. (21).

$$C_i = \frac{d_i^-}{d_i^+ + d_i^-}, 1 \leq i \leq M \tag{21}$$

6. **Ranking of Alternatives** - The relative closeness C_i are utilized to rank the alternatives A_i.

The experiments and results are presented in the next section.

4 Experiments

4.1 Dataset

A dataset of 70 ontologies of the biomedical domain downloaded from the BioPortal [28] repository is used in this study. Furthermore, eight complexity metrics (Eqs. (1) to (8)) of each ontology in the dataset is computed to constitute the decision matrix for the MCDM algorithms. The ontologies are indexed with $O_i, 1 \leq i \leq 70$ to facilitate the reference to them in the paper.

4.2 Computer and Software Environments

A computer with the following characteristics was used to carried out the experiments: 64-bit Genuine Intel (R) Celeron (R) CPU 847, Windows 8 release preview, 2 GB RAM and 300 GB hard drive. The computation of the complexity metrics of the ontologies in the dataset was done in Java Jena API library configured in Eclipse Integrated Development Environment (IDE) Version 4.2.

4.3 Experimental Results

An application was developed in Jena API to compute the complexity metrics of ontology in Eqs. 1 to 8 to form the decision matrix of the MCDM algorithms with the ontologies or alternatives (lines) and the complexity metrics or criteria (columns). To improve the quality and scale of values in the decision matrix, it was further normalized for WSM and WPM algorithms using a procedure adopted from [13] to obtain the normalized decision matrix. Similarly, the decision matrix was normalized and weighted for the TOPSIS algorithm with Eqs. (15) and (16).

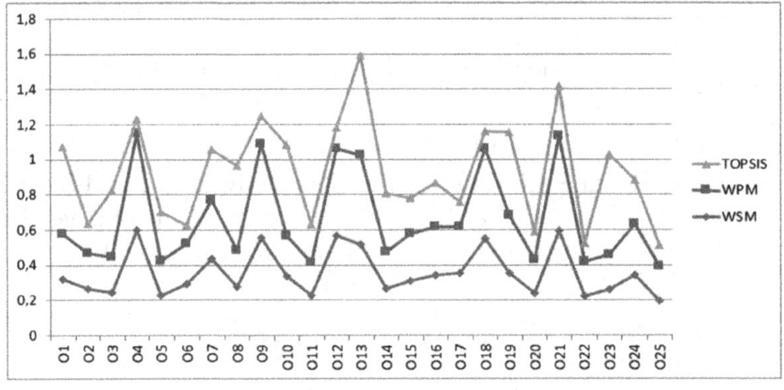

Fig. 1. Ranking scores of algorithms on Part I of dataset

The normalized decision matrix obtained for WSM and WPM was directly used to compute the score of each alternative as in Eqs. (13) and (14). With regards to the TOPSIS algorithm, its normalized decision matrix was used to get the positives and negatives ideal solutions with Eqs. (17) and (18) as well as the distances between these solutions as in Eqs. (19) and (20). Finally, the score of each alternative was computer based on the distances from positive to negative ideal solutions using Eq. (21).

In the following, the ranking scores are split into three parts due to the large number of ontologies in the dataset (70) and to improve the quality of the presentation and discussion. In fact, the scores of the three algorithms are small numbers and yielded low quality results when plotted for the 70 ontologies in the dataset. The first part includes the ontologies encoded $O_i, 1 \leq i \leq 25$ (Part I), the second part the ontologies encoded $O_j, 26 \leq j \leq 50$ (Part II) and the third or last part the ontologies $O_k, 51 \leq k \leq 70$ (Part III). Figures 1, 2 and 3 show the curves of ranking scores of the WSM, WPM and TOPSIS algorithms on the first, second and third parts of the dataset, respectively.

It is shown in Figs. 1, 2 and 3 that the curves of the ranking scores of the three algorithms follow the same patterns and that the TOPSIS scores are higher than that of WSM and WPM algorithms, whereas, the WPM scores are slightly higher than those of WSM algorithm. Furthermore, Fig. 1 portrays that the ranking scores of TOPSIS and WPM algorithms for a number of ontologies are very closed (O_4, O_9, O_{12}, O_{17}, O_{18}, O_{20}, O_{21}, O_{22}, O_{25}); and for the same ontologies, the ranking scores of the WSM algorithm are slightly low compared to that of WPM and TOPSIS. The remaining parts of the curves in Fig. 1 (middle to right) show some gaps between the ranking scores of the three algorithms.

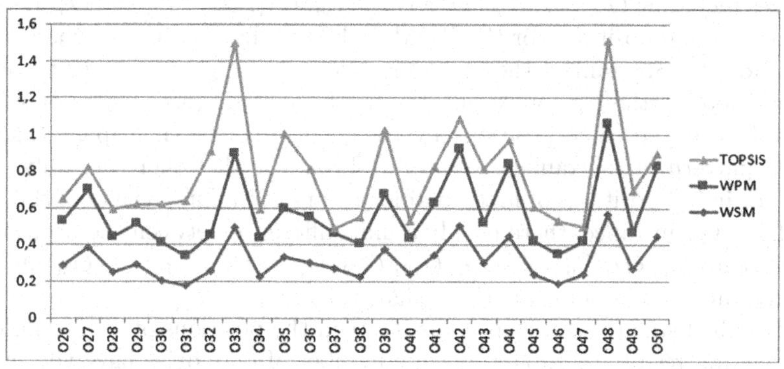

Fig. 2. Ranking scores of algorithms on Part II of dataset

A similar pattern is displayed in Fig. 2 where the ranking scores of TOPSIS and WPM algorithms are closed for most of the ontologies and slightly distanced on the rest of the ontologies. Once more, in Fig. 2, there remains a constant gap between the scores of the WSM algorithm compared to the scores of TOPSIS and WPM algorithms. Lastly, Fig. 3 maintains a constant gap between the ranking scores of the three algorithms across the entire curves, despite some closeness between the ranking scores of TOPSIS and WPM for a few number of ontologies. Overall, Figs. 1, 2 and 3 display a correlation between the ranking scores of the three algorithms where the scores are closed to each other for some ontologies and slightly distanced for others.

Table 1 displays the ranking results of all the three algorithms on the 70 ontologies in the dataset. It can be noticed that only the ontology O_8 received the same rank for all the three algorithms. However, many ontologies have been ranked at the same positions by pairs of algorithms. In fact, the pair of algorithms WSM and WPM has ranked the ontologies O_1, O_4, O_6, O_{13}, O_{29}, O_{46}, O_{49}, O_{56},

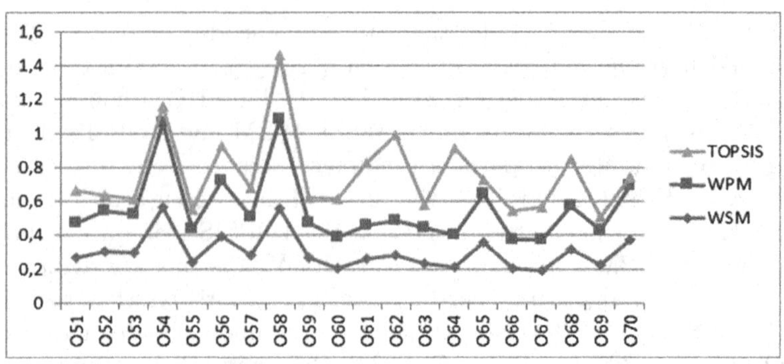

Fig. 3. Ranking scores of algorithms on Part III of dataset

O_{57} and O_{61} at the same positions in the ranking results in Table 1. Similarly, the ontologies including O_5, O_{19}, O_{25}, O_{26}, O_{28}, O_{39}, O_{44}, O_{50}, O_{53} and O_{58} received the same ranking positions for the WSM and TOPSIS algorithms, whereas, the WPM and TOPSIS ranked the ontologies O_{21}, O_{33}, O_{37}, O_{38} and O_{54} at the same positions in the ranking results. Apart from these cases where ontologies were ranked at the same positions by the algorithms taken in pairs, another common pattern in the ranking results in Table 1 is that there are only slight differences in the positions of many ontologies in the ranking results of the three algorithms. For instance, there is only a unit difference between the positions of the ontologies O_1, O_4, O_6, O_9, O_{15}, O_{19}, O_{29}, O_{37}, O_{44}, O_{46}, O_{50}, O_{59} and O_{68} in the ranking results of all the three algorithms.

Although there are significant gaps between the positions of some ontologies in the ranking results in Table 1, a large number of ontologies have very closed ranking positions in all the three algorithms. These findings suggest a close correlation between the ranking results of the three algorithms. These findings were further confirmed with the Pearson correlation coefficients of $\rho = 0,947835387$ on the ranking results of WSM and WPM, $\rho = 0,990346397$ for WSM and TOPSIS and $\rho = 0,87918235$ for WPM and TOPSIS. The strong correlation between the ranking results of the three algorithms indicates that they are all good candidates MADM methods for the ranking of semantic web ontologies.

The ranking patterns of the WSM, WPM and TOPSIS MCDM algorithms presented above is also observed in the ranking results of ELECTRE I&III and AKTiveRank in [5]. This shows that the three algorithms have successfully ranked the ontologies in the dataset based on their level/degree of complexity. Moreover, the robustness of the results obtained with a large dataset of 70 ontologies compared to 12 and 17 ontologies in [5] and [4], respectively, provides more insights on the application of decision making in ontologies ranking.

Table 1. Ranking results of WSM, WPM and TOPSIS algorithms

Index	WSM	WPM	TOPSIS	Index	WSM	WPM	TOPSIS	Index	WSM	WPM	TOPSIS
O_1	44	44	42	O_{26}	51	55	51	O_{51}	31	10	28
O_2	27	25	29	O_{27}	23	26	22	O_{52}	39	43	40
O_3	21	24	20	O_{28}	17	13	17	O_{53}	14	4	14
O_4	70	70	69	O_{29}	69	69	68	O_{54}	54	52	52
O_5	12	21	12	O_{30}	10	19	11	O_{55}	20	15	19
O_6	3	3	2	O_{31}	28	27	25	O_{56}	49	49	47
O_7	36	38	48	O_{32}	24	17	27	O_{57}	61	61	59
O_8	57	57	57	O_{33}	48	50	50	O_{58}	43	50	43
O_9	32	29	31	O_{34}	4	16	6	O_{59}	52	51	49
O_{10}	5	7	4	O_{35}	35	40	34	O_{60}	40	35	37
O_{11}	64	68	63	O_{36}	55	53	53	O_{61}	59	59	55
O_{12}	46	39	54	O_{37}	29	30	30	O_{62}	52	54	56
O_{13}	11	11	15	O_{38}	30	32	32	O_{63}	19	5	18
O_{14}	68	63	67	O_{39}	33	36	33	O_{64}	53	56	58
O_{15}	6	8	7	O_{40}	22	12	24	O_{65}	56	58	60
O_{16}	62	65	64	O_{41}	34	37	36	O_{66}	2	1	3
O_{17}	26	33	23	O_{42}	38	34	35	O_{67}	58	62	65
O_{18}	42	47	39	O_{43}	7	28	5	O_{68}	65	67	66
O_{19}	8	9	8	O_{44}	1	2	1	O_{69}	18	6	21
O_{20}	47	48	45	O_{45}	25	14	26	O_{70}	67	64	70
O_{21}	50	46	46	O_{46}	60	60	61				
O_{22}	9	20	10	O_{47}	37	41	38				
O_{23}	15	22	13	O_{48}	13	31	9				
O_{24}	63	66	62	O_{49}	45	45	44				
O_{25}	16	23	16	O_{50}	41	42	41				

5 Conclusion

This study have applied three MCDM algorithms, namely, WSM, WPM and TOPSIS in ontologies ranking. The decision matrix of these algorithms was built with as list of ontologies considered as the alternatives and a number of metrics that measure the design complexity of these ontologies. The complexity metrics constituted the attributes of the ontologies or alternatives. A large dataset of 70 ontologies of the biomedical domain was used in the experiment and the ranking scores and results of the three algorithms were analyzed and compared. The results revealed that although there were significant gaps between the positions of some ontologies in the ranking results, a large number of ontologies have very closed ranking positions in all the three algorithms. These findings suggested a strong correlation between the ranking results of the three algorithms and recorded that the three algorithms have successfully ranked the ontologies in the dataset based on their level/degree of complexity. Moreover, the robustness of the results obtained when compared to previous studies, provided more insights on the application of decision making in ontologies ranking. The future direction of research would be to experiment the ELECTRE family of MCDM methods on the dataset used and compare their ranking results to that obtained in this study.

References

1. Trokanas, N., Cecelja, F.: Ontology evaluation for reuse in the domain of process systems engineering. Comput. Chem. Eng. **85**, 177–187 (2016)
2. Lonsdale, D., Embley, D.W., Ding, Y., Xu, L., Hepp, M.: Reusing ontologies and language components for ontology generation. Data Knowl. Eng. **69**, 318–330 (2010)
3. Bontas, E.P., Mochol, M., Tolksdorf, R.: Case studies on ontology reuse. In: 5th International Conference on Knowledge Management (I-Know 2005), Graz, Austria (2005)
4. Groza, A., Dragoste, I., Sincai, I., Jimborean, I., Moraru, V.: An ontology selection and ranking system based on the analytical hierarchy process. In: The 16th International Symposium on Symbolic and Numerical Algorithms for Scientific Computing, Timisoara, Romania (2014)
5. Esposito, A., Zappatore, M., Tarricone, L.: Applying multi-criteria approaches to ontology ranking: a comparison with AKtiveRank. Int. J. Metadata Semant. Ontol. **7**, 197–208 (2012)
6. Alani, H., Brewster, C., Shadbolt, N.: Ranking ontologies with AKTiveRank. In: Cruz, I., Decker, S., Allemang, D., Preist, C., Schwabe, D., Mika, P., Uschold, M., Aroyo, L.M. (eds.) ISWC 2006. LNCS, vol. 4273, pp. 1–15. Springer, Heidelberg (2006). https://doi.org/10.1007/11926078_1
7. Naskar, D., Dutta, B.: Ontology and ontology libraries: a study from an ontofier and an ontologist perspective. In: Proceedings of 19th International Symposium on Electronic Theses and Dissertations (ETD 2016 "Data and Dissertations"), Lille, France, pp. 1–12 (2016)
8. d'Aquin, M., Noy, N.F.: Where to publish and find ontologies? A survey of ontology libraries. Web Semant. Sci. Serv. Agents World Wide Web **11**, 96–111 (2012)
9. Ensan, F., Du, W.: A semantic metrics suite for evaluating modular ontologies. Inf. Syst. **38**, 745–770 (2013)
10. Zhang, H., Li, Y.F., Tan, H.B.K.: Measuring design complexity of semantic web ontologies. J. Syst. Softw. **83**, 803–814 (2010)
11. Liao, L., Shen, G., Huang, Z., Wang, F.: Cohesion metrics for evaluation semantic web ontologies. Int. J. Hybrid Inf. Technol. **9**, 369–380 (2016)
12. Leea, H.C., Chang, C.T.: Comparative analysis of MCDM methods for ranking renewable energy sources in Taiwan. Renew. Sustain. Energy Rev. **92**, 883–896 (2018)
13. Chou, J.R.: A weighted linear combination ranking technique for multi-criteria decision analysis. S. Afr. J. Econ. Manag. Sci. Spec. **16**, 28–41 (2013)
14. Park, J., Ohb, S., Ahn, J.: Ontology selection ranking model for knowledge reuse. Expert. Syst. Appl. **38**, 5133–5144 (2011)
15. Sridevi, K., Umarani, R.: Ontology ranking algorithms on semantic web: a review. Int. J. Adv. Res. Comput. Commun. Eng. **2**, 3471–3476 (2013)

16. Butt, A.S., Haller, A., Xie, L.: DWRank: learning concept ranking for ontology search. Semant. Web **7**, 447–461 (2016)
17. Yu, W., Cao, J., Chen, J.: A novel approach for ranking ontologies on the semantic web. In: 1st International Symposium on Pervasive Computing and Applications, Urumchi, Xinjiang, China, pp. 608–612 (2006)
18. Yu, W., Chen, J.: Ontology ranking for the semantic web. In: 3rd International Symposium on Intelligent Information Technology Application, NanChang, China, pp. 573–574 (2009)
19. Jones, M., Alani, H.: Content-based ontology ranking. In: 9th International Protégé Conference, Stanford, CA, USA, pp. 1–4 (2006)
20. Subhashini, R., Akilandeswari, J., Haris, S.: An integrated ontology ranking method for enhancing knowledge reuse. Int. J. Eng. Technol. (IJET) **6**, 1424–1431 (2014)
21. Vrandečić, D., Sure, Y.: How to design better ontology metrics. In: Franconi, E., Kifer, M., May, W. (eds.) ESWC 2007. LNCS, vol. 4519, pp. 311–325. Springer, Heidelberg (2007). https://doi.org/10.1007/978-3-540-72667-8_23
22. Duque-Ramos, A., Boeker, M., Jansen, L., Schulz, S., Iniesta, M., Fernandez-Breis, J.T.: Evaluating the good ontology design guideline (GoodOD) with the ontology quality requirements and evaluation method and metrics (OQuaRE). PLoS One **9**, 1–14 (2014)
23. Zavadskas, E.K., Turskis, Z., Kildiene, S.: State of art surveys of overviews on MCDM/MADM methods. Technol. Econ. Dev. Econ. **20**, 165–179 (2014)
24. Taha, R.A., Daim, T.: Multi-criteria applications in renewable energy analysis, a literature review. Res. Technol. Manag. Electr. Ind. **8**, 17–30 (2013)
25. Kumara, A., et al.: A review of multi criteria decision making (MCDM) towards sustainable renewable energy development. Renew. Sustain. Energy Rev. **69**, 596–609 (2017)
26. Mulliner, E., Malys, N., Maliene, V.: Comparative analysis of MCDM methods for the assessment of sustainable housing affordability. Omega **59**, 146–156 (2016)
27. Balcerzak, A.P., Pietrzak, M.B.: Application of TOPSIS method for analysis of sustainable development in European Union countries. In: Proceedings of 10th International Days of Statistics and Economics, Prague, Czech Republic, pp. 82–92 (2016)
28. Ochs, C., et al.: An empirical analysis of ontology reuse in BioPortal. J. Biomed. Inform. **71**, 165–177 (2017)

Method of Decision-Making Logic Discovery in the Business Process Textual Data

Nina Rizun[1], Aleksandra Revina[2](✉), and Vera Meister[3]

[1] Gdansk University of Technology, 80-233 Gdansk, Poland
nrizun@zie.pg.gda.pl

[2] Technical University of Berlin, 10623 Berlin, Germany
revina@tu-berlin.de

[3] Brandenburg University of Applied Sciences, 14770 Brandenburg an der Havel, Germany
vera.meister@th-brandenburg.de

Abstract. Growing amount of complexity and enterprise data creates a need for novel business process (BP) analysis methods to assess the process optimization opportunities. This paper proposes a method of BP analysis while extracting the knowledge about Decision-Making Logic (DML) in a form of taxonomy. In this taxonomy, researchers consider the routine, semi-cognitive and cognitive DML levels as functions of BP conceptual aspects of Resources, Techniques, Capacities, and Choices. Preliminary testing and evaluation of developed method using data set of entry ticket texts from the IT Helpdesk domain showed promising results in the identification and classification of the BP Decision-Making Logic.

Keywords: Business process management · Decision-making · Robotic Process Automation · Natural Language Processing · Text Mining

1 Introduction

A strong market-driven digitization trend opens many opportunities for organizations, such as cost savings and performance increase achieved by the process automation, but at the same time puts some barriers, such as how to assess enterprise processes for an optimal utilization of digitization opportunities offered on the market and in the open source communities. One of these increasingly discussed process automation technologies is Robotic Process Automation (RPA) [1]. RPA software can automatically take over the execution of routine repetitive tasks of a human employee. Today, it is also possible to augment simple RPA with so-called "cognitive" functions based on scalable Natural Language and Image Processing technologies equipped with Machine Learning capabilities. Here, marketing and consulting specialists use the term Intelligent Process Automation (IPA) [2]. However, independently from the terminology, key challenges for an organization remain the following: (1) how to identify process activities that are suitable for automation and (2) how to identify the achievable degree or form of automation (for example, RPA vs. IPA).

W. Abramowicz and R. Corchuelo (Eds.): BIS 2019, LNBIP 353, pp. 70–84, 2019.
https://doi.org/10.1007/978-3-030-20485-3_6

In this paper, motivated by the above-mentioned challenges, the authors set the research objective to develop a novel method of extracting knowledge regarding the decision-making nature of processes in a form of Decision-Making Logic (DML) taxonomy from the process textual data. Under DML, the researchers understand "cognition" level of a decision-making process, i.e. perceived processing complexity of tasks to be performed within the process by a process worker and related task automation possibility in the context of existing rules, available information for task execution and automation costs [3–5]. To the extent of the authors' knowledge, there is no other approach combining the same methodological and technological setup, thus, no other approach reveals the same merits as the method presented in the paper. The researchers look into the IT ticket texts considering diverse conceptual aspects to discover the DML, i.e. parts of speech organized as per ***Resources*** (nouns), ***Techniques*** (verbs and verbal nouns), ***Capacities*** (adjectives), and ***Choices*** (adverbs) (RTCC). With the help of the RTCC framework, the authors can correctly capture the various aspects of the DML levels hidden in the entry ticket texts.

The rest of the paper is organized as follows. Section 2 introduces the related work. Section 3 presents the method for the DML identification, discusses its major steps and preliminary evaluation. Finally, Sect. 4 concludes the paper and outlines the future work.

2 Related Work

The authors suggest to structure the related work section into (1) the work related to knowledge extraction in general and the BP textual data specifically, (2) the sources necessary for understanding the method and applied technologies, i.e. Natural Language Processing (NLP) and Text Mining, and (3) subjects closely related to the present research, i.e. decision-making and taxonomies in the process context.

2.1 Approaches of Knowledge Extraction

The topic related to the extraction and representation of knowledge from texts is rather varied in the sense of approaches and especially results interpretation and formalization. Thus, in semantic technologies, widespread RDF (Resource Description Framework) approach [6] describes the data about (web) resources as subject-predicate-object triples [7]. Here, the subject must be an entity, whereas the object may also be a textually named literal. Approaches such as [8] use logical-linguistic models that consider the knowledge as sentences in the form of subject-predicate-object triplets. Another group of scientists deeply working on the knowledge extraction from the BP textual data highlight two research application areas [9]: (1) analysis of natural language inside process models [10]; (2) techniques that analyze the natural language captured in textual process descriptions [11]. One of the most relevant research publications in the context of the current paper deals with the identification of RPA candidate tasks in the BP textual descriptions [9]. The scientists suggest a three-step Machine Learning-based approach to automatically detect the degree of automation of tasks (manual, user, automatic) described in the textual process descriptions. However, the formalization approach of the extracted knowledge in the routine-cognitive classification context is missing. The authors of the present paper while building up upon the findings of [3, 9] suggest a novel approach of knowledge extraction, interpretation and formalization with the DML taxonomy of the BP activities.

2.2 Technologies Applied for Knowledge Extraction

In the paper, the main source of knowledge are the IT ticket texts coming either per email or directly entered in the ticketing system of the case study. Thus, the technologies for knowledge extraction are related to NLP and Text Mining. One of these knowledge types connected with semantic aspects hidden in BP texts is Latent Semantic Relations (LSR), which can be found both inside the documents and between them and are used to identify the context of the analyzed document and to classify a group of documents based on their semantic proximity. Specifically, a mathematical model of a text collection describing the words or documents is associated with a family of probability distributions on a variety of topics [12]. The aim of the LSR analysis is to extract "semantic structure" from the collection of information flows and automatically expand them into the underlying topic. A variety of approaches, such as discriminative Latent Semantic Analysis (LSA) [13] or probabilistic Latent Dirichlet Allocation (LDA) [12], evolved in this field with the time. While using the mentioned state-of-the-art technologies, the researchers aim to experiment on improving their quality and finding the ways to eliminate the limitations.

2.3 Decision-Making and Taxonomies in the Process Context

There is a number of research studies devoted to the decision-making processes in the business context. One group addresses the Theory of Decision-Making in general and related open questions [14–16], second group of studies discusses the challenges and opportunities of decision-making in an enterprise context, such as context-aware group decision-making [17], the criteria and approaches in decision-making support [18], or text mining-based extraction of decision elements out of project meeting transcripts [19]. Nonetheless, to the best of the authors' knowledge, the approach of the DML discovery suggested in the present paper is not researched so far.

Decision Mining in the context of Process Mining represents another relevant research direction. Process Mining [20] is a technique that aims to extract facts out of the event log. Hereby, the analysis of the latter can provide important knowledge that can help organizations to improve their decision-making processes. Specifically, Decision Mining aims at the detection of data dependencies that affect the event-based routing within a process model [21]. Regardless of this fact, if compared to the proposed research, Decision Mining primarily focuses on the analysis of event logs generated by the machines and not on the natural language texts generated by human workers within the process.

Being a widespread method of knowledge structuring and management [22], taxonomies find also their application in the decision-making related classifications [23, 24], nonetheless taxonomies applied on the levels of DML extracted from BP texts are not researched so far.

To sum up, these studies do represent a significant contribution to the development of science. However, first, the approach of the DML discovery in the unstructured BP textual data is not well represented, and, second, taxonomies applied to the DML levels extracted from BP texts are not researched so far. Furthermore, the present research makes a valuable contribution with its focus on the RPA bringing it into the new DML context. As [9] fairly state, automation in BPM is not a recent development. Research on RPA, not to mention the most recent IPA technology, by contrast, is still scarce [1].

3 Method for the Identification of the Decision-Making Logic

The work is based on the Design Science Research guidelines by Hevner et al. [25] and uses common methods of Information Systems research, such as case study, computer experiments, interviews, and observations.

The business need for novel process analysis methods in the context of emerging technologies has been identified while observing the R&D achievements (see Related Work Section), growth of the market offerings, increasing enterprise process complexity and costs (see Introduction Section). The envisioned artifact, i.e. method for process analysis, aims at classifying the BPs into simple (routine) and complex (cognitive) ones from the perspective of decision-making complexity of the worker responsible for the BP processing. At present research stage, the method uses unstructured textual data triggering the process and generated in a natural language by a process participant. This data can be received via different communication channels – email, chat, phone call, or directly entered in a specific task management system in a natural language form. In future work, the routine BPs can be automated with mentioned RPA technologies, and for the cognitive BPs – diverse levels of support for process workers can be proposed. In the paper, the artifact is evaluated using the case study qualitative survey approach. In this regard, the following research questions, which are proved in the evaluation, are raised:

RQ1: What measurements can enable the identification of the BP "cognition" level based on the unstructured BP textual data triggering the process?

RQ2: Does the semantic knowledge extracted from the unstructured BP textual data using the proposed method provide valuable information to identify the specific DML level?

The researchers use the methodological triangulation approach, which is based on multiple data sources in an investigation to produce rich and well-developed understanding for the research artifact [26]. Currently, the researchers implemented three major phases: (1) literature review, recent research and market observations to conceptualize the DML taxonomy; (2) DML taxonomy vocabulary population and experimental set-up based on the case study BP texts; (3) evaluation of the approaches (1) and (2) via qualitative survey with the case study process workers.

In the subsections below, the mentioned phases as well as detailed research artifact description (see Fig. 1) are presented.

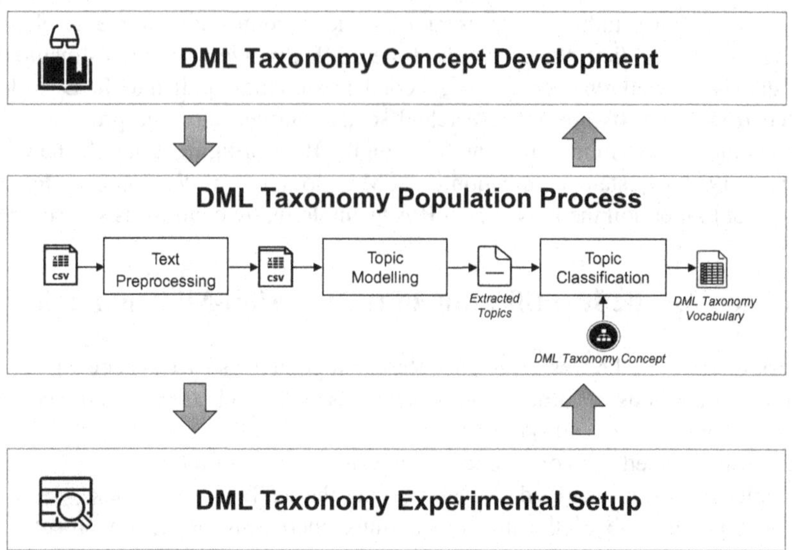

Fig. 1. Steps of method for the DML identification

3.1 DML Taxonomy Concept Development

In the first phase of the methodological triangulation, aiming to address the RQ1, i.e. measurements that could enable the identification of the BP "cognition" level, the researchers developed an understanding of the DML levels and using the systematic literature analysis enhanced with recent research and market observations distinguished three classes – *routine, semi-cognitive* and *cognitive*. Herewith, the following definitions have been accepted: (1) *routine* DML level activities or tasks are those expressible in rules so that they are easily programmable and can be performed by computers at economically feasible costs [4, 5]; (2) *semi-cognitive* DML level activities or tasks are those where no exact rule set exists and there is a clear need of information acquisition and evaluation [3, 4]. Here, computer technology cannot substitute but increases the productivity of employees [27] by partial task processing; (3) *cognitive* DML level activities or tasks are the most complex ones where not only information acquisition and evaluation is required, but also complex problem solving [4]. Computers can offer only a minimal support.

In order to measure the DML levels, a taxonomy-based approach is suggested based on the following principles: (1) consideration of a sentence as a tuple of parts of speech compound of nouns, verbs, verbal nouns, adjectives, and adverbs; (2) assumptions that DML effects of a BP in the form of *routine, semi-cognitive* and *cognitive* can be largely understood as functions of their (BP) conceptual aspects ***Resources***, ***Techniques***, ***Capacities***, and ***Choices*** (referred by the authors as RTCC semantic tagging framework). Hereby, ***Resources*** (nouns) indicate the specificity of business process task items affected by the decision-making activity; ***Techniques*** (verbs and verbal nouns) represent knowledge and information transformation activities by which the decision-making process affects existing resources; ***Capacities*** (adjectives) describe situational specificity of decision-making techniques or resources; and finally, *Choices* (adverbs) determine the selection of the required set of decision-making techniques or resources in the course of DML.

Furthermore, using a systematic literature approach and previous work [15, 16], the researchers drafted a set of indicators, or contextual variables, based on which the classification into the three DML levels according to the RTCC framework takes place (see Table 1).

In the subsection below, while populating the DML taxonomy with the contextual values of variables, the authors aim to develop a set of domain attributes and characteristics for each of the DML levels.

3.2 DML Taxonomy Population Process

In the second phase of the methodological triangulation, the case study is introduced. The data set in a form of entry ticket texts comes from the IT Change Management (ITIL Framework[1]) ticket-processing department of a big enterprise with more than 200,000 employees worldwide. The tickets can be opened in the system using an incoming email text from the customer or directly by a professional process worker who already understands what needs to be done having received the request via different channel. While developing the DML taxonomy vocabulary, the researchers extracted the topics with descriptive key words out of the case study data set, entry ticket description texts, using mentioned LD/SA approach [28, 29]. The vocabulary (see Table 1) was developed based on the available data set processed and converted into a CSV-formatted text corpus with more than 1,000,000 documents (text entries) of English, German and English-German ticket texts created in the period of 2015–2018. After removing duplicates and selecting English texts, the final case study data sample comprised 28,157 entries.

The architecture of the DML taxonomy population process (DML_TPP) is visualized on Fig. 1 as a part of the method for the DML identification. The unstructured use case BP textual data in natural language, entry ticket texts, served as an input. As a result, the researchers obtained domain dependent attributes in the form of descriptive key words extracted from the textual data (separate parts of speech – nouns, verbs, verbal nouns, adjectives, adverbs) and classified them with the help of the contextual variables, grouped based on the conceptual aspects of RTCC structure, into *routine*, *semi-cognitive* and *cognitive* DML levels.

The first step of DML_TPP is to pre-process, parse the text entries and build the document term matrices for the separate parts of speech. It is an important step in order to be able to perform topic modeling in the corpus and afterwards populate the DML taxonomy. The result of this step is cleaned textual process data with separate parts of speech. The second step of DML_TPP is to create topics with descriptive key words over the complete preprocessed data set. In particular, the created document term matrices for each part of speech (nouns, verbs, verbal nouns, adjectives, adverbs) are processed using the combination of LD/SA topic modeling methods [28, 29]. In the

[1] IT Infrastructure Library Framework, www.axelos.com/best-practice-solutions/itil.

Table 1. DML taxonomy vocabulary with exemplary key words

Contextual variables	Decision-making logic levels		
	Routine	*Semi-cognitive*	*Cognitive*
	Conceptual aspects		
	Resources		
	22%	8%	2%
Problem processing level	user, task, user request, interface, tool	team, leader, project, colleague, production	management, CAB, measure, server farm
Indeterminacy	time, application, product, name, ID	description, environment, requirement, solution, problem	risk
Information	server, database, file, location, dataset	requestor, case, rule, outage, power-supply	impact, approval
	Techniques		
	16%	6%	2%
Experience	send, note, deploy, document, decommission	check, assign, increase, create, modify	approve, delegate, define
Action alternative	follow, start, stop, monitor, run	implement, deploy, require, classify, process	propose
Effort	cancel, delete, activate, finish, mount	perform, support, plan, verify, migrate	freeze
	Capacities		
	12%	9%	5%
Specificity	additional, attached, online, virtual, same	separate, specific, technical, minor, successful	major, high, big, small, strong
Decisions formulation	new, old, preinstalled, fixed, ready	available, necessary, important, significant, successful	possible, desired, related, different, multiple
Predictability	actual, full, current, valid, same	temporary, normal, previous, similar, standard	random, randomized, expected
	Choices		
	11%	5%	2%
Precision	automatically, manually, internally, instead, there	normally, well, shortly, enough, recently	approximately, properly
Time	current, still, now, often, daily	newly, immediately, later, urgently	soon
Ambiguity	consequently, completely, never, simultaneously, accordingly	successfully, however, usually, temporarily, previously	randomly, likely, maybe

third step of DML_TPP, the extracted topics with descriptive key words are classified based on the contextual variables into the suggested DML levels of *routine, semi-cognitive* and *cognitive*. Here, the involvement of the process workers being familiar with the context is essential for the right key words classification. In Table 1, the exemplary extracted key words classified according to the three DML levels based on the RTCC framework are presented. The majority of the identified key words belong to the ***Resources*** counting up to 71 key words in the *routine* DML level. Due to the size limits of the paper, the researchers provided up to five exemplary key words per each RTCC element and DML level. As the key word relative distribution based on the total count of 324 (100%) vocabulary key words shows (see also Table 1), the majority of 61% belongs to the class *routine*, 28% to the *semi-cognitive* and very few 11% to the *cognitive*, what can be explained by: (1) the specificity of the data set domain, IT ticket processing, and (2) the fact that the entry texts of routine tickets contain a substantial level of details explaining every single step to be done and not ambiguous generic words that can imply a lot of action options.

The mentioned experiments of data preprocessing and topic modeling were performed using Python 3.4.3. The vocabulary population process is to be executed manually involving the process workers familiar with the context, what is essential for the right key words classification. At this research stage, the researchers performed the population and related classification based on the available process documentation, findings obtained within the workshop with process workers and qualitative survey based findings (see Sect. 3.4). To sum up, based on the developed DML levels taxonomy vocabulary, the authors suggest the following interpretation (characteristics of the DML levels of the data set):

- *Routine* DML level is characterized by prevailingly ***Resources*** indicating the specificity of BP task items followed by ***Techniques*** of knowledge and information transformation. This characteristic can be interpreted as an intensive and clear naming of the process resources, i.e. exact names of servers, databases, configuration items (CIs) related to the ticket.
- Compared with the routine DML level, *semi-cognitive* DML level is increasingly described with ***Capacities*** describing situational specificity of decision-making techniques and ***Techniques*** themselves. The accent shifts from ***Resources*** to the description of the situational specificity of the ***Techniques,*** what can be explained by the absence of the simple and exact rule set to be followed while processing the ticket.
- Compared with routine and semi-cognitive levels, *cognitive* DML level is not well represented in the taxonomy vocabulary. The rare findings, however, show the growing "cognition" within the cognitive DML level key words relative distribution, i.e. increased relative numbers of ***Capacities*** describing situational specificity and ***Choices*** describing the selection of the ***Resources*** or ***Techniques***. In this case, the process workers need to act based on their "gut" feeling performing mental simulations of the situation, acquiring and evaluating further information via various channels. All these complicates further ticket processing.

3.3 DML Taxonomy Experimental Setup

In the experimental setup step of the method, the developed DML taxonomy vocabulary was tested (using Python 3.4.3) on the final case study data sample with 28,157 entries (processed and English language based) to analyze the specific distributions of the DML vocabulary words in the tickets (all parts of speech). As shown in Table 2, the following analytical findings can be derived: (1) there are only 20.23% of *pure routine, cognitive* or *semi-cognitive* tickets in the data sample with the clear majority of *pure routine* tickets 18.87%, only 1.30% of *pure semi-cognitive* and scarcely 0.07% of *pure cognitive*[2]; (2) no key words were identified in the group of tickets 2.82%. These tickets are characterized by the 100% presence of unique names of process ***Resources.*** Therefore, the authors classify them as *pure routine* ones (see the interpretation of *routine* DML level above); (3) the prevailing amount of tickets in the data sample 76.95% is the mixed one. In this group, the majority of tickets 33.31% contains >=50% of *routine* key words, <=50% of *semi-cognitive* and 0% of *cognitive*. The second representative subgroup with 19.00% comprises >50% of *routine* key words, <=25% of *semi-cognitive* and <=25% of *cognitive*. Thus, also in the mixed group, the *routine* amount of key words prevails. The obtained analytical findings provide an understanding about the "cognition" level of the tasks the process workers are dealing with in the case study department. The practical value of these findings will be investigated in the future work, i.e. Recommender System approach: (1) *pure routine* tickets can be processed automatically while implementing existing templates; (2) mixed and *pure semi-cognitive* groups of tickets should be studied in more detail, and the recommendation with drop-down lists of possible choices can be provided to the process workers; (3) processing of *pure cognitive tickets* or tickets with relative high amount of the *cognitive* key words (mixed group) should be supported by the provision of the history of similar tickets.

3.4 DML Taxonomy Evaluation

In order to evaluate the approaches presented in Sects. 3.1, 3.2 and 3.3, in the third phase of the methodological triangulation, the researchers developed a qualitative survey in a form of a questionnaire. The survey was conducted in the period of January-February 2019 with the process workers responsible for the ticket processing in the case study IT Change Management department. The sample comprised 13 process managers. The respondents were asked to critically evaluate and provide their own (practical) view on: (1) the definitions of the three DML levels suggested by the researchers (Sect. 3.1); (2) the developed DML taxonomy vocabulary (Sect. 3.2); (3) the ticket classification examples (see Table 3) according to the three DML levels performed by the researchers using sentence-by-sentence approach based on the randomly selected tickets.

As one can conclude from Table 3, the underlying decision-making processes of the process worker in the exemplary anonymized ticket are predicted to be 77%

---

[2] The relative distributions were calculated based on the presence of the DML taxonomy vocabulary key words in a ticket and not on the overall count of words in a ticket.

Table 2. DML taxonomy vocabulary key word distribution in tickets

Routine key words in a ticket, %	Semi-cognitive key words in a ticket, %	Cognitive key words in a ticket, %	Number of tickets with such a distribution, %
100	0	0	18.87
0	100	0	1.30
0	0	100	0.07
0	0	0	2.82
>=50	<=50	0	33.31
>50	<=25	<=25	19.00
The rest			24.64

Table 3. Anonymized ticket classification example

Ticket (sentences)	Values	Contextual semantic aspect	Word DML level	Ticket DML summary
Please stop-start XYZ databases mentioned below: XYZ1, XYZ2, XYZ3, XYZ4	stop	verb (technique)	routine	77% routine 15% semi-cognitive 8% cognitive
	start	verb (technique)	routine	
	database	noun (resource)	routine	
server: xyzxyz	server	noun (resource)	routine	
Please check mentioned databases if were stop-start successfully, and if applications after start running properly	check	verb (technique)	semi-cognitive	
	database	noun (resource)	routine	
	stop	verb (technique)	routine	
	start	verb (technique)	routine	
	successfully	adverb (choice)	semi-cognitive	
	application	noun (resource)	routine	
	start	verb (technique)	routine	
	run	verb (technique)	routine	
	properly	adverb (choice)	cognitive	

routine, 15% *semi-cognitive* and only 8% *cognitive* (the second characteristic subgroup of the mixed group with 19.00%, see Table 2). Such ***Techniques*** as "stop", "start", "run" in the case study context are based on the simple decision-making processes implying a straightforward action (routine DML) while the activity or task "check" demands a certain amount of experience on the level of direct habits to be performed correctly (semi-cognitive DML). Such process ***Resources*** as "database", "server", "application" have exact names indicating high accuracy, certainty and complete information for the process worker (routine DML). However, such ***Choices*** determining the selection of the required set of ***Techniques*** as "successfully" (semi-cognitive DML) or "properly" (cognitive DML) have an implicit meaning and in the majority of the cases are based on the experience-gained "gut" feelings.

The results of the qualitative survey and their implications for the current research are presented below:

(1) the accepted DML levels definitions were complemented with the following new contextual aspects (see Summary Table 4). Thus, for example, the DML *routine* level can be characterized from the process worker perspective with the time, frequency, effort, and impact aspects, while the researchers first considered the theoretical perspective of rules, information, and automation based on the literature, recent research and market observations.
(2) the researchers enhanced the DML taxonomy vocabulary with context-based key words received from the respondents, such as: *routine* – "firewall", "user request", "rundown", "decommission" and *cognitive* – "big measures", "server farm", "freeze". As a result, the DML taxonomy vocabulary has been specified with the contextual key words.
(3) the majority of respondents agreed with the provided examples of ticket classification (sentence-by-sentence approach based on the identified descriptive key words) performed by the researchers using the DML taxonomy vocabulary.

The evaluation-based findings mentioned above enabled the researchers to answer the research questions posed at the beginning of the Sect. 3. Hence, while approaching the RQ1 in Sects. 3.1 and 3.2, the researchers suggested (and experimentally tested in Sect. 3.3) a set of measurements for the BP "cognition" level identification based on the unstructured BP textual data triggering the process. These measurements are the DML levels definitions, RTCC structure and the DML taxonomy vocabulary. In the evaluation phase, the proposed set of measurements was specified with the context-based definitions and key words provided by process workers. While approaching the RQ2, the researchers provided ticket classification examples for the evaluation by the process workers. Prevailingly positive evaluation results showed the plausibility of the method (RQ2).

Table 4. Summary of DML levels definitions

Theoretical definitions	Context-based definitions
Routine	
Rules: simple **Information:** complete **Automation:** easily programmable at economically feasible	**Time:** less than 5 min **Frequency:** daily occurred work **Effort:** few mouse clicks **Impact:** no impact
Semi-cognitive	
Rules: no exact rule set **Information:** need for information acquisition and evaluation **Automation:** partial task processing to increase the productivity of employees	**Number of tasks:** many **Number of CIs:** many **Impact:** with clear impact
Cognitive	
Rules: complex **Information:** arguable information demanding complex problem solving **Automation:** minimal possible	**Challenging** **Multi-solution** Thinking of **what?, where?, how?**

4 Conclusion and Future Work

In this paper, the researchers presented a method of extracting knowledge about the Decision-Making Logic ("cognition") of business processes in a form of DML taxonomy from the unstructured textual data. In contrast to the most of existing approaches in process and text analysis, the proposed method is based on the novel combination of methodological and technological approaches, offering new merits for process analysis in the automation context. Following the methodological triangulation approach, the researchers present the method in three main phases: (1) literature review, recent research and market observations to conceptualize the DML taxonomy; (2) DML taxonomy vocabulary population and experimental set-up based on the case study BP texts, IT entry ticket texts; (3) evaluation of the approaches (1) and (2) via qualitative survey with the case study process workers.

The main contribution of the paper is finding answers to the research questions suggested by the authors. The results of the evaluation phase provided additional (contextual) information on the DML definitions and taxonomy vocabulary. Furthermore, the evaluation demonstrated that the method in general and DML taxonomy vocabulary in particular are able to deliver plausible results in the classification of the entry ticket texts according to *routine, semi-cognitive* and *cognitive* DML levels. The discovered topics with descriptive key words appeared to be coherent and informative for DML taxonomy vocabulary building.

However, in the presented research, the authors performed an in-depth analysis of unstructured textual data using only one criterion – specific semantics of an entry ticket text. The researchers will include additional criteria to analyze the mentioned unstructured texts, for example length of the texts and their stylistic characteristics.

Furthermore, such event log data as time stamps, number of tasks and CIs per ticket, responsible groups will be also included into analysis to verify the semantically based DML level findings.

Besides, the researchers will address the demonstration of the research practical value. The proposed DML method can be formalized and automated with the support of an OMG framework for a rule-based decision modeling DMN [30]. The researchers plan to formalize the case study process using BPMN [31] for modeling the overall process with strict procedures while preferring CMMN [32] in case of wide range of free plannable activities depending on the situational context. Hereby, the experimental model of the case study process (Recommender System approach mentioned in Sect. 3.3) will be built using DMN extension to measure the entry ticket DML level as an attempt of partial automation. To enable a functioning prototype of such a multi-level Recommender System, such technologies as robust graph databases, ML algorithms for exact match search, RPA to automate front- and back-end rule-based tasks will be studied and evaluated in detail.

References

1. Bourgouin, A., Leshob, A., Renard, L.: Towards a process analysis approach to adopt robotic process automation. In: 15th International Conference on e-Business Engineering, Xi'an, pp. 46–53. IEEE (2018)
2. Lowes, P., Cannata, F.R.S., Chitre, S., Barkham, J.: The Business Leader's Guide to Robotic and Intelligent Automation. Automate This (2016). Deloitte: https://www2.deloitte.com/content/dam/Deloitte/nl/Documents/operations/deloitte-nl-operations-the-business-leaders-guide-to-robotic-and-intelligent-automation-automate-this-report.pdf. Accessed 1 Jan 2019
3. Koorn, J.J., Leopold, H., Reijers, H.A.: A task framework for predicting the effects of automation. In: Twenty-Sixth European Conference on Information Systems Proceedings, Portsmouth (2018)
4. Autor, D.H., Levy, F., Murnane, R.J.: The skill content of recent technological change: an empirical exploration. Q. J. Econ. **118**, 1279–1333 (2003)
5. Levy, F., Murnane, R.: With what skills are computers a complement? Am. Econ. Rev. **86**(2), 258–262 (1996)
6. Jetschni, J., Meister, V.G.: Schema engineering for enterprise knowledge graphs - a reflecting survey and case study. In: Conference on Intelligent Computing and Information Systems Proceedings, Cairo, pp. 271–277. IEEE (2017)
7. RDF Scheme. https://www.w3.org/TR/rdf-schema/. Accessed 1 Jan 2019
8. Khairova, N.F., Petrasova, S., Gautam, A.P.S.: The logical-linguistic model of fact extraction from English texts. In: Dregvaite, G., Damasevicius, R. (eds.) ICIST 2016. CCIS, vol. 639, pp. 625–635. Springer, Cham (2016). https://doi.org/10.1007/978-3-319-46254-7_51
9. Leopold, H., van der Aa, H., Reijers, H.A.: Identifying candidate tasks for robotic process automation in textual process descriptions. In: Gulden, J., Reinhartz-Berger, I., Schmidt, R., Guerreiro, S., Guédria, W., Bera, P. (eds.) BPMDS/EMMSAD -2018. LNBIP, vol. 318, pp. 67–81. Springer, Cham (2018). https://doi.org/10.1007/978-3-319-91704-7_5

10. Leopold, H., Mendling, J.: Automatic derivation of service candidates from business process model repositories. In: Abramowicz, W., Kriksciuniene, D., Sakalauskas, V. (eds.) BIS 2012. LNBIP, vol. 117, pp. 84–95. Springer, Heidelberg (2012). https://doi.org/10.1007/978-3-642-30359-3_8
11. Leopold, H., van der Aa, H., Pittke, F., Raffel, M., Mendling, J., Reijers, H.A.: Searching textual and model-based process descriptions based on a unified data format. Softw. Syst. Model. **18**(2), 1179–1194 (2017)
12. Blei, D.M.: Probabilistic topic models. Commun. ACM **55**(4), 77–84 (2012)
13. Dumais, S.T., Furnas, G.W., Landauer, T.K.: Using latent semantic analysis to improve access to textual information. In: Conference on Human Factors in Computing Proceedings, pp. 281–285. ACM, New York (1988)
14. Xinghua, L., Zhixin, M.: Research on the problem of the intuitive decision making. In: The 2nd International Conference on E-Education, E-Business and E-Technology Proceedings, Beijing, pp. 192–196. ACM (2018)
15. Rizun, N., Shmelova, T.: Decision-making models of the human-operator as an element of the socio-technical systems. In: Batko, R., Szopa, A. (eds.) Strategic Imperatives and Core Competencies in the Era of Robotics and Artificial Intelligence, pp. 167–204. IGI Global, Hershey (2017)
16. Rizun, N., Taranenko, Y.: Simulation models of human decision-making processes. Manage. Dyn. Knowl. Econ. **2**(5), 241–264 (2014)
17. De Maio, C., Fenza, G., Loia, V., Orciuoli, F., Herrera-Viedma, E.: A framework for context-aware heterogeneous group decision-making in business processes. Knowl.-Based Syst. **102**, 39–50 (2016)
18. Pérez-Álvareza, J.M., Maté, A., Gómez-López, M.T., Trujillo, J.: Tactical business-process-decision support based on KPIs monitoring and validation. Comput. Ind. **102**, 23–39 (2018)
19. Chibelushi, C., Thelwall, M.: Text mining for meeting transcript analysis to extract key decision elements. In: International MultiConference of Engineers and Computer Scientists Proceedings, IMECS, Hong Kong (2009)
20. van der Aalst, W.: Process Mining. Springer, Heidelberg (2016). https://doi.org/10.1007/978-3-662-49851-4
21. Rozinat, A., van der Aalst, W.: Decision mining in ProM. In: Dustdar, S., Fiadeiro, J.L., Sheth, A.P. (eds.) BPM 2006. LNCS, vol. 4102, pp. 420–425. Springer, Heidelberg (2006). https://doi.org/10.1007/11841760_33
22. Guttmann, C.: Towards a taxonomy of decision making problems in multi-agent systems. In: Braubach, L., van der Hoek, W., Petta, P., Pokahr, A. (eds.) MATES 2009. LNCS (LNAI), vol. 5774, pp. 195–201. Springer, Heidelberg (2009). https://doi.org/10.1007/978-3-642-04143-3_19
23. Scherpereel, C.M.: Decision orders: a decision taxonomy. Manag. Decis. **44**(1), 123–136 (2006)
24. Gibcus, P., Vermeulen, P.A., de Jong, J.P.: Strategic decision making in small firms: a taxonomy of small business owners. Entrepreneurship Small Bus. **7**(1), 74–91 (2009)
25. Hevner, A.R., March, S.T., Park, J., Ram, S.: Design science in information systems research. MIS Q. **78**(1), 75–105 (2004)
26. Patton, M.Q.: Enhancing the quality and credibility of qualitative analysis. Health Serv. Res. **34**(5), 1189–1208 (1999)
27. Spitz-Oener, A.: Technical change, job tasks, and rising educational demands: looking outside the wage structure. J. Labor Econ. **24**(2), 235–270 (2006)

28. Rizun, N., Taranenko, Y., Waloszek, W.: The algorithm of modelling and analysis of latent semantic relations: linear algebra vs. probabilistic topic models. In: Różewski, P., Lange, C. (eds.) KESW 2017. CCIS, vol. 786, pp. 53–68. Springer, Cham (2017). https://doi.org/10.1007/978-3-319-69548-8_5
29. Rizun, N., Taranenko, Y., Waloszek, W.: Improving the accuracy in sentiment classification in the light of modelling the latent semantic relations. Information **9**(12), 53–68 (2018)
30. OMG: Decision Model and Notation (DMN) – Version 1.1. Object Management Group, Needham (2016)
31. OMG: Business Process Model and Notation (BPMN) – Version 2.0.2. Object Management Group, Needham (2013)
32. OMG: Case Management Model and Notation (CMMN) – Version 1.1. Object Management Group, Needham (2016)

Prediction of Productivity and Energy Consumption in a Consteel Furnace Using Data-Science Models

Panagiotis Sismanis(✉)

Sidenor Steel Industry SA, 15125 Athens, Greece
psismanis@sidenor.vionet.gr

Abstract. The potential to predict the productivity and the specific electric-energy furnace consumption is very important for the economic operation and performance of a Consteel electric-arc furnace. In this work, these two variables were predicted based on specific operating parameters with the use of machine learning. Actually, three different algorithms were tested for this study: the BRF method of support vector machine (SVM), the light gradient boosting method (lightGBM), and the Keras system with TensorFlow as backend. The results appear to be good enough for production scheduling, and are presented and discussed in this work.

Keywords: Gradient boosting method · Support vector machine · Keras · TensorFlow · Productivity · Energy · Consteel

1 Introduction

The need for the potential of predicting the productivity, and specific electrical-energy consumption based on process parameters has been always the desire in steelmaking plants worldwide. It facilitates procurement and results in a more rational approach in the schedule of actions. Budget development is then performed in a faster and more reliable basis. In a monumental work, Koehle [1] had come up with a very compact formula based on regression analysis from data collected from a relatively large number of electric-arc furnaces (EAF) worldwide. This equation had taken under consideration the specific electric-energy reduction from installations with scrap preheating, like Consteel. On the other hand, in a more recent work, Memoli [2] elaborated data from Consteel installations and pointed out the effect of liquid heel upon productivity in addition to the installed and operating electric-arc power. However, each Consteel installation has its own peculiarities depending upon scrap quality, furnace condition, and operating personnel skills. In this study, process parameters that are recorded daily per heat were selected and an attempt to predict the expected productivity and specific electric-furnace energy consumption was carried out. Machine learning was applied in order to derive supervised models that could be used in off-line scheduling from the scrap yard till the secondary metallurgy treatment of liquid steel.

W. Abramowicz and R. Corchuelo (Eds.): BIS 2019, LNBIP 353, pp. 85–99, 2019.
https://doi.org/10.1007/978-3-030-20485-3_7

2 Preparation for the Computations

2.1 EAF Installation

The SOVEL plant, which is part of the VIOHALCO/SIDENOR group of companies, is located at a seacoast area of Almyros next to the city of Volos in the middle of Greece. In 2006, SOVEL decided to convert to a Consteel operation in order to increase production capacity and decrease electrical energy consumption. The furnace is a 3-phase EBT EAF with a 120 MVA transformer, and 600-mm-diameter electrodes.

Table 1. Considered operating parameters (independent variables).

No.	Symbol	Description
1	*YIELD*	Yield (t of good billet per t of scrap)
2	*POW_ON*	Power on time (min)
3	*TAP_TAP*	Tap-to-tap time (min)
4	*POWER_AVG*	Average power per heat (MW)
5	*SPEC_OX*	Specific oxygen consumption (Nm^3/t)
6	*SPEC_LIME*	Specific lime consumption (kg/t)
7	*SPEC_MGO*	Specific MgO consumption (kg/t)
8	*SPEC_CHRG_CARBON*	Specific charged carbon (kg/t)
9	*SPEC_INJ_CARBON*	Specific injected carbon (kg/t)
10	*TEMP*	Tapping temperature (°C)
11	*ppmO*	Oxygen content (ppm)
12	*PC_HMS_1*	Scrap type HMS #1 (percentage)
13	*PC_HMS_2*	Scrap type HMS #2 (percentage)
14	*PC_SHREDDED*	Shredded scrap (percentage)
15	*PC_BUSHELLING*	Busheling scrap (percentage)
16	*PC_RETURNS*	Returns (percentage)
17	*PC_TURNINGS*	Turnings (percentage)
18	*PC_PIG_IRON*	Pig iron (percentage)

It has the necessary modules to supply chemical energy as well as to inject carbon units in order to retain the appropriate volume of foaming slag. The meltshop has a ladle furnace (LF) for the secondary metallurgy treatment of liquid steel, and a 6-strand continuous caster (CCM) that mainly casts 140 × 140 mm × mm billets. The furnace-tapped weight is 130 t and the annual capacity is around 1 million tons of steel. The interested reader may refer to [3] for a more detailed description of the furnace installation.

2.2 Parameters Considered in the Computations

Two dependent parameters were analyzed: productivity in t/h (*TON_PER_HR*), and specific electrical-energy consumption in kWh/t good billet (*SPEC_ENRGY_BILLET*).

However, since these two parameters are very critical with respect to the economics policy of a company, only scaled data/results were presented in this study. Nevertheless, the degree of correlation achieved in the analysis is mostly important to the data scientist and not the actual values.

Table 1 presents the 18 independent operating parameters considered as the important factors that were supposed to have an influence upon the two dependent variables. The main parameters related to the electrical energy consumption are the average power of the heat (*POWER_AVG*), and the time duration in which the energy was supplied (*POW_ON*). The tap-to-tap time (*TAP_TAP*) influences productivity and electrical energy consumption; it reflects the proper condition of a furnace not only with respect to operations but with respect to maintenance, as well. The specific oxygen consumption (*SPEC_OX*) plays an important role not only to the selected dependent variables but also to the chemistry of the tapped liquid steel. The specific addition of lime (*SPEC_LIME*) and magnesia (*SPEC_MGO*) play a paramount role in the proper chemical composition of the slag for foaming. In practice, we aim for $FeO \approx 20\%$, $MgO \approx 6\%$, and $B_3 \approx 2.0$; the index B_3 is given by the following equation [4]:

$$B_3 = \frac{CaO}{SiO_2 + Al_2O_3} \tag{1}$$

We wanted to keep the derived supervised results as close as possible to the operating parameters that we controlled more easily in practice. For that reason, we decided to employ only the specific additions of the slag fluxes than including the slag chemical analyses into the computations. The liquid steel temperature (*TEMP*) and oxygen (*ppmO*) as measured by the CELOX probe [5] have an effect on energy consumption and liquid steel cleanliness; the yield (*YIELD*) plays a great role on specific consumptions and cost-effective meltshop operation. Finally, seven scrap parameters related to scrap mix were taken under consideration (Table 1, 12–18). One may realize that scrap mix is the top factor that greatly influences the two selected dependent parameters, so it was impossible to proceed in this type of study without considering it. Nevertheless, a meltshop facility may not have the desired scrap mix all year around. On this basis, proper management may help eliminate extreme cases and still make production at desired costs.

2.3 Computational Approach

Pieces of software were developed in order to tackle the process of data, and tune the selected machine-learning algorithms. Python (version 3.5.4, 64-bit, Anaconda installation [6]) was the deployed language. The total number of cases (heats) that were selected for the analysis were 9148 in total, and they belonged in the production period from January 2016 until July 2018, that is about 1.2 million tons of produced liquid steel. In fact, the initial number of heats for that period was more but almost one-third of heats were filtered out in order to compensate for ambiguous or missing data. The library of pandas [7] was used for data manipulation, and the scikit-learn library [8] was used for validation and tuning of the models. Furthermore, the kernel RBF (radial basis function) from the support vector machine (SVM) library of the scikit-learn package was deployed as one of the three selected algorithms for machine learning. SVM was selected, as it is a powerful and widely used machine-learning algorithm. It generally constructs a hyperplane or set of hyperplanes in a multi-dimensional space which can be used for regression by selecting the hyperplane that has the largest distance to the nearest training-data point of any class (margin), since in general the larger the margin, the lower the generalization error of the derived supervised model. The second selected algorithm was lightGBM [9] (gradient boosting method) that appears to be reliable, simple, and flexible. LightGBM is a gradient boosting framework that uses histogram-based algorithms, which bucket continuous feature (attribute) values into secret bins. This speeds up training and reduces memory usage. Finally, the last algorithm was Keras [10, 11] with TensorFlow [12] as backend. Keras uses neural networks for the derivation of the supervised models. Keras is an open-source neural-network library written in Python. Designed to enable fast experimentation with deep neural networks, it focuses on being user-friendly, modular, and extensible. It supports standard, convolutional, and recurrent neural networks. An attempt to apply machine learning in a Python environment was the triggering mechanism to work with the three aforementioned machine-learning algorithms after having published another study in an environment under R [13]. The software was run in a DELL Alienware laptop with the Intel i7-6700HQ CPU (8 cores) @2.6 GHz, 16 GB RAM, running under a 64-bit Windows 10 Professional O/S. The data were scaled before further processing. Finally, two important functions from the scikit-learn package were deployed: the StandardScaler function was used in order to scale the data so that errors coming from computations amongst too big and too low values to be minimized; furthermore, the GridSearchCV function was called in order to fine-tune the machine learning algorithms.

3 Results and Discussion

3.1 Principal Component Analysis

At first, a search was held in order to verify whether the independent variables exhibited some patterns. For this purpose, the KMeans method from scikit-learn was applied after scaling in order to identify the optimum number of clusters in which the independent variables might be included; the average-silhouette-width [14] criterion was put into effect and proved that the optimum number of clusters was two. Furthermore, applying a principal component analysis (PCA) after filtering the scaled data through the GradientBoostingRegressor (all scikit-learn [8] functions) for the two optimum clusters, Fig. 1 was created showing the results.

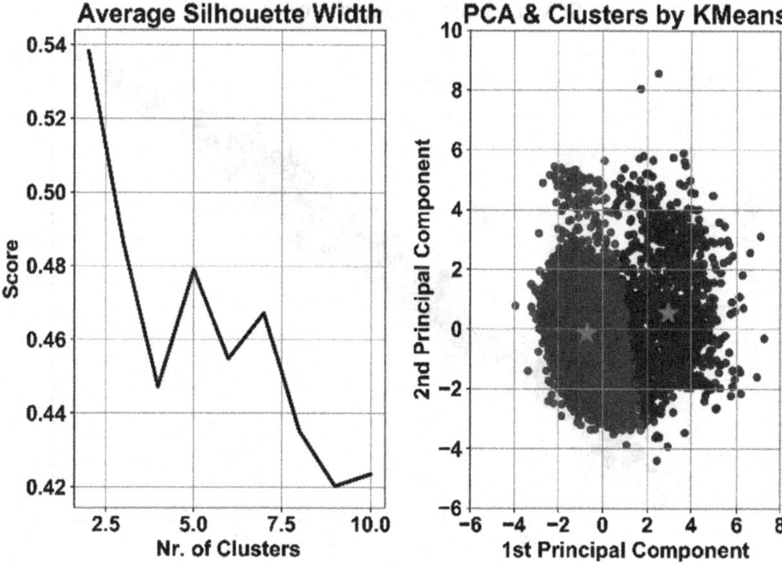

Fig. 1. Clusters and PCA analysis.

The two stars in Fig. 1 illustrate the centroids for the two clusters. It seems that the two clusters are related to the heats that were produced at larger average-power values (left-hand-side centroid), and smaller average-power values (right-hand-side centroid).

3.2 Productivity Analysis-SVM

Figure 2 depicts actual versus predicted scaled-productivity values. One may realize that there is a relatively good prediction upon productivity. In fact, the analysis of variance (ANOVA) between actual and predicted values showed that the root-mean-squared-error (RMSE) was 0.150, with a correlation coefficient (R2_squared) equal to 0.9778. However, in order to deduce these relatively good results with the RBF kernel of the support vector machine we had to tune the model. Two tuning parameters were required [15, 16] in order to improve the accuracy of the model: C, and gamma; actually, after tuning using a grid search (GridSearchCV), the optimum values were C = 100, and gamma = 0.01.

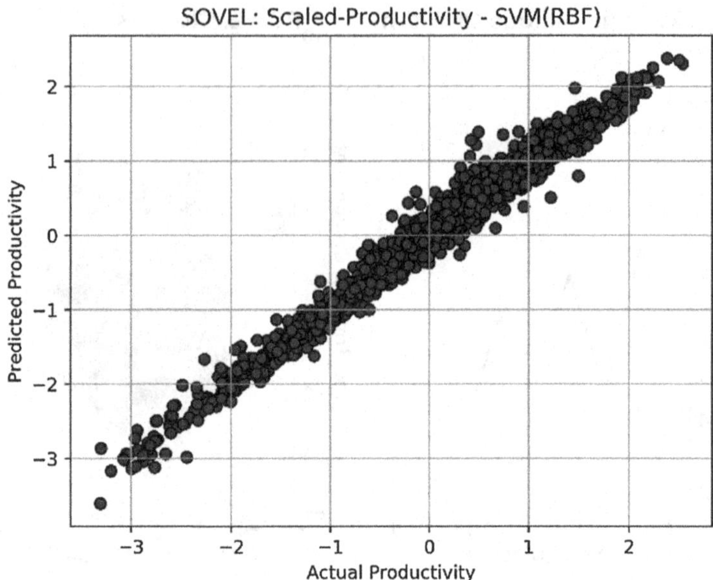

Fig. 2. Actual vs. predicted scaled-productivity values using the RBF kernel of the support vector machine.

3.3 Productivity Analysis-LightGBM

Figure 3 illustrates actual versus predicted scaled-productivity values by the lightGBM model. Here, the ANOVA showed that the RMSE value was 0.098, with R2_squared equal to 0.9905. Furthermore, a grid search was held in order to tune the appropriate parameters. In this case, the optimum parameters were learning_rate = 0.01; n_estimators = 10000.

3.4 Productivity Analysis-Keras

Keras was found more difficult to tune and the attained optimum parameters gave adequate values for the RMSE = 0.156, and R2_squared = 0.9760. Figure 4 illustrates the attained training and validation accuracy versus epochs. One may notice that the error minimizes flattening at epochs equal to 700. Actually, the optimum parameters were found to be batch_size = 150, and epochs = 700. Figure 5 depicts the actual versus predicted productivity values by Keras. Two main sequential models were applied the first with 64 units, and the second 18 units; the activation selected was of 'relu' type that is a good choice for regression models, with a 2% dropout percentage, and an 'l2' regularizer kernel equal to 0.1%. The 'rmsprop' type was selected as the routine for the mean-squared-error optimizer.

As a concluding remark, it seems that productivity can be predicted to a reasonable extent by the operating parameters from all three selected models.

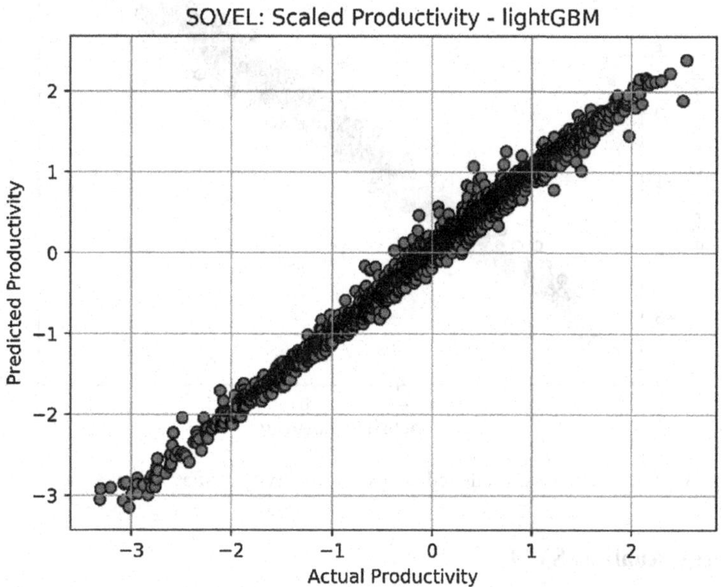

Fig. 3. Actual vs. predicted scaled-productivity values using lightGBM.

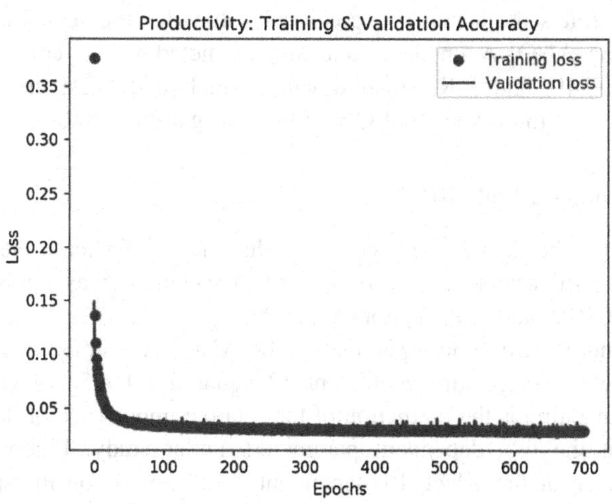

Fig. 4. Training and validation accuracy vs. epochs during grid-search tuning of the Keras model for productivity.

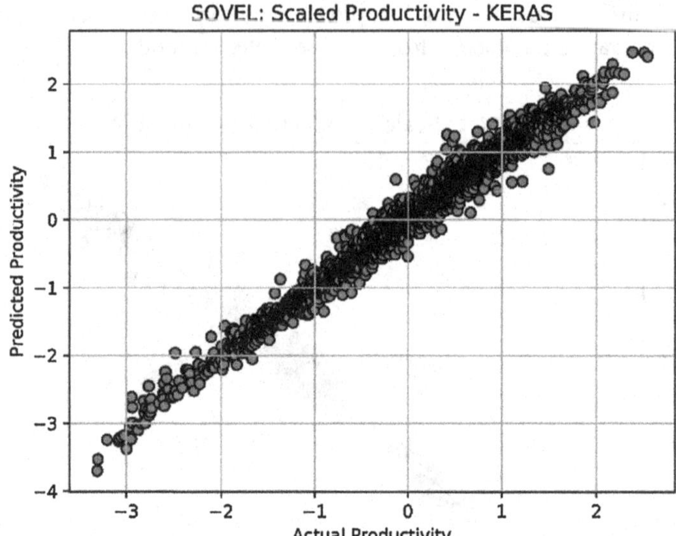

Fig. 5. Actual vs. predicted scaled-productivity values using Keras.

3.5 Energy Analysis-SVM

The scaled specific electrical-energy consumption (*SPEC_ENRGY_BILLET*) was predicted by the RBF kernel of the support vector machine; a grid search was applied in order to quantify the tuning parameters (C, gamma) and optimize the mean-squared-error. Figure 6 depicts the actual and predicted values for the scaled specific-energy consumption. The ANOVA on the actual and predicted values computed a RMSE value equal to 0.240, and a R2-squared value equal to 0.9421; these values were calculated for the optimum values of C = 1000, and gamma = 0.01.

3.6 Energy Analysis-LightGBM

Figure 7 illustrates the actual and predicted values for the specific electrical energy consumption. A grid search was also performed calculating as optimum values a learning_rate = 0.010, and n_estimators = 10000.

For Fig. 7 under the optimum parameters the ANOVA resulted in a RMSE value equal to 0.1858 with a correlation coefficient R2_squared = 0.9654. Nevertheless, what appears to be interesting is the prediction of the relative importance of the independent parameters upon the two dependent parameters under study. Figure 8 shows the variable importance of the selected independent parameters upon the specific energy consumption; it is worth saying that the same relative importance of the independent parameters appears true for productivity, as well. One may notice that the independent parameters related to power and time play a paramount importance upon energy consumption, as well as productivity. Indeed, *POWER_AVG*, *POW_ON*, and *TAP_TAP* seem to be very important, as expected from real practice experience. Yield (*YIELD*) cannot be underestimated and energy-inducing factors like oxygen (*SPEC_OX*) and carbon (*SPEC_CHRG_CARBON, SPEC_INJ_CARBON*) are important as well.

We should notify that we operate the furnace at low natural-gas consumptions; actually, we use natural gas only for the proper operation of modules and not as burners. Lime is an important slag constituent (*SPEC_LIME*), and tapping temperature (*TEMP*) and oxygen content (*ppmO*) could not be excluded from the estimating picture. The influence of the scrap mix is more-or-less expected according to the experience from our practice. However, the low effect from pig iron (*PC_PIG_IRON*) comes from the fact that we do not add more than 3% in most cases, for the heats that we do add it.

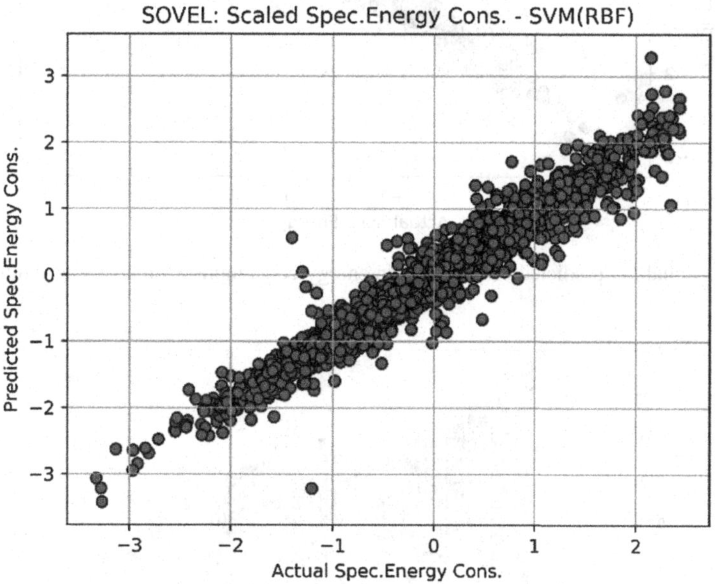

Fig. 6. Actual vs. predicted scaled specific-energy-consumption values using the RBF kernel of the support vector machine.

3.7 Energy Analysis-Keras

Figure 9 illustrates the training and validation accuracy in the tuning process.

Applying a grid search to tune the Keras model it was found that the batch_size = 100, at epochs = 600 gave rise to the best possible optimization of the mean-squared-error. In fact, the ANOVA gave the value of RMSE = 0.238, at a multiple correlation coefficient R2_squared = 0.9432. Neural networks seem to be hard to tune in order to attain the best desirable values. Again in this case, two main sequential units were applied with the first being 64 dense units, and 18 the second; the optimizer selected was the type 'rmsprop' at a 'relu' type of activation. A 2% dropout was applied, too.

Figure 10 illustrates the actual and predicted specific energy-consumption values from this analysis with Keras.

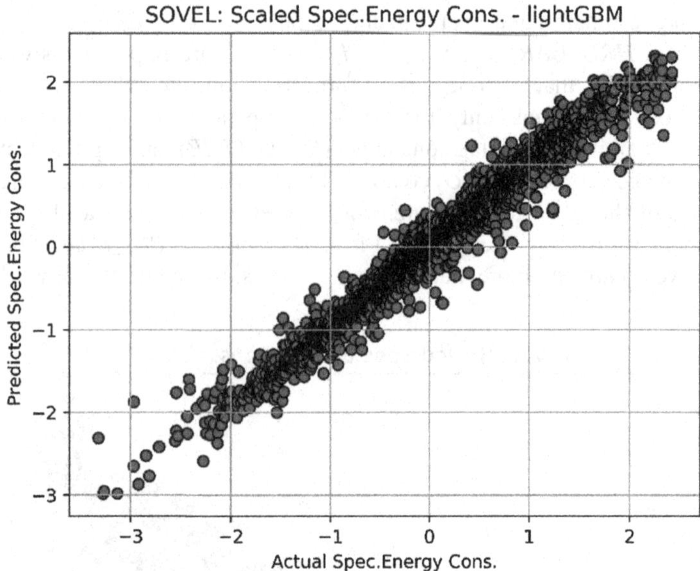

Fig. 7. Actual vs. predicted scaled specific-energy-consumption values using lightGBM.

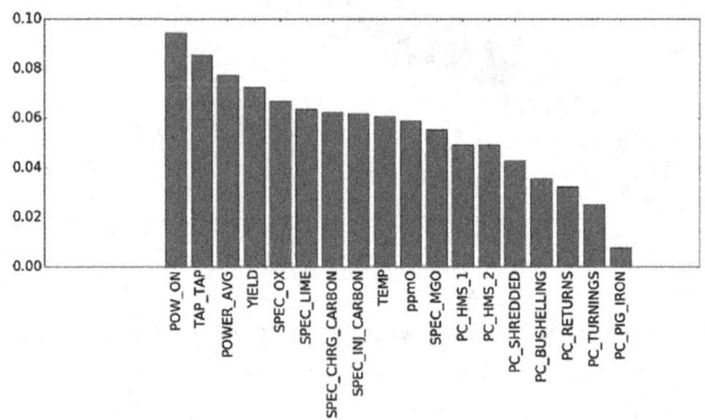

Fig. 8. Relative importance of the independent parameters upon the specific energy consumption (the same holds true for productivity).

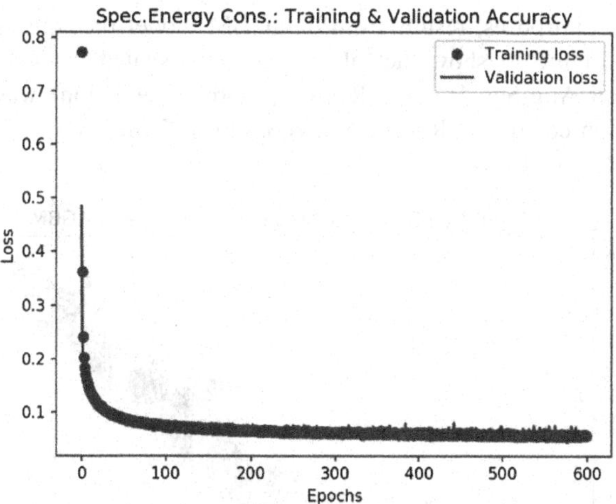

Fig. 9. Training and validation accuracy vs. epochs during grid-search tuning of the Keras model for specific energy consumption.

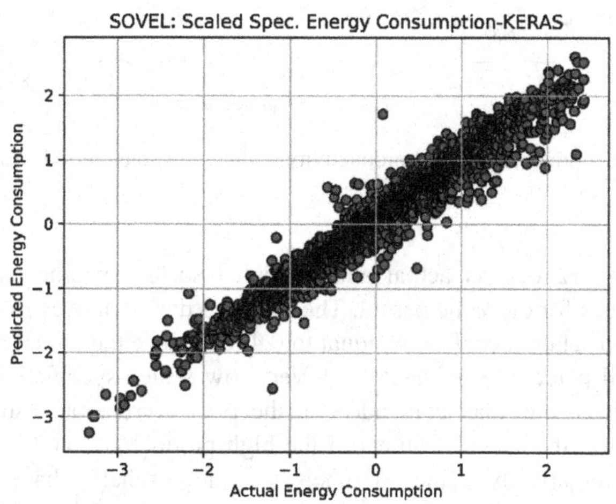

Fig. 10. Actual vs. predicted scaled specific-energy-consumption values using Keras.

3.8 Energy Analysis-Verification by Practice

After the summer-2018 maintenance period, the scrap yard collected enough quantity of relatively good scrap with minimal gangue content. Consequently, we experienced a good September-2018 month with high productivities and small specific electrical-energy consumptions. For this reason, testing the supervised model predictions for productivity and energy consumption based on the past set of data for the independent variables would be a good verification about the reliability of these models.

For illustration purposes, results from lightGBM model predictions were selected for presentation. Figure 11 shows actual and predicted scaled-productivity values for the "good" period Aug-Sep'18. The RMSE (standard deviation) was 0.303 with a multiple correlation coefficient R2_squared equal to 0.9108.

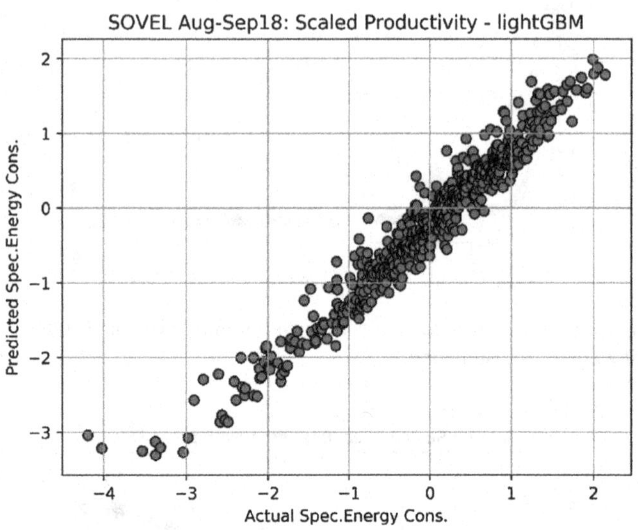

Fig. 11. Actual vs. predicted scaled-productivity values for the period Aug-Sep'18 using lightGBM.

Similarly, Fig. 12 depicts actual and predicted scaled specific electrical-energy consumption values for the same period. The standard deviation was in this case 0.699 with a multiple correlation coefficient equal to 0.4463. Since the model had been tuned with a very small percentage of heats with very low scaled-specific electrical-energy consumption values, one may conclude that the poor correlation results were to be expected. However, the salient features of the high productivity and low energy consumption were remarkably achieved albeit to a less reliable base. Nevertheless, including the Aug-Sep'18 data in the tuning database of the models it was expected that new and better-supervised models could be computed that will behave much better in similar cases in the future. Since statistical analysis has exhibited much better results for the prediction of productivity for the limited data of the Aug-Sep'18 period, emphasis was given to energy analysis for that period by encompassing these data into the initial database and deriving a new lightGBM supervised model in order to check any potential improvement.

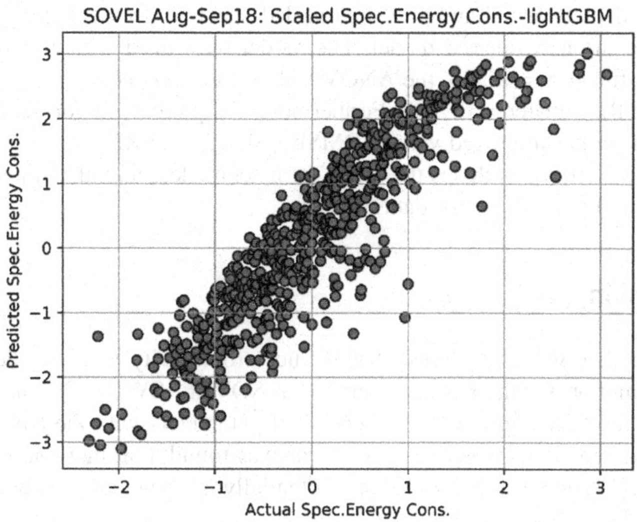

Fig. 12. Actual vs. predicted scaled specific-energy-consumption values for the period Aug-Sep'18 using lightGBM

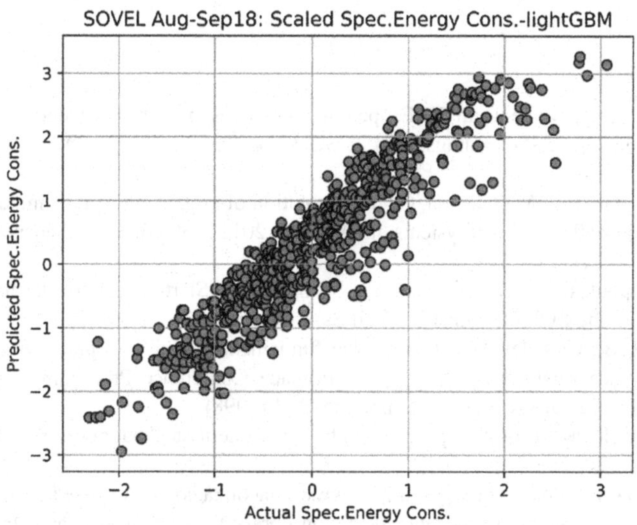

Fig. 13. Actual vs. predicted scaled specific-energy-consumption values for the period Aug-Sep'18 using the new derived lightGBM-model by including the data of this period

Figure 13 depicts actual and predicted scaled specific electrical-energy consumption values for the new derived model. The model was tuned with a learning_rate = 0.01, and n_estimators = 10000; the ANOVA showed improved model values, RMSE = 0.1266, and R2_squared = 0.9833. Furthermore, the predictions for the Aug-Sep'18 data exhibited further improved values, RMSE = 0.573, and R2_squared = 0.628. We would like to point out that this is the 'heart' of machine learning: the system is trained to behave better by getting fresh data by time.

4 Conclusion

Supervised models were developed that predict productivity and specific electrical-energy consumption for the Consteel furnace at SOVEL. SVM and lightGBM based models predicted adequately well the dependent parameters with the latter exhibiting the best correlation characteristics, while Keras was found a bit more difficult to tune. As a further work, one may propose that a potentially online model can be deduced by incorporating Level 2 automation data.

Acknowledgment. The author is grateful to the top management for the continuous support on this type of studies.

References

1. Koehle, S.: Improvements in EAF operating practices over the last decade. In: Electric Furnace Conference Proceedings, Iron & Steel Society, Pittsburgh, PA, vol. 57, pp. 3–14 (1999)
2. Memoli, F., Guzzon, M., Giavani, C.: The evolution of preheating and the importance of hot heel in supersized Consteel® systems. In: AISTech 2011 Proceedings, Indianapolis, IN, vol. I, pp. 823–832 (2011)
3. Bouganosopoulos, B., Papantoniou, V., Sismanis, P.: Start-up experience and results of Consteel® at the SOVEL meltshop. Iron Steel Technol. (2), 38–46 (2009)
4. Pretorius, E.B., Carlisle, R.C.: Foamy slag fundamentals and their practical application to electric furnace steelmaking. In: Electric Furnace Conference Proceedings, Iron & Steel Society, New Orleans, LA, vol. 56, pp. 275–292 (1998)
5. Maes, R.: Celox® for on-line process control in modern steelmaking, (brochure), Heraeus Electro-Nite (2012)
6. Continuum/Anaconda Homepage. https://www.anaconda.com/. Accessed Sept 2017
7. McKinney, W.: pandas: a foundational Python library for data analysis and statistics. http://pandas.pydata.org/. Accessed 20 Aug 2018
8. Pedregosa, F., et al.: Scikit-learn: machine learning in Python. J. Mach. Learn. Res. **12**, 2825–2830 (2011)
9. Ke, G., et al.: LightGBM: a highly efficient gradient boosting decision tree. In: 31st Conference on Neural Information Processing Systems (NIPS 2017), Advances in Neural Information Processing Systems 30, Microsoft Research (2017)
10. Chollet, F.: Deep Learning with Python. Manning Publications, Shelter Island (2018)
11. Brownlee, J.: Develop Deep Learning Models on Theano and TensorFlow Using Keras. Deep Learning with Python. Jason Brownlee, Melbourne (2018)

12. TensorFlow. https://www.tensorflow.org/. Accessed Aug 2018
13. Sismanis, P.: Analysis of rolled-plates' mechanical properties with a machine-learning software. In: AISTech 2018 Proceedings, Philadelphia, PA, AIST, pp. 2273–2286 (2018)
14. Zumel, N., Mount, J.: Practical Data Science with R. Manning Publications, Shelter Island (2014)
15. Avila, J., Hauck, T.: scikit-learn Cookbook, 2nd edn. Packt Publishing, Birmingham (2017)
16. Garreta, R., Moncecchi, G., Hauck, T., Hackeling, G.: scikit-learn: Machine Learning Simplified. Packt Publishing, Birmingham (2017)

Understanding Requirements and Benefits of the Usage of Predictive Analytics in Management Accounting: Results of a Qualitative Research Approach

Rafi Wadan(✉) and Frank Teuteberg(✉)

Universität Osnabrück, Osnabrück, Germany
Rafi.wadan@gmail.com,
frank.teuteberg@uni-osnabrueck.de

Abstract. The accuracy of a forecast affects the financial result of a company. By the improvement of Management Accounting (MA) processes, the introduction of advanced technology and additional skills is prognosticated. Even though companies have increasingly adopted Predictive Analytics (PA), the impact on MA overall has not been investigated adequately. This study investigates this problem through a single case study of a German company. The interview results provide an overview of requirements and benefits of PA in MA. In the future, Management Accountants will be able to focus on business partnering, but require advanced statistical knowledge to fully benefit from PA.

Keywords: Predictive Analytics · Management Accounting · Forecasting · Competencies

1 Introduction

Over the years, the role of management accountants (MAs) has significantly changed. The increasing relevance of data is revolutionizing business analytics and is a key technology trend for MAs. Competition in business has increased tangentially with technology development, the scope of MA has also expanded from historical value reporting towards usage of real time reporting and predictive reporting [1]. Recently, Predictive Analytics (PA) software solutions have gained considerable attention in organizations [1]. PA can be considered as part of BI that focuses on future forecasting to enable better planning and decision-making, whereas BI focuses on reporting and analyzing historical data. MAs now could utilize data analytics techniques to answer the question regarding a potential sales development. In a highly competitive business, especially accurate sales forecasts are essential for managing risks by predicting sales figures [2]. Forecast accuracy affects the efficiency of the company planning process, the degree of goal achievement, total costs and the level of customer needs fulfillment [3]. Failed forecast processes which have led to wrong sales anticipations have been documented widely in the literature [4]. However, the current literature also shows that, in addition to the overpowering of large infrastructure issues, adaptations in business management are necessary [5]. The increase in complexity due to comprehensive

W. Abramowicz and R. Corchuelo (Eds.): BIS 2019, LNBIP 353, pp. 100–111, 2019.
https://doi.org/10.1007/978-3-030-20485-3_8

networking and digitization requires new forms of MA to successfully meet tomorrow's challenges. Seufert [5] identified in their literature review that significant effort is needed in the development of methodological skills in the context of Big Data and Advanced Analytics. Lee et al. [6] suggest that realizing the potential of PA requires not just investing in supporting IT systems, but also making a strong commitment to new ways of managerial thinking. Empirical research in order to understand the predictive forecast methods on MA are also rare [7]. Despite the growing body of research in practice [7] and in science [8] on the socio-technical phenomenon predictive analytics, empirical research on how predictive analytics can be used to the advantage of companies and what requirements companies are facing is missing [8]. This study aims to further develop an understanding about the impact and the requirements on MAs through the usage of PA in order to address current and future MA professionals. We apply the socio-technical system (STS) theory as conceptual lens for our case study in order to broaden the view towards a holistic social-technical perspective on a phenomenon that is technologically discussed so far [9]. Drawing on the results of a revelatory case study as well as on the socio-technical systems theory we present requirements for the adoption of PA. Our research addresses the following research questions:

RQ: Which requirements and which benefits arise for Management Accounting (MA) from the integration of Predictive Analytics (PA) into sales forecasting?

The remainder of this article is structured as follows. Next, we describe the theoretical background providing a conceptualization of PA. Subsequently, in Sect. 3 we present our methodological approach to answer the research questions. Thereafter, we represent the results supported by interview quotations. The last chapter of the study concludes with a discussion and conclusion for future research and limitations.

2 Theoretical Background

2.1 The Social-Technical System Perspective on Predictive Analytics

A socio-technical system (STS) perspective is useful as a lens to examine organizational changes and individuals' behaviors toward new PA implementations [10]. The STS model examines the co-development of people and artefacts, socio and technical or human actors and non-human actors – to avoid the situation where one outcome is supported or privileged by one element over another. The work system as sociotechnical system consists of the technical system and the social system. First, the social system comprises (i) people (e.g., project participants and stakeholders) and their characteristics and attributes (ii) structure which represents institutionalized rules and arrangements. Second, the technical system comprises (iii) tasks which refer to what and how work is accomplished; and (iv) technology including hardware, software and tools [10]. Leavitt states that components of an organizations system are interdependent; the change of one component affects other components and leads to organizational change [10]. The successful use of PA requires the implementation of technology, which is able to process, store and collect a vast amount of data with respect to data variety, variability, velocity, and value [11]. These technologies enable the task to uncover previously unknown patterns, correlations and information from diverse and unexploited data sources to enhance competitiveness in terms of profits and efficiency, speed and service, products through timely and more wise decisions [12]. An organization utilizing PA would have invested significant resources to collect, process, prepare, and eventually analyze it and consequently expects deeper insights and knowledge as results. Essential for any type of data, beyond being big or not, is being of high quality. High quality data is complete, precise, valid, accurate, relevant, consistent, and timely [13]. The people need to develop a supportive culture, appropriate capabilities as well as knowledge, whereas the structures (i.e., collaboration and organizational department structures) need to support PA initiatives [14]. Particularly, the collaboration of IT-related and business departments plays an important role for PA as organizations ought to combine technology and domain knowledge [15].

2.2 Related Work

The impact of technological developments on MA has been widely analyzed at a normative level internationally in recent years [16]. Recent studies show that, although digitization affects the entire company, new potentials and challenges for MA [5]. For example, new technical toolboxes for MAs are presented in which the link to profile-specific, process-related and organizational prerequisites is missing [16]. However, the academic literature regarding PA in the accounting context is almost non-existent. Schwegmann, Matzner, and Janiesch [17] designed a predictive analytics tool combining business intelligence and real-time process monitoring for a maintenance application scenario. By following an event-driven approach, this tool is able to reduce the lag between event observation and the decision-maker's response. Breuker et al. [18] integrated predictive modelling techniques to streamline operational business processes. Huikku, Hyvönen and Järvinen [19] investigated the role of PA project initiator in the integration of financial and operational sales forecast. They identified that MA functions are paying more responsiveness to the assimilation than the representatives from other departments. Granlund [20] identified in his literature review a clear gap in the BI research. In line with this view, Elbashir, Collier and Sutton [21] invited researchers to explore the diffusion of BI technologies and specifically to enhance our current scarce understanding about the impact of this diffusion on the roles of MAs in organizations. Taipaleenmäki and Ikäheimo [15] further suggested that it would be a benefit for research to investigate technology-driven changes in accounting. Seufert [5] highlighted that MAs still do not have enough statistical knowledge in the context of Big Data. Following current literature, it can be stated that some qualitative empirical contributions on analytics tool and their impact on the role of MAs exist, but in that context in the field of the use of PA, research is still nascent. Therefore, by means of this study the effects of the use of a PA system on MA in a large company should be investigated. In particular, the research questions regarding requirements and benefits through the use of PA have not yet been researched. Furthermore, it can be stated that case studies from practice in order to investigate the impact through the PA tool on MA have not yet taken place.

3 Research Method

3.1 Methodological Foundations

The data of this study was collected with the aid of a qualitative empirical research method that is particularly suited for the chosen object through its fall orientation. Since the topic of digitalization impacting the MA as well as the related changes in roles is a theoretically poorly ordered and empirically under-researched subject of research, it was decided against a quantitative survey and evaluation method with subsequent testing of hypotheses. For this purpose, a survey was selected by semi-standardized interviews as a primary data collection tool. A qualitative approach based on a case study research [22] was adopted to gain a deep understanding of the requirements and benefits of the usage of PA. The case study research methodology is particularly well-suited for investigating organizational issues and can provide a deep insight into complex and social phenomena [22]. A single case study can contribute to scientific development through a deep understanding of the context and by capturing experiences [23]. The aim is a comprehensive description of the case with problem-centered interviews (the interviews were conducted in German and translated into English) as a method of data collection [22]. In the present study, the case selection was carried out with the aim of identifying the broadest possible basis for the practicalities associated with the use of PA and benefits and thus contributing to answering the research question. The chosen approach was based on the recommendation of Yin [22] to first create a selection matrix based on various criteria, into which potential case study can be subsequently classified. Furthermore, such an approach allows a higher generalizability of the results. An intensity criterion and a homogeneity criterion has been defined, which must be fulfilled by every expert eligible for the survey. In addition, an additional criterion was defined to guarantee the highest possible heterogeneity of the final sample. In order to reach the intensity criterion, the observed experts should have been making effective use of a PA for a period of more than one year. This duration is of particular importance for this study, as other studies already found that the longer a system is used in a company, the greater its effects on MA and on the role of MAs [24]. This criterion is intended to reinforce the generation of practical experience. The requirement for the experts of having used PA for more than a year, increases the likelihood of identifying experts, who participated in the implementation process of the PA tool and thus can provide the researcher with a general overview. With regard to the homogeneity criterion, a restriction of the experts participating in the forecast process was laid. The selected experts are participating in the forecast process, which make them able to give relevant insights for the participation in the investigation. The heterogeneity criterion includes that the experts differ in their professional experience from each other. This criterion was meant to obtain heterogeneous perspectives and knowledge from different disciplines. As a suitable company, Company X could be gained, which is an international operating company based in Germany with over 1,000 employees worldwide. The investigated company had to be a production-driven company, which is in our case a manufacturing company. The reason for this is the forecast process, which is initiated by the production department. To maintain sustainable innovation, manufacturing companies are demanding the ability to be proactive and profitable in their production. Thus, it can be assumed that MAs in this industry are affected by the changes in their activities. The company is particularly challenged by the obstacles of predicting the demand as precisely as possible. Prior to the PA implementation, Company X worked with the ERP system SAP and used spreadsheets in Microsoft Excel as their forecast solution. Since 2016, the PA tool from SAS has also been used for forecasting purposes.

3.2 Data Collections/Analysis

In this study, the data collection resorted to multiple sources of evidence, which allowed us to increase the validity of our constructs [22]. To obtain in-depth qualitative data, explorative interviews with analyst, consultants and managers were conducted as primary source for data collection. At least six experts for case analysis should serve as the source of the survey. Thus, the different perspectives based on functions within the company and job experience can be pointed out (in this case a Financial Analyst/2 years, Chief Information Officer (CIO)/5 years job experience, Chief Financial Officer (CFO)/2 years, Financial Consultant/1 year, Production Planner/12 years and Quality Process Engineer/4 years). In addition to the criteria of expert selection, a decision on the number of interviews to be examined is made when selecting the examination design. In the literature, a benchmark of four to ten interviews to be included in the analysis is recommended, until a representative statement and thus a theoretical saturation occurs [25]. For this reason, the integration of other experts would provide little or no new input. Prior to conducting the interviews, an interview guideline was developed following the guidelines by Laforest [26]. Based on the respective knowledge of the interviewees and the interview context, additional questions are asked. The interviews lasted between 30–45 min and were held from January 2018 to November 2018. They were transcribed based on the audio recordings resulting in a Microsoft Excel database. The results of the case studies were extracted based on a one-step process and all characteristic content related to the benefits and requirements were extracted. The categories that were obtained from coding are separated into benefits and requirements whereby requirements consist of four dimensions, i.e., technology, people, tasks and structure. In the course of the interviews it was important that the openness of the answers was preserved. This requires the interviewer to openly oppose information and not be categorized [26].

4 Results

There were several objectives that the company wanted to meet with the implementation of PA. All in all, the following general theses on the requirements and benefits of Company X using a PA tool in the context of the MA can be established.

With the use of PA, an enhanced process in order to reflect potential sales figure is required from the system. Besides, the interviewees and the literature [5, 16, 20, 21] emphasized that an ERP and additionally a BI system as supportive tool is required. Nevertheless, Microsoft Excel is the basic system in Company X for analyzing and reviewing the results from the PA.

"The creation of the reporting is currently partly done via SAS, but mostly via Excel. Previously, a ready-made template was used, which generated a reporting from the data. Meanwhile, we rely on the BI and SAS reporting tools more." CFO, 2017.

Proposition 1: PA needs to be supported with an integrated BI.

In order to use the output of the PA tool, management needs also to trust the figures [8]. In Company X, the quality of the data did not change a lot, but continuous improvements and data quality checks were an essential step to stabilize the PA system and increase trust within the company and management. As well, all interviewees agreed that the trust in data and openness to use the tool constitute crucial factors.

"By using seasonal factors, our sales forecast can be determined very well. As a result, the trust of the management is quite high. [...] *Trust in false information, however, can lead to system failure. However, there must be an openness to such a system to live the system and then adapt it to our business"*. CFO, 2017.

Proposition 2: Management needs to support and trust the PA data.

In order to use and understand the PA tool with his comprehensive functions, the MAs must have knowledge in statistical methods, but not necessarily expand IT skills [5, 8]. Previous MAs in Company X were not willing to support the implementation of the PA tool and decided to leave the company.

"Programming languages are also helpful in controlling today. Not only for predictive analytics, but also for macros and databases. But a pronounced numerical understanding should nevertheless be present. I also do not see any end of the use of Excel, so in general Excel is the basic tool for graduates in accounting." CIO, 2017.

Nevertheless, the interviewees are not completely of an opinion as to which software should be considered elementary for MAs. It should be highlighted, that all interviewees mentioned the importance of having knowledge in statistical methods.

Proposition 3: MAs need intermediate statistical knowledge in order to use PA.

With the use of PA, the forecast process activities of MAs partially have been taken over by other departments or employees [5, 8, 19]. After PA implementation, the MA team of Company X needs to exchange more often with other teams, e.g. the collaboration between the other departments and the MAs has changed.

"We work closely with Sales, Planning and IT. Above all, the exchange with the engineers has intensified more strongly. [...] This helps us in addition in terms of knowledge sharing and many progressive questions." Financial Analyst, 2017.

Proposition 4: Collaboration with other teams needs to be increased.

Although other departments are considered in the decision-making process, the update process of the data is still located in the MA department. The literature also mentions it [8, 19]. All interviewees mentioned the MA department as single source of truth regarding PA.

"We cooperate more closely with sales. It ensures a faster reaction to the market. We can discuss new events with the Sales Department." Production Planner, 2018.

Among the changes in the interaction of the MAs with other employees or departments of the company through the use of PA tool, based on the case study analysis, the following theses can be generated.

Proposition 5: MA department owns the PA tool and is single source of truth.

Nevertheless, it is of high importance that the Sales and Production Planning Department are updating the tool. That leads to the desired effect of producing a real time forecast [19].

"[...] Nevertheless, it is assumed that all colleagues such as Sales and Production Planning regularly update the PA tool." Financial Consultant, 2017.

Proposition 6: Sales and Production have to increase their data input more regularly.

According to the statements of the interviewees and literature, two key benefits of using the PA tool were of particular importance. The first benefit is the reduction of the operational risks through improving forecasts [5, 8, 15, 16]. Regarding the reduction of the operation business, the desire for a supporting system, which facilitates predictive figures of sales, was mentioned.

"[...] Faster reaction in production and more efficient production design and transparency." Quality Process Engineer, 2018.

Proposition 7: PA will reduce the operational risk through rolling forecast and real time scenario calculations.

The other key objective is the changing role of MAs. It should be achieved by the independency from the creation of previous forecast and reducing the time for analyzing output of data. Hereby MAs could change their role towards business partnering the management.

"The predictable information enables me to adjust the data so that I can support the management with measures and make recommendations." Financial Analyst, 2017

In general, the access to real time data is mentioned as the strongest impact through PA [5, 8, 16]. With the integrated PA tool and the comprehensive data access, the transparency within the company could be increased significantly. Especially the consideration of the time series and thereby the opportunity to generate timely information is expressed by all interviewees.

Proposition 8: PA will allow MAs to spend more time in business partnering and update less in updating processes.

An overview of the reviewed literature (Sect. 2.2) matched to the proposition is represented in Table 1.

Table 1. Results of the literature analysis.

References	Propositions	Research question	Main statement
A Seufert (2016)	P1, P3, P4, P7, P8	Which methodic competences are required through PA?	To meet the challenge of digitizing big data, controllers need to gain more statistical methodological skills
B Baesens et al. (2016)	P2, P3, P4, P5, P7, P8	How are the technical and managerial issues of PA?	The era of PA is upon us and is changing the world dramatically
C Taipaleenmäki et al. (2013)	P7	How is the convergence between MA and IT?	MA elements are often intertwined with IT and IT has a crucial Role in PA
D Warren et al. (2015)	P1, P7, P8	How will PA change accounting?	PA will improve the quality and relevance of MA information
E Breuker et al. (2016)	P7	How to design a new predictive modeling technique based on weaker biases?	Predictive modeling provide way to streamline operational business processes
F Huikku et al. (2017)	P4, P5, P6	What is role of PA in the integration of financial and operational sales forecasts?	The initiators from MA tend to pay more attention to the integration than other functions
G Granlund (2011)	P1	How could we proceed to understand this relationship between IT and MA?	The mainstream tradition tends to ignore the design and implementation processes
H Elbashir et al. (2011)	P1	How is the influence of organizational controls related to KM?	Organizational absorptive capacity is critical to assimilating Bl systems for organizational benefit

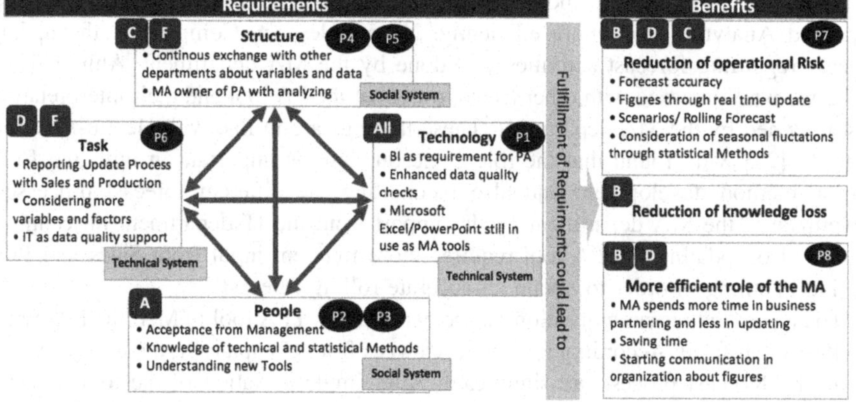

Fig. 1. Explanatory model based on the socio-technical system model [10]

The advanced approach of the STS model in Fig. 1 is added with the benefits based on the with the references from the literature review (e.g. A) and the derived propositions (e.g. P1).

5 Discussion

With regard to the **requirements**, it can be stated that SAP and as an additional tool BI were the basic supportive systems in order to use PA. In congruence with this view Elbashir, Collier and Sutton [21] maintain that only BI systems that are effectively integrated with the data warehouses of organizations can create value for the business. The opponents of data integration refer to inherent challenges that are related to the different purposes of different forecasts, i.e., problems related to a simultaneous use of forecasts for challenging targets and realistic estimates [27]. Taken all together, companies today seem not to be totally content with the ability of their ERPs or BI to support sales forecasting, but they acquire PA tools on top of their systems for these processes. Previous work already has shown that the acceptance of employees is an important success factor for an implementation of systems [28]. A boycotted system change can lead to difficulties or even to a failure of a PA tool implementation and application. For business practice therefore, it is of particular importance to promote the management and employee acceptance for a PA system. The CFO of Company X mentioned the relevance of knowledge regarding statistical method skills. The results also confirm earlier empirical findings, e.g. Seufert [5] explicitly points to emerging knowledge in the areas of statistics and information technology. All interviewees pointed out that the interaction between them has changed, because the involvement of MAs in decision processes changed. The CIO of Company X mentioned that e.g. IT was needed as support in SAP questions. According to this qualitative empirical study, MAs now are more involved in projects and processes and are part of several interdepartmental teams. This finding partially confirms previous research, e.g. by Grabski, Leech and Schmidt [28], who also argue in terms of the involvement of MAs in company-wide projects. In general, the conclusion could be stated, that the accounting area will have to work even more closely with the IT department in order to use Big Data and Analytics tools more efficiently. Nevertheless, in Company X the update process regarding forecast activities was done by the MA department. Although the exchange was increased with other teams, owner of the PA tool and their interpretation is still owned by the MA department. These findings are in line with Henttu-Aho and Järvinen [27], who found that the role of accountants is important in initiating forecasting method development and also reconciling the different forecast processes. Nevertheless, the MA department needs support from the IT department in technical issues. Also updating the PA tool requires more frequent input from Sales and Production Planning in order to obtain an accurate rolling forecast.

To answer the research question on the **benefits** for a PA tool in MA, the following results can be stated. According to the interviewees from Company X, the benefits using PA in the forecast process are significant. After implementation of the tool the MA department started to work more on analysis tasks and less on manual reporting update. Technical tasks are exposed to this transformation, declining more and more, whereas data analysis tasks take now majority of the working time. In the development of the various roles, a shift from the analyst to the business partner is shown, which is accompanied by an increase in competence requirements [16, 29]. Indeed, literature argues that rolling forecasts are necessary, especially in uncertain and competitive business environments [3, 16]. It can be concluded that the output of real time sales figure could lead to further management activities in order to enhance the financial result. In addition to that, the interviewees mentioned that the usage of PA allows an organization an independence from MAs. By relying on the systematic forecasting methodology, the loss of knowledge due to employee loss is no longer so strong. From the research results of this study, several recommendations for business practice can be derived.

6 Conclusion

6.1 Main Contribution

This study provides theoretical as well as practical contributions to the change of MAs role through PA. A correlation could be shown between the use of PA and the impact on risk reduction as well as an efficient role of MA. Particularly, the results indicate which requirements in detail are relevant for the social (barrier No. 2, 3, 4, 5) as well as for the technical system (barrier No. 1, 6) of a company's work system. For practice, they provide guidance for companies whose financial result depends on productions and need to be tackled to leverage the business value of PA tools. Thus, they provide a first direction for companies thinking of investing in PA tools. Additionally, we argue that the successful implementation of PA tools requires fundamental organizational changes before a PA technology is put in place. In company X, the PA tool helped to improve sales forecasts. Furthermore, MA employees were able to focus on analytical topics in their activities. Through this change, in the future companies could make a stronger use of PA in practice. Furthermore, the results from the literature and case study of company X showed that the use of PA tools require increasing knowledge in statistic methodology. The study was rounded off by the findings from Company X, which show that although the MA department still has to update the PA tool, there is an increasing exchange between MA and other teams. However, the boundaries of the work are not disregarded and should be presented as links for future research in this field.

6.2 Limitation

Like any research, this study has some limitations resulting from the chosen research method. In this qualitative empirical research, only six people of different level were interviewed. As a consequence, the results partly refer to the subjective assessments of the interviewees [30]. This leads to a limited depth of this work. Furthermore, the work offers a limited generalizability with regard to the size of the company, since only a large company was investigated. Typically, large companies are the "early adopters" of new technologies or the leaders. Small and medium-sized enterprises (SMEs), on the other hand, are much more hesitant. They often do not have budget to implement technology projects and cannot afford to make mistakes.

6.3 Future Work

We acknowledge that further research in the area of MAs roles and change through the PA is required to study further aspects of competencies as well as define how MAs could change in future practice. The conducted qualitative empirical research of this study sheds light on the effects of the use of PA tools on MA in a large German company. Since the results of this work only relate to a specific company, they cannot be generalized and there is still a great need for research in companies. In addition to that, an investigation of the usage of analytics tools like PA in relation to other countries than Germany could be interesting. Accordingly, for future research it would be conceivable investigating the case in a longitudinal manner, i.e. analyzing the perception of people regarding the PA tool over time to increase the objectivity of the results.

References

1. Cokins, G.: Top 7 trends in management accounting. Strateg. Financ. **95**(6), 21–30 (2013)
2. Hofer, P., Eisl, C., Mayr, A.: Forecasting in Austrian companies; do small and large Austrian companies differ in their forecasting processes? J. Appl. Acc. Res. **16**(3), 359–382 (2015)
3. Fabianová, J., Kačmáry, P., Molnár, V., Michalik, P.: Using a software tool in forecasting: a case study of sales forecasting taking into account data uncertainty. Open Eng. **6**(1) (2016). https://doi.org/10.1515/eng-2016-0033
4. Kerkkänen, A., Korpela, J., Huiskonen, J.: Demand forecasting errors in industrial context: measurement and impacts. Int. J. Prod. Econ. **118**(1), 43–48 (2009)
5. Seufert, A.: Die Digitalisierung als Herausforderung für Unternehmen: Status Quo, Chancen und Herausforderungen im Umfeld BI & Big Data. In: Fasel, D., Meier, A. (eds.) Big Data. EHMD, pp. 39–57. Springer, Wiesbaden (2016). https://doi.org/10.1007/978-3-658-11589-0_3
6. Lee, J., Elbashir, M.Z., Mahama, H., Sutton, S.G.: Enablers of top management team support for management control systems innovations. Int. J. Acc. Inf. Syst. **15**, 1–25 (2014)
7. Henke, N., et al.: The age of analytics: competing in a data-driven world. McKinsey Global Institute (2016)
8. Baesens, B., Bapna, R., Marsden, J.R., Vanthienen, J., Zhao, J.L.: Transformational issues of big data and analytics in networked business. MIS Q. **40**(4), 807–818 (2016)
9. Yoo, Y.: It is not about size. J. Inf. Technol. **30**(1), 63–65 (2015)
10. Leavitt, H.J.: Applied organizational change in industry: structural, technological and humanistic approaches. In: Handbook of Organizations, pp. 2976–3045 (2013)
11. Van den Broek, T., Van Veenstra, A.F.: Modes of governance in inter-organizational data collaborations. In: Proceedings of the 24th European Conference of Information Systems, Münster, Germany (2015)
12. Woerner, S., Wixom, B.H.: Big data: extending the business strategy toolbox. J. Inf. Technol. **30**(1), 60–62 (2015)
13. Redman, T.C.: Data Driven: Profiting from Your Most Important Business Asset. Harvard Business Press (2013). ISBN 978-1-4221-6364-1
14. Chatfield, A., Reddick, C., Al-Zubaidi, W.: Capability challenges in transforming government through open and big data: tales of two cities. In: Proceedings of the 36th International Conference on Information Systems, Fort Worth, USA (2015)

15. Taipaleenmäki, J., Ikäheimo, S.: On the convergence of management accounting and financial accounting – the role of information technology in accounting change. Int. J. Acc. Inf. Syst. **14**, 321–348 (2013)
16. Warren Jr., J.D., Moffitt, K.C., Byrnes, P.: How big data will change accounting. Account. Horiz. **29**(2), 397–407 (2015)
17. Schwegmann, B., Matzner, M., Janiesch, C.: A method and tool for predictive event-driven process analytics. In: 11. Internationale Tagung Wirtschaftsinformatik (WI), Merkur, pp. 721–735 (2013)
18. Breuker, D., Matzner, M., Delfmann, P., Becker, J.: Comprehensible predictive models for business processes. MIS Q. **40**(4), 1009–1034 (2016)
19. Huikku, J., Hyvönen, T., Järvinen, J.: The role of a predictive analytics project initiator in the integration of financial and operational forecasts. Baltic J. Manage. **12**(4), 427–446 (2017)
20. Granlund, M.: Extending AIS research to management accounting and control issues: a research note. Int. J. Acc. Inf. Syst. **12**, 3–19 (2011)
21. Elbashir, M.Z., Collier, P.A., Sutton, S.G.: The role of organizational absorptive capacity in strategic use of business intelligence to support integrated management control system. Acc. Rev. **86**, 155–184 (2011)
22. Yin, R.K.: Case Study Research: Design and Methods, 4th edn. Thousand Oaks Zahra (2009)
23. Flyvbjerg, B., Budzier, A.: Why your IT project may be riskier than you think. Harvard Bus. Rev. **89**(9), 601–603 (2011)
24. Booth, P., Matolcsy, Z., Wieder, B.: The impacts of enterprise resource planning systems on accounting practice - the Australian experience. Aust. Acc. Rev. **10**(3), 4–18 (2000)
25. Lamnek, S.: Methoden und Techniken. Qualitative Sozialforschung (Band 2), 2. Auflage, Benz, Weinheim, p. 102 (1993)
26. Laforest, J.: Safety diagnosis tool kit for local communities. In: Guide to Organizing Semi-Structured Interviews with Key Informants. Institut national de sante publ., Quebec (2009)
27. Henttu-Aho, T., Järvinen, J.: A field study of the emerging practice of beyond budgeting in industrial companies: an institutional perspective. Eur. Acc. Rev. **22**, 765–785 (2013)
28. Grabski, S., Leech, S., Schmidt, P.: A review of ERP research: a future agenda for accounting information systems. J. Inf. Syst. **25**(1), 37–78 (2011)
29. Brands, K., Holtzblatt, M.: Business analytics: transforming the roles of management accountants. Manage. Acc. Q. **16**(3), 1–12 (2015)
30. King, M.F., Bruner, G.C.: Social desirability bias: a neglected aspect of validity testing. Psychol. Mark. **17**(2), 79–103 (2000)

Deriving Corporate Social Responsibility Patterns in the MSCI Data

Zina Taran[1(✉)] and Boris Mirkin[2,3]

[1] Delta State University, 1003 W Sunflower Road, Cleveland, MS 38733, USA
ztaran@deltastate.edu
[2] Birkbeck University of London, Malet Street, London WC1 7HX, UK
mirkin@dcs.bbk.ac.uk
[3] National Research University Higher School of Economics, Moscow, Russian Federation

Abstract. Empirical research effort over Corporate Social Responsibility (CSR) is typically concentrated on a limited number of aspects. We focus on the whole set of CSR activities to find out if there is a structure in those. We take data on the four major dimensions of CSR: environment, social & stakeholder, labor, and governance, from the MSCI database. To find out the structure hidden under almost constant average values, we apply a modification of K-means clustering with its complementary criterion. This method leads us to discover an impressive process of change in patterns that we predict will continue in the future.

Keywords: Corporate social responsibility · Pattern · K-means · Anomalous clusters

1 Introduction. Corporate Social Responsibility

Corporate Social Responsibility (CSR) and sustainability have been recognized as topics of importance by both academics and business practitioners [1, 8, 11, 13, 27].

Corporate social responsibility (CSR) is usually defined as a company's "actions that appear to further some social good, beyond the interests of the firm and that which is required by law" [22, 28]. Such definition of CSR is based on the idea of stakeholders, that are groups that have interest in the way the company does business as well as to its outcomes [9, 10, 14] different stakeholder groups have different concerns that may at times conflict with one another [6]. CSR is understood as a multi-dimensional construct [20, 31, 33] to include a variety of actions and principles directed at satisfying society-related concerns of non-shareholder stakeholders.

A popular set of aggregate dimensions for CSR are:

(1) Social dimension as directed at the local community and society at large,
(2) Labor dimension as directed at own employees,
(3) (Natural) Environment, and
(4) Governance [2, 27].

We refer to these four as to the "Four dimensions" (4D) list.

W. Abramowicz and R. Corchuelo (Eds.): BIS 2019, LNBIP 353, pp. 112–121, 2019.
https://doi.org/10.1007/978-3-030-20485-3_9

Let us refer to the set of grades of a corporation over all four dimensions in the 4D List as its CSR profile. Such CSR profiles may vary on a continuum between two extreme pattern types, an "even" pattern and a "focused" one. An even CSR profile has about the same ratings along each of the four dimensions. On the other end of the continuum is a company which concentrates its all CSR efforts in just one of the dimensions, which we are going to refer to as a focused CSR pattern. Discerning such patterns from the aggregate data is all but impossible in most cases.

A great many research projects study the effect of CSR on various aspects of company performance (company reputation, returns, etc.), as well as factors moderating and mediating such effect [see, for example, 7, 15, 17, 18, 32]. Yet, to the best of our knowledge, no publication has analyzed the structure of corporate efforts among various dimensions of CSR as a whole. Meanwhile, an insight into the structure of the relationship could provide both a context to companies and other bodies concerned and a background reference to facilitate their analysis, planning and control of CSR activities.

So, this paper asks: What are the patterns in the CSR activities? Are even profiles more prevalent than focused ones? Do they change over time?

To answer these questions, we are using company ratings along the 4D list dimensions above that are provided by the MSCI Inc. which took over the popular Kinder, Lydenberg, and Domini (KLD) database. This paper takes 1850 companies in 2007 and 2012 MSCI database and looks for patterns in CSR activities and their changes over time by using K-means clustering [16, 24] with a modified, but equivalent, criterion.

The remainder includes a description of our take on K-means (Sect. 2), and a presentation of the found patterns and their evolution (Sect. 3). Section 4 discusses the found patterns and their possible evolution. In the Conclusion, main results of this study are pointed out, along with possible implications for scholars and practitioners.

2 K-means Clustering: Classic and Modified

A popular clustering method, K-means partitioning, seems especially suitable for our goals [16, 24].

This technique is intuitive and computationally convenient. However, the method requires the user to pinpoint initial cluster seeds or, if the user cannot, generates them randomly thus leading to results that can be inadequate [3]. One more user-specified parameter required by the method, the number of clusters, may be difficult to specify at times too. In the literature, there has been a number of proposals developed to automate this [see, for example, 19, 21, 23, 24, 29]. This paper pursues an approach based on a complementary criterion for K-means. The complementary criterion is mathematically equivalent to the original K-means criterion, but it leads to a different rationale to clustering. According to this complementary criterion, the goal is to find big anomalous clusters, which leads to a simple heuristic for building "anomalous" clusters one by one, thus making the choice of the number of clusters much easier. In this way, the complementary criterion serves as a substantiation of the so-called anomalous cluster initialization heuristic [5, 12].

The cluster structure in K-Means is specified by a partition S of the entity set in K non-overlapping clusters, $S = \{S_1, S_2, \ldots, S_K\}$, and cluster centroids, $c_k = (c_{k1}, c_{k2}, \ldots, c_{kV})$, $k = 1, 2, \ldots, K$.

$$D(S,c) = \sum_{k=1}^{K} \sum_{i \in S_k} \sum_{v \in V} (y_{iv} - c_{kv})^2 = \sum_{k=1}^{K} \sum_{i \in S_k} d(y_i, c_k) \tag{1}$$

Given K, the problem is to find such a partition $S = \{S_1, S_2, \ldots, S_K\}$ and cluster centroids $c_k = (c_{k1}, c_{k2}, \ldots, c_{kV})$, $k = 1, 2, \ldots, K$, that minimize the square error criterion where $d(y_i, c_k)$ is the squared Euclidean distance between data point y_i and cluster center c_k.

To derive the complimentary criterion, let us do elementary transformations of the criterion in (1):

$$D(S,c) = \sum_{k=1}^{K} \sum_{i \in S_k} \sum_{v=1}^{V} (y_{iv} - c_{kv})^2 = \sum_{k=1}^{K} \sum\nolimits_{i \in S_k} \sum_{v=1}^{V} (y_{iv}^2 - 2y_{iv}c_{kv} + c_{kv}^2)$$
$$= \sum_{i=1}^{N} \sum_{v \in V} y_{iv}^2 - \sum_{k=1}^{K} |S_k| \sum_{v \in V} c_{kv}^2$$

where $|S_k|$ is the number of elements in cluster S_k. The last equation above holds because $c_{kv}|S_k| = \sum_{i \in S_k} y_{iv}$.

Let us denote $T(Y) = \sum_{i=1}^{N} \sum_{v \in V} y_{iv}^2$, referred to as the data scatter, and

$$F(S,c) = \sum\nolimits_{k=1}^{K} |S_k| \sum\nolimits_{v \in V} c_{kv}^2 = \sum\nolimits_{k=1}^{K} |S_k| < c_k, c_k >, \tag{2}$$

where $< c_k, c_k >$ is the inner product of c_k by itself, the squared Euclidean distance from c_k to 0. Then the derived equation above can be reformulated as

$$T(Y) = F(S,c) + D(S,c) \tag{3}$$

The complementary criterion in (2) is the sum of contributions by individual clusters, $f(S_k, c_k) = |S_k| < c_k, c_k >$; each is the product of the cluster's cardinality and the squared distance from the cluster's center to the origin, 0. Provided that the origin preliminarily is shifted into the point of 'norm', i.e. the gravity center, the meaning of the complementary criterion is this: find as numerous and as anomalous clusters as possible, to maximize F(S, c). In contrast to the square-error criterion $D(S, c)$ which does not depend on the location of the space origin, 0, the criterion $F(S, c)$ pertains to the origin, as its items $< c_k, c_k >$ heavily depend on that. Therefore, it is recommended, when using the complementary criterion, to not skip a data preprocessing option, the subtraction of the point of "norm" from all the data points.

An option for finding big and anomalous clusters would be to begin by building anomalous clusters independently so that each cluster S and its center c maximize the contribution

$$f(S,c) = |S| < c, c > \tag{4}$$

A locally optimal solution for this can be obtained as follows. Assume an initial cluster *S* to be a singleton. To maximize (4) then, one has to put it into the point which is furthest away from the origin, 0. This, unlike at the conventional K-means, gives us a reasonable initialization to the clustering process. To move further, we attend to the same alternating minimization scheme which is utilized in the conventional K-means algorithm. Given cluster *S*, its center *c* is computed as the average:

$$c = c(S) = \frac{\sum_{i \in S} y_i}{|S|}$$

where y_i is a row of the data matrix corresponding to observation $i \in I$. Given *c*, an optimal update of cluster S should be computed according to the following rule CUR:

Cluster Update Rule (CUR):
Given a cluster *S*, remove $i \in S$ from *S* if $f(S, c) > 2|S| <c, y_i> - <y_i, y_i>$, or add $i \notin S$ to *S* if $f(S, c) < 2|S| <c, y_i> + <y_i, y_i>$.

Algorithm EXTAN (EXTracting an ANomalous cluster).
Input: A data matrix.
Output: List of observations *S* and its center *c*.

1. **Initialization.** Find an observation maximally distant from 0 and make it the initial center, *c*, of the anomalous cluster being built.
2. **Anomalous cluster update.** Given *c*, update *S* according to CUR rule above.
3. **Anomalous center update.** Given S, update the center as the within-*S* mean *c'*.
4. **Test.** If $c' \neq c$, assign $c = c'$ and go back to step 2. Otherwise, move on to Step 5.
5. **Output.** Output the list *S* and its center *c*.

Using EXTAN as a subroutine, we utilize the following one-by-one algorithm for greedily maximizing the complementary criterion *F(S, c)* in (2).

Algorithm BANCO (Big Anomalous Clusters One-by-One).
Input: A data matrix and a user-defined integer *t* – the minimum cluster size (and, possibly, the point of norm, *g*).
Output: A partition of the set of observations *S* in *K* clusters (*K* is determined by *t*) and cluster centers $c_1, c_2, \ldots, c_K$.

1. **Data preprocessing.** Centering: Take the input point g if provided or, if not, compute the grand mean, the vector of average values of the features, and take it as *g*. Subtract *g* from all the data matrix rows. Optionally, normalize data features. Set counter of clusters $k = 1$. Define I_k the set of all the observations.
2. **Iterative EXTAN.** At a given *k*, apply EXTAN to the data matrix over the set of observations I_k to output cluster S_k and its center c_k. Define $I_{k+1} = I_k - S_k$. If $I_{k+1} \neq \emptyset$, set $k = k + 1$ and start step 2 again. Otherwise, move to the next step.
3. **Small cluster removal.** Consider all the sets S_k obtained, and remove those satisfying $|S_k| \leq t$. Define *K* the number of remaining clusters S_k ($k = 1, 2, \ldots, K$).
4. **Output the K anomalous clusters and their centers.**

The BANCO's output serves as the input to a run of K-means.

Here, the number of clusters *K* is determined by another user-defined quantitative parameter, *t*, the minimum number of observations in a cluster. We will show, in the next section, how this can be used for determining the "right" *K*.

3 Empirical Analysis

3.1 Aggregate Data Characteristics

The "MSCI ESG" database evaluates company performance based on each of the four dimensions of the 4D list described above [25]. Each of the scores is a weighted summative score of subfactors scored by the MSCI experts. We use the 1850 companies that have ratings in both, March 2012 and March 2007, data.

We included into our analysis the major components in 4D List. These variables form the so-called Social and Eco ratings: Strategic Governance Factor v1, Human Capital Factor v2, Environment Factor v3, and Stakeholder Capital Factor v4. Table 1 provides descriptive statistics for the variables.

Table 1. Basic statistics

		2012		2007	
		Mean	St.Dev	Mean	St.Dev
Strategic Governance Factor	v1	5.27	1.64	5.44	1.88
Human Capital	v2	5.66	1.90	5.53	1.72
Environment Factor	v3	5.06	1.90	4.90	1.72
Stakeholder Capital	v4	4.93	1.67	5.29	1.85

Table 2. Correlation coefficients between the variables

N = 1850		v1	v2	v3
		2007		
Strategic Governance Factor	v1			
Human Capital	v2	0.680***		
Environment Factor	v3	0.692***	0.563***	
Stakeholder Capital	v4	0.724***	0.640***	0.628***
		2012		
Strategic Governance Factor	v1			
Human Capital	v2	0.252**		
Environment Factor	v3	0.193**	0.192**	
Stakeholder Capital	v4	0.141**	0.160**	0.245**

**Correlations statistically significant at 0.01 level
***Correlations statistically significant at 0.001 level

Table 1 shows that all the mean levels are close to 5.0 whereas the levels of variation are about 30–35% of the mean. This pattern does not much change from 2007 to 2012.

Looking at the pattern of correlations between the CSR components (Table 2), we can see a great difference between the time points. In 2007, correlations between all the four dimensions are much greater than those in 2012. The level of correlations in 2007 is at the level of 0.65–0.70, whereas in 2012 they are around 0.15–0.25. This shows that the starting point of the CSR activity process is not something that happened a long time ago. The starting period, however early CSR discussions started, did cover year 2007 and probably later. But we can safely claim that a new phase already started by 2012. The dramatic changes in the patterns of CSR activities manifest changes in the patterns of activity.

To take a closer look at the structure of CSR activity patterns, let us move on to finding and analyzing CSR similarity clusters of corporations.

3.2 The Structure of Clusters: Comparative Analysis of CSR Activities

Method BANCO applied to datasets of 2007 and 2012 produced six clusters presented in Tables 3, for 2007, and 4, for 2012. The number $K = 6$ is derived from the distributions of the sequential anomalous clusters. At both datasets, 2007 and 2012, a significant drop in cluster cardinalities occurs after the 6th cluster: from more than a hundred to fifty or less. In the Tables, clusters are represented by their centers in the natural scales of MSCI scores, as well as in their relation to the grand mean as the relative difference.

Table 3. Clusters 2007

k	N_k	Cluster center				Relation to Grand Mean (% over/under Grand Mean)			
		Environment	Strat. Governance	Human Capital	Stakeholder Capital	Environment	Strat. Governance	Human Capital	Stakeholder Capital
1	258	6.99	8.11	7.8	7.97	42.5	49	41	50.7
2	347	6.37	6.74	6.29	6.35	29.9	23.9	13.7	20.1
3	329	4.38	5.86	6.33	5.77	−10.6	7.8	14.4	9.1
4	337	5.45	4.98	4.75	4.57	11.2	−8.4	−14.1	−13.6
5	333	3.38	4.1	4.78	4.3	−31	−24.6	−13.6	−18.7
6	246	2.65	2.69	3.1	2.67	−45.9	−50.6	−43.9	−49.5
Total	1850	4.90	5.44	5.53	5.29	0	0	0	0

One can see, from the relative part on the right of Table 3, that many clusters-2007 indeed manifest even patterns. Clusters 1 and 2 (totaling to about 600 companies) perform much better than the grand mean values. In contrast, clusters 5 and 6 (totaling to 579 companies) exhibit profiles that are lower than the grand mean. Clusters 5 and 6 are uniformly underperforming by about 15–30% and 40–50% respectively. In general, in 2007 about two thirds of the companies (1184) exhibit uniform profiles.

The 2012 clusters in Table 4 give a rather different picture. The thoroughly overperforming and underperforming clusters still are present, but they cover a much smaller part of the set: only cluster 1 (258 companies), on the plus side, and cluster 6 (287 companies), on the minus side, fall within this category. This represents a sharp decline of the balanced effort: from 1184 down to 545 companies.

Table 4. Clusters 2012

k	N_k	Cluster center				Relation to Grand Mean (% over/under Grand Mean)			
		Environment	Strat. Governance	Human Capital	Stakeholder Capital	Environment	Strat. Governance	Human Capital	Stakeholder Capital
1	258	6.15	7.36	6.74	6.28	21.6	39.5	19.1	27.3
2	297	5.74	4.82	8.16	5.07	13.6	−8.5	44.1	2.8
3	346	7.09	5.02	4.82	5.35	40.1	−4.8	−14.9	8.4
4	492	3.91	5.21	5.42	5.66	−22.7	−1.1	−4.3	14.7
5	170	4.22	6.44	5.92	2.07	−16.5	22.1	4.6	−58
6	287	3.38	3.58	3.39	3.53	−33.1	−32.1	−40.1	−28.3
Total	1850	5.06	5.27	5.66	4.93	0	0	0	0

Also, the levels of deviation from the grand mean values are smaller in 2012. Other clusters do not manifest even patterns at all.

4 Discussion

Overall, the found clusters provide the following answers to our questions: in 2007, prevailing pattern was to uniformly outperform or underperform on all four dimensions. By 2012, while less than 1/3 of the companies still exhibited the even pattern, the prevailing pattern changed to that of single-focus one. This turn to more focused patterns shows a tendency which is likely to prevail in the future developments.

Most likely, the clusters of “staunchly uniform” over- and under-performers will remain, although at further reduced sizes. The clusters with single focus patterns will be more numerous and better defined. Moreover, probably, as CSR activities mature, the single focus groups could be further enhanced in the direction of expanding CSR activities to embrace two or more dimensions. Therefore, a greater number of double-focus CSR clusters should be expected in the future data.

The inherently exploratory nature of cluster analysis may be viewed as a limitation of the current study. Specifically, the issue of the “right” number of clusters has received no universally adequate approach so far [see, for example 4, 24, 26, 34]. In this study, the other characteristic of granularity with which the EXTAN algorithm operates – the minimum number of objects in a cluster – has received an empirical support.

Also, one should notice that the concept of cluster anomality, much important in our study, is not postulated but rather derived from the K-means square error criterion (1). The anomalous cluster approach, in a slightly different version, was experimentally validated in [5, 12].

5 Conclusion

In this paper, we use the four dimensions - environmental, social/stakeholder, labor and governance – in the 4D list to represent the structure of the CSR process, albeit in a somewhat aggregate form. We use a set of 1850 world companies that covers two separate time moments. To find out structures of prevailing patterns, we develop a version of K-means method, exposing more flexibility regarding the issue of number of clusters than the conventional K-means. To this end, we consider a clustering criterion emerging in the context of data scatter decomposition in two items, the square error and complementary criterion. This complementary criterion explicitly states the goal of clustering as that of finding big anomalous clusters, which leads us to one-by-one finding the clusters and, in this way, to reasonably determining the number of clusters.

The cluster structures found for 2007 and 2012 datasets lead us to conclusions that can be stated, in brief, as follows.

An unexpected phenomenon is observed that as recently as of 2007, the overall CSR process on the level of a single corporation was as yet at an early stage of the CSR efforts, more or less uniformly distributed over all the dimensions in the 4D List, both on high and low levels.

We can see a rather impressive process of change from the predominantly uniform patterns of CSR activities in 2007 to the predominantly single-focus patterns of CSR activities in 2012. We predict that this process will continue into the future.

The research is somewhat limited by the nature of the ratings provided by the MSCI [30].

Directions for further research lie in comparing the method we suggested with other methods in their ability to disclose useful patterns in such data. Additionally, researching the effects of choosing the CSR profile should be studied.

To scholars and analysts studying the impact of various CSR activities, this paper adds understanding of CSR as a multi-dimensional process. In particular, CSR profiles may provide important variables mediating and moderating financial and reputational impact of CSR. Additionally, the paper describes a useful and practical modification of K-means method that gives a different angle to efforts in finding the right number of clusters and may serve as an example at which the number of clusters is not defined ad hoc but rather derived from the data.

To marketing practitioners, we suggest looking at profiles when benchmarking against competition. We also provide them with an easy to implement tool to do so.

References

1. Adam, A.M., Shavit, T.: How can a ratings-based method for assessing corporate social responsibility (CSR) provide an incentive to firms excluded from socially responsible investment indices to invest in CSR? J. Bus. Ethics **82**(4), 899–905 (2008)
2. Albinger, H.S., Freeman, S.J.: Corporate social performance and attractiveness as an employer to different job seeking populations. J. Bus. Ethics **28**(3), 243–254 (2000)

3. Arthur, D., Vassilvitskii, S.: K-means++: the advantages of careful seeding. In: Proceedings of the Eighteenth Annual ACM-SIAM Symposium on Discrete Algorithms. Society for Industrial and Applied Mathematics, pp. 1027–1035 (2007)
4. de Amorim, R.C., Hennig, C.: Recovering the number of clusters in data sets with noise features using feature rescaling factors. Inf. Sci. **324**, 126–145 (2015)
5. de Amorim, R.C., Makarenkov, V., Mirkin, B.: A-Wardpβ: effective hierarchical clustering using the Minkowski metric and a fast K-means initialisation. Inf. Sci. **370**, 343–354 (2016)
6. Betts, S.C., Taran, Z.: Conflicting issues and corporate social responsibility: aligning organizational efforts with stakeholder interests. J. Int. Manag. Stud. **11**(3), 39–46 (2011)
7. Block, J.H., Wagner, M.: The effect of family ownership on different dimensions of corporate social responsibility: evidence from large us firms. Bus. Strategy Environ. **23**(7), 475–492 (2014)
8. Bosch-Badia, M.T., Montllor-Serrats, J., Tarrazon, M.A.: Corporate social responsibility from Friedman to Porter and Kramer. Theor. Econ. Lett. **3**(3A), 11–15 (2013)
9. Carroll, A.B.: Corporate social responsibility: evolution of a definitional construct. Bus. Soc. **38**(3), 268–295 (1999)
10. Clarkson, M.B.E.: A stakeholder framework for analyzing and evaluating corporate social performance. Acad. Manag. Rev. **20**(1), 92–117 (1995)
11. Chen, R.Y., Chen-Hsun, L.: Assessing whether corporate social responsibility influence corporate value. Appl. Econ. **49**(54), 5547–5557 (2017). https://doi.org/10.1080/00036846.2017.1313949
12. Chiang, M., Mirkin, B.: Intelligent choice of the number of clusters in K-Means clustering: an experimental study with different cluster spreads. J. Classif. **27**(1), 3–40 (2010)
13. Cochran, P.L.: The evolution of corporate social responsibility. Bus. Horiz. **50**(3), 449–454 (2007)
14. Fassin, Y.: The stakeholder model refined. J. Bus. Ethics **84**(1), 113–135 (2009)
15. Harjoto, M.A., Jo, H.: Corporate governance and CSR nexus. J. Bus. Ethics **100**(1), 45–67 (2011)
16. Hartigan, J.A., Wong, M.A.: Algorithm AS 136: a K-means clustering algorithm. J. Roy. Stat. Soc.: Ser. C (Appl. Stat.) **28**(1), 100–108 (1979)
17. Jones, E.: Bridging the gap between ethical consumers and corporate social responsibility: an international comparison of consumer-oriented CSR rating systems. J. Corp. Citizsh. **2016**(65), 30–55 (2017)
18. Krüger, P.: Corporate goodness and shareholder wealth. J. Financ. Econ. **115**(2), 304–329 (2015)
19. Lord, E., Willems, M., Lapointe, F.J., Makarenkov, V.: Using the stability of objects to determine the number of clusters in datasets. Inf. Sci. **393**, 29–46 (2017)
20. Martinez, F.: Corporate strategy and the environment: towards a four-dimensional compatibility model for fostering green management decisions. Corp. Gov. **14**(5), 607–636 (2014)
21. Matlab: kmeans. https://www.mathworks.com/help/stats/kmeans.html (2018). Accessed 26 July 2018
22. McWilliams, A., Siegel, D., Wright, P.M.: Corporate social responsibility: a theory of the firm perspective. Acad. Manag. Rev. **26**(1), 117–127 (2011)
23. Mirkin, B.G.: A sequential fitting procedure for linear data analysis models. J. Classif. **7**(2), 167–195 (1990)
24. Mirkin, B.: Choosing the number of clusters. WIREs Data Min. Knowl. Discov. **1**(3), 252–260 (2011)
25. MSCI: User Guide and ESG Ratings Definition (2011). http://msci.com. Accessed 25 Oct 2015

26. Mur, A., Dormido, R., Duro, N., Dormido-Canto, S., Vega, J.: Determination of the optimal number of clusters using a spectral clustering optimization. Expert Syst. Appl. **65**, 304–314 (2016)
27. Peloza, J., Shang, J.: How can corporate social responsibility activities create value for stakeholders? A systematic review. Acad. Mark. Sci. J. **39**(1), 117–135 (2011)
28. Porter, M.E., Kramer, M.R.: Creating shared value. Harvard Bus. Rev. **89**(1), 2–17 (2011)
29. Rodriguez, A., Laio, A.: Clustering by fast search and find of density peaks. Science **344** (6191), 1492–1496 (2014)
30. Schendler, A., Toffel, M.: The factor environmental ratings miss. MIT Sloan Manag. Rev. **53** (1), 17–18 (2011)
31. Schreck, P.: Reviewing the business case for corporate social responsibility: new evidence and analysis. J. Bus. Ethics **103**(2), 167–188 (2011)
32. Sen, S., Bhattacharya, C.B.: Does doing good always lead to doing better? Consumer reactions to corporate social responsibility. J. Mark. Res. **38**(2), 225–243 (2001)
33. Weber, J., Gladstone, J.: Rethinking the corporate financial–social performance relationship: examining the complex, multistakeholder notion of corporate social performance. Bus. Soc. Rev. **119**(3), 297–336 (2014)
34. Zhou, S., Xu, Z., Liu, F.: Method for determining the optimal number of clusters based on agglomerative hierarchical clustering. IEEE Trans. Neural Netw. Learn. Syst. **28**(12), 3007–3017 (2017). https://doi.org/10.1109/tnnls.2016.2608001

Profile Inference from Heterogeneous Data
Fundamentals and New Trends

Xin Lu, Shengxin Zhu(✉), Qiang Niu(✉), and Zhiyi Chen

Department of Matheamtics, Xi'an Jiaotong Liverpool University, Suzhou, China
{Xin.Lu15,Zhiyi.Chen15}@student.xjtlu.edu.cn,
{Shengxin.Zhu,Qiang.Niu}@xjtlu.edu.cn

Abstract. One of the essential steps in most business is to understand customers' preferences. In a data-centric era, profile inference is more and more relaying on mining increasingly accumulated and usually anonymous (protected) data. Personalized *profile* (preferences) of an anonymous user can even be recovered by some data technologies. The aim of the paper is to review some commonly used information retrieval techniques in recommendation systems and introduce new trends in heterogeneous information network based and knowledge graph based approaches. Then business developers can get some insights on what kind of data to collect as well as how to store and manage them so that better decisions can be made after analyzing the data and extracting the needed information.

Keywords: User profile · Heterogeneous data · Recommendation systems · Information network · Similarity

1 Introduction

Recommendations based on customers' preferences usually lead to better products or service promotion. Knowing customers' preferences is usually the first step to develop a successful business. The description of what information is of interest to a user is commonly referred to as a *user profile*. Parts of users' accurate profile can be obtained through their direct subscriptions, while the remaining are often hidden in their online activities and thus need to be inferred. At the heart of the profile inference is the so called recommendation system. Conventional recommendation systems are usually based on structured rating data, such approaches have been thriving and dominating for a long time until recently

The research is supported by Natural Science Foundation of China (NSFC. 11501044), Jiangsu Science and Technology Basic Research Programme (BK20171237), Key Program Special Fund in XJTLU (KSF-E-21), Research Development Fund of XJTLU (RDF-2017-02-23), Research Enhance Fund of XJTLU (REF-18-01-04) and partially supported by NSFC (No. 11571002, 11571047, 11671049, 11671051, 61672003, 11871339).

W. Abramowicz and R. Corchuelo (Eds.): BIS 2019, LNBIP 353, pp. 122–136, 2019.
https://doi.org/10.1007/978-3-030-20485-3_10

more *heterogeneous* data are available. For example for the MovieLens data [11], we have not only the rating data as a matrix, but also cast list as a graph or network, and movie reviews as text. As more and more companies view data as new resource like fuel, various kinds of data are collected and interconnected. Like fossil oil need to be refined, these primary data need to be mined for better business development. Further techniques are demanded, which brings new research problems, technique challenges as well as new business opportunities.

The aim of this paper is to give a review on the fundamentals of mathematical profile inferences based on mathematical similarities, and then introduce two popular techniques for personalized profile retrieval, i.e. recovers users preferences as possible as one can. We aim at making the paper serve as a readable tutorial to explain the rationale behind them for more people (other than only data scientists) with a business minds so that they can get some insights for their enterprise and business development. For more technique details, the reader is directed to more technique reports.

The remaining of this paper is organised as follows. In Sect. 2, we first introduce some commonly used similarities. According to the saying that things of a kind come together, people of a mind fall into the same group. Similarity is the most important concept for preference reference, which serves as the mathematical fundamentals for most recommendation techniques. In Sect. 3 we introduce two approaches for heterogeneous data. Finally, we give some discussion and remarks in Sect. 4.

2 Similarity for Profile Inference

There are various algorithms for profile inference or recommendation. Despite some traditional filtering algorithms such as demographic filtering and content based filtering, Collaborative Filtering (CF) is a popular method used in recommender systems. The CF techniques can be categorized into memory-based, model-based and hybrid recommender [1]. CF assumes that if two persons have similar preferences on some items, then they will be more likely to have similar preferences on other items. For memory-based CF, a data set of items and users' ratings will be used to analyze users' preferences and then make inference. In this regard, a CF algorithm is either user-based CF or item-based CF. The user-based CF aims at matching similar-minded people to a given user, and using their rating data to predict given user's potential preferences. The rationale of item-based CF is that two products are considered to be similar if many customers buy them together or give them similar ratings, and thus the more people conformably doing these, the more similar two items are. Then, the system predicts users' potential preferences basing on similarities between items they have bought and further buy. Amazon implements an efficient item-based CF technique with the idea that X and Y are related items if people are unusually likely to buy item Y given that X has been bought, which means the probability that you bought Y after buying X when they are related will be significantly greater than the probability when X and Y are not related at all [8,14].

Clearly, determining similarity is one of the most important issues in CF. Memory-based approaches use user rating data to compute the similarity between users or items. The three main categories are vector cosine based similarity, correlation based similarity and conditional probability based similarity, summarized by Su [15]. The intuition of computation of user similarity and item similarity are almost the same, but one uses the row vectors of the user-item rating matrix and the other uses the column vectors. Some typical user similarity computation methods are listed below and the corresponding item similarities can be derived correspondingly.

2.1 Vector Cosine Based Similarity

$$sim(x,y) = \cos(\boldsymbol{x},\boldsymbol{y}) = \frac{\boldsymbol{x}\cdot\boldsymbol{y}}{\|\boldsymbol{x}\|_2\,\|\boldsymbol{y}\|_2} = \frac{\sum\limits_{s\in S_{xy}} r_{x,s}r_{y,s}}{\sqrt{\sum\limits_{s\in S_{xy}} r_{x,s}^2}\sqrt{\sum\limits_{s\in S_{xy}} r_{y,s}^2}}. \tag{1}$$

The cosine-based metric [4] measures the user similarity by calculating the cosine value of their rating vectors. In this formula, S_{xy} is defined as the set of all items co-rated by user x and y, and $r_{x,s}$ represents the rating of user x on item s. Some commentators argue that this method is exposed to a proportional problem which will mislead the result to show high similarity of two totally different people [12]. For example, user 1 rated two movies as $(1,1)$ while user 2 rated them as $(5,5)$, and cosine method will get 1 showing the two people have the same taste. Although it is difficult to get proportional scores as the number of co-rated items increases, this problem is still not negligible when sparse data situation occurs.

2.2 Correlation Based Similarities

a. Pearson correlation coefficient (PCC)

$$sim(x,y) = \frac{\sum\limits_{s\in S_{xy}} (r_{x,s}-\bar{r}_x)(r_{y,s}-\bar{r}_y)}{\sqrt{\sum\limits_{s\in S_{xy}} (r_{x,s}-\bar{r}_x)^2}\sqrt{\sum\limits_{s\in S_{xy}} (r_{y,s}-\bar{r}_y)^2}}. \tag{2}$$

The Pearson correlation coefficient was first developed by Karl Pearson to measure linear correlation between two variables. This was introduced in GroupLens study [11] as a metric of user similarity because the linear correlation between two rating vectors could reflect to what extent two users may agree with each other in aspect of ratings. The $\bar{r}_x$ and $\bar{r}_y$ represent the average rating of user x and y respectively. It was argued that the PCC method can reduce the severity of problems presented in the Cosine method, because the Pearson correlation metric carries out a normalization process with the average of each user's ratings, which then adjusts the influence of individual tendency to rate very positively or very negatively [12]. However, the proportional problem still exists. What's more, one controversial issue is that what the subtracted mean values should be. Some people use the average ratings of users to calculate but others prefer to subtract the average ratings of items instead. Also, the Pearson correlation is subject to some assumptions [2]. It requires linear relationships between X and Y, which are supposed to be two continuous normally distributed variables.

b. Constrained Pearson correlation coefficient

$$sim(x,y) = \frac{\sum\limits_{s \in S_{xy}} (r_{x,s} - r_{med})(r_{y,s} - r_{med})}{\sqrt{\sum\limits_{s \in S_{xy}} (r_{x,s} - r_{med})^2} \sqrt{\sum\limits_{s \in S_{xy}} (r_{y,s} - r_{med})^2}}. \quad (3)$$

Ringo music recommender [13] introduced constrained Pearson correlation method by replacing the average ratings in Pearson correlation method with the midpoint of rating scale r_{med} of the system, which was claimed to have a better performance according to their experiments. However, they provided no theoretical explanation to the rationale behind this.

c. Spearman rank correlation

$$sim(x,y) = \frac{\sum\limits_{s \in S_{xy}} (rank_{x,s} - \overline{rank}_x)(rank_{y,s} - \overline{rank}_y)}{\sqrt{\sum\limits_{s \in S_{xy}} (rank_{x,s} - \overline{rank}_x)^2} \sqrt{\sum\limits_{s \in S_{xy}} (rank_{y,s} - \overline{rank}_y)^2}}. \quad (4)$$

Spearman's rank coefficient does not require model assumptions compared with the Pearson correlation method and it is suitable for both continuous and discrete ordinal variables. This is a rougher metric because it only assesses monotonic relationships whereas Pearson metric measures linear relationships. The major difference between them is that Spearman rank method uses rank values of the rating scores when the data are sorted. It was pointed out that a great amount of tied rankings could lead to a degradation of the accuracy of Spearman correlations. To illustrate this argument, strategies for assigning rankings to equal values, which are said to be tied, should be discussed first. Two common methods are standard competition ranking (1224 ranking) which assigns the smaller rank value to all the equal items and fractional ranking (1 2.5 2.5 4 ranking) where each tied observation receives the average of the ordinary ranks they would have if they were slightly different. Therefore, if the rating scale of movies is 1 to 5 discretely, a large number of rating scores will have same rank as there are only 5 distinct rating values.

d. Kendall's tau correlation

$$\tau = \frac{(number\ of\ concordant\ pairs) - (number\ of\ discordant\ pairs)}{n(n-1)/2}. \quad (5)$$

Kendall's tau is a rank correlation measure as well according to herlocker [22]. Different from the Spearman's metric, Kendall uses the relative orders of the ranks rather than their true values. The logic of this method is to count the number of pairs which move in the same direction, and the more such pairs, the stronger the positive correlation. The kendall's has many types of formulas and what is given above is the one with a visual representation where n is the total number of pairs. For any pair of observations (x_i, y_i) and (x_j, y_j), they are said to be concordant if both $rank(x_i) > rank(x_j)$ and $rank(y_i) > rank(y_j)$ or if both $rank(x_i) < rank(x_j)$ and $rank(y_i) < rank(y_j)$. The kendall's tau has a good property because the distribution of τ tends to normality for both large and low values of n, whereas the distribution of ρ in Spearman's rank appears to present peculiar features even though it also has asymptotic normal distribution when n is large [23].

2.3 Conditional Probability-Based Similarity

The conditional probability-based similarity method is specifically designed for item-based similarity. This metric was developed in [5] based on a basic idea that the similarity between each pair of items i and j can be measured by the conditional probability of purchasing item j given that item i has already been purchased, denoted as $P(j|i)$. It is calculated through dividing the number of customers who purchase both items by the total number of customers who purchased item i. The result can be presented in a matrix form:

$$P(j|i) = \frac{Freq(ij)}{Freq(i)} = \frac{\mathbf{G}_{(i,j)}}{\mathbf{G}_{(i,i)}}, \tag{6}$$

where $\mathbf{G}_{(i,j)}$ denotes the $(i, j)th$ element of item-item matrix $\mathbf{G}$, which is the number of customers who purchase both items i and j. Given the user-item matrix $\mathbf{X}$, where the $(u, i)th$ element is 1 when user u have bought item i, and 0 otherwise. The item-item matrix can be derived by the matrix multiplication, which is $\mathbf{G} = \mathbf{X}^T\mathbf{X}$.

To address the influence of popular items which are being purchased frequently, $Freq(j)$ is supplemented to the denominator with a scaling parameter α, because if j is a popular item, $P(j|i)$ will always be high for any item i. In addition, due to the common belief that customers who buy fewer goods might be more reliable indicators for item similarities, the measure is further modified by adding a normalization process to the user-item matrix, where each row of the matrix $\mathbf{X}$ is normalized to be of unit length and the normalized matrix is denoted as $\mathbf{R}$. Then redefine item-item matrix $\mathbf{G}$ as: $\mathbf{G} = \mathbf{R}^T\mathbf{R}$. The similarity metric then becomes:

$$P(j|i) = \frac{\mathbf{G}_{(i,j)}}{\mathbf{G}_{(i,i)}(\mathbf{G}_{(j,j)})^{\alpha}}. \tag{7}$$

2.4 Other Similarities

a. Mean Squared Difference (MSD)

$$MSD(x,y) = \sum_{s \in S_{xy}} (r_{x,s} - r_{y,s})^2,$$
$$sim(x,y) = \frac{L - MSD(x,y)}{L}. \tag{8}$$

A less common method introduced in the Ringo system [13] named MSD is a basic numerical metric based on the geometrical principles of the Euclidean distance. L is a threshold to scale down the result. Some people found it not as efficient as the PCC [6] while some others conclude that this metric can produce good accuracy in prediction except its major drawback in coverage [12].

b. Jaccard similarity

$$sim(x,y) = \frac{|N(x) \cap N(y)|}{|N(x) \cup N(y)|}. \tag{9}$$

The Jaccard method [7] calculates the proportion between the number of co-rated items and the total number of different items that rated by at least one of user x and user y. It is a qualitative metric, which takes only the amount of items into account without considering the relative values of ratings. Nevertheless, it could to some extent reflect user similarity because people with similar preferences tend to choose the same products before rating them and many researchers use it or its modified version as a part of their recommendation model.

2.5 New Metrics Proposed by Researchers

Based on those basic metrics, many new similarity computation methods were proposed and developed to improve the accuracy of prediction in recommender systems. Herlocker et al. proposed a weighted Pearson correlation coefficient method by adding a correlation significance weighting factor and an item-variance weight factor, taking the influence of popular items into consideration [6]. Bobadilla et al. developed a new similarity metric called JMSD based on Jaccard and Mean squared-difference (MSD), achieving better performances compared to the Pearson correlation method in aspects of mean absolute error (MAE) and perfect predictions on MovieLens and NetFlix database, but it failed to outperform PCC when applied to FilmAffinity database [2]. Combining rating comparison and sigmoid-function, Wu and Huang created a new method named SigRA [17]. Moreover, some similarity measures are proposed in machine learning framework, such as metrics base on genetic algorithms [3] and Artificial Neural Network [9].

2.6 Comparison of Similarities

To compare the accuracy in rating predictions of those traditional similarity methods, a basic recommendation system is established. The movieLens ***ml-latest-small*** dataset was used in the experiment, which contains 100,836 ratings by 610 users on 9,742 movies. The dataset was randomly divided into 80% and 20% for train set and test set respectively.

Firstly, we use the similarity computation methods to calculate user similarity scores. Then, the predicted ratings of a target user in the test set will be calculated based on the top k-neighbors who are more similar to the target user. The predicted rating p_x^i of user x on item i is calculated as [9]:

$$p_x^i = \bar{r}_x + \frac{\sum_{n=1}^{k}[sim(x, u_n)(r_{u_n}^i - \bar{r}_{u_n})]}{\sum_{n=1}^{k} sim(x, u_n)}, \tag{10}$$

where $\bar{r}_x$ is the average rating of user x, and u_n represents the nth neighbor whose rating on item i and average rating are $r_{u_n}^i$ and $\bar{r}_{u_n}$ respectively. This aggregation approach is used to alleviate the problem that users are different in rating scale. Some picky users tend to have a low average rating score and some others may have a rating style to rate everything higher.

After making predictions for all users in the test set, the Mean Absolute Error (MAE) of the prediction system can be obtained through the Eq. (11).

$$MAE = \frac{\sum_{i,u \in T} |p_u^i - r_u^i|}{|T|}. \tag{11}$$

T represents the test set and $|T|$ is the total number of existing ratings r_u^i in the test set. The results in the Fig. 2 show that Constrained Pearson, MSD and Cosine have relative low MAE's. The k value should not be set too large, otherwise some dissimilar users will be included, which would reduce the accuracy. It seems reasonable for those rank correlation methods to have relatively high MAE's because those methods may lose some information when transferring the true observed values into their ranks. The results could reflect that those similarity measures for collaborative filtering are kind of thorough and we could already use only the rating data to make predictions in recommender systems with a satisfactory accuracy as long as we get enough data. The state of the art system should not merely focus on achieving low MAE because it will not help so much by reducing the MAE by 0.1 if the MAE is already relatively low at, for example, 0.6. The aim now should be to use other information to help improve the system. Those heterogeneous data could help us with profile inference, classifying users and items and making predictions. This trend may imply that the cold start and sparse data problems would somehow be alleviated. We could make recommendations for a user even if his historical rating data is unknown to us. As long as we build his basic profile, his preference is predictable to some extent, and better predictions could be made with some more complete information.

3 Similarity Based on Heterogeneous Data

Memory-based similarity computation methods only take the rating data into consideration, the recommender system cannot give explicit explanation on the reason why some items are recommended, neither their latent feature associations. Heterogeneous Information Network and Knowledge Graph Embedding are two advanced methods [9,10]. Their multi-relational structures include not only the users rating scores but also abundant attributes of users and items. Exploring these multi-relational graphs would help us dig out the latent semantic information and thus use this tacit knowledge to predict unknown facts. For example, one attribute of a movie is genre and the fact that a person likes a movie may implicitly signal that he likes the genre of this movie, then we can use this information to predict the user's preference and make further recommendation [19–21]. Both Knowledge Graph and Information Network are based on the concept of graph in mathematics. A similarity model based on collaborative filtering and knowledge graph representation learning was proposed and achieved better prediction accuracy in dealing with both general situation and cold start problem [10]. Similarly, based on heterogeneous information network, a personalized entity recommendation method is designed but binary user feedback is used which only reflects whether a user has viewed or bought an item or not, with no information about the degree of preference [18].

Knowledge Graph has brought successful application in many real-world scenarios for its great power in data representation and information retrieval. Triplets in the format of $(headentity, relation, tailentity)$, also called *fact*, are used to represent relations. The core technique is the so called Knowledge Graph Embedding, which can be regarded as the procedure to imitate the true graph using learned embeddings. Therefore, for some given data (i.e. triplets or facts), one can derive more information base on them to predict those unknown facts. For a general KG embedding technique, observed data are stored in a collection of triplets $D^+ = \{(h, r, t)\}$, where h and t belong to the set of entities and r belongs to the set of relations. The entities and relations should be interpreted in a low-dimensional continuous vector space for simplicity, without losing any peculiarity of each entity and relation. Later, a scoring function $f_r(h, t)$ should be defined to measure the plausibility of each fact, which means that for a fact (h, r, t), the possibility that h and t does have the relation r. Observed facts should have higher scores. One possible scoring function can be the negative distance between two vectors, say, $\mathbf{h} + \mathbf{r}$ and $\mathbf{t}$. And a modified one is the distance between their projections to some relation-specific hyperplanes. After the scoring function is defined, the last part can be considered as an optimization problem that maximizes the total plausibility of observed facts subjects to some constraints. More approaches and applications for KG embedding are surveyed and summarized by Wang et al., as well as Goyal [24,25]. The rest of this paper focuses on another technique, which is called heterogeneous information network.

3.1 Heterogeneous Information Network

Information Network. Similar to [16], an information network can be defined as a directed graph $G = (V, E)$ with V and E respectively representing the set of entities and the set of edges. An edge can also be called the relation of two entities. Assume the set of entity types is $\mathcal{A}$ and the set of relation types is $\mathcal{R}$, then there will be an entity type mapping function $\phi : V \rightarrow \mathcal{A}$ such that $\phi(v) \in \mathcal{A}$ and an relation type mapping function $\psi : E \rightarrow \mathcal{R}$ such that $\psi(l) \in \mathcal{R}$. Through these mapping functions, each entity instance $v \in V$ can be mapped to a type in $\mathcal{A}$ and each edge instance $l \in E$ can be mapped to a relation type in $\mathcal{R}$. Two edges are considered to have the same type if and only if the head and tail entities linked by them have the same type respectively.

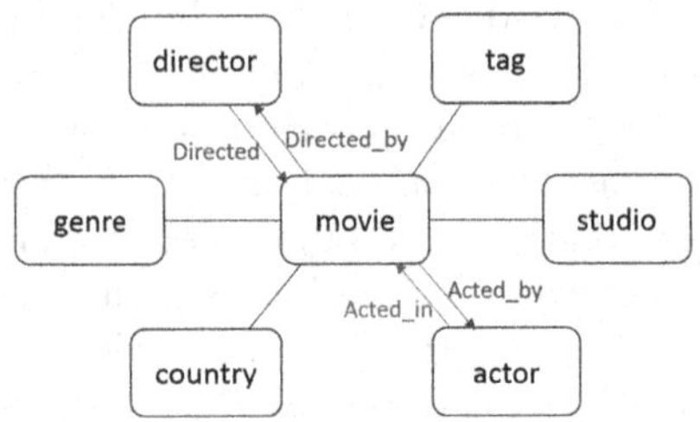

Fig. 1. Network schema (an example)

Network Schema. The *network schema*, denoted by $G_T = (\mathcal{A}, \mathcal{R})$, can be regarded as an abstract graph representing the entity and relation type restrictions in an information network. Figure 1 provides a possible network schema for the movie recommendation scenario.

Heterogeneous Information Network. An information network is called a *Heterogeneous Information Network* when $|\mathcal{A}| > 1$ or $|\mathcal{R}| > 1$, which means there are more than one type of entities or more than one type of relations.

Meta-Path. Intuitively, meta-paths can be regarded as the types of paths in information networks. A meta-path is defined as

$$\mathcal{P} = A_0 \xrightarrow{R_1} A_1 \xrightarrow{R_2} ... \xrightarrow{R_k} A_k$$

in a network schema $G_T = (\mathcal{A}, \mathcal{R})$ where $A_i \in \mathcal{A}$ and $R_i \in \mathcal{R}$ for $i = 0, ..., k$. In a recommender system, the main purpose is to discover potential connections between users or items. Symmetric meta-paths are useful from this point of view, as such paths can represent links between two entities of the same type. A path $\mathcal{P}$ is symmetric if it can be written in the form of $\mathcal{P} = (\mathcal{P}_l \mathcal{P}_l^{'})$, where $\mathcal{P}_l$ is an arbitrary path and $\mathcal{P}_l'$ is its reverse. For example, $movie \xrightarrow{Acted\_by} actor$ is a short meta-path with length 1 and its inverse is $actor \xrightarrow{Acted\_by^{-1}} movie$, then the path:

$$movie \xrightarrow{Acted\_by} actor \xrightarrow{Acted\_by^{-1}} movie$$

is a symmetric meta-path. Note that the semantic meaning of "*Acted_by*$^{-1}$" is just "*Acted_in*". This meta-path help establish linkages between two movies in aspect of actors. Therefore, we can use this meta-path to calculate the similarity between two movies, with "actor" as the criterion. The more common actors two movies have, the more similar the two movies are.

3.2 PathSim Similarity

PathSim is a similarity measure based on symmetric meta-paths which returns the similarity scores between two entities of the same type. Given a symmetric meta-path $\mathcal{P}$, the PathSim score between two entities x and y is:

$$s(x,y) = \frac{2 \times |\{p_{x \rightsquigarrow y} : p_{x \rightsquigarrow y} \in \mathcal{P}\}|}{|\{p_{x \rightsquigarrow x} : p_{x \rightsquigarrow x} \in \mathcal{P}\}| + |\{p_{y \rightsquigarrow y} : p_{y \rightsquigarrow y} \in \mathcal{P}\}|}, \tag{12}$$

where $p_{x \rightsquigarrow y}$ is a path instance between x and y and $|\{p_{x \rightsquigarrow y} : p_{x \rightsquigarrow y} \in \mathcal{P}\}|$ is the number of path instances $p_{x \rightsquigarrow y}$ between x and y following $\mathcal{P}$. We can get the number of path instances from matrix operations. To do the calculation, we first need to identify adjacency matrices and commuting matrices.

The adjacency matrix between type A_i and A_j is denoted by $W_{A_iA_j}$ and each element in this matrix is the interaction between the corresponding two entities, one from type A_i and the other from type A_j. A binary value can be used in the ***movie-actor*** example where the $(i,j)th$ element has value 1 if the ith movie is acted by jth actor and 0 if not. The type of values in the adjacency matrix depends on how the interaction between these two entities is measured. For example, in a ***movie-user*** matrix, each venue has a real value ranging from 0 to 5, which is the corresponding rating score.

Then, based on the definition of adjacency matrix, the commuting matrix M for a meta-path $\mathcal{P} = (A_1A_2...A_k)$ can be defined as:

$$M = W_{A_1A_2}W_{A_2A_3}...W_{A_{k-1}A_k}. \tag{13}$$

If each adjacency matrix $W_{A_iA_j}$ only has binary values for relation representation, then the value in element $M(i,j)$ will be the number of paths instances between entity $x_i \in A_1$ and entity $y_j \in A_k$, following the meta-path $\mathcal{P} = (A_1A_2...A_k)$. Thus, given a symmetric meta-path $\mathcal{P} = (A_1A_2...A_2A_1)$, we can calculate the similarity between entity $x_i \in A_1$ and $x_j \in A_1$ under path $\mathcal{P}$. The PathSim formula becomes:

$$sim(x_i,x_j) = \frac{2M_{ij}}{M_{ii} + M_{jj}}. \tag{14}$$

3.3 Recommendation Model with Heterogeneous Information Network

User Preference Score and Predicted Rating. Assume we can get item similarities through the PathSim measurement, then we need to add a new entity type, *User*, to the symmetric path mentioned above. For example, the meta-path ***movie-actor-movie*** is extended to ***user-movie-actor-movie*** after adding one node at the beginning. The latent semantic description of this type of meta-path can be that users watch movies cast by certain actors. If a user picks actors when choosing movies, then he or she will be likely to explore new movies following this meta-path.

Given the set of users $U = \{u_1, u_2 ..., u_m\}$ and the set of items $I = \{e_1, e_2, ..., e_n\}$ in a recommendation system, we can combine the users' ratings and the item similarities obtained from PathSim method to predict the user preference score. $R_{u_i,e}$ is the rating of user u_i on item e, which can be found in the user-item adjacency matrix. Then the preference score of user u_i on item e_j can be calculated as:

$$s(u_i, e_j) = \sum_{e \in I} R_{u_i,e} sim(e, e_j).$$

However, this is a rough measure and we only make recommendations by sorting these scores to find some top items. We can make further modification to the formula so that it can return a prediction of the rating of user u_i on item e_j:

$$s(u_i, e_j) = \frac{\sum_{e \in I} R_{u_i,e} sim(e, e_j)}{\sum_{e \in I} R'_{u_i,e} sim(e, e_j)}, \tag{15}$$

where $R'_{u_i,e}$ is the binary representation of user's rating, with value 1 if user u_i has rated item e_j and 0 if not. Therefore, it can be regarded as a weighted average of all the ratings of this user and the weights are related to item similarities. Higher weights are assigned to items that are more similar to the aimed item e_j which we want to make prediction on.

Given a meta-path $\mathcal{P} = (A_1 A_2 ... A_k)$, we define one of its sub-path as $\mathcal{P}' = (A_2 ... A_k)$ where the first node is omitted. Then the predicted rating for user i on item j along meta-path $\mathcal{P}$ can be defined as:

$$s(u_i, e_j | \mathcal{P}) = \frac{\sum_{e \in I} R_{u_i,e} sim(e, e_j | \mathcal{P}')}{\sum_{e \in I} R'_{u_i,e} sim(e, e_j | \mathcal{P}')}, \tag{16}$$

where M is the commuting matrix for meta-path $\mathcal{P}'$.

User Predicted Rating Matrix. This predicted rating score is specifically designed for meta-paths in the format of $\boldsymbol{user - item - * - item}$, where the third node can be any attribute of items that we want to take into consideration when making recommendation. In the movie recommendation, some meta-paths can be like:

- $\mathcal{P}_1$: $user - movie - actor - movie$,
- $\mathcal{P}_2$: $user - movie - director - movie$,
- $\mathcal{P}_3$: $user - movie - genre - movie$,

  ...

- $\mathcal{P}_L$: $user - movie - country - movie$.

Assume we want to take L attributes into consideration, then there will be L meta-paths in the system. For each meta-path, by calculating all the predicted rating scores between m users and n items in the system, we can generate a user predicted rating matrix $\tilde{R}^{(l)} \in \mathbb{R}^{m \times n}$. Thus there will be L predicted rating matrices in total and we denote them as $\tilde{R}^{(1)}, \tilde{R}^{(2)}, ..., \tilde{R}^{(L)}$, where $\tilde{R}^{(l)}$ is represented as:

$$\tilde{R}^{(l)} = \begin{bmatrix} s(u_1, e_1|\mathcal{P}_l) & \cdots & s(u_1, e_n|\mathcal{P}_l) \\ \vdots & \ddots & \vdots \\ s(u_m, e_1|\mathcal{P}_l) & \cdots & s(u_m, e_n|\mathcal{P}_l) \end{bmatrix}, \quad for\ l = 1, 2, ..., L.$$

For systems with big data set, we must consider the storage space because as m and n growing, the storage size for $\tilde{R}^{(l)}$ will grow significantly and the whole procedure will be time and speed expensive. Therefore, we need to use some matrix transformation techniques to alleviate this problem. We use matrix factorization techniques to approximate the predicted rating matrix $\tilde{R}^{(l)}$ with the products of two low-rank matrices as:

$$\tilde{R}^{(l)} \approx \hat{U}^{(l)} \hat{V}^{(l)\mathbf{T}}, \tag{17}$$

where $d < min(m, n)$ and $\hat{U}^{(l)} \in \mathbb{R}^{m \times d}$ can be regarded as the latent feature for users along path $\mathcal{P}_l$ with $\hat{U}_i^{(l)}$, the ith row of $\hat{U}^{(l)}$, representing the feature of user i specifically. $\hat{V}^{(l)} \in \mathbb{R}^{d \times n}$ represents the latent feature for items in the same manner (Fig. 3).

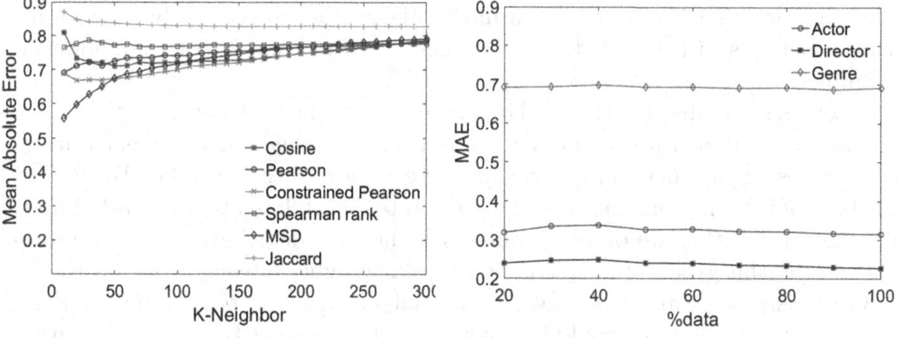

Fig. 2. Comparison between different similarity metrics

Fig. 3. MAE's under different mental paths

The Proposed Recommendation Model. We can get a global predicted rating matrix by combining all the $\tilde{R}^{(l)}$'s, each assigned a weight w_l:

$$\tilde{R} = \sum_{l=1}^{L} w_l \tilde{R}^{(l)}. \tag{18}$$

With the matrix factorization techniques, the final predicted score of user u_i on item e_j can be given as:

$$r(u_i, e_j) = \tilde{R}(i, j) = \sum_{l=1}^{L} w_l^i \hat{U}_i^{(l)} \hat{V}_i^{(l)\mathbf{T}}. \tag{19}$$

Then we can use this model to make predictions on the users' preferences according to personalized w_l^i's.

4 Conclusion

The results in Fig. 2 shows that the basic similarity computation methods based on structured rating data have already achieved relative high accuracy in rating predictions when we choose 50 or less neighbors. The proposed model get a general performance with MAE around 0.7, failing to surpass some conventional methods but it can dig out the potential linkage between users and movies, making recommendations explicable. Also, it can serve as a basic demonstration of the new technology. It can be found that the meta-path containing genre as the central node has the lowest MAE, but the one with actor has the highest MAE. That may because there are only 18 genres for movies but there are 14,832 different actors for 9,310 movies in this data set, and the later leads to a very sparse adjacency matrix. It may be a little bit difficult to find many useful links to help make predictions if there are many disjoint relations. The information network will efficiently help users find some other movies acted by his or her favorite actors but the point is that, according to the model, two movies will have higher similarity if their cast lists contain more common actors. That suggests that if a user gives a movie high rating score mainly because his favorite actor is in the cast, this *actor* meta-path method is not that powerful at finding such latent semantic.

There is no unique solution on how to determine the best weights w_l^i's of each mental-path for each user. Different users may have different judgment criteria when choosing products and that should be a part of their profile. To dig out this kind of information, one should incorporate data related to users' attributes. A rough idea is that we divide people with the same attributes into one group and assume this group of people assign similar weights to each meta-paths.

Obviously, more and more data are available, which are tend to be heterogeneous. More information should be incorporated to enhance the recommendation system. The heterogeneous information network approach or knowledge graph based approach are promising for preferences inference from heterogeneous data sets if they are well designed and properly used. It is an very active research field which deserves further research and It is a promising tool to develop new business products or services.

References

1. Adomavicius, G., Tuzhilin, A.: Toward the next generation of recommender systems: a survey of the state-of-the-art and possible extensions. IEEE Trans. Knowl. Data Eng. **17**(6), 734–749 (2005)
2. Bobadilla, J., Serradilla, F., Bernal, J.: A new collaborative filtering metric that improves the behavior of recommender systems. Knowl. Based Syst. **23**, 520–528 (2010)
3. Bobadilla, J., Ortega, F., Hernando, A., Alcal, J.: Improving collaborative filtering recommender system results and performance using genetic algorithms. Knowl. Based Syst. **24**, 1310–1316 (2011)
4. Breese, J.S., Heckerman, D., Kadie, C.: Empirical analysis of predictive algorithms for collaborative filtering. In: Proceedings of the Fourteenth Conference on Uncertainty in Artificial Intelligence, UAI 1998, pp. 43–52. Morgan Kaufmann Publishers Inc., San Francisco (1998)
5. Deshpande, M., Karypis, G.: Item-based top-N recommendation algorithms. ACM Trans. Inf. Syst. **22**(1), 143 (2004)
6. Herlocker, J.L., Konstan, J.A., Borchers, A., Riedl, J.: An algorithmic framework for performing collaborative filtering. In: Proceedings of the 22nd Annual International ACM SIGIR Conference on Research and Development in Information Retrieval, SIGIR 1999, pp. 230–237. ACM, New York (1999)
7. Koutrika, G., Bercovitz, B., Garcia-Molina, H.: Flexrecs: expressing and combining flexible recommendations. In: Proceedings of the 2009 ACM SIGMOD International Conference on Management of Data, SIGMOD 1909, pp. 745–758. ACM, New York (2009)
8. Linden, G., Smith, B., York, J.: Amazon.com recommendations: item-to-item collaborative filtering. IEEE Internet Comput. **7**(1), 76–80 (2003)
9. Mannan, N.B., Sarwar, S.M., Elahi, N.: A new user similarity computation method for collaborative filtering using artificial neural network. In: Mladenov, V., Jayne, C., Iliadis, L. (eds.) EANN 2014. CCIS, vol. 459, pp. 145–154. Springer, Cham (2014). https://doi.org/10.1007/978-3-319-11071-4_14
10. Mu, R., Zeng, X.: Collaborative filtering recommendation algorithm based on knowledge graph. Math. Probl. Eng. **2018**, 1–11 (2018). Article ID 9617410
11. Resnick, P., Iacovou, N., Suchak, M., Bergstrom, P., Riedl, J.: GroupLens: an open architecture for collaborative filtering of netnews. In: Proceedings of the 1994 ACM Conference on Computer Supported Cooperative Work, CSCW 1994, pp. 175–186. ACM, New York (1994)
12. Sanchez, J.L., Serradilla, F., Martinez, E., Bobadilla, J.: Choice of metrics used in collaborative filtering and their impact on recommender systems. In: 2008 2nd IEEE International Conference on Digital Ecosystems and Technologies, pp. 432–436, February 2008
13. Shardanand, U., Maes, P.: Social information filtering: algorithms for automating "word of mouth". In: Proceedings of the SIGCHI Conference on Human Factors in Computing Systems, CHI 1995, pp. 210–217. ACM Press/Addison-Wesley Publishing Co., New York (1995)

14. Smith, B., Linden, G.: Two decades of recommender systems at amazon.com. IEEE Internet Comput. **21**(3), 12–18 (2017)
15. Su, X., Khoshgoftaar, T.M.: A survey of collaborative filtering techniques. Adv. Artif. Intell. **2009**, 4:2 (2009)
16. Sun, Y., Han, J., Yan, X., Yu, P.S., Wu, T.: PathSim: meta path-based top-k similarity search in heterogeneous information networks. In: VLDB 2011 (2011)
17. Wu, X., Huang, Y.: SigRA: a new similarity computation method in recommendation system. In: 2017 International Conference on Cyber-Enabled Distributed Computing and Knowledge Discovery (CyberC), pp. 148–154, October 2017
18. Yu, X., et al.: Personalized entity recommendation: a heterogeneous information network approach. In: Proceedings of the 7th ACM International Conference on Web Search and Data Mining, WSDM 2014, pp. 283–292. ACM, New York (2014)
19. Gao, B., Zhan, G., Wang, H., Wang, Y., Zhu, S.: Learning with linear mixed model for group recommendation. In: Proceedings of the 11th International Conference on Machine Learning and Computing, ICMLC 2019, Zhu Hai, Guangdong, China (2019)
20. Zhu, S., Gu, T., Xu, X., Mo, Z.: Information splitting for big data Analytics. In: CyberC 2016, Chengdu, China, pp. 294–302 (2015)
21. Zhu, S.: Fast calculation of restricted maximum likelihood methods for unstructured high-throughput data. In: Proceeding of 2nd IEEE International Conference on Big Data Analysis, ICBDA 2017, Beijing, China, pp. 40–43. IEEE (2017)
22. Herlocker, J., Konstan, J., Terveen, K., Riedl, J.: Evaluating collaborative filtering recommender systems. ACM Trans. Inf. Syst. **22**(1), 5–53 (2004)
23. Kendall, M.G.: A new measure of rank correlation. Biometrika **30**, 81–93 (1938)
24. Wang, Q., Mao, Z., Wang, B., Guo, L.: Knowledge graph embedding: a survey of approaches and applications. IEEE Trans. Knowl. Data Eng. **29**(12), 2724–2743 (2017)
25. Goyal, P., Ferrara, E.: Graph embedding techniques, applications, and performance: a survey. CoRR, abs/1705.02801 (2017)

Triggering Ontology Alignment Revalidation Based on the Degree of Change Significance on the Ontology Concept Level

Adrianna Kozierkiewicz and Marcin Pietranik(✉)

Faculty of Computer Science and Management,
Wroclaw University of Science and Technology,
Wybrzeze Wyspianskiego 27, 50-370 Wroclaw, Poland
{adrianna.kozierkiewicz,marcin.pietranik}@pwr.edu.pl

Abstract. Following a common definition, ontologies can be seen as a formal specification of a conceptualisation. However, it cannot be expected that there will be no changes applied to them. Obviously, any application build on top of some ontology needs to adjust to the introduced alterations. For example, a mapping designated between two ontologies (also called an ontology alignment) is valid only if participating ontologies are fixed. In this paper we present a function that can indicate, whether or not, the aforementioned alignment needs updating, in order to follow modifications done to participating ontologies, and to avoid mapping them again from scratch.

Keywords: Ontology evolution · Ontology alignment · Knowledge management · Consensus theory

1 Introduction

Ontologies, following a common definition, can be seen as a formal specification of a conceptualisation, which translates to a formal description of some selected area of knowledge using a set of concepts and relationships that hold between them. They can be used to enrich data mining tools, algorithms of fraud detecting and semantic publishing. Their biggest downside is a problem of heterogeneity, which entails that between two independently created ontologies there is no possibility to develop any framework which could assert a consistency between them.

One of the approaches to overcoming this difficulty is finding which parts of ontologies define the same or similar parts of the aforementioned selected area of knowledge. This issue is a widely investigated topic, and in a literature, designating such mappings, is referred to as an ontology alignment [4]. Most of the available sources emphasise the complexity of this procedure. Therefore, it is not possible to relaunch a selected aligning algorithm whenever a mapped

W. Abramowicz and R. Corchuelo (Eds.): BIS 2019, LNBIP 353, pp. 137–148, 2019.
https://doi.org/10.1007/978-3-030-20485-3_11

ontology has changed in order to revalidate that the available mapping is still valid.

A frequent simplification of this issue is based on an assumption that ontologies do not change in time and their authors do not update their contents. However, it is obvious that in modern applications it is impossible to build any kind of flexible knowledge base build on such fixed foundations. In our current research we want to concentrate on managing alterations applied to ontologies on a concept level, and to investigate how they may influence an alignment initially established between them. We claim that not all modifications that appear during the ontology's lifespan are significant enough to entail invalidation of ontology alignment that has been previously designated. For example, note, that small changes concerning some concept's label are not equally important and influential as a major update of its structure.

Formally, the research task can be described as follows: *For a given ontology O in its two consecutive states in time, denoted as $O^{(m)}$ and $O^{(n)}$, one should determine a function Ψ_C representing the degree of significancy to which concepts within it have been changed in time.* Informally speaking - our main goal is developing a function that could be used as an indicator of having to check if the alignment of concepts at hand may potentially need revalidating, in the light of changes applied to maintained ontologies. Such measure can be confronted with some accepted threshold in order to ascertain the necessity of updating ontology alignment.

The article is structured as follows. In Sect. 2 an overview of related researches is given. Section 3 includes a mathematical foundation for our work (which involves basic definitions etc.). The main contribution can be found in Sect. 4 which contains a description of the developed function Ψ_C. Experimental evaluation can be found in Sect. 5. The paper ends in Sect. 6 with a summary and brief overview of our upcoming research plans.

2 Related Works

An ontology can be understood as a structure which allows to store and process some knowledge. If our knowledge is distributed in many sources, then an ontology integration process should be applied in order to allow reasoning about the whole available knowledge. For this task an alignment between input ontologies is required to conduct such integration process (also referred to as merging). However, knowledge stored in ontologies could be out of date and therefore, an update process is required. The modification of ontologies may entail changes in the existing alignment. To the best of our knowledge, problems referring to the ontology alignment evolution are not well investigated so far.

Zablith and others [17] divide an ontology evolution process into five subproblems:

1. Detecting the need for evolution which initiates the ontology evolution process by detecting a need for change.

2. Suggesting change representations and suggests changes to be applied to the ontology
3. Validating changes filters out those changes that should not be added to the ontology as they could lead to an incoherent or inconsistent ontology, or an ontology that does not satisfy domain or application-specific constraints.
4. Assessing impact measures the impact on external artefacts that are dependent on the ontology or criteria such as costs and benefits of the proposed changes.
5. Managing changes applies and records changes and keeps track of the various versions of the ontology.

To assert a proper ontology evolution management, all of the mentioned above subtasks have to be solved. However, most of the research available in the literature focus on detecting and managing changes implemented in an ontology. For example, [2] is devoted to a repository of large ontologies. Authors propose an algorithm called the Ontology Version Detector, which implements a set of rules analysing and comparing URIs of ontologies to discover versioning relations between ontologies.

The paper [13] is especially addressing a problem of detecting changes between versions of the same knowledge bases. Authors propose a formal framework that was used for the definition of their language of changes. The detection semantics of the defined language is used as the basis for a change detection algorithm.

The practical idea of an ontology version management and change detection is raised in [8]. Authors designed a system OntoView which is able to store an ontology, provide a transparent interface to different versions, specify relations between versions of ontologies, identify scheme for ontologies, and finally helps users to manage changes in online ontologies.

The ontology evolution management system is also designed by Khattak and others [9]. Authors describe a change history management framework for evolving ontologies. The paper addresses several subproblems such as ontology versioning (also covered in [7]), tracking a change's provenance, a consistency assertion, a recovery procedure, a change representation and a visualisation of the ontology evolution. Experimental results show that the proposed system has better accuracy against other existing systems.

In [18] authors propose a framework called temporal OWL 2 (τOWL), which supports a temporal schema versioning, by allowing changing these components and by keeping track of their evolution through the conventional schema versioning and annotating document versions, respectively. Some tools for managing a temporal versions of an ontology could be found also in [5,16].

The problem of the ontology evolution involves measuring and managing changes applied within ontologies, and assessing its impact on mappings between such ontologies. However, in many papers the alignment evolution is pushed to a background and not considered properly. Authors of [15] noticed that alignments originally established between ontologies can become stale and invalid when certain changes have been applied to maintained ontologies. Thus, they propose a preliminary algorithm for a revalidation and preserving correctness of alignment of two ontologies.

In [3] an original method for identifying the most relevant subset of concept's attributes, which is useful for interpreting the evolution of mappings under evolving ontologies is designed. Such solution aims at facilitating a maintenance of mappings based on the detected attributes.

The alignment evolution is addressed from the different point of view in [10]. In this paper, a solution that allows query answering in data integration systems under evolving ontologies without mapping redefinition is provided. It is achieved by rewriting queries among ontology versions and then forwarding them to the underlying data integration systems to be answered. The changes among ontology versions using a high level language of changes are detected and described. They are interpreted as sound global-as-view mappings, and are used in order to produce equivalent rewritings among ontology versions.

COnto-Diff [6] is a rule-based approach which detects high-level changes according to a dedicated language of changes. The detection process is coupled with a mapping between the elements (concepts, properties) of two ontology versions. The application of the detected modifications (and their inverses) is also considered in this work.

In many real approaches dedicated to the ontology evolution, when maintained ontologies change, the mappings between them are recreated from scratch or are adjusted manually, a process which is known to be error-prone and time-consuming. In this paper we propose a function representing the degree of change significancy, which allows detecting outdated alignments. In consequence it will provide us with an ability of automatic alignments revalidation.

3 Basic Notions

In our research, we assume a following formal definition of an ontology:

$$O = (C, H, R^C, I, R^I) \tag{1}$$

where C is a finite set of concepts; H denotes a concepts' hierarchy; R^C is a finite set of relations between concepts $R^C = \{r_1^C, r_2^C, ..., r_n^C\}$, $n \in N$, such that every $r_i^C \in R^C$ ($i \in [1, n]$) is a subset of $C \times C$; I represents a set of instances' identifiers; $R^I = \{r_1^I, r_2^I, ..., r_n^I\}$ denotes a set of relations between concepts' instances.

By "a real world" we call a pair *(A, V)*, where A denotes a set of attributes and V is a set of valuations of these attributes (their domains). A concept's $c \in C$ structure from the *(A, V)*-based ontology is defined as:

$$c = (id^c, A^c, V^c, I^c) \tag{2}$$

where: id^c is it's unique identifier, A^c denotes a set of its attributes ($A^c \subseteq A$) with their domains included in the set V^c (formally: $V^c = \bigcup_{a \in A^c} V_a$ where V_a is a domain of an attribute a taken from the set V), and I^c is a set of assigned instances. For simplicity, we write $a \in c$ to denote that an attribute a belongs to the concept c (formally: $a \in c \iff a \in A^c$).

To ascribe a meaning to attributes included in some concept, we assume an existence of a sub-language of the sentence calculus (denoted as L_s^A) and a function $S_A : A \times C \to L_s^A$, which assigns logic sentences to every attribute from a concept c. For example, an attribute $DateOfBirth$ within a concept $Person$ obtains the following semantics: $S_A(DateOfBirth, Person)$: $birthYear \wedge birthMonth \wedge birthDay \wedge age$.

The overall meaning of a concept (further referred to as its context) is defined as a conjunction of semantics of each of its attributes. Formally, for a concept c, such that $A^c = \{a_1, a_2, ..., a_n\}$, its context is as follows $ctx(c) = S_A(a_1, c) \wedge S_A(a_2, c) \wedge ... \wedge S_A(a_n, c)$.

Due to the fact, that in this article we focus only on the concept level of ontologies, we do not provide detailed definitions of remaining elements from the Eq. 1. For broader explanations, please refer to our previous publications, such as [11].

In order to track changes applied to ontologies, we accept a notion of a universal timeline, which can be understood as an ordered set of discrete moments in time: $\overline{TL} = \{t_n | n \in N\}$. $TL(O)$ denotes a subset of this timeline for a selected ontology - it contains only those elements of $\overline{TL}$ for which the ontology O has changed. By using a superscript $O^{(m)} = (C^{(m)}, H^{(m)}, R^{C(m)}, I^{(m)}, R^{I(m)})$ we denote the ontology O in a given moment in time $t_m \in TL(O)$. We also introduce the notion $O^{(m-1)} \prec O^{(m)}$ which represents a fact that $O^{(m)}$ is a later version of O than $O^{(m-1)}$. For simplicity we extend this notation for particular elements of the given ontology, e.g. $c^{(m-1)} \prec c^{(m)}$ denotes that a concept c has at least two versions, and $c^{(m-1)}$ is earlier than $c^{(m)}$. On top of these definitions, we define a repository of an ontology O, which is an ordered set of its subsequent versions in time, formally defined as $Rep(O) = \left\{ O^{(m)} | \forall m \in TL(O) \right\}$.

Assuming an existence of two independent, (A, V)-based ontologies, O and O', an alignment on a concept level between them is defined as a finite set $Align(O, O')$ containing tuples of the form:

$$(c, c', \lambda_C(c, c'), r) \tag{3}$$

where: c and c' are concepts from O and O' respectively, $\lambda_C(c, c')$ is a real value representing a degree to which the concept c can be mapped into the concept c', and r is one of types of relation that connects c and c' (equivalency, generalisation or contradiction). $\lambda_C(c, c')$ can be calculated using one of the similarity methods taken from a very broad literature concerning the ontology alignment. A robust overview of a current state of the art in this field can be found in [1]. Due to its simplicity and flexibility, we can use this notion also for time-tracked ontologies. For example, $Align(O^{(m)}, O'^{(n)})$ is an alignment of the ontology O in a state it had in a moment m, and the ontology O' in a state from a moment n. Obviously, both $m, n \in \overline{TL}$.

4 Ontology Change Significance on the Concept's Level

In order to compare two states of a single ontology O we introduce a function $diff_C$ which, when fed with its two successive states $O^{(m-1)}$ and $O^{(m)}$ (such that $O^{(m-1)} \prec O^{(m)}$), generates three sets containing concepts added, deleted and somehow altered. Formally, these sets are defined below:

$$diff_C(O^{(m-1)}, O^{(m)}) = \Big\langle new_C(C^{(m-1)}, C^{(m)}), \\ del_C(C^{(m-1)}, C^{(m)}), \\ alt_C(C^{(m-1)}, C^{(m)}) \Big\rangle \quad (4)$$

where:

1. $new_C(C^{(m-1)}, C^{(m)}) = \Big\{ c \Big| c \in C^{(m)} \wedge c \notin C^{(m-1)} \Big\}$
2. $del_C(C^{(m-1)}, C^{(m)}) = \Big\{ c \Big| c \in C^{(m-1)} \wedge c \notin C^{(m)} \Big\}$
3. $alt_C(C^{(m-1)}, C^{(m)}) = \Big\{ (c^{(m-1)}, c^{(m)}) | c^{(m-1)} \in C^{(m-1)} \wedge c^{(m)} \in C^{(m)} \wedge c^{(m-1)} \prec c^{(m)} \wedge (A^{c^{(m-1)}} \neq A^{c^{(m)}} \vee V^{c^{(m-1)}} \neq V^{c^{(m)}} \vee I^{c^{(m-1)}} \neq I^{c^{(m)}}) \vee ctx(c^{(m-1)}) \neq ctx(c^{(m)}) \Big\}$

The first two descriptors in the definition above are self-explanatory. The last one represents alterations applied to concepts from $O^{(m-1)}$, as a set of pairs of concepts' versions, that have been neither added nor deleted, but differ structure-wise or in terms of their contexts.

The function $diff_C$ describes changes applied to a certain ontology, however, it does not show how significant they were. For an ontology $O = (C, H, R^C, I, R^I)$ in its two subsequent states $O^{(m-1)}$ and $O^{(m)}$, such that $O^{(m-1)} \prec O^{(m)}$, and a concept difference function $diff_C$ defined above, a function calculating **a degree of change significance on the level of concepts** has a following signature:

$$\Psi_C : C^{(m-1)} \times C^{(m)} \to [0, 1] \quad (5)$$

Such function must meet a following two postulates:

- **P1.** $\Psi_C(C^{(m-1)}, C^{(m)}) = 0 \iff diff_C(C^{(m-1)} \times C^{(m)}) = \Big\langle \phi, \phi, \phi \Big\rangle$
- **P2.** $\Psi_C(C^{(m-1)}, C^{(m)}) = 0 \iff del_C(C^{(m-1)}, C^{(m)}) = C^{(m-1)} \wedge$ $\wedge\, alt_C(C^{(m-1)}, C^{(m)}) = \phi$

P1 states that the change significance is minimal if no alterations on the concept level have been applied. Namely, no new concepts have appeared, no concepts have been removed, no concepts have been changed.

P2 describes that the change significance is maximal if the ontology has been completely modified, meaning, every concept from the earlier state has been deleted, every concept in a later state is new (or nothing has been added), and therefore no concepts have been altered.

Having the above postulates in mind, we define the function Ψ_C as follows:

$$\Psi_C(C^{(m-1)}, C^{(m)}) = \frac{|new_C(C^{(m-1)}, C^{(m)})| + |del_C(C^{(m-1)}, C^{(m)})|}{|C^{(m)}| + |del_C(C^{(m-1)}, C^{(m)})|} + \\ + \frac{\sum_{(c_1,c_2)\in alt_C(C^{(m-1)},C^{(m)})} d_s(ctx(c_1), ctx(c_2))}{|C^{(m)}| + |del_C(C^{(m-1)}, C^{(m)})|} \tag{6}$$

The function above is build from three components. The first two are cardinalities of sets describing new and removed concepts. The last component utilises a function d_s which calculates a distance between two logic formulas - we initially transform the passed formulas (concepts' contexts) to a conjunctive normal form and incorporate the Jaccard's measure to calculate the distance between them. For details please refer to our previous publication [14].

In the next section we will describe an experiment that we designed and conducted in order to verify a usefulness of the developed function Ψ_C along with an analysis of obtained results.

5 Experimental Verification

5.1 Experiment's Setup and Procedure

The ontology alignment is a frequently covered topic. Ontology Alignment Evaluation Initiative (OAEI) is an organisation which annually organises a campaign aiming at assessing strengths and weaknesses of ontology matching systems and comparing their performances [1]. Participants of these campaigns designate mappings between preprepared ontologies that, for logistical reasons, are grouped into groups called tracks. Within every track, for every ontology pair, OAEI provides a reference alignment, with which a collected mappings are compared using a variety of different measures.

In order to verify the usefulness of a function Ψ_C in detecting a necessity of potential revalidating an alignment that became stale due to the ontology evolution, we needed a robust dataset and an independent ontology alignment tool. We chose "a Conference Track" consisting of 16 ontologies describing the domain of organising conferences, that was used in the OAEI'2017 campaign. We also decided to base our experiment on LogMap [12], which is an ontology alignment and alignment repair system. It is a highly scalable ontology matching solution with an integrated reasoning and inconsistency repair capabilities. It is capable of extracting mappings between concepts, relations and instances. More importantly, LogMap earned high positions in subsequent OAEI campaigns.

The experiment was divided into two parts. The first one was aimed at showing how different modifications of an ontology that may appear during its evolution can affect its alignments. It consisted of following phases:

1. Select a source ontology (called *confOf*) and a target ontology (*CMT*) from a Conference Track of OAEI'2017 campaign.
2. Generate a base alignment between the two selected ontologies using LogMap.
3. Apply random modifications to the source ontology according to every alteration scenario from Table 1
4. For the two versions of the source ontology, calculate a value of Ψ_C.
5. Using LogMap generate a new alignment between a modified version of the source ontology and the target ontology.
6. Calculate a Dice coefficient measure between the base and the new alignment, in order to illustrate differences between alignments of ontologies that have changes over time. This measure has been chosen as the very intuitive functions allowing to compare dissimilarity between two sets. This value can clearly show changes within ontologies (which significancy can be calculated using Ψ_C function) affect mappings between ontologies.

The goal of the second part was showing how the developed function, calculating the degree of concepts' change significancy, behaves when different ontologies and their evolutions are processed. This phase of the experiment had a following steps:

1. Select a source ontology (*confOf*)
2. Using LogMap, generate base alignments between the source ontology and every other ontology from the track (presented in Table 2)
3. Apply a random modification relying on adding and removing 5 related concepts to the source ontology and calculate the value of Ψ_C. Obviously, in this part of the experiment, the value of Ψ_C is constant and equals 0.128.
4. Using LogMap, generate new alignments between the modified source ontology and every other ontology from the track (except *confOf*).
5. Calculate a Dice coefficient measures between the base alignment and the new alignments between the source ontology and every other ontology from the track collected in the previous step.

Results collected during both parts of the experiment, along with their statistical analysis, can be found in a next section of the paper.

5.2 Results of the Experiment

As it was mentioned in the previous section, the experiment has been divided into two parts. In the first one, we prepared some scenarios which apply all possible ontology changes on the concept level like: adding, removing or modifying some or all concepts.

Let us suppose that there exists some correlation between the degree of significancy to which concepts within maintained ontologies have been changed in time and a Dice distance between the base and the new alignment. We confirmed this hypothesis using a statistical analysis of gathered data obtained using a procedure described in the previous section and presented in the Table 1.

Before selecting a proper statistical test, we analysed the distribution of obtained data using a Shapiro-Wilk test. Because for both samples $p - values$ are greater than $\alpha = 0.05$, we rejected the null hypothesis and claim that samples do not come from a normal distribution. Next, we calculated Spearman's rank correlation coefficient. Comparing the obtained $p - value$ equals 0.00642 with the assumed significance level α, we could draw a conclusion that there is a monotonic dependence between Ψ_C and calculated values of the Dice measure (Fig. 1). This dependence is directly proportional. The Spearman's rank correlation coefficient was equal 0.71 and can be interpreted as a strong, monotonic relation between the examined samples. It allows us to claim that, the developed function Ψ_C can serve as a trigger of alignment revalidation in case of significant change that may appear during the ontology evolution.

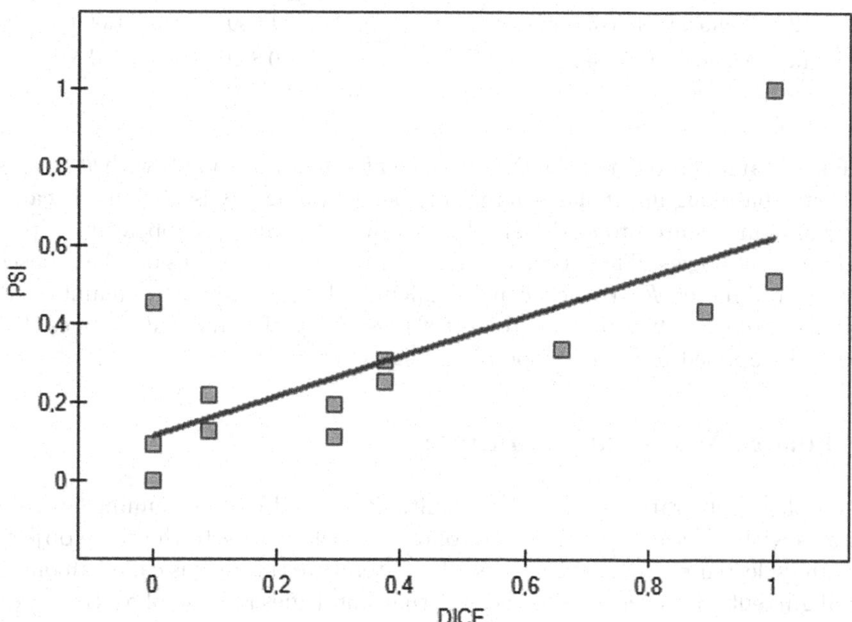

Fig. 1. A graphical representation of a monotonic dependence between Ψ_C and calculated values of the Dice measure

Based on the results presented in the Table 1, we observed that LogMap does not consider attributes and their modifications in its alignment determination process. Thus, for the second part of our experiment we focused only on adding and removing concepts for the chosen ontology where the Ψ_C was equal 0.294. Then, according to the experimental procedure described earlier, we calculated values of the Dice measure. Obtained results are shown in the Table 2.

Table 1. Different scenarios for a single pair of ontologies, 39 concepts in the base ontology *confOf*

No.	Description	Dice measure	Ψ_C
1	No changes	0	0
2	Removing 5 concepts	0.091	0.128
3	Adding 5 related concepts	0.294	0.114
4	Modifying 5 concepts	0	0.093
5	Adding and removing 5 concepts	0.375	0.256
6	Adding and modifying 5 concepts	0.294	0.196
7	Modifying and removing 5 concepts	0.091	0.221
8	Adding, removing and modifying 5 concepts	0.375	0.309
9	Removing 20 concepts	1.000	0.513
10	Adding 20 concepts	0.657	0.339
11	Modifying 20 concepts	0	0.455
12	Removing 40 concepts	1.000	1.000
13	Adding 40 concepts	0.889	0.435

For all target ontologies, the Dice coefficient value corresponds with Ψ_C, especially if a base alignment has a large number mappings. It is obvious, because larger alignments are proportionally less sensitive for changes appearing in participating ontologies. Therefore, we can draw a general conclusion that, based only on a value of Ψ_C (which can be calculated by processing a maintained, evolving ontology). It is possible to detect a necessity of significant changes that need to be applied in the corresponding alignments.

6 Future Works and Summary

An ontology integration task is a difficult, time-, and cost-consuming process. It starts with designating elements of ontologies that relate to the same objects from the selected universe of discourse. In a literature this task is called an ontology alignment. However, due to the fact that ontologies are complex structures, and in modern days it is not possible to assume that they won't change in time, therefore, predesignated mappings between two or more ontologies may become obsolete.

In this paper we have presented a component of an ontology alignment evolution framework. The developed tool can be used to check, whether or not, the aforementioned situation in which the mapping between ontologies is no longer valid. This trigger is build on top on an analysis of changes that appeared within ontologies over time. Its usefulness was proved on a basis of a statistical analysis of experimental results gathered from a procedure utilising a broadly accepted OAEI datasets, created for validating ontology alignment tools.

Table 2. The same modification scenario for different pairs of ontologies

Target ontology	Number of mapping in a base alignment	Dice measure
CMT	5	0.375
cocus	9	0.660
conference	14	0.226
confious	5	0.429
crs	6	0.375
edas	13	0.241
ekaw	15	0.278
iasted	4	0.385
linklings	6	0.474
micro	9	0.440
myReview	7	0.368
OpenConf	6	0.294
paperdyne	11	0.417
pcs	6	0.294
sigkdd	5	0.330

Our upcoming research plans are twofold. At first, we want to develop an algorithm that based solely on an analysis of an evolving ontology will be capable of updating existing mappings with other ontologies when such necessity occurs. Secondly, we will extend the created framework for other elements available within ontologies - relations and instances.

Acknowledgement. This research project was supported by grant No. 2017/26/D/ST6/00251 from the National Science Centre, Poland.

References

1. Achichi, M., et al.: Results of the ontology alignment evaluation initiative 2017. In: OM 2017–12th ISWC Workshop on Ontology Matching, pp. 61–113 (2017). No commercial editor
2. Allocca, C., d'Aquin, M., Motta, E.: Detecting different versions of ontologies in large ontology repositories. In: Proceedings of IWOD 2009, Washington, D.C., USA (2009)
3. Dinh, D., Dos Reis, J.C., Pruski, C., Da Silveira, M., Reynaud-Delaître, C.: Identifying relevant concept attributes to support mapping maintenance under ontology evolution. Web Semant.: Sci. Serv. Agents World Wide Web **29**, 53–66 (2014)
4. Euzenat, J.: Revision in networks of ontologies. Artif. Intell. **228**, 195–216 (2015)
5. Grandi, F.: Multi-temporal RDF ontology versioning. In: Proceedings of IWOD 2009, Washington, D.C., USA (2009)
6. Hartung, M., Groß, A., Rahm, E.: COnto-Diff: generation of complex evolution mappings for life science ontologies. J. Biomed. Inform. **46**(1), 15–32 (2013)

7. Heflin, J., Pan, Z.: A model theoretic semantics for ontology versioning. In: McIlraith, S.A., Plexousakis, D., van Harmelen, F. (eds.) ISWC 2004. LNCS, vol. 3298, pp. 62–76. Springer, Heidelberg (2004). https://doi.org/10.1007/978-3-540-30475-3_6
8. Klein, M., Fensel, D., Kiryakov, A., Ognyanov, D.: Ontology versioning and change detection on the web. In: Gómez-Pérez, A., Benjamins, V.R. (eds.) EKAW 2002. LNCS (LNAI), vol. 2473, pp. 197–212. Springer, Heidelberg (2002). https://doi.org/10.1007/3-540-45810-7_20
9. Khattak, A.M., Latif, K., Lee, S.: Change management in evolving web ontologies. Knowl.-Based Syst. **37**, 1–18 (2013)
10. Kondylakis, H., Plexousakis, D.: Ontology evolution without tears. J. Web Semant. **19**, 42–58 (2013)
11. Kozierkiewicz, A., Pietranik, M.: The knowledge increase estimation framework for integration of ontology instances' relations. In: Lupeikiene, A., Vasilecas, O., Dzemyda, G. (eds.) DB&IS 2018. Communications in Computer and Information Science, vol. 838, pp. 172–186. Springer, Cham (2018). https://doi.org/10.1007/978-3-319-97571-9_15
12. Jiménez-Ruiz, E., Cuenca Grau, B.: LogMap: logic-based and scalable ontology matching. In: Aroyo, L., et al. (eds.) ISWC 2011. LNCS, vol. 7031, pp. 273–288. Springer, Heidelberg (2011). https://doi.org/10.1007/978-3-642-25073-6_18
13. Papavassiliou, V., Flouris, G., Fundulaki, I., Kotzinos, D., Christophides, V.: High-level change detection in RDF(S) KBs. ACM Trans. Database Syst. **38**(1), 1–42 (2013)
14. Pietranik, M., Nguyen, N.T.: A Multi-atrribute based framework for ontology aligning. Neurocomputing **146**, 276–290 (2014). https://doi.org/10.1016/j.neucom.2014.03.067
15. Pietranik, M., Nguyen, N.T.: Framework for ontology evolution based on a multi-attribute alignment method. In: CYBCONF 2015, pp. 108–112 (2015). https://doi.org/10.1109/CYBConf.2015.7175915
16. Sassi, N., Jaziri, W., Gargouri, F.: Z-based formalization of kits of changes to maintain ontology consistency. In: Proceedings of KEOD 2009, pp. 388–391 (2009)
17. Zablith, F., et al.: Ontology evolution: a process-centric survey. Knowl. Eng. Rev. **30**(1), 45–75 (2015)
18. Zekri, A., Brahmia, Z., Grandi, F., Bouaziz, R.: τOWL: a systematic approach to temporal versioning of semantic web ontologies. J. Data Semant. **5**(3), 141–163 (2016)

Challenging SQL-on-Hadoop Performance with Apache Druid

José Correia[1,2(✉)], Carlos Costa[1,3], and Maribel Yasmina Santos[1]

[1] ALGORITMI Research Centre, University of Minho, Guimarães, Portugal
{josecorreia, carlos.costa, maribel}@dsi.uminho.pt
[2] NATIXIS, on Behalf of Altran Portugal, Porto, Portugal
[3] Centre for Computer Graphics - CCG, Guimarães, Portugal

Abstract. In Big Data, SQL-on-Hadoop tools usually provide satisfactory performance for processing vast amounts of data, although new emerging tools may be an alternative. This paper evaluates if Apache Druid, an innovative column-oriented data store suited for online analytical processing workloads, is an alternative to some of the well-known SQL-on-Hadoop technologies and its potential in this role. In this evaluation, Druid, Hive and Presto are benchmarked with increasing data volumes. The results point Druid as a strong alternative, achieving better performance than Hive and Presto, and show the potential of integrating Hive and Druid, enhancing the potentialities of both tools.

Keywords: Big Data · Big Data Warehouse · SQL-on-Hadoop · Druid · OLAP

1 Introduction

We are living in a world increasingly automated, where many people and machines are equipped with smart devices (e.g. smartphones and sensors) all integrated and generating data from different sources at ever-increasing rates [1]. For these characteristics (volume, velocity and variety), Big Data usually arises with an ambiguous definition. However, there is a consensus defining it as data that is "too big, too fast or too hard" to be processed and analyzed by traditional techniques and technologies [1–3]. The organizations that realize the need to change their processes to accommodate adequate decision-making capabilities, supporting them with Big Data technologies, will be able to improve their business value and gain significant competitive advantages over their competitors.

The Big Data concept also impacts the traditional Data Warehouse (DW), leading to a Big Data Warehouse (BDW) with the same goals in terms of data integration and decision-making support, but addressing Big Data characteristics [4, 5] such as massively parallel processing; mixed and complex analytical workloads (e.g., ad hoc querying, data mining, text mining, exploratory analysis and materialized views); flexible storage to support data from several sources or real-time operations (stream processing, low latency and high frequency updates), only to mention a few. Also, SQL-on-Hadoop systems are increasing their notoriety, looking for interactive and low

W. Abramowicz and R. Corchuelo (Eds.): BIS 2019, LNBIP 353, pp. 149–161, 2019.
https://doi.org/10.1007/978-3-030-20485-3_12

latency query executions, providing timely analytics to support the decision-making process, in which each second counts [6]. Aligned with the research trends of supporting OLAP (Online Analytical Processing) workloads and aggregations over Big Data [7], this paper compares Apache Druid, which promises fast aggregations on Big Data environments [8], with two well-known SQL-on-Hadoop systems, Hive and Presto.

This paper is organized as follows: Sect. 2 presents the related work; Sect. 3 describes Druid and the experimental protocol; Sect. 4 presents the obtained results; and Sect. 5 discusses the main findings and concludes with some future work.

2 Related Work

Several SQL-on-Hadoop systems have been studied to verify their performance, supporting interactive and low latency queries. The work of [9] benchmarks different SQL-on-Hadoop systems (Hive, Spark, Presto and Drill) using the Star Schema Benchmark (SSB), also used in [10], testing Hive and Presto using different partitioning and bucketing strategies. In [6], Drill, HAWQ, Hive, Impala, Presto and Spark were benchmarked showing the advantages of in-memory processing tools like HAWQ, Impala and Presto. Although the good performance of these in-memory processing tools, this work also shows the increase in the processing time that is verified when these tools do not have enough Random Access Memory (RAM) and activate the "Spill to Disk" functionality, making use of secondary memory. In terms of scalability, HAWQ showed the worst result when taking into consideration querying response time with increased data volumes (in the same infrastructure), while Spark, Presto and Hive showed good scalability. The results obtained with Spark point that this technology is appropriate when advanced analysis and machine learning capabilities are needed, besides querying data, and that Hive can perform similarly to Presto or Impala in queries with heavy aggregations. Although other benchmarks are available, to the best of our knowledge, they do not evaluate such diverse set of tools.

For Druid, although a very promising tool, few works have studied this technology and most of them do not use significant data volumes [8, 11] or do not compare the results against other relevant technologies, typically used in OLAP workloads on Big Data environments [8]. The work of [12] contributed to this gap by using the SSB to evaluate Druid, pointing out some recommendations regarding performance optimization. The obtained results were impressive in terms of processing time, but Druid was not compared with other systems. Thus, this paper seeks to fulfil this gap by comparing Druid against two well-known SQL-on-Hadoop systems, Hive and Presto, a work of major relevance for both researchers and practitioners concerned with low latency in BDW contexts. The two SQL-on-Hadoop systems were selected based on their advantages, shown in the literature, such as the robustness of Hive with increased data volumes and the good overall performance of Presto. Moreover, these tools are here benchmarked using the same infrastructure and the same data as in [10], allowing the comparison of the results.

3 Testing Druid Versus SQL-on-Hadoop

Druid (http://druid.io) is an open source column-oriented data store, which promises high-performance analytics on event-driven data, supporting streaming and batch data ingestion. Its design combines search systems, OLAP and timeseries databases in order to achieve real-time exploratory analytics [8, 12]. Druid has several features, being important to explain the ones addressed in this work: (i) segment granularity is the granularity by which Druid first partition its data (e.g. defining this as month generates 12 segments per year); (ii) query granularity specifies the level of data aggregation at ingestion, corresponding to the most detailed granularity that will be available for querying; (iii) hashed partitions are responsible for further partitioning the data, besides segment granularity; (iv) cache stores the results of the queries for future use; (v) memory-mapped storage engine deals with the process of constantly loading and removing segments from memory [8, 12].

3.1 Technological Infrastructure and Data

This work compares the obtained results with some of the results available in [12] and [10], reason why the same technological infrastructure is used. This infrastructure is based on a 5-node cluster, including 1 HDFS NameNode (YARN ResourceManager) and 4 HDFS DataNodes (YARN NodeManagers). Each node includes: (i) 1 Intel Core i5, quad-core, clock speed ranging between 3.1 and 3.3 GHz; (ii) 32 GB of 1333 MHz DDR3 Random Access Memory; (iii) 1 Samsung 850 EVO 500 GB Solid State Drive with up to 540 MB/s read speed and up to 520 MB/s write speed; (iv) 1 Gigabit Ethernet card connected through Cat5e Ethernet cables and a gigabit Ethernet switch. The several nodes use CentOS 7 with an XFS file system as the operative system. Hortonworks Data Platform 2.6.4 is used as the Hadoop distribution, with the default configurations, excluding the HDFS replication factor, which was set to 2. Druid 0.10.1 is used with its default configurations. Taking into consideration that the main goal of this paper is to compare the performance of Druid against other SQL-on-Hadoop technologies under similar circumstances, the SSB [13] must be used, as in [12] and [10]. The SSB is based on the TPC-H BenchmarkTM (http://www.tpc.org/tpch), but following the principles of dimensional data modeling with a star schema [14]. This is a reference benchmark often used to measure the performance of database systems that process large volumes of data, supporting Data Warehousing applications [13].

Since Druid does not support joins, the SSB was denormalized, originating two different flat tables: (i) *A Scenario*, including all the attributes from the SSB; (ii) *N Scenario*, containing the attributes strictly needed to answer the queries of the benchmark. The goal is to compare Druid, Hive and Presto in the A Scenario and evaluate how the number of attributes affects the processing time of Druid in the *N Scenario*. Moreover, Druid's features, and their impact on performance, are evaluated whenever possible in both scenarios. Besides this, the 13 SSB queries were also used in their denormalized version. In order to obtain rigorous results and allow replicability, several scripts were coded running each query four times and calculating the final time as the average of the four runs (scripts available at https://github.com/jmcorreia/Druid_SSB_Benchmark).

3.2 Experimental Protocol

Figure 1 depicts the experimental protocol, aiming to evaluate Druid's performance in different scenarios and comparing these results with the ones obtained by Hive and Presto under similar circumstances. Three Scale Factors (SFs) are used to evaluate the performance of the tools for different workloads (30, 100 and 300 GB), using data and queries from SSB.

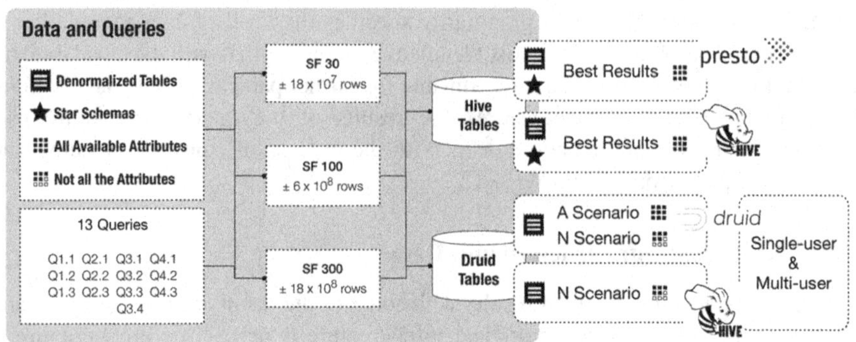

Fig. 1. Experimental protocol.

Besides studying the impact of different data volumes on query performance, other Druid features were explored, namely segment granularity, query granularity and hashed partitions. Some of the obtained results and main conclusions regarding these properties were retrieved from [12] and are used in this work to: (i) analyze Druid's best results [12] with Hive and Presto's best results [10]; (ii) compare the results obtained by Druid in scenarios that do not optimize performance to its maximum potential with the best results obtained by Hive and Presto [10]; and, (iii) study the potential of integrating Hive and Druid, theoretically and practically, on single and multi-user environments.

4 Results

This section presents the results for the defined experimental protocol. In Druid, the several stored tables follow this notation: *S{Segment Granularity}_Q{Query Granularity}_PHashed{Number of Partitions}*, with the different configurations of segments, query granularity and hashed partitions [12]. In the *A Scenario*, the query granularity property is not used as the existing keys exclude the possibility of data aggregation when storing the data (all rows are different). In the *N Scenario*, as it only uses the attributes required for answering the queries, keys are not present and data aggregation is possible (using query granularity). In this paper, the goal is to compare Druid's potential with other technologies and not to explore Druid properties in deeper detail, as this was done in [12], reason why here the tables with the best results, for the tested features, are used exploring the results of [12]. To this end, Fig. 2 summarizes the selected tables and why they were selected and used in the analysis of the following subsections. The results presented in this section, if not mentioned otherwise, include the total average processing time for the execution of the 13 queries.

	Table	Section	Reason
SF 30	SQuarter_PHashed3	4.1	Best result for the *A Scenario*
	SQuarter_QMonth_PHashed3	4.1	Best result for the *N Scenario*
	SQuarter	4.2	Best result for the *A Scenario*, without using hashed partitions
	SMonth	4.2	Best result for the *N Scenario*, without using query granularity or hashed partitions
	SMonth_QWeek	4.2	Best result for the *N Scenario*, without using hashed partitions
SF 100	SQuarter_PHashed6	4.1	Best result for the *A Scenario*
	SMonth_QWeek_PHashed2	4.1	Best result for the *N Scenario*
	SQuarter	4.2	Best result for the *A Scenario*, without using hashed partitions
	SMonth	4.2	Best result for the *N Scenario*, without using query granularity or hashed partitions
	SQuarter_QMonth	4.2	Best result for the *N Scenario*, without using hashed partitions
SF 300	SQuarter_QMonth_PHashed10	4.1	Best result for the *N Scenario*
	SMonth	4.2	Best result for the *N Scenario*, without using query granularity or hashed partitions
	SQuarter_QMonth	4.2	Best result for the *N Scenario*, without using hashed partitions

Fig. 2. Considered Druid tables.

4.1 Druid Versus Hive and Presto

In this subsection, Druid's best results are compared with Hive and Presto's best results, retrieved from [10], where the authors explored different partition and bucketing strategies, in order to identify the most effective strategies to enhance performance. The main purpose is to assess if Druid constitutes an alternative to Hive and Presto, two well-known SQL-on-Hadoop systems. Figure 3 presents the best results obtained by Hive and Presto, without results for the denormalized table in SF 300, as in [10] the authors were not able to execute this test, due to memory constraints of the infrastructure.

	Hive			Presto		
Scale Factor (SF)	30	100	300	30	100	300
Star Schema	349s	865s	982s	77s	256s	452s
Denormalized	256s	424s	---	33s	90s	---

Fig. 3. Best results for Hive and Presto. Processing time based on [10].

Analyzing Fig. 3, we can verify that the denormalized table achieves better processing time than the star schema for all the evaluated SFs, both for Hive and Presto. It is also noticeable that Presto outperforms Hive in all the scenarios. Considering these aspects and the fact that Druid does not support joins, Fig. 4 considers the processing time obtained by Presto for the denormalized table, except for the SF 300, in which we considered the performance obtained for the star schema (because there is no result for the denormalized version). Figure 4 also presents the best results obtained by Druid for the scenarios *A* and *N*, showing that Druid is the fastest technology for all SFs, presenting improvements between 93.2% and 98.3%. For the SF 300, Druid was only tested for the *N Scenario*, as in [12] the SFs 30 and 100 were considered representative enough in the *A Scenario* for the analysis of Druid capabilities. So, it is important to have in mind that the comparison for the SF 300 is made considering a star schema for Hive and Presto, and a denormalized table with a subset of the attributes for Druid.

		Time (s)	Difference (%)
SF 30	**Presto**	33	---
	Druid (SQuarter_PHashed3)	2.09	-93.7%
	Druid (SQuarter_QMonth_PHashed3)	1.35	-95.9%
SF 100	**Presto**	90	---
	Druid (SQuarter_PHashed6)	6.12	-93.2%
	Druid (SMonth_QWeek_PHashed2)	3.72	-95.9%
SF 300	**Presto**	452	---
	Druid (SQuater_QMonth_PHashed10)	7.6	-98.3%

Fig. 4. Druid and Presto best results. Processing time based on [10, 12].

With this context, in Fig. 4 it is noticeable that the tables of the *N Scenario* achieved better processing times than the *A Scenario*. This happens because this scenario uses a model with less attributes, reducing storage needs and enabling the application of query granularity (which aggregates data, also reducing storage needs). This way, the data can be easily loaded into memory, because the segments use less memory. In addition, as the data is aggregated, the answers for the queries may need less calculations. However, users must be aware of the trade-off between using query granularity and the limitations this will impose on the ability to query the data, because the query granularity represents the deepest level of detail for querying data. These good results are obtained due to the optimized way used by Druid to store data, enhancing its efficiency with an advanced indexing structure that allows low latencies. Besides this, the number of attributes and the use of query granularity also impact ingestion time, as the tables of the *N Scenario* take less time to be ingested. For the SF 100, for example, the tables of the *A Scenario* spent 4 h and 14 min (on average) and the tables of the *N Scenario* spent 1 h and 22 min on average (68% less time). Regarding storage needs, the *N Scenario* and the use of query granularity were able to reduce the needed space by more than 85%. For performance, Druid cache mechanisms and its memory-mapped storage engine also have a relevant impact, as detailed later in this paper.

4.2 Suboptimal Druid Versus Hive and Presto

In the previous subsection, Druid revealed significantly faster processing time compared to Presto and Hive. However, we used Druid's best results, which were obtained using the tables with hashed partitions, an advanced property that should only be used if it is strictly necessary to optimize Druid's performance to its maximum potential. In most cases, tables without hashed partitions achieve results that are good enough to satisfy the latency requirements [12]. Therefore, this subsection does not use Druid's best results, but the better ones obtained without hashed partitions. Figure 5 shows the time obtained by running the 13 queries and the difference between the results obtained by Druid and Presto. In this case, even with less concerns regarding optimization, Druid achieves significantly faster processing time when compared to Presto. In the worst case, Druid was able to use less 90.3% of the time needed by Presto. This result considers the table *SQuarter* belonging to the *A Scenario*, which uses the model with all the attributes and does not aggregate data during its ingestion, meaning that even with raw data, Druid performs very well, outperforming Hive and Presto.

		Time (s)	Difference (%)
SF 30	**Presto**	33	---
	Druid (SQuarter)	3.21	-90.3%
	Druid (SMonth)	2.22	-93.3%
	Druid (SMonth_QWeek)	1.42	-95.7%
SF 100	**Presto**	90	---
	Druid (SQuarter)	8.08	-91.0%
	Druid (SMonth)	7.02	-92.2%
	Druid (SQuarter_QMonth)	5	-94.4%
SF 300	**Presto**	452	---
	Druid (SMonth)	20.02	-95.6%
	Druid (SQuater_QMonth)	8.99	-98.0%

Fig. 5. Druid suboptimal vs. Presto. Processing time based on [10, 12].

Figure 5 also reveals that the most significant performance differences between Presto and Druid were obtained for the tables belonging to the *N Scenario*, as expected, and using the query granularity property to aggregate data during its ingestion, also expected. As previously mentioned, this happens because this scenario uses a model with the attributes needed to answer the queries, taking less storage space. Besides, this model enables the use of the query granularity feature, aggregating data during its ingestion, which also reduces storage needs and improves performance for the same reasons mentioned above (see Subsect. 4.1). The table *SQuarter_QMonth* (in the SF 300), for example, was able to obtain a processing time 98.0% lower than Presto, taking 8.99 s to execute all the queries. In this case, data is segmented by quarter and aggregated to the month level, meaning that this table corresponds to a view with less rows than the ones analyzed by Presto. Although, in this case, Presto uses raw data while Druid uses aggregated data, this comparison is important to understand how Druid's characteristics can be used to enhance data processing efficiency, as in similar conditions (the same number of rows) Druid outperforms Presto. Looking to the overall performance, Druid seems to be a credible alternative to well-established SQL-on-Hadoop tools in scenarios when interactive querying processing is needed, even with less optimization concerns.

4.3 Scalability

For querying vast amounts of data, the processing tools are designed to be scalable, meaning that as the data size grows or the computational resources change, the tools need to accommodate that growth/change. Each tool usually applies specific approaches to scale, either in terms of data size or in terms of the used computational resources. In this paper, and due to the limited hardware capacity of the used cluster, the scalability of the tools is analyzed looking into the time needed by the tools to process the data, as the data size grows. The 30 GB, 100 GB and 300 GB SFs were used, because we had baseline results for Hive and Presto, under similar circumstances, available in the literature. Higher SFs were not used due to limitations of the infrastructure and because there would be no results available in the literature to make a fair comparison with Druid. In this case, we believe that the three different SFs used provide a fair analysis of the scalability of the tools.

Looking first into Hive and Presto best results, Fig. 6 shows that in the denormalized version, Hive increases in 1.7 times the time needed to process the data when the dataset grows from 30 to 100 GB, while Presto increases this value in 2.7 times. In the star schema, these values are 2.5 and 3.3 times, respectively. In this data model, moving from 100 from 300 GB has almost no impact on Hive, with a marginal increase in the needed overall time (1.1), while Presto almost double the needed processing time (1.8). For Druid, looking both into the best and suboptimal results, the time needed to process the data seems to increase around 3 times when the data volume increases from 30 to 100 GB, and from 2 to 3 times when the data volume increases from 100 to 300 GB. Datasets higher than 300 GB are needed to have a more detailed perspective on the scalability of the tools, but Hive seems to be the tool that better reacts to the increase of data volume, although taking more time than the other two to process the same amount of data.

	Time (s)				Increase along SF (x times)			
	Denormalized		Star Schema		Denormalized		Star Schema	
	Hive	Presto	Hive	Presto	Hive	Presto	Hive	Presto
SF 30	256	33	349	77				
SF 100	424	90	865	256	1.7	2.7	2.5	3.3
SF 300	---	---	982	452	---	---	1.1	1.8
	Denormalized		Denormalized		Denormalized		Denormalized	
	Druid Best	Druid Suboptimal	Druid Best	Druid Suboptimal	Druid Best	Druid Suboptimal	Druid Best	Druid Suboptimal
SF 30	2.09	3.21	1.35	2.22				
SF 100	6.12	8.08	3.72	7.02	2.9	2.5	2.8	3.2
SF 300	---	---	7.60	20.02	---	---	2.0	2.9

Fig. 6. Hive, Presto and Druid scalability analysis.

4.4 Hive and Druid: Better Together?

The previous subsection showed Druid as an alternative to Presto and Hive. However, new versions of Hive include a novel feature named Live Long and Process (LLAP) [15], allowing the integration between Hive and Druid [16]. Hence, this subsection studies the relevance of this integration. In theory, the integration of both technologies adequately combines their capabilities, providing some advantages, such as: (i) querying and managing Druid data sources via Hive, using a SQL (Structured Query Language) interface (e.g., create or drop tables, and insert new data); (ii) efficiently execute OLAP queries through Hive, as Druid is well suited for OLAP queries on event data; (iii) use more complex query operators not natively supported (e.g. joins) by Druid; (iv) using analytical tools to query Druid, through ODBC/JDBC Hive drivers; (v) indexing the data directly to Druid (e.g. a new table), instead of using MapReduce jobs [16, 17].

Next, in order to study the real impact of integrating Hive and Druid, Fig. 7 shows the processing time obtained by Hive querying a Druid data source, compared with the time of Druid itself. The results were obtained querying the table *SMonth_QDay* from the SF 300. This table was selected because it is a table of the higher SF used in this work and it does not use hashed partitions, which is not a trivial property and, as so, may not be frequently used in real contexts by most users. Besides, the configuration of the segment granularity "month" and query granularity "day" seems to simulate well a real scenario, in which users do not want to aggregate data more than the day level, avoiding losing too much detail and potential useful information. As can be seen, in terms of performance, it is preferable to query Druid directly, as it reduces processing time by an average value of 76.4%, probably because Druid tables are optimized to be queried by Druid. Considering only the first run of the queries (ignoring Druid's cache effect), this reduction is of 65.5%. Nevertheless, the results obtained when querying a Druid data source through Hive are also satisfactory (55.03 s on average), being better than the results obtained only using Hive (982 s) or Presto (452 s) when querying Hive tables (Fig. 3). Using Druid to query data stored in Druid was expected to achieve better results than being queried through Hive, however, Druid does not allow joins or other more complex operations, thus, in this type of scenario, Hive would achieve increased performance querying Druid, in a scenario that takes advantages from the best of both tools.

	Hive (querying Druid data)		Druid	
	Run 1	Average (Runs)	Run 1	Average (Runs)
Time (s)	67.34	55.03	23.22	13.01
Difference (%)	---	---	-65.5%	-76.4%

Fig. 7. Hive and Druid integration: single-user.

4.5 Hive and Druid in Multi-user Environments

In a real-world scenario, systems are not just queried by a single user. Thus, it is also relevant to study these technologies in multi-user environments, in order to verify their behavior compared with the previously obtained single-user results. Therefore, we included a scenario wherein four users simultaneously query Druid data sources (table *SMonth_QDay* from the SF 300, as this was the table used in the Subsect. 4.4), directly or through Hive. As can be seen in Fig. 8, it is still preferable to query Druid directly (not using Hive), as it reduces 66.6% of the time on average. However, the difference in terms of performance was reduced from 76.4% (Fig. 7) to 66.6%, meaning that Hive was less affected than Druid in multi-user environments. With Hive, from single-user to multi-user, the processing time increased 49.1% on average, while Druid increased 110.8% (which is still a satisfactory result, as an increase below 300% means that the system was able to execute the queries faster than executing the queries four times each in a single-user environment). This scenario also points out the robustness of Hive, a characteristic already mentioned.

	Hive (querying Druid data)		Druid	
	Run 1	Average (Runs)	Run 1	Average (Runs)
Time (s)	94.59	82.06	39.69	27.42
Difference (%) (from Hive to Druid)	---	---	-59.1%	-66.6%
Increase (s) (from Single to Multi-user)	40.5%	49.1%	66.6%	110.8%

Fig. 8. Hive and Druid integration: multi-user.

Figure 9 shows the processing time obtained by *User 1* and *User 2* for the table *SMonth_QDay* (SF 300), executing queries with and without Druid's cache mechanism.

	Druid with Cache				Druid without Cache			
	User 1 (s)		User 2 (s)		User 1 (s)		User 2 (s)	
	Run 1	Average	Run 1	Average	Run 1	Average	Run 1	Average
Q1.1	18.68	4.70	0.02	0.02	27.53	9.92	2.84	2.76
Q1.2	2.36	0.61	0.02	0.02	1.47	1.36	1.33	1.31
Q1.3	0.60	0.18	0.02	0.02	0.38	0.65	0.30	0.36
Q2.1	7.13	4.65	17.98	6.32	4.70	4.85	17.88	6.61
Q2.2	1.63	2.51	3.13	3.51	5.14	5.70	7.51	3.77
Q2.3	1.59	2.33	3.80	3.25	1.26	2.00	2.57	5.32
Q3.1	7.39	4.73	3.85	5.37	8.50	5.30	3.71	3.44
Q3.2	1.08	1.22	1.79	1.45	0.67	1.41	3.27	2.65
Q3.3	0.83	0.51	1.00	1.11	1.00	1.58	1.58	1.13
Q3.4	1.37	1.05	0.28	0.73	0.26	1.21	0.16	1.03
Q4.1	4.92	3.86	1.89	3.03	1.42	1.61	4.95	4.01
Q4.2	1.23	0.95	2.33	1.51	0.46	0.45	1.77	1.96
Q4.3	0.74	0.90	1.17	1.14	0.16	0.19	0.60	1.31
Total	49.55	28.20	37.28	27.48	52.95	36.23	48.47	35.66

Fig. 9. Results by user, with and without cache.

It is relevant to highlight that *User 1* started executing *Q1.1* and finished with *Q4.3*, while *User 2* started with *Q2.1* and finished with *Q1.3*. In general, the results show that the use of Druid's cache mechanism increases performance, although for some cases the difference is not very noteworthy. This is observable looking into the time of *Run 1*, compared with the average time of the four runs. The average total time of the four runs is always inferior to the total time of the *Run 1* of the queries. Even more interesting is the processing time observed among different queries within the same group. Subsequent queries take less time than the previous ones (e.g. *Q1.2* takes less time than *Q1.1*). This is caused by Druid's memory-mapped storage engine, which maintains recent segments in memory, while the segments less queried are paged out. With this feature, subsequent queries benefit from the fact that some of the needed segments are already in main memory, avoiding reading data from disk. This storage engine seems to have more impact than the cache mechanism, since without this mechanism the results were not so good, but the average time of the four runs remained inferior to the time of the *Run 1*. Druid's cache mechanism and its memory-mapped storage engine allow different users in a multi-user environment to benefit from each other.

5 Discussion and Conclusion

This paper presented several results comparing the performance obtained by Druid, Hive and Presto, running a denormalized version of the 13 queries of the SSB for different SFs. The results show Druid as an interesting alternative to the well-known SQL-on-Hadoop tools, as it always achieved a significant better performance than Hive and Presto. Even with less concerns regarding optimization, Druid was more efficient in terms of processing time. This paper also investigated the performance of Hive querying Druid tables, which showed to be more efficient than Hive and Presto querying Hive tables. The integration of Hive and Druid enhances Druid with relevant features, such as the possibility to execute more complex query operators (e.g. joins), or the opportunity to manage Druid data sources through Hive's SQL interface, among others, avoiding some possible drawbacks in Druid's adoption. Moreover, this work also analyzed Hive and Druid in multi-user environments, showing the impressive influence on performance from Druid's memory-mapped storage engine and cache mechanism. Besides this, both the integration of Hive and Druid, or Druid alone, showed an adequate behavior in this environment.

In conclusion, this work highlighted Druid's capabilities, but also focused the integration between this technology and Hive, challenging SQL-on-Hadoop technologies in terms of efficiency. Although Druid achieved significantly better performance than Hive and Presto, other aspects rather than data processing performance need to be considered in the adoption of a specific technology, such as the technology maturity, the available documentation and the resources (time, people, etc.) to learn and implement a new technology. However, the Big Data world is characterized by several distinct technologies and the community is used to change and to adopt new technologies. Allied to this fact, the integration of Druid in the Hadoop ecosystem and, in particular, with Hive, can facilitate the adoption of this technology.

As future work, it will be relevant to use higher data volumes and to further explore: (i) Druid's memory-mapped and cache mechanisms; (ii) Hive and Druid's integration, benchmarking complex query operators (e.g. joins); (iii) the possibility to use Presto to query Druid tables, verifying if Presto performance can be improved querying data stored in Druid rather than in Hive.

Acknowledgements. This work is supported by COMPETE: POCI-01-0145- FEDER-007043 and FCT – *Fundação para a Ciência e Tecnologia* within Project UID/CEC/00319/2013 and by European Structural and Investment Funds in the FEDER component, COMPETE 2020 (Funding Reference: POCI-01-0247-FEDER-002814).

References

1. IBM, Zikopoulos, P., Eaton, C.: Understanding Big Data: Analytics for Enterprise Class Hadoop and Streaming Data, 1st edn. McGraw-Hill Osborne Media (2011)
2. Ward, J.S., Barker, A.: Undefined by data: a survey of big data definitions. CoRR, abs/1309.5821 (2013)
3. Madden, S.: From databases to big data. IEEE Internet Comput. **16**(3), 4–6 (2012)
4. Krishnan, K.: Data Warehousing in the Age of Big Data, 1st edn. Morgan Kaufmann Publishers Inc., San Francisco (2013)
5. Costa, C., Santos, M.Y.: Evaluating several design patterns and trends in big data warehousing systems. In: Krogstie, J., Reijers, H.A. (eds.) CAiSE 2018. LNCS, vol. 10816, pp. 459–473. Springer, Cham (2018). https://doi.org/10.1007/978-3-319-91563-0_28
6. Rodrigues, M., Santos, M.Y., Bernardino, J.: Big data processing tools: an experimental performance evaluation. Wiley Interdisc. Rev. Data Min. Knowl. Discov. **9**, e1297 (2019)
7. Cuzzocrea, A., Bellatreche, L., Song, I.-Y.: Data warehousing and OLAP over big data: current challenges and future research directions. In: Proceedings of the Sixteenth International Workshop on Data Warehousing and OLAP, New York, USA, pp. 67–70 (2013)
8. Yang, F., Tschetter, E., Léauté, X., Ray, N., Merlino, G., Ganguli, D.: Druid: a real-time analytical data store. In: Proceedings of the 2014 ACM SIGMOD International Conference on Management of Data, pp. 157–168 (2014)
9. Santos, M.Y., et al.: Evaluating SQL-on-Hadoop for big data warehousing on not-so-good hardware. In: ACM International Conference Proceeding Series, vol. Part F1294, pp. 242–252 (2017)
10. Costa, E., Costa, C., Santos, M.Y.: Partitioning and bucketing in hive-based big data warehouses. In: Rocha, Á., Adeli, H., Reis, L.P., Costanzo, S. (eds.) WorldCIST'18 2018. AISC, vol. 746, pp. 764–774. Springer, Cham (2018). https://doi.org/10.1007/978-3-319-77712-2_72
11. Chambi, S., Lemire, D., Godin, R., Boukhalfa, K., Allen, C.R., Yang, F.: Optimizing druid with roaring bitmaps. In: ACM International Conference Proceeding Series, 11–13 July 2016, pp. 77–86 (2016)
12. Correia, J., Santos, M.Y., Costa, C., Andrade, C.: Fast online analytical processing for big data warehousing. Presented at the IEEE 9th International Conference on Intelligent Systems (2018)
13. O'Neil, P.E., O'Neil, E.J., Chen, X.: The Star Schema Benchmark (SSB) (2009)
14. Kimball, R., Ross, M.: The Data Warehouse Toolkit: The Definitive Guide to Dimensional Modeling. Wiley, Hoboken (2013)

15. LLAP - Apache Hive - Apache Software Foundation. https://cwiki.apache.org/confluence/display/Hive/LLAP. Accessed 07 Nov 2018
16. Druid Integration - Apache Hive - Apache Software Foundation. https://cwiki.apache.org/confluence/display/Hive/Druid+Integration. Accessed 07 Nov 2018
17. Ultra-fast OLAP Analytics with Apache Hive and Druid - Part 1 of 3, Hortonworks, 11 May 2017. https://hortonworks.com/blog/apache-hive-druid-part-1-3/. Accessed 07 Nov 2018

The Impact of Imbalanced Training Data on Local Matching Learning of Ontologies

Amir Laadhar[1(✉)], Faiza Ghozzi[2], Imen Megdiche[1], Franck Ravat[1], Olivier Teste[1], and Faiez Gargouri[2]

[1] Institut de Recherche en Informatique de Toulouse, Toulouse, France
amir.laadhar@irit.fr

[2] MIRACL, Sfax University, Sakiet Ezzit, Sfax, Tunisia

Abstract. Matching learning corresponds to the combination of ontology matching and machine learning techniques. This strategy has gained increasing attention in recent years. However, state-of-the-art approaches implementing matching learning strategies are not well-tailored to deal with imbalanced training sets. In this paper, we address the problem of the imbalanced training sets and their impacts on the performance of the matching learning in the context of aligning biomedical ontologies. Our approach is applied to local matching learning, which is a technique used to divide a large ontology matching task into a set of distinct local sub-matching tasks. A local matching task is based on a local classifier built using its balanced local training set. Thus, local classifiers discover the alignment of the local sub-matching tasks. To validate our approach, we propose an experimental study to analyze the impact of applying conventional resampling techniques on the quality of the local matching learning.

Keywords: Imbalanced training data · Machine learning · Ontology matching · Semantic web

1 Introduction

Biomedical ontologies such as SNOMED CT, the National Cancer Institute Thesaurus (NCI), and the Foundational Model of Anatomy (FMA) are used in the biomedical engineering systems [23]. These ontologies are developed based on different modeling views and vocabularies. The integration of these knowledge graphs requires efficient biomedical ontology matching systems [7,22]. The mapping between heterogeneous ontologies enables data interoperability.

Ontology mapping becomes a challenging and time-consuming task due to the size and the heterogeneity of biomedical ontologies. In this domain, Faria et al. [8] identified different challenges such as handling large ontologies or exploiting background knowledge. In addition to these problems, we highlight the importance of ensuring good matching quality while aligning large ontologies. Among the cited strategies to deal with large ontologies, we find search-space reduction techniques

W. Abramowicz and R. Corchuelo (Eds.): BIS 2019, LNBIP 353, pp. 162–175, 2019.
https://doi.org/10.1007/978-3-030-20485-3_13

encompassing two sub-strategies: partitioning and pruning [7]. The partitioning approach divides a large matching task into a set of smaller matching tasks, called partitions or blocks [23]. Each partition focuses on a specific context of the input ontologies. The ontology alignment process consists of aligning similar partition-pairs. The partitioning process aims to decrease the matching complexity of a large matching task. A line of work performs the partitioning process of pairwise ontologies to align the set of identified similar partitions (e.g., [1,4,9,11,26]). The state-of-the-art partitioning approaches use global matching settings (e.g., matchers choice, thresholds and weights) for all extracted partition-pairs [23]. Therefore, they do not employ any local tuning applied to each partition-pair to maximize the matching quality. Despite the existing work applying global matching process, we perform a local matching process over each extracted partition-pair. The local matching resolved by machine learning techniques, known as "local matching learning", requires a set of local training sets. Imbalanced training sets are one of the main issues occurring while dealing with ontology matching learning. Imbalanced data typically refers to a classification problem where the number of observations per class is not equally distributed [18]. Current matching learning work does not consider the resampling process to resolve the problem of imbalanced matching learning training data. Nonetheless, resampling is essential to deliver better matching learning accuracy.

In this paper, the main novelty is studying the impact of the imbalanced training data issue in the case of aligning large biomedical ontologies. Unlike state-of-the-art approaches, we automatically derive a local training set for each local matching learning classifier based on external biomedical knowledge resources. We automatically generate a local training set for each classifier without the use of any gold standard. Furthermore, we employ existing resampling methods to balance local training sets of local classifiers, and to align each partition-pair. Then, we evaluate the matching accuracy after applying different resampling techniques on the local matching tasks. To the best of our knowledge, there is a lack of works that automatically generate and resample matching learning training data. In sum, the contributions of this paper are the following:

- Local matching learning of biomedical ontologies;
- Automatically generating labeled local training sets, which are inferred from external biomedical knowledge bases to build local-based classifiers;
- Comparative study of the resampling methods to balance the generated local training sets.

The remainder of this paper is organized as follows: the next section presents the related work. Section 3 introduces preliminaries for the local matching approach and the local training set. Section 4 presents an overview of the proposed local matching architecture. In Sect. 5, we present our method for local matching learning. In Sect. 6, we perform a comparative study of the state-of-the-art resampling techniques in order to resample imbalanced local training sets. Finally, Sect. 7 concludes this paper and gives some perspectives.

2 Related Work

Faria et al. [8] identified the different challenges to align large biomedical ontologies. Ensuring good quality alignments while aligning these ontologies is challenging. To cope with these issues, we propose to employ matching learning techniques in order to fully automate the ontology matching process. Therefore, matching learning automates the alignment process while ensuring quality alignment independently from the matching context. In this section, we review the state-of-the-art matching learning strategies as well as the resampling techniques. Resampling methods could be applied to balance imbalanced training sets.

2.1 Ontology Matching Learning

There has been some relevant work dealing with supervised matching learning [6, 10, 19, 20]. Machine learning approaches for ontology alignment usually follow two phases [7]:

1. In the training phase, the machine learning algorithm learns the matching settings from a training set. This training set is usually created from the reference alignments of the same matching task.
2. In the classification phase, the generated classifier is applied over the input ontologies to classify the candidate alignments of the input ontologies.

Eckert et al. [6] built a meta-learner strategy to combine multiple learners. Malform-SVM [10] constructed a matching learning classifier from the reference alignments through a set of element level and structural level features. Nezhadi et al. [19] presented a machine learning approach to aggregate different types of string similarity measures. The latter approach is evaluated through a relatively small bibliographic matching track provided by the OAEI benchmark. Yam++ [20] defined a decision tree classifier based on a training set with different similarity measures. The decision tree classifier is built from the reference alignments and evaluated through the matching tasks. Nkisi-Orji et al. [21] proposed a matching learning classifier based on 23 features. Wang et al. [16], feed a neural network classifier with 32 features covering commonly used measures in the literature. A global classifier is built based on all the 32 features. Both approaches [16, 21] do not mention if the training set is balanced or not. Moreover, the training set is generated from the reference alignments.

Existing matching learning approaches build their machine learning classifiers from the reference alignments or derive it manually from a particular matching task [7]. We automatically generate a local training set for each sub-matching task. We do not use any reference alignments or user interactions to build the local training sets. Each local machine learning classifier is based on its local training set, which provides adequate matching settings for each sub-matching context.

2.2 Training Set Resampling

Classification problems often suffer from data imbalance across classes. This is the case when the size of instances from one class is significantly higher or lower relative to the other classes. A small difference often does not matter [18]. However, if there is a modest class imbalance in training data like 4:1, it can cause misleading classification accuracy. Imbalanced data refers to classification problems where we have unequal instances for different classes.

Most of machine learning classification algorithms are sensitive to imbalanced training data. An imbalanced training data will bias the prediction classifier towards the more common class. This happens because machine learning algorithms are usually designed to improve accuracy by reducing the error. Thus, they do not take into account the class distribution of classes. Different methods have been proposed in the state-of-the-art for handling the data imbalance [5]. The common approaches [3] to generate a balanced dataset from an imbalanced one are undersampling, oversampling, and their combination:

- Undersampling approach balances the dataset by reducing the size of the abundant class by keeping all samples in the rare class and randomly selecting an equal number of samples in the abundant class;
- Oversampling approach is used when the quantity of data is insufficient. It tries to balance dataset by increasing the size of rare samples. Rather than removing of abundant samples, new rare samples are generated;
- Performing a combination of oversampling and undersampling can yield better results than either in isolation.

Most of the state-of-the-art matching learning approaches neglect the problem of the imbalanced dataset. Existing work does not give any importance to resampling. However, the resampling method can strongly affect the obtained accuracy by the matching learning strategy. In this paper, we propose to resample training data in the context of local matching learning of biomedical ontologies. We apply the state-of-the-art resampling techniques on the local training data. Then, we study the impact of the applied resampling techniques on the local matching accuracy.

3 Preliminaries

In this section, we briefly present the fundamental definitions used in our work.

Definition 1 (Ontology partition). *An ontology partition* $p_{i,k}$ *of an ontology* $\mathcal{O}_i$ *is a sub-ontology denoted by* $p_{i,k} = (\mathcal{V}_{i,k}, \mathcal{E}_{i,k})$*, such as:*

- $\mathcal{V}_{i,k} = \{e_{i,k,1}, ..., e_{i,k,m_k}\}$, $\mathcal{V}_{i,k} \subseteq \mathcal{V}_i$. *is a finite set of classes of the ontology partition,*
- $\mathcal{E}_{i,k} = \{(e_{i,k,x}, e_{i,k,y}) \mid (e_{i,k,x}, e_{i,k,y}) \in \mathcal{V}_{i,k} \ \exists \ (e_{i,k,x}, e_{i,k,y}) \in \mathcal{E}_{i,k}\}, \mathcal{E}_{i,k} \subseteq \mathcal{E}_i$. *is a finite set of edges, where an edge encodes the relationship between two classes into an ontology partition*

Definition 2 (Set of ontology partitions). *An ontology $\mathcal{O}_i$ can be divided into a set of ontology partitions $\mathcal{P}_i = \{p_{i,1}, .., p_{i,s_i}\}$.*

- $\forall k \in [1..s_i], \mathcal{V}_{i,k} \neq \emptyset$;
- $\bigcup_{k=1}^{s_i} \mathcal{V}_{i,k} = \mathcal{V}_i$;
- $\forall\ k \in [1..s_i], \forall\ l \in [1..s_i],\ k \neq l,\ \mathcal{V}_{i,k} \cap \mathcal{V}_{i,l} = \emptyset$.

Definition 3 (Local-based training Set). *For each local matching $lm_{ij,q}$ of $\mathcal{LM}_{ij}$, we automatically generate a local training set denoted $ts_{ij,q}$. A local training set $ts_{ij,q}$ for a local matching task $lm_{ij,q}$ of $\mathcal{LM}_{ij}$ is denoted by $ts_{ij,q} = \{f_{i_1}, ..., f_{n_{i_q}}\}$. Each local training set $ts_{ij,q}$ contains a set of features associated with a prediction class attribute (match/not match). We denote $ts_{ij,q} = \{f_{ijq,1}, ..., f_{ijq,r}, c_{ijq}\}$. Since we are dealing with a binary classification task, $c_{ijq} \in \{0, 1\}$. A set of local machine learning classifiers is generated from a set of local training sets $\mathcal{T}s_{ij}$, denoted $\mathcal{T}s_{ij,q} = \{ts_{ij,1}, ..., ts_{ij,q}\}$.*

Hypothesis 1. *For a local matching $\mathcal{LM}_{ij}$ between two ontologies $\mathcal{O}_i$ and $\mathcal{O}_j$ associated with a set of imbalanced local training sets $\mathcal{T}s_{ij} = \{ts_{ij,1}, ..., ts_{ij,q}\}$, performing the adequate resampling technique improves the accuracy of the local matching learning.*

4 Local Matching Learning Architecture Overview

In Fig. 1, we depict an architectural overview of the local matching workflow. This architecture follows three modules: (i) ontology indexing and partitioning, (ii) local matching learning and (iii) alignment evaluation. We participated in the Ontology Alignment Evaluation Initiative (OAEI)[1] of 2018 using this architecture [14]. In the following, we describe the different modules of our proposal.

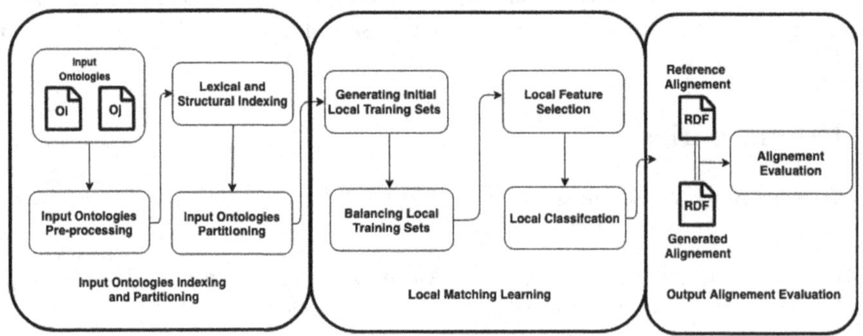

Fig. 1. Local matching architecture overview

[1] http://oaei.ontologymatching.org/2018/.

Input Ontologies Indexing and Partitioning. The input ontologies indexing is composed of two sequential steps: input ontologies pre-processing, and lexical and structural indexing. In the first step, we pre-process the lexical annotations. Thus, we apply the Porter stemming [15] and a stop word removal process over the extracted lexical annotations. In the second step, those lexical annotations are indexed. Moreover, we use structural indexing to store all the relationships between entities.

We then perform the partitioning of the input ontologies based on the approach of [13]. This partitioning approach is based on the Hierarchical Agglomerative Clustering (HAC) [17] to produce a set of partition-pairs with a sufficient coverage ratio and without producing any isolated partitions. Partitions with only one entity are considered as isolated. This partitioning process follows these steps:

1. Ontology partitioning processing: the HAC algorithm generates a list of structural similarity scores between all the entities of each input ontology. The structural similarity measure computes the structural relatedness between every pair of entities of one ontology.
2. Dendrogram construction: the HAC approach receives as an input the list of structural similarities of every input ontology. The HAC generates a dendrogram for every ontology. A dendrogram represents the structural representation of an input ontology.
3. Dendrogram cut: we cut the two generated dendrograms. A single cut of a dendrogram can result in a set of large partitions. To cope with this issue, we perform an automated multi-cut strategy of every resulted dendrogram. The multi-cut strategy results in a set of partitions for each ontology.
4. Finding similar partition-pairs: we find the set of similar partition-pairs between the two ontologies. These partition-pairs represent the set of local matching tasks, which will be employed by the next module.

The employed partitioning approach to generate local matching tasks is explained in depth in our previous research work [13].

Local Matching Learning. The local matching approach is based on generating a local classifier for each sub-matching task. The local classifiers are generated based on a set of local training sets composed of element and structural level features. Each local classifier is based on adequate features to align its sub-matching task. These features are automatically selected based on a feature selection. The local matching approach is composed of the following steps:

1. Generating Initial Local Training Set: Initial local training set are generated for each local matching task. These local training sets are not balanced.
2. Resampling of the Local Training Data: During this step, local training sets are balanced using conventional resampling techniques. Resampling aims to balance local training sets for better classification accuracy.

3. Wrapper Local Feature Selection: We apply wrapper feature selection over each resampled local training set. This local feature selection aims to choose the adequate features for each local matching task.
4. Local Classification: Local classification aims to classify the candidate correspondences of a local matching task to be aligned or not. This process is based on a set of local classifiers generated from the set of local training sets.

We will provide an in-depth description of these steps in the next section.

Output Alignment Evaluation. The generated output correspondences for every local matching task $lm_{ij,q}$ are unified to generate the final alignment file for the whole ontology matching task. The alignment file is compared to the reference alignment to evaluate the overall local matching $\mathcal{LM}_{ij}$ accuracy.

5 Local Matching Learning

In this section, we present the different steps of the local matching learning module depicted in the architecture of Fig. 1.

Generating Initial Local Training Set: Each local matching task has its own specific context. Therefore, a local matching task should be aligned based on its adequate matching settings, such as the weight of each matcher and its threshold. To cope with this issue, a local based classifier should be built for each local matching task. Therefore, we automatically construct a supervised training set $ts_{ij,q}$ for each local matching task $lm_{ij,q}$. These training sets serve as the input for each local classifier. Labeled data for the class attribute c_{ijq} are usually hard to acquire. Existing work construct the labeled data either from the reference alignments or by creating it manually [20]. However, the reference alignments commonly do not exist. We automatically generate the local training sets without any user manual involvement or any reference alignments. We derive the positive mappings samples (minority class) of the class attribute c_{ijq} by combining the results of two methods: cross-searching and cross-referencing. This combination allows the enrichment of the local training sets in order to cover a wide range of biomedical ontologies.

For a given local matching $lm_{ij,q}$, we generate:

- Positive samples $ps_{ij,q} = \{(e_i, q, x, e_{j,q,y})\}$, the total number of these positive samples is $\mathcal{N} = |ps_{ij,q}|$.
- Negative samples $ns_{ij,q} = \{(\mathcal{V}_{i,q} \times \mathcal{V}_{j,q}) \setminus ps_{ij,q}\}$. Therefore, the total number of negative samples is $\mathcal{M} = \mathcal{N}(\mathcal{N} - 1)$.

In the following, we present the cross-searching and the cross-referencing methods:

Cross-searching: Cross-searching employs external biomedical knowledge sources as a mediator between local matching tasks in order to extract bridge alignments. A bridge alignment is extracted if a similar annotation is detected

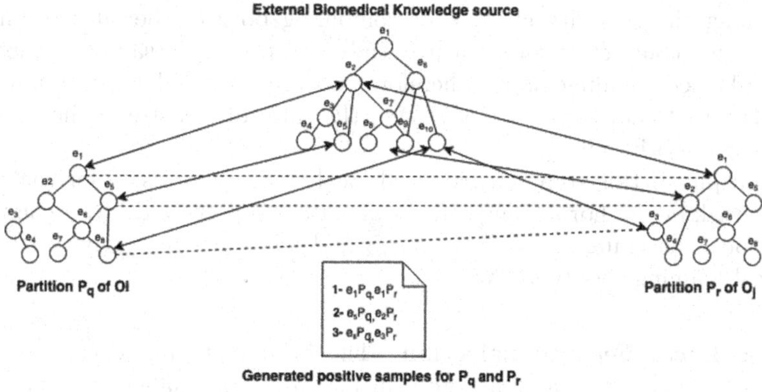

Fig. 2. Local training set extraction

between an entity of the bridge ontology and two entities of a local matching task. We consider bridge alignments as the positive samples. For example, in Fig. 2, we extracted the following positive samples $PS_{ij,q} = \{(\mathrm{e}_1\mathcal{P}_q, \mathrm{e}_1\mathcal{P}_r), (\mathrm{e}_8\mathcal{P}_q, \mathrm{e}_2\mathcal{P}_r), (\mathrm{e}_5\mathcal{P}_q, \mathrm{e}_3\mathcal{P}_r)\}$, with $\mathcal{N} = 3$. These labeled classes are generated by cross-searching the two partitions and an external biomedical knowledge base. Therefore, we deduce that the number of negative samples $\mathcal{M} = \mathcal{N}\ (\mathcal{N} - 1) = 6$ for the partition-pair $\mathcal{P}_q$ and $\mathcal{P}_r$ respectively from $\mathcal{O}_i$ and $\mathcal{O}_j$. *Cross-referencing*: we employ Uberon as an external biomedical knowledge source in order to derive positive samples for each local training set. Uberon is an integrated cross-species ontology covering anatomical structures and includes relationships to taxon-specific anatomical ontologies. Indeed, we explored the property "hasDbXref", which is mentioned in almost every class of Uberon. This property references the classes URI of external biomedical ontologies. We align every two entities of a given local matching task in case if one of their entities are both referenced in a single entity of Uberon. For example, the UBERON ontology includes references to different biomedical ontologies (via annotation property "hasDbXRef"). For instance, the class UBERON_0001275 ("pubis") of Uberon references the FMA class 16595 ("pubis") and NCI class C33423 ("pubic bone"). Therefore, the later entities construct a positive sample of a local training set [8].

Resampling of the Local Training Data: The training set is not balanced since the number of the negative samples $\mathcal{M}$ is higher than the number of positive samples $\mathcal{N}$. Therefore, we initially undersample each local training set $ts_{ij,q}$ by a heuristic method which consists of removing all the negative samples (majority class) having at least one element level feature equal to zero. The result of this initial treatment is not enough to balance the local training data, due to the high number of negative samples $\mathcal{M}$ (majority class) compared to the number of positive samples $\mathcal{N}$ (minority class). Hence, an additional sampling method is required to result in a balanced training set, we employ the state-of-the-art resampling methods to perform the undersampling of the majority class $\mathcal{M}$, oversampling the minority class $\mathcal{N}$ or combining both of the later technique. In Sect. 5, we conduct a comparative study of applying resampling methods over imbalanced training data. Therefore, we obtain a balanced training set ($\mathcal{N} = \mathcal{M}$). We Denote by $r = \|\mathcal{N}\|/\|\mathcal{M}\|$ the ratio of the size of the minority class to the majority class.

The output of this step is a balanced local training set $ts_{ij,q}$ for each local matching task $lm_{ij,q}$. For instance, if a local matching process $\mathcal{LM}_{ij}$ is composed of three local matching tasks: $lm_{ij,1}$ $lm_{ij,2}$ and $lm_{ij,3}$, we respectively result in three local training sets $ts_{ij,1}$ $ts_{ij,2}$ $ts_{ij,3}$.

Wrapper Local Feature Selection. The local matching $\mathcal{LM}_{ij}$ approach splits a large ontology matching problem into a set of smaller local matching tasks $lm_{ij,q}$. Each local matching task focuses on a specific sub-topic of interest. Therefore, it should be aligned based on its suitable features. We employ wrapper feature selection approaches in order to determine the suitable features for each local matching task $lm_{ij,q}$ among 23 structural-level and element-level features. Element level features consider intrinsic features of entities such as their textual annotations. Element level features refer to well-known similarity measure matchers, which can be classified into four groups: edit-distance, character-based, term-based and subsequence-based [13]. Structure level features consider the ontological neighborhood of entities in order to determine their similarity. We previously introduced all these features in a previous research work [13]. Feature selection is performed over each local training set $ts_{ij,q}$ in order to build local classifiers. The later identifies the local alignments of a local matching task $lm_{ij,q}$. For example for a given local training sets: $ts_{ij,1}$, $ts_{ij,2}$ and $ts_{ij,3}$, we separately perform the feature selection over these three local training sets.

Local Classification. Candidate correspondences of each local matching task $lm_{ij,q}$ are determined through performing the Cartesian product between the entities $\mathcal{V}_{i,q}$ and $\mathcal{V}_{j,q}$ of $lm_{ij,q}$. A local classifier classifies each candidate correspondence into a true or a false alignment. We build local classifiers using the Random Forest algorithm. We have selected Random Forest after comparing the efficiency of several machine learning algorithms for ontology matching learning [13].

6 Evaluation and Comparative Study

In this section, we evaluate the Hypothesis 1 "Resampling improves the accuracy of the local matching learning" by conducting a set of experiments. All experiments have been implemented in Java (weka library for classification) and Python (imblearn library for training sets resampling) on a MacOs operating system with 2.8 GHz Intel I7-7700HQ (4 cores) and 16 GB of internal memory. In the following subsections, we compare the accuracy of the commonly employed resampling methods (undersampling, oversampling and their combination) then we discuss the results. Our experiments are performed based on the dataset of the Evaluation Initiative of 2018, in particular, the ontology matching track of Anatomy.

6.1 Impact of Undersampling on Local Training Data

We evaluate four different methods of the widely employed undersampling method in order to balance the class distribution of the local training data. These methods are described as follows:

- **Random undersampling** is a non-heuristic method that aims to balance the class distribution through the random elimination of instances belonging to the majority class.
- **Tomek links** [24] removes unwanted overlap between classes where majority class links are removed until all minimally distanced nearest neighbor pairs are on the same class.
- **One-sided selection (OSS)** [12] aims at creating a training dataset composed only by "safe instances". This technique removes instances that are noisy, redundant, or near to the decision border. Similar to the other undersampling techniques, OSS removes only instances from the majority class.
- **Edited Nearest Neighbors** [25] method removes the instances of the majority class with prediction made by the K-means method is different from the majority class. Therefore, if an instance has more neighbors of a different class, this instance will be removed.

In Table 1, we depict the results of each undersampling method in terms of the obtained accuracy. The Edited Nearest Neighbors resulted in the highest F-Measure of 85.3%.

Table 1. Local matching accuracy for each undersampling method

Undersampling method	Precision	Recall	F-Measure
Random Undersampling	65.9%	86.4%	74.8%
Tomek links	93.7%	77.4%	84.8%
One-sided selection	93.7%	77.1%	84.6%
Edited Nearest Neighbors	93.4%	78.4%	85.3%

6.2 Impact of Oversampling on Local Training Data

In this section, we perform the oversampling of the minority class instead of performing the undersampling of the majority class. There are several oversampling methods used in typical classification problems. The most common techniques are SMOTE [2] (Synthetic Minority Oversampling Technique) and random oversampling method:

- **SMOTE** oversamples the minority class by taking each positive instance and generating synthetic instances along a line segments joining their k nearest neighbors

- **Random oversampling** is a non-heuristic method that aims to balance class distribution through the random elimination of instances belonging to the minority class.

In the following Table 2, we depict the results of the local matching $\mathcal{LM}_{ij}$ after performing the oversampling of the local training sets for each local matching task $lm_{ij,q}$ of $\mathcal{LM}_{ij}$. We show below the results of applying SMOTE with $k = 5$ as the number of nearest neighbors in order to achieve a ratio $r = 1$. We also perform the random oversampling with a ratio $r = 1$ We deduce from Table 2 that SMOTE outperforms the random oversampling method in terms of precision and F-Measure. We argue this result due to the randomly generated instances by the random oversampling method.

Table 2. Local matching accuracy for each oversampling method

Oversampling method	Precision	Recall	F-Measure
Random oversampling	70.8%	82.8%	76.3%
SMOTE	86.7%	78.8%	82.6%

6.3 Combination of Oversampling and Undersampling on Local Training Data

It is possible to combine oversampling and undersampling techniques into a hybrid strategy. Common state-of-the-art methods [3,18] include the combination of SMOTE and Tomek links, SMOTE and Edited Nearest Neighbors (ENN) or SMOTE and Random Undersampling. SMOTE is employed for the oversampling with a ratio r = 0.5 and each of the latter techniques is employed for undersampling. We evaluate these three methods by resampling the local training sets, then we perform the local matching process based on the generated local classifiers. In the following Table 3, we depict the results of each combination method. We deduce that the combination of SMOTE and Tomek Link results in the best accuracy in terms of F-Measure and Precision. The combination of SMOTE and Random Under sampling results in the best recall value due to the random nature of this approach. Therefore, it returns the highest number of alignments with the lowest precision compared to the other combination methods.

Table 3. Combining oversampling and undersampling techniques

Hybrid method	Precision	Recall	F-Measure
SMOTE + Random Undersampling	87.8%	81.3%	84.4%
SMOTE + ENN	88.4%	80.5%	84.3%
SMOTE + Tomek Link	92.0%	79.0%	85.0%

6.4 Discussion

According to the achieved results, we highlight the following points:

- The random undersampling and oversampling methods are unstable. A more deep study on the convergence of these methods can be investigated to argue their usage.
- We can observe in Table 3 that combining SMOTE with undersampling methods decreases the matching accuracy.

We deduce that for the matching learning context, undersampling methods outperform the other resampling methods. We argue this result since the undersampling method removes redundant instances rather than creating new synthetic instances like the oversampling of the minority class. We conclude that the undersampling using ENN [25] yields to the best Precision and F-Measure. ENN removes instances of the majority class with prediction made by K-means method which are different from the majority class. We mention that we combined the use of cross-searching and cross-referencing with external resources in order to construct local training sets. The impact of each method on the resulted matching quality can be investigated. We tend to validate the hypothesis that the quality could depend on the use of methods used in the construction of the training data sets.

7 Conclusion

Ontology matching based on machine learning techniques has been an active research topic during recent years. In this paper, we focus on the problem of imbalanced learning training sets in the case of Local Matching Learning. To the best of our knowledge, there is a lack of works that automatically generate and resample local matching learning training sets. To perform the resampling of the local training sets, we evaluate the common state-of-the-art resampling techniques in order to improve the classification performance by employing the best technique. Our comparative study shows that the undersampling methods outperform the oversampling methods and their combination.

In future work, we tend to automate the choice of the resampling method for each local matching task. We will also evaluate the impact of the external knowledge resource on the global quality of our output alignments. Finally, we plan to experiment on different domain-based ontologies.

References

1. Algergawy, A., Babalou, S., Kargar, M.J., Davarpanah, S.H.: SeeCOnt: a new seeding-based clustering approach for ontology matching. In: Morzy, T., Valduriez, P., Bellatreche, L. (eds.) ADBIS 2015. LNCS, vol. 9282, pp. 245–258. Springer, Cham (2015). https://doi.org/10.1007/978-3-319-23135-8_17
2. Chawla, N.V., et al.: SMOTE: synthetic minority over-sampling technique. J. Artif. Intell. Res. **16**, 321–357 (2002)

3. Chawla, N.V.: Data mining for imbalanced datasets: an overview. In: Maimon, O., Rokach, L. (eds.) Data Mining and Knowledge Discovery Handbook. Springer, Boston (2009). https://doi.org/10.1007/978-0-387-09823-4_45
4. Chiatti, A., et al.: Reducing the search space in ontology alignment using clustering techniques and topic identification. In: ICKC. ACM (2015)
5. de Souto, M.C.P., Bittencourt, V.G., Costa, J.A.F.: An empirical analysis of undersampling techniques to balance a protein structural class dataset. In: King, I., Wang, J., Chan, L.-W., Wang, D.L. (eds.) ICONIP 2006. LNCS, vol. 4234, pp. 21–29. Springer, Heidelberg (2006). https://doi.org/10.1007/11893295_3
6. Eckert, K., Meilicke, C., Stuckenschmidt, H.: Improving ontology matching using meta-level learning. In: Aroyo, L., et al. (eds.) ESWC 2009. LNCS, vol. 5554, pp. 158–172. Springer, Heidelberg (2009). https://doi.org/10.1007/978-3-642-02121-3_15
7. Euzenat, J., Shvaiko, P.: Ontology Matching, vol. 1. Springer, Heidelberg (2007). https://doi.org/10.1007/978-3-540-49612-0
8. Faria, D., Pesquita, C., Mott, I., Martins, C., Couto, F.M., Cruz, I.F.: Tackling the challenges of matching biomedical ontologies. JBS **9**(1), 4 (2018)
9. Hu, W., Qu, Y., Cheng, G.: Matching large ontologies: a divide-and-conquer approach. DKE **67**(1), 140–160 (2008)
10. Ichise, R.: Machine learning approach for ontology mapping using multiple concept similarity measures. In: 7th IEEE/ACIS (2008)
11. Jiménez-Ruiz, E., et al.: We divide, you conquer: from large-scale ontology alignment to manageable subtasks. In: Ontology Matching (2018)
12. Kubat, M., Matwin, S.: Addressing the curse of imbalanced training sets: one-sided selection. In: ICML 1997 (1997)
13. Laadhar, A., Ghozzi, F., Megdiche, I., Ravat, F., Teste, O., Gargouri, F.: Partitioning and local matching learning of large biomedical ontologies. In: ACM SIGAPP SAC, Limassol, Cyprus (2019, to appear)
14. Laadhar, A., Ghozzi, F., Megdiche, I., Ravat, F., Teste, O., Gargouri, F.: OAEI 2018 results of POMap+. In: Ontology Matching, p. 192 (2018)
15. Porter, M.F.: Snowball: a language for stemming algorithms (2001)
16. Wang, L.L., et al.: Ontology alignment in the biomedical domain using entity definitions and context (2018)
17. Müllner, D.: Modern hierarchical, agglomerative clustering algorithms. arXiv preprint arXiv:1109.2378 (2011)
18. More, A.: Survey of resampling techniques for improving classification performance in unbalanced datasets. arXiv preprint arXiv:1608.06048 (2016)
19. Nezhadi, A.H., Shadgar, B., Osareh, A.: Ontology alignment using machine learning techniques. IJCSIT **3**, 139 (2011)
20. Ngo, D., Bellahsene, Z.: Overview of YAM++—(not) Yet Another Matcher for ontology alignment task. Web Semant.: Sci. Serv. Agents World Wide Web **41**, 30–49 (2016)
21. Nkisi-Orji, I., Wiratunga, N., Massie, S., Hui, K.-Y., Heaven, R.: Ontology alignment based on word embedding and random forest classification. In: Berlingerio, M., Bonchi, F., Gärtner, T., Hurley, N., Ifrim, G. (eds.) ECML PKDD 2018. LNCS (LNAI), vol. 11051, pp. 557–572. Springer, Cham (2019). https://doi.org/10.1007/978-3-030-10925-7_34
22. Shvaiko, P., Euzenat, J., Jiménez, E., Cheatham, M., Hassanzadeh, O.: OM 2017. In: International Workshop on Ontology Matching (2017)

23. Stuckenschmidt, H., Parent, C., Spaccapietra, S. (eds.): Modular Ontologies. LNCS, vol. 5445. Springer, Heidelberg (2009). https://doi.org/10.1007/978-3-642-01907-4
24. Tomek, I.: Two modifications of CNN. IEEE Trans. Syst. Man Cybern. **6**, 769–772 (1976)
25. Wilson, D.L.: Asymptotic properties of nearest neighbor rules using edited data. IEEE Trans. Syst. Man Cybern. **2**(3), 408–421 (1972)
26. Xue, X., Pan, J.-S.: A segment-based approach for large-scale ontology matching. Knowl. Inf. Syst. **52**(2), 467–484 (2017)

A DMN-Based Method for Context-Aware Business Process Modeling Towards Process Variability

Rongjia Song[1,2(✉)], Jan Vanthienen[2], Weiping Cui[3], Ying Wang[1], and Lei Huang[1]

[1] Department of Information Management, Beijing Jiaotong University, Beijing, China
{rjsong, ywangl, lhuang}@bjtu.edu.cn
[2] Department of Decision Sciences and Information Management, KU Leuven, Leuven, Belgium
{rongjia.song, jan.vanthienen}@kuleuven.be
[3] State Grid Energy Research Institute, State Grid Corporation of China, Beijing, China
cuiweiping@sgeri.sgcc.com.cn

Abstract. Business process modeling traditionally has not paid much attention to the interactive features considering the dynamism of the environment in which a business process is embedded. As context-awareness is accommodated in business process modeling, decisions are still considered within business processes in a traditional way. Moreover, context-aware business process modeling excessively relies on expert knowledge, due to a lack of a methodological way to guide its whole procedure. Lately, BPM (Business Process Management) is moving towards the separation of concerns paradigm by externalizing the decisions from the process flow. Most notably, the introduction of DMN (Decision Model and Notation) standard provides a solution and technique to model decisions and the process separately but consistently integrated. The DMN technique supports the ability to extract and operationalize value from data analytics since the value of data analytics lies in improving decision-making. In this paper, a DMN-based method is proposed for the separate consideration of decisions and business processes, which allows to model context into decisions as context-aware business process models for achieving business process variability. Using this method, the role of analytics in improving some part of the decision making can also be integrated in the context-aware business process modeling, which increases the potential for using big data and analytics to improve decision-making. Moreover, a formal presentation of DMN is extended with the context concept to set the theoretical foundation for the proposed DMN-based method.

Keywords: Decision modeling · Context-aware business process · Process modeling · Process variability

W. Abramowicz and R. Corchuelo (Eds.): BIS 2019, LNBIP 353, pp. 176–188, 2019.
https://doi.org/10.1007/978-3-030-20485-3_14

1 Introduction

A business process model captures a set of logically related tasks to achieve a particular goal. BPM (Business Process Management) continues receiving significant attention, and building business processes accommodated to dynamic context is considered as a major challenge. Recent studies have explored context-awareness as a new paradigm and principle in designing and managing a business process [1, 2]. In order to model a context-aware business process towards variability, relevant contextual data needs to be integrated in a traditional process model as a key aspect of its data perspective, and its consequential process variants need to be modeled as well. Then the research question can be presented as "*What data constitutes the context of a business process and how does the context lead to business process variability?*".

Some researchers have attempted to answer this research question by proposing methods to organize different adaptation variants in a business process model, such as methods based on process invariants [3], context-based process variants [4], adaptation components [5] and instance fragments [6]. Nevertheless, relevant contextual data and resulting process variants, i.e., process fragments that possibly occur, have been determined according to expert knowledge. There is a lack of a methodological method to determine relevant data that constitutes the context of a business process and to bridge the gap between the dynamic context and consequent process variants.

Lately, BPM is moving towards the separation of concerns paradigm by externalizing decisions from the process flow. Most notably, the introduction of DMN (Decision Model and Notation) standard [7], provides a solution and technique to model decisions and the process separately but consistently integrated. Moving towards the context-aware BPM, decisions has been still considered within business processes in a traditional way, which impairs the maintainability, scalability, and flexibility of both processes and decisions, as well as the analytics capability of business processes. In this paper, a DMN-based method for context-aware business process modeling is proposed to provide a systematic solution of the research question, which allows to model context into decisions to achieve process variability.

2 Motivation and Related Work on Context-Aware Business Process Modeling

Up to date, many researchers have realized that context is vital to agile BPM and have paid attention to the content and characteristics of context-aware business processes [1, 8], approaches to model context-aware business processes [9–11], tools for supporting the integration of contextual data and business processes [12] and some prototype cases were presented as well. In particular, a number of research efforts have been undertaken to integrate contextual elements for extending the traditional notion of a business process to pursue flexible business processes [13]. Process flexibility is referred to as the capability to cope with externally triggered changes by modifying only parts of a business process instead of replacing it [14], without losing its "identity" [3]. Capturing variability in business process models can provide process flexibility. Typically, for a particular process type, a multitude of process variants exists, each of them being valid in a particular context [15]. Moreover, the context of a business process is identified as the extrinsic driver for process flexibility [14], which needs to be considered in the process modeling towards variability in BPM.

Moreover, Since the early 1960s, the context concept has been modeled and exploited in many areas of informatics [16]. Many researchers have tried to define the context concept in their own work. Context seems to be a slippery concept, which keeps to the periphery, and slips away when one attempts to define it [17]. Researchers understand context and the context of a business process in different ways according to their background. Consequently, context-aware process modeling relies on expert knowledge excessively, due to a lack of a methodological way to guide the whole procedure of context-aware business process modeling.

Context-aware business processes gather contextual information of a user and a business process and adapt their behavior accordingly [4]. A workaround that is often observed in modeling practices of a context-aware business process is that a contextual variable becomes an explicit condition of control flow leading to a decision point such as "check if the process occurs in the holidays". Moreover, rule-based approaches are commonly used for automating decision making to enable a business process to be aware of context since rules usually derivate from dynamic changes of context [9]. However, decisions are still considered within context-aware business process modeling, which impairs the capability of decisions on context-aware BPM. Hence, in this paper, decision modeling is utilized for modeling the context-aware business process in a methodological way with less reliance on experts.

3 DMN Modeling Technique

DMN is designed as a declarative decision language to model the decision dimension of a business process, which consists of two levels that are to be used in conjunction [7]. Figure 1 depicts key elements involved in the DMN Technique. One is the decision requirements level, represented by the DRD (Decision Requirements Diagram), which depicts requirements and dependencies between data and sub-decisions involved in the decision model. Figure 4 depicts an example of a DRD. These input data can be static or dynamic, which may be extracted directly from databases, sensors and IoT devices, or generated by data fusion, data analytics and machine learning. The other one is the decision logic level, presented by the Business Knowledge Model (BKM), which encapsulates business know-how in the form of decision rules or a decision table. Figure 5 is an example of the decision logic presentation. BKMs origin from some knowledge sources of authorities that can be guidelines, regulations or analytics systems. Analytics capability is therefore able to be integrated to improve the decision-making of a business process.

Organizations are increasingly investing in data-driven analytics to improve their business results, deepen customer understanding and better manage risk. The value of these analytics lies in improving decision-making. In other words, unless a decision is improved as a result of analytics it is hard to argue that the analytics have any value [18]. Explicit decisions from a business process can improve using analytics. If decisions are identified, modeled and understood, the potential for analytics to improve it would be much clearer. Data analytics can be used to add support for BKMs or data requirements of decisions. More specifically, outcomes of data analytics can be provided as parameters of decision logic (i.e., decision tables or decision rules) or data inputs for decision models. The DMN technique provides an effective solution for the separate consideration of the decision dimension from a business process and supports the ability to extract and operationalize value from data analytics.

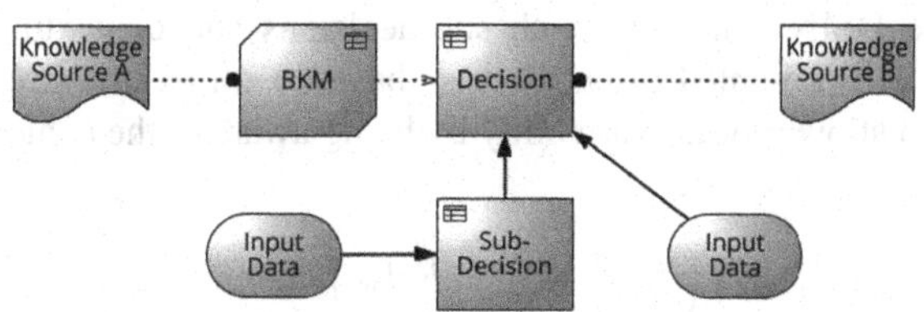

Fig. 1. Key elements of the DMN technique.

4 DMN Formal Presentation Extended with the Context Concept

We adopt the formal presentation of DMN constructs provided in [19] and extend it with sub-decisions and contextual data as the theoretical preliminaries of the proposed DMN-based method for modeling context-aware business processes.

In the DMN constructs, a decision in a business process can be formally defined as follows:

A decision in a process model, $d^a \in D_{dm}$ is a tuple (I_{da}, O_{da}, L_{da}), where $a \subseteq A_d$, $I_{da} \subseteq I(a)$, $O_{da} \subseteq O(a)$, and $L_{da} \subseteq L$.

The symbols in the formal presentation of a decision in a business process:

- d^a refers to a decision of decision type activities in a business process.
- D_{dm} refers to a finite non-empty set of decision nodes of a DRD, and an example is depicted in Fig. 4. DRD is a tuple (D_{dm}, ID, IR), consisting of decision nodes D_{dm}, a finite non-empty set of input data nodes ID, and a finite non-empty set of directed edges IR representing the information requirements such that $IR \subseteq (D_{dm} \cup ID) \times D_{dm}$, and $(D_{dm} \cup ID, IR)$ is a directed acyclic graph (DAG).
- $d^a \in D_{dm}$ is a tuple (I_{da}, O_{da}, L_{da}), presents that a decision d^a is one of decision nodes in a DRD D_{dm}, where $I_{da} \subseteq ID$ is a set of input symbols, O is a set of output symbols, and L is the decision logic defining the relation between symbols in I_{da} and symbols in O_{da}.
- A refers to a finite non-empty set of activities in a business process and $A_o \cup A_a \cup A_d = A$, where A_o is a finite non-empty set of operational activities ((no) inputs, no outputs), A_a is a finite non-empty set of administrative activities (no inputs, outputs), and A_d is a finite non-empty set of decision activities (inputs, outputs), which serve a decision purpose by transforming inputs into an outcome.

- $a \subseteq A_d$ presents that a is a type of decision activity and $I(a) \neq \emptyset \wedge O(a) \neq \emptyset$ indicates that there are inputs and outputs for the decision activity a.
- $I_{da} \subseteq I(a), O_{da} \subseteq O(a)$, respectively present the inputs and outputs of the decision d^a.
- $L_{da} \subseteq L$, presents the decision logic of the decision d^a. A Decision Table (DT) is commonly used for presenting the decision logic in DMN. In this case, a commonly used reasoning construct in decision models, L_{da} is the DT of decision d^a demonstrating decision rules, and I_{da} and O_{da} contain the names of the input and output elements respectively.

To extend the DMN constructs with sub-decisions and contextual data, the following theorems are presented as the formal basis to utilize the decision modeling technique DMN to allow process variability by being aware of the context of a business process.

Theorem 1. $\left(d^{a'} \in D_{dm}\right) \wedge \left(\left(d^{a'} \leftrightarrow d^a\right) = False\right)$

Note that not only data is the input of a decision activity, but sub-decisions are another source of input, such that, if $d^{a'}$ is also a decision node of the DRD of d^a, and $d^{a'}$ is not the same decision with d^a, then $d^{a'}$ is a sub-decision of d^a.

Theorem 2. $IR \subseteq (D_{dm} \times D_{dm}) \cup (D_{dm} \times ID)$

Moreover, $IR \subseteq (D_{dm} \cup ID) \times D_{dm}$ is equal to $IR \subseteq (D_{dm} \times D_{dm}) \cup (D_{dm} \times ID)$. More specifically, $D_{dm} \times ID$ presents a decision node has a data input, and $D_{dm} \times D_{dm}$ presents a decision node has a sub-decision input.

Theorem 3. $DI_{da} \cup DecI_{da} = I_{da}$

The inputs I_{da} of the decision d^a consists of data inputs DI_{da} and decision inputs $DecI_{da}$ (i.e., the sub-decision $d^{a'}$ of the decision d^a).

Theorem 4. $(IR_{da} \neq \emptyset) \wedge (DI_{da} \neq \emptyset)$

Any decisions including sub-decisions need data inputs, which means these decisions are not leaf (end) nodes of a DRD.

Theorem 5. $(MI_{da} \cup C_{da}) = DI_{da}$

MI_{da} is provided manually or intentionally, other leaf inputs of decisions (i.e., data inputs) are collected from the context C_{da} including raw contextual data (e.g., IoT sensors, GIS) or contextual variables that are obtained by aggregating and processing various sensor data.

5 A Proposed DMN-Based Method for Context-Aware Business Process Modeling

The main objectives of the proposed DMN-based method is to provide a mechanism and methodology for correctly modeling business processes that fit their contexts and thus are able to properly execute across different situations by using decision models. Figure 2 outlines the methodology of the proposed DMN-based method, which consists of three phases including: (a) integrated modeling of the base process, (b) modeling the process variants, and (c) modeling the context of the business process. In particular, context leads to process variability by affecting decisions of a business process, since contextual data is used to make decisions in gateways or calculate in decision activities. Hence, in a business process without decisions, context-awareness is normally not needed for the business process.

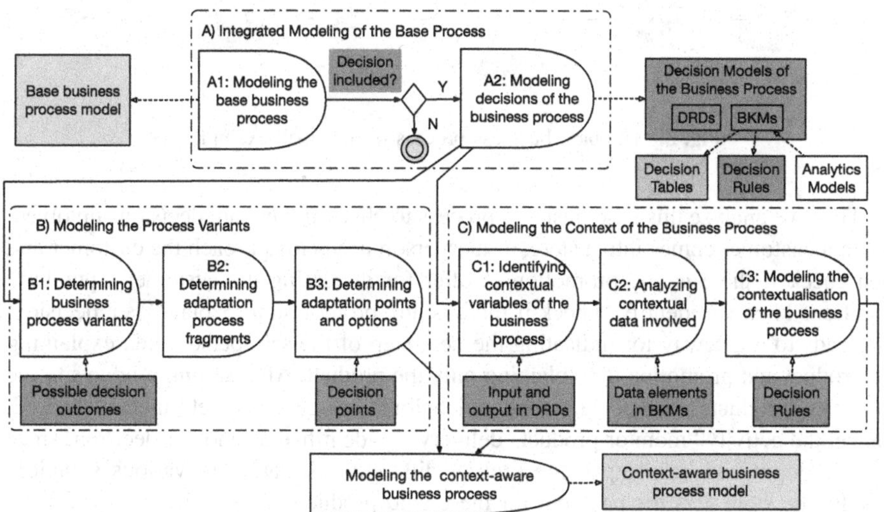

Fig. 2. The methodology of DMN-based method to model context-aware business processes.

5.1 Integrated Modeling of the Base Process

Firstly, the base process should be modeled in an integrated way, which means modeling decisions separately from the process flow and then obtaining a process-decision model for the initialization. In this way, decisions are not hidden in a business process as intricate control flows that can result in cascading gateways. The DMN technique is used to model the decision dimension of processes as decision models, including information requirements and the decision logic. After identifying and modeling decisions of a business process, the resulting model must be subsequently integrated with the process model. Consequently, an integrated process-decision model consists of the process workflow, DRDs presenting information requirements and the BKMs presenting the decision logic, which is the base to include context afterwards.

A1: Modeling the Base Business Process

Initially, the base business process is modeled in a traditional way. We propose a process of a customer buying a product as an example to depict how this DMN-based method works in applications. Figure 3 presents the base business process of this small example. In this case, we model the business process based on BPMN (Business Process Model and Notation) in a procedural way, but CMMN (Case Management Model and Notation) can also be used to combine with the DMN technique to model the decision-process integrated model in a declarative way, especially for flexible business processes.

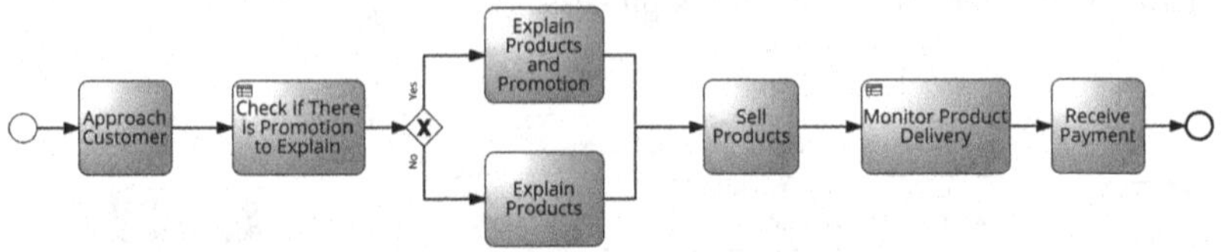

Fig. 3. The base business process of the small example.

Then we analyze this base business process to check if there are decisions involved. When a customer comes into a store, a salesperson needs to approach the customer and then check if there is any promotion to offer for increasing the customer's purchase intention. Then we identify "check if there is any promotion to explain" is a decision, and leads to a gateway for indicating the next step of the salesperson, i.e., explaining the product and promotion or explaining only the product. After selling products to the customer, products need to be monitored before the customer gets them delivered. Hence, the activity "monitor products delivery" is identified as another decision, since it needs input and has output constituting different calculation in various situation. Finally, the store gets the payment for these sold products.

A2: Modeling Decisions of the Base Business Process

After identifying decisions involved in the base business process, we need to model the decision dimension which consists of information requirements and the decision logic. Figure 4 depicts the DRDs of the base business process, which presents the information requirements level.

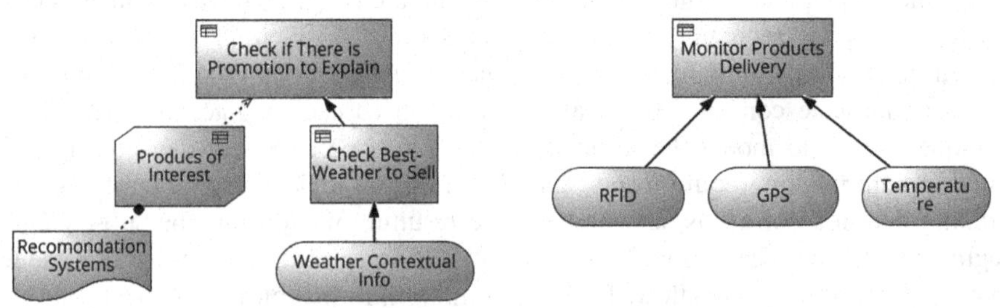

Fig. 4. The DRDs of the example.

The decision "monitor products delivery" needs "RFID", "GPS" and "temperature" as inputs to execute. More specifically, RFID can be used to invoke the identity information of products; GPS can be used to track the location of the delivery; and temperature can be needed to monitor the status of the delivery especially for fresh products such as milk or seafood. The other decision "check if there is promotion to explain" depends on products that the customer is interested in and bestsellers of different weather states which needs certain contextual information of weather as inputs. Furthermore, the BKM of "products of interest" is needed to provide customer's preference, which depends on the recommendation system. Outcomes of the recommendation system provide inputs to determine the products of interest in terms of a customer. The capability of data analytics can then be utilized for decision making in this business process.

Then the decision logic needs to be modeled, which includes the BKM of "products of interest", the decision table of Best-WeatherToSell, and decision rules for checking promotions. Figure 5 depicts the decision logic representation involving data analytics. More specifically, the BKM of "products of interest" uses data analytics techniques, which could include user-based filtering and content-based filtering to provide recommendation for products that the given customer could be interested. The Best-WeatherToSell needs follow several decision rules that are presented in a decision table. Making the decision of checking promotion needs the decision rule to connect all elements involved in the decision logic. In addition, the decision logic of "monitor products delivery" could be a calculation to produce an array to output the delivery situation.

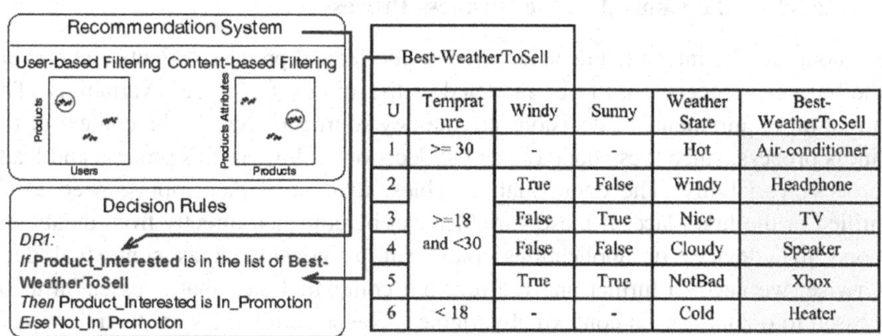

U	Temprat ure	Windy	Sunny	Weather State	Best-WeatherToSell
1	>= 30	-	-	Hot	Air-conditioner
2	>=18 and <30	True	False	Windy	Headphone
3		False	True	Nice	TV
4		False	False	Cloudy	Speaker
5		True	True	NotBad	Xbox
6	< 18	-	-	Cold	Heater

Fig. 5. The representation of the decision logic involving data analytics.

After these two steps, a process-decision model is obtained with a separate consideration of decisions and the business process.

5.2 Modeling the Process Variants

When the decision dimension of a business process is set, possible decision outcomes can be used to determine the process variability. Then possible process variants need to be modeled.

B1: Determining Business Process Variants

Different decisions outcomes lead to process variants. Especially strategic decisions, such as different requirements of production costs, quality control and safety monitoring, could trigger process changes. In this example, two process variants can be anticipated according to the outcomes of decision "check promotion", that are "promotions to explain" or "no promotion to explain" if we take the context of promotion into consideration.

B2: Determining Adaptation Process Fragments
Then some process fragments need to be modeled to optimize or complete the base business process for responding to different situations towards process flexibility, personalized services or knowledge-intensive tasks. In this case, the activity of checking promotion and the subsequent activity of explain products and the promotion is modeled as a process fragment.

B3: Determining Adaptation Points and Options
Decisions can lead to partial process splits as a decision point, or result in different calculations of a decision activity. Moreover, these process splits result in process variants of one process goal. Hence, we need analyze possible process splits of the business process to determine how to configure the base business process and process fragments due to different decision outcomes. The decision points are the key indicators of adaptation points. Moreover, the adaptation options could be "insert before", "insert after", "around", "delete", etc.

5.3 Modeling the Context of the Business Process

After modeling the integrated decision-process model and process variants, the context of the business process need to be modeled as the stimuli for process variability. The information requirements of decisions are the key source to identify the context of the business process, since these data affects the decision making in this process and leads to process variability. The contextual variables, i.e., direct data inputs, need to be identified in the first place. If these variables can be collected directly from databases, sensors, IoT devices or applications, they can be considered as contextual data. Otherwise, we need to further analyze relevant contextual data that is essential to be processed to obtain certain contextual variables. The semantic rules are then needed to be presented for context reasoning, which must be consistent with decision rules to guarantee the correctness of context interpretation. These identified contextual data and semantic rules need to be organized in an extendible way. Hence, context models may be needed especially in which the business process embedded is complicated.

B1: Identifying Contextual Variables of the Business Process
The contextual variables can be identified according to data inputs in DRDs of a business process. These contextual variables are required as parameters for executing gateways to choose a process branch or calculating in decision activities.

According to the inputs of the DRDs, we can identify relevant contextual variables of the base business process, which are "weather contextual info", "weather state", "products of interest", "best-weather to sell", "RFID", "GPS" and "temperature" in this example.

B2: Analyzing Contextual Data Involved
After identifying contextual variables, we need to examine if these variables can be directly collected. However, some contextual variables can only be obtained by context fusion or context reasoning, so we need to further identify contextual data of lower

semantic level needed. Thus, the decision logic is the key source to identify contextual data involved, since these rules are the logic to get the intermediate variables to make decisions in the business process.

In this example, "weather contextual info" and data required in the BKM of "products of interest" is still not clear. Through analyzing the decision table of Best-WeatherToSell, we know that contextual weather information including "temperature", "windy or not" and "sunny or not" is needed for the contextualization of the business process, i.e., components of the context of the business process. The contextual data needs also to be analyzed if the analytical model of the BKM of "products of interest" is provided.

B3: Modeling the Contextualization of the Business Process
After determining the context of a business process, we need to organize contextual data and their relations in a context model, especially when a business process is embedded in a complicated context. Since semantic hierarchies exist in the context, a reference technique is also needed for modeling context. Different techniques have been proposed until now to model context [20]. Although all techniques have advantages and disadvantages, the ontology technique is one of the best choices to model context, which also allows the reasoning technique embedded. Note that the context modeling technique needs to be compatible with the process-decision modeling technique in order to ensure the usability of the context model. Since the context of this example is simple, we don't provide details for the context model in this paper.

After these three steps, the context of the business process is identified and organized using the decision models.

5.4 Modeling the Context-Aware Business Process

In this small example, the process variants are simple as a split of two exclusive activities "explain products and promotion" and "explain products". However, the design of process variants can be complicated in real world.

In this case, context affects the business process in two points of decision activities. One in the decision "check if there is promotion to explain", which leads to a gateway for flow variants in the business process. The other one is the decision "monitor products delivery", which leads to different performance variants in the business process. Through applying the DMN-based method, we finally obtained the context-aware business process, which allows process variability. Figure 6 depicts the final context-aware business process model of this example.

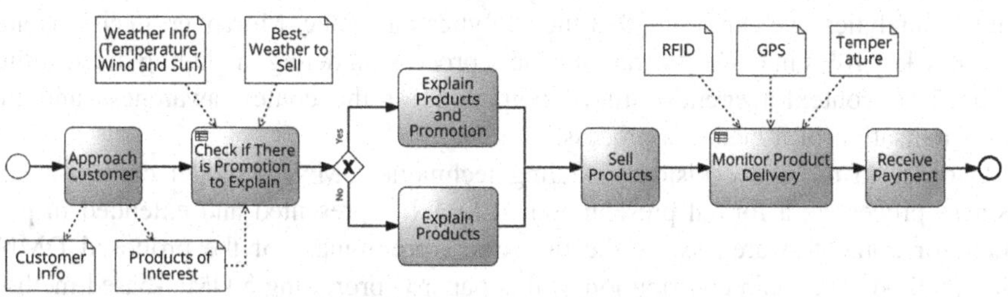

Fig. 6. The context-aware business process model of the example.

6 Discussion

Context is taken into consideration to model a complete and optimized business process at design time. Business process designers use decision models to integrate contextual data and anticipate possible situations for modeling a flexible or personalized business process.

Different types of context-dependent decisions could influence the context-aware business process from different perspectives, such as process flexibility, personalized services and knowledge-intensive tasks. As decision activities are more sensitive to context compared to operational activities and administrative activities in a business process, it is of importance to separately consider decisions from business processes, rather than considering decisions as a hidden and intrinsic flow in a traditional way of business process modeling. The DMN-based method proposed in this paper takes the separate consideration of decisions and processes which provides advantages of allowing process variability depending on dynamic context. It provides the methodology for the whole procedure of the context-aware business process modeling.

Context influences business processes and leads to process variability in both the flow and the performance levels. However, it is challenging to identify relevant contextual data of a given business process, especially in this data explosion world (ubiquitous and pervasive computing). This DMN-based method also provides a methodological way to guide the contextualization of a given business process, which can be used in any application domain. This also presents a straightforward path for researchers if they want to use decisions and rules to model context-awareness. Moreover, the capability of analytics can be integrated in the context-aware business processes and directly contributes to decision making based on the proposed DMN-based method.

7 Conclusion and Future Work

This work provides a DMN-based method for the separate consideration of decisions and processes, which allows to model context into decisions for achieving process variability.

Traditionally, decisions have been modeled as the hidden and intrinsic part of a business process, also in the context-aware modeling field. Such an approach impairs the maintainability, scalability, and flexibility of both processes and decisions, as well as the analytics capability of a business process. Only if a decision is improved as a result of analytics, we can argue that the analytics has value. Moreover, decisions and rules play key roles in context-aware business process modeling, including determining the need of context-awareness, the anticipation for the context-awareness and the contextualisation of a business process.

In order to use the decision modeling technique DMN to model context-aware business processes, a formal presentation of DMN is presented and extended in particular for context-awareness, as the theoretical preliminary of the proposed DMN-based method. The main contribution of this paper is proposing a DMN-based method to model context-aware business processes systematically with less reliance on experts.

Using this method, the role of analytics in improving parts of decision making can also be integrated in context-aware business process modeling. The more specific role for analytics increases the potential for using big data and analytics, which leads to decision-making improvement and the business process with that decision-making involved.

In future endeavours we will investigate a real-world case for applying the proposed method in more detail. Especially in the IoT paradigm, achieving context-awareness in business process modeling for process variability is significant. Moreover, context modeling techniques such as the ontology technique are interesting to involve in the future solution.

Acknowledgments. This research is supported by the Natural Science Foundation of China under Grant 71502010.

References

1. Rosemann, M., Recker, J., Flender, C.: Contextualisation of business processes. Int. J. Bus. Process Integr. Manag. **3**, 47–60 (2008)
2. Viaene, S., Schmiedel, T., Recker, J., vom Brocke, J., Trkman, P., Mertens, W.: Ten principles of good business process management. Bus. Process Manag. J. **20**, 530–548 (2014)
3. Regev, G., Bider, I., Wegmann, A.: Defining business process flexibility with the help of invariants. Softw. Process: Improv. Pract. **12**, 65–79 (2007)
4. Hallerbach, A., Bauer, T., Reichert, M.: Context-based configuration of process variants. In: 3rd International Workshop on Technologies for Context-Aware Business Process Management (TCoB 2008) Barcelona, Spain (2008)
5. Hermosillo, G., Seinturier, L., Duchien, L.: Creating context-adaptive business processes. In: Maglio, P.P., Weske, M., Yang, J., Fantinato, M. (eds.) ICSOC 2010. LNCS, vol. 6470, pp. 228–242. Springer, Heidelberg (2010). https://doi.org/10.1007/978-3-642-17358-5_16
6. Bucchiarone, A., Marconi, A., Pistore, M., Raik, H.: Dynamic adaptation of fragment-based and context-aware business processes. In: 2012 IEEE 19th International Conference on Web Services (ICWS), pp. 33–41. IEEE (2012)
7. OMG: Decision Model and Notation 1.2. (2018)
8. Recker, J., Flender, C., Ansell, P.: Understanding context-awareness in business process design. In: ACIS 2006 Proceedings, p. 79 (2006)
9. Bernal, J.F.M., Falcarin, P., Morisio, M., Dai, J.: Dynamic context-aware business process: a rule-based approach supported by pattern identification. In: Proceedings of the 2010 ACM Symposium on Applied Computing, Sierre, Switzerland, pp. 470–474. ACM (2010)
10. de la Vara, J.L., Ali, R., Dalpiaz, F., Sánchez, J., Giorgini, P.: COMPRO: a methodological approach for business process contextualisation. In: Meersman, R., Dillon, T., Herrero, P. (eds.) OTM 2010. LNCS, vol. 6426, pp. 132–149. Springer, Heidelberg (2010). https://doi.org/10.1007/978-3-642-16934-2_12
11. Frece, A., Juric, M.B.: Modeling functional requirements for configurable content-and context-aware dynamic service selection in business process models. J. Vis. Lang. Comput. **23**, 223–247 (2012)
12. Serral, E., De Smedt, J., Snoeck, M., Vanthienen, J.: Context-adaptive Petri nets: supporting adaptation for the execution context. Expert Syst. Appl. **42**, 9307–9317 (2015)

13. Nunes, V.T., Werner, C.M.L., Santoro, F.M.: Dynamic process adaptation: a context-aware approach. In: 2011 15th International Conference on Computer Supported Cooperative Work in Design (CSCWD), pp. 97–104. IEEE (2011)
14. Rosemann, M., Recker, J.C.: Context-aware process design: exploring the extrinsic drivers for process flexibility. In: The 18th International Conference on Advanced Information Systems Engineering. Proceedings of Workshops and Doctoral Consortium, pp. 149–158. Namur University Press (2006)
15. Hallerbach, A., Bauer, T., Reichert, M.: Capturing variability in business process models: the Provop approach. J. Softw. Maint. Evol.: Res. Pract. **22**, 519–546 (2010)
16. Coutaz, J., Crowley, J.L., Dobson, S., Garlan, D.: Context is key. Commun. ACM **48**, 49–53 (2005)
17. Dourish, P.: What we talk about when we talk about context. Pers. Ubiquit. Comput. **8**, 19–30 (2004)
18. Taylor, J., Fish, A., Vanthienen, J., Vincent, P.: Emerging standards in decision modeling. An introduction to decision model notation, pp. 133–146 (2013)
19. De Smedt, J., Hasić, F., vanden Broucke, S.K.L.M., Vanthienen, J.: Towards a holistic discovery of decisions in process-aware information systems. In: Carmona, J., Engels, G., Kumar, A. (eds.) BPM 2017. LNCS, vol. 10445, pp. 183–199. Springer, Cham (2017). https://doi.org/10.1007/978-3-319-65000-5_11
20. Perera, C., Zaslavsky, A., Christen, P., Georgakopoulos, D.: Context aware computing for the internet of things: a survey. IEEE Commun. Surv. Tutor. **16**, 414–454 (2014)

The Shift from Financial to Non-financial Measures During Transition into Digital Retail – A Systematic Literature Review

Gültekin Cakir(✉), Marija Bezbradica, and Markus Helfert

School of Computing, Dublin City University, Dublin, Ireland
{gueltekin.cakir,marija.bezbradica, markus.helfert}@dcu.ie

Abstract. Researchers in the retail domain today propose that, in particular, complex and non-financial goals such as 'customer experience' represent the new imperative and leading management objective in the age of Digital Retail, questioning the role of conventional financial measures such as revenue. However, there is no evidence in research showing the corresponding and necessary shift from financial measures to non-financial measures as subject of interest in recent years. This article aims to reveal the development of financial versus non-financial metrics used in retail research in the last ten years and thus highlight the transition from conventional retail into Digital Retail from a metrics perspective. A systematic literature review, conducted on the basis of 80 high quality journals, serves as the research method of choice and sheds light on the various range of metrics used in the last ten years of retail research. More importantly, the results still show a major focus on financial measures despite indicating a rising awareness of non-financial, more complex and intangible measures such as 'customer experience' or 'customer satisfaction'. While this finding supports proposed shift towards non-financial measures in current retail research in one side, it also shows a lack of research focusing on non-financial objectives in retail, in comparison to financial measures.

Keywords: Omni channel retail · Key performance indicator · KPI · Customer experience

1 Introduction

With a total revenue generation of around US$14,1 trillion per year (2017), the significance of today's retail industry is still apparent as it always has been and moreover, it even grew steadily to US$14,9 trillion in 2018 (Euromonitor 2018).

At the same time, retail is undergoing a major transformation in the process of becoming Digital Retail. Digital Retail can be understood as the digitalisation in the retail industry through digital technologies (Hansen and Sia 2015). Researchers argue that during this transformation, 'customer experience' resembles the new narrative and leading goal for retailers (Briel 2018; Lemon and Verhoef 2016; Pickard 2018; Verhoef et al. 2015) and is considered as a major differentiator for competitive advantage in Digital Retail (Verhoef et al. 2009). However, compared to the conventional retail

W. Abramowicz and R. Corchuelo (Eds.): BIS 2019, LNBIP 353, pp. 189–200, 2019.
https://doi.org/10.1007/978-3-030-20485-3_15

objectives such as revenue, sales growth or profit, customer experience represents a non-financial, non-monetary and complex objective to measure. Consequently, the transformation in retail also entails a potential transition from financial to non-financial measures. Ideally, this change would be reflected in retail research through emergence of more non-financial measures as subject of interest. In literature, there is no evidence showing this corresponding potential shift and thus providing support for the emergence of more intangible measures, underpinning the change in the retail industry. Objective of this paper is to investigate the development and use of financial versus non-financial measures throughout retail research in the last ten years. This is done through a systematic literature review on the basis of 80 high quality journals. The outcome will show the relation and significance of financial and non-financial measures in ten years of retail research and reveal changes regarding their significance. Moreover, a systematic overview of measures used in retail and retail research will be provided.

This paper is structured as follows: after a brief background and context description in section two, section three outlines how the systematic literature review was conducted. Descriptive results follow and a comprehensive discussion of findings is presented in section four. A conclusion, summarised contributions and future research as well as limitations complete the article in the last section.

2 Background

This section discusses and classifies financial and non-financial measures, followed by the related work in this field.

2.1 Financial and Non-financial Measures

In order to manage a business effectively and efficiently, the concept of performance measurement is utilized (Melnyk et al. 2014). This is done through measures quantifying and indicating progress and actions (Neely et al. 1995). Hereby, indicators can be distinguished by their measurability between financial and non-financial measures. Measures are considered as a financial metric in case of its financial nature, or can be measured directly on a monetary basis. For example, ‘revenue’ can simply be broken down to ‘price times goods sold’, or ‘revenue growth’ in percentage (relative measure) can be translated into actual sales (absolute measure). On the other side, non-financial measures are often described as abstract objectives and comprise of intangible, non-monetary and, in many cases, complex goals such as ‘customer experience’, ‘customer loyalty’ or ‘customer satisfaction’. They are difficult to measure because of their multidimensional nature (Kranzbühler et al. 2018; Lemon and Verhoef 2016) and are not able to be captured fully by current accounting measures (Ittner and Larcker 1998). Nevertheless, they are also relevant to evaluate progress of actions and to formulate necessary goals for retailers.

Further distinction can be made between strategic and operational measures. If used on top management decision level only, measures are considered as *strategic indicators* and are characterized by their long-term scope (e.g. year-to-year revenue growth, market share, employee size, change in profitability between time periods, etc.). Strategic measures are abstract and viewed as independent from industry (Zentes et al. 2017) to allow suitable performance comparison for shareholders. *Operational indicators* can be summed up as measures for evaluating day to day progress of actions such as in inventory management (e.g. stock accuracy, shelf capacity), fulfillment (delivery times, order times, etc.) or store metrics (store profit, salesperson net profit, etc.) (Gunasekaran et al. 2001) as well as metrics used in an e-commerce context (conversion rates, web visits, cost per click, etc.) (Tsai et al. 2013). The classification is shown in the following Table 1, along with examples.

Table 1. Classification of metrics.

Measurability	Management level	Metric example
Financial	Strategic	Retailer's shareholder value (e.g. Kashmiri et al. 2017)
	Operational	Cross space elasticity & shelf space (Schaal and Hübner 2018)
Non-financial	Strategic	Customer satisfaction, customer loyalty, service quality (e.g. Lin et al. 2016)
	Operational	Purchase Likelihood (Ho et al. 2014)

2.2 Related Work

Related work, covering development or classification of measures in a retail context is scarce. Nevertheless, some authors discuss metrics and metrics classifications.

Table 2. Related work.

Authors	Year	Context	Classification of metrics
Gunasekaran et al.	2001	Supply chain	Financial, non-financial, strategical, tactical, operational
Anand and Grover	2013	Supply chain	Transport, inventory, IT and resources
Glanz et al.	2016	Retail food store	Qualitative and quantitative indicators
Kumar et al.	2017	Retail	Market-related, firm-related, store-related, customer-related

For example, from a supply chain perspective, Gunasekaran et al. (2001) developed a framework, comprising strategical, tactical and operational level metrics. The authors argue that there is a need for more balance between financial and non-financial measures, as well as clear distinction between application level (strategic, tactical, operational). However, despite relation to retail, the focus of the study remains strongly oriented towards supply chain. Anand and Grover (2015) also identified and classified retail supply chain key performance indicators, providing a comprehensive overview of measures in the dimensions of transport, inventory, IT and resources. The broad overview gives valuable insights into supply chain operations and measures, however, the focus is not completely retail-related and there is no systematic development about measure evolvement. Glanz et al. (2016) conducted a review about retail food store environment measures concluded with specific and static collection of suitable measures. Although, similar research approach was applied here, the scope and focus are too specific to deduct conclusion for Digital Retail. Kumar et al. (2017) propose an organizing framework for retailers with four dimensions (Market, Firm, Store and Customer) where they map studies related to profitability and list applied measures from those articles. The article shows measures in different manifestations along the above mentioned dimensions (e.g. sales, ROI, MVA, CLV, customer experience or loyalty), however, despite providing an implicit strategic and operational differentiation of measures (strategy and actions), there is no explicit distinction between financial and non-financial measures and a specific timeline comparison or element. Related work is summed up in Table 2.

3 Method – Systematic Literature Review

A systematic literature review represents a vital method to extract knowledge and conduct critical reflection considering the vast amount and variety of sources (books, journals, conferences, etc.). Following the review approach as proposed by Webster & Watson, ensuring high quality and rigour in the reviewing process (Webster and Watson 2002), answers to the following research questions are pursued: (1) "How evolve financial and non-financial retail measures as subject of interest throughout the years in retail research?" and (2) "What measures were subject of interest throughout the years in retail research?".

This section is structured into two parts: (1) review approach and scope, explaining search strategy and process and (2) analysis framework, discussing the concept matrix used for article analysis.

3.1 Review Approach and Scope

This review is conducted in four steps (Fig. 1). We first identified relevant journals on the basis of Scopus, the abstract and citation database of Elsevier. Source material is retrieved from high quality journals covering a wide field of disciplines. We filtered those journals within the 5% citescore percentile of the database to ensure high level quality, measured by citation quality. Second, we applied a search string on each journal ('search in title and abstract'). The search string utilized "AND" and "OR" Boolean operators to link keywords and was set as follows: "TITLE-ABS-KEY(retail* AND measure* OR performance* OR goal* OR metric* OR KPI OR objective* OR indicator*)" – applied to the subject areas "Business Management Accounting", "Decision Science" and "Information Systems" of the Scopus database to include a wide and relevant coverage of disciplines. The search string covered the keywords "measure" and appropriate synonyms (KPI, metric, indicator). Additionally, it included the terms "performance", "goal" and "objective" to address a context of "performance-goal-relationship" in retail research.

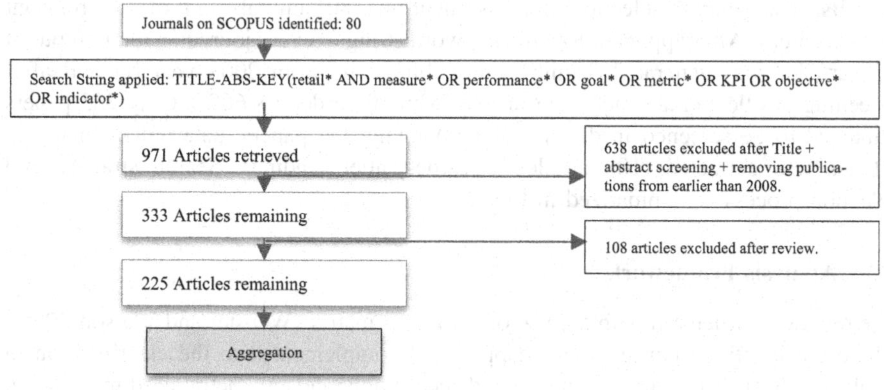

Fig. 1. Search strategy and selection process.

Third, all articles published earlier than 2008 were excluded – the review has a scope of ten years to cover an appropriate period of time for reflection. Also, non-English written articles and articles without any relevance on retail industry as the field of research or articles not providing a retail-performance measure-relationship were excluded as well. Excluded was for example a study about customer decision context ("shopper's goals") with no relation and discussion about subsequent retail performance (Guha et al. 2018). In the same step, all abstracts of the remaining articles were screened regarding context. The last step comprised all remaining articles for review and data extraction. Here, articles with no relevance to the research objectives were excluded as well. On- or offline retail focus in articles was treated equally, so there was no distinction between different retail channels in the reviewing process since Digital retail is defined by the merging of on- and offline channels into "omnichannel" to provide a seamless customer experience (Lemon and Verhoef 2016; Verhoef et al. 2015). Research in travel and bank retail was also considered. Inclusion and exclusion criteria are summed up in Table 3.

Table 3. Inclusion and exclusion criteria for the literature review.

Inclusion criteria	Exclusion criteria
• Article has retail metric(s) as subject of research interest, uses retail metric(s) in research design or conceptualizes a novel metric • Keywords: retail, measure, performance, goal, metric, KPI, objective, indicator • Top5 citescore percentile of journals on Scopus • Multidisciplinary sources • Research context regardless of on- or offline retail	• Non-English publication • Publication earlier than 2008 • No relation to retail industry in research context • Not providing a relationship between measures and performance

Results. The journal selection process identified 80 journals out of the relevant research areas. After application of the keyword string to every journal, a total output of 971 articles was generated. Limiting to publications from 2008 year on and after screening of title and abstracts, 638 articles were discarded (~66%). Only 333 papers remained to be screened in depth. After reviewing the papers, data out of final 225 articles was extracted (108 articles discarded after reading). Search strategy and selection process is summarized in Fig. 1.

3.2 Analysis Framework

The review is structured with the use of a concept matrix (Webster and Watson 2002), where the identified concepts are mapped and complemented to the classification in Table 1 in Sect. 2.1. Articles are screened accordingly and any metric used in research, as a measure of interest, a measure validating a framework or conceptualizing a new measure are looked at and classified accordingly to the criteria in Table 1.

The link between these types of measures lies in the cause-effect-relationship between non-financial and financial drivers. Non-financial drivers can be driven by financial drivers but can also lead to general strategic financial indicators such as Return on Investment or impact on Shareholder Value (Ittner and Larcker 2003). For example, long delivery times of an online retailer can affect customer satisfaction negatively and be reflected by declining sales and thus poor return on investment on shareholder perspectives. The concept matrix is shown in Table 4.

Table 4. Concept matrix.

Metric	Financial	Non-financial	Strategic	Operational
Author(s) X	X			X
Author(s) Y		X	X	
…				

4 Findings and Discussion

This section discusses the findings of the review in three parts, beginning with overviews of strategic and operational metrics and following with dynamic results (evolvement of financial/non-financial metrics throughout retail research).

4.1 Strategic Metrics

Findings. The majority of strategic financial indicators used in retail research is represented by 'sales'/'revenue', 'profit' and 'return on [investment]-measures' (e.g. return on investment, return on equity) and thus reflect a focus on accounting measures (see Table 5). Some articles use indicators such as 'cash flow' (e.g. in Kalaignanam et al. 2018; Lado et al. 2008) or 'market share' (e.g. in Shee et al. 2018; Srinivasan et al. 2013) and differentiate themselves by applying a liquidity or market related perspective.

Regarding non-financial indicators, trust, loyalty, satisfaction and other customer related measures resemble the major part of measures of interest. In addition, quality and image indicators are discussed consistently throughout retail research in the last ten years.

Discussion. The major focus on strategic measures lies in accounting measures. This is probably due to their simple measurability and ease to use and acquire. Non-financial indicators such as "Customer Channel Awareness" (Bell et al. 2018) reflect awareness for Digital Retail peculiarities (different channels in retail). Additionally, a surprising wide variety of articles discuss employee related indicators, showing significance of the organisational element in retail (e.g. leadership (Lee et al. 2011), racial diversity (Richard et al. 2017) or employee orientation (Tang et al. 2014)).

Table 5. Overview of strategic metrics used in retail research from 2008 to 2018.

	Financial metrics	Non-financial metrics
Strategic	• Sales/Revenue (total/annual/average/target/per channel/per store/per square foot/per salesperson/per customer/growth/return on/elasticities) • Profit (gross/margin/target/growth/per store/per salesperson/per product/per channel/per partnership) • Return on Sales (ROS), Return on Investment (ROI), Return on Equity (ROE), Return on Assets (ROA) • Return on advertising/Marketing spending/Ad spending • Cost of Goods Sold (COGS), Cost of investments, cost efficiency • (Relative) market share, market share growth • Tobin's Q, shareholder value (e.g. market to book ratio, stock returns) • Cash flow (net/volatility) • Order size, firm size/Employee size • Working capital turnover • Customer Lifetime Value (CLV) • Share of wallet • Stores (amount/growth/closing ratio)	• Consumer trust, loyalty (Consumer/Employee), customer engagement, customer experience, online experience, satisfaction (Customer/Employee) • Behaviour (Customer/Employee), customer orientation • Service and e-service quality, service image • Brand image, brand awareness, brand equity, brand profitability, corporate reputation • Price premium • Customer channel awareness • Competitiveness • Employee racial diversity, employee identification, social responsibility, leadership • Word of Mouth (WOM) and Electronic Word of Mouth (eWOM)

4.2 Operational Metrics

Findings. In terms of operational metrics, a variety of financial and non-financial measures were revealed (see Table 6). The majority of the indicators have productivity and/or efficiency purposes. Financial measures are for example represented by inventory performance (e.g. inventory inaccuracy, inventory variance or inventory costs), customer related measures (e.g. complaint rates, customer spending) or e-commerce measures (impression rates, web visits or e-mail response rates). Non-financial measures comprise mainly customer oriented, subjective indicators such as purchase intentions (Berry et al. 2015; Wen and Lurie 2018), likelihoods (Ho et al. 2014) or perceived value (Karpen et al. 2015).

Discussion. The wide variety signifies the complex nature of operations in retail. Besides established productivity indicators in the domain of supply chain (inventory, delivery, logistics), also web and store performance measures play a role, as well as indicators assessing organisational performance. However, it becomes evident that also non-financial indicators are well established in retail research on operational level.

Table 6. Overview of operational metrics used in retail research from 2008 to 2018.

	Financial metrics	Non-financial metrics
Operational	• Order cycle time • (Product) return rate • Productivity (per store/per salesperson) • Category share, category sales, share of shelf • Shelf space capacity, cross space elasticity • Store traffic • Purchase frequency per customer • Click through rate, impression rates, web visits, email response rates, SEA Price/Click • Customer complaints • Customer spending (shopping basket size), coupon redemption rate, customer trip revenue, customer recency rate, customer cross buying rate • Inventory costs, inventory inaccuracy, inventory variance ratio, inventory average • Promotion costs • Stock capacity, logistics costs (transportation/warehousing/ procurement) • Demand forecast, demand variance, forecasting period, lead time average • Order variance ratio • Salesperson salary, share of wages, employee absenteeism, labour hours	• (Re)purchase intentions • In-store experience • Store image, perceived showrooming, perceived product value • Customer referral behaviour • Consumer goodwill, convenience • Ease of use (Web) • Purchase uncertainty, purchase likelihood, product return likelihood • Order fulfillment quality • Service behaviour • Organizational climate, employee retention, employee motivation, employee compliance behaviour, salesperson emotional intelligence

4.3 Evolvement of Metrics

Findings. Overall, the development of metrics along the timeline from 2008 to 2018 shows a general growth in research with retail measures as subject of interest. Here, the increase of non-financial measures from 2015 to 2018 can be characterized as the most significant one. Also noteworthy is a peak in 2009, showing a year of high interest for retail measures. The relation between financial and non-financial measures can be described as unequally distributed. The focus in retail research lies in financial measures throughout the scope of the review, shown by a relation of ca. 55% to 45% in favour of financial metrics. However, it is observable that the gap is shrinking: in 2017, the interest in non-financial measures rose even more than in financial measures.

Discussion. On the one hand, the development shows a general rising emphasise on articles, incorporating retail measures, but on the other hand, a rising awareness for non-financial measures, compared to earlier years of publication, which is most significantly shown in 2018, is clearly visible. Consequently, findings show indeed a continuous shift towards more non-financial measures in the last ten years (Fig. 2).

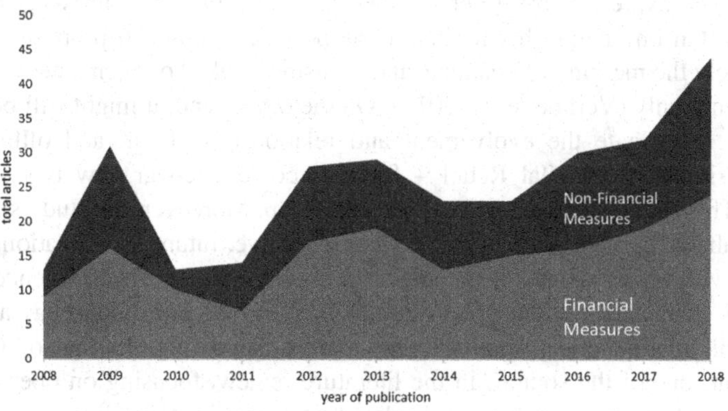

Fig. 2. Distribution and frequency of financial and non-financial metrics used in retail research from 2008 to 2018.

5 Conclusion

Following a systematic literature approach, this article investigated the development and forms of financial and non-financial measures throughout retail research in the last ten years to show a shift from financial to non-financial measures. Findings revealed an increase since 2015 in non-financial measures as subject of interest in retail research and support the proposition of authors, arguing the leading and rising role of non-financial measures, represented primarily by 'customer experience', in the transformation process of Digital Retail. This finding is also encouraged by rising awareness from the industry perspective. Here, it is observable that, for example, the British multinational retailer TESCO now incorporates the strategic non-financial measure

"Group Net Promoter Score" and measures assessing "employee trust" and "recommendation" in their annual reports (Tesco PLC 2018) – these indicators were not communicated before 2014. Another interesting indicator is proposed by KPMG: a new metric, conceptualised as "Experience per Square Foot", emphasises the rising role of 'customer experience' in Digital Retail (Pickard 2018). Regarding the second objective of this study, a wide range of different measures used in empirical or conceptual studies were uncovered, synthesised and classified according to strategic and operational level, showing an almost balanced variety between financial and non-financial measures. However, financial measures remain still a focus in retail research so far. A noticeable finding is also given by the emphasise on rather organisational metrics, for example, evaluating employee performance on different dimensions, e.g. satisfaction, service quality, behaviour or simply net profit. Also, other organisational elements such as climate, social diversity or compliance behavior are metrics of interest. This might appear as an interesting finding, considering the rising role of cultural mindset and organisational capabilities in Digital Retail which are also argued by several authors (Briel 2018; Homburg et al. 2017).

As with similar studies, this study involves limitations. To begin with, this study did not distinguish between on- and offline retail, considering the dissimilarity between retail channels, expressed by different operations and different measures (Beck and Rygl 2015). On one hand, this limitation supports the approach proposed in Digital Retail, namely the merging of channels and focusing on the "omnichannel perspective" as one channel only (Verhoef et al. 2015). On the other hand, it might still be a fruitful research to investigate the evolvement and relationship of on- and offline metrics during transition into Digital Retail – findings could uncover new type of metrics emerging. This study can be considered as a first step. Moreover, the study solely relied on one database. Despite Scopus being comprehensive, future investigations could be conducted with more variety of databases to support (e.g. Web of Science). Further limitation is characterised by the fact that this literature review did not pay attention to the potential dilution effect, caused by journal subjects, e.g. Journal of Operations Management, one of the sources in the literature review, focusing on operations only and thus providing mostly operational-related measures in its studies.

Acknowledgements. ■ This project has received funding from the European Union's Horizon 2020 research and innovation programme under the Marie Skłodowska-Curie grant agreement No. 765395, and was also supported, in part, by Science Foundation Ireland grant 13/RC/2094 and co-funded under the European Regional Development Fund through the Southern & Eastern Regional Operational Programme to Lero – the Irish Software Research Centre (www.lero.ie).

References

Anand, N., Grover, N.: Measuring retail supply chain performance - theoretical model using key performance indicators (KPIs). Benchmarking Int. J. **22**(1), 135–166 (2015)

Beck, N., Rygl, D.: Categorization of multiple channel retailing in multi-, cross-, and omni-channel retailing for retailers and retailing. J. Retail. Consum. Serv. **27**, 170–178 (2015)

Bell, D.R., Gallino, S., Moreno, A.: Offline showrooms in omnichannel retail: demand and operational benefits. Manage. Sci. **64**(4), 1629–1651 (2018)

Berry, C., Mukherjee, A., Burton, S., Howlett, E.: A COOL effect: the direct and indirect impact of country-of-origin disclosures on purchase intentions for retail food products. J. Retail. **91** (3), 533–542 (2015)

Von Briel, F.: The future of omnichannel retail: a four-stage Delphi study. Technol. Forecast. Soc. Chang. **132**, 217–229 (2018)

Euromonitor: Euromonitor Retail Market Sizes. Euromonitor International (2018)

Glanz, K., Johnson, L., Yaroch, A.L., Phillips, M., Ayala, G.X., Davis, E.L.: Measures of retail food store environments and sales: review and implications for healthy eating initiatives. J. Nutr. Educ. Behav. **48**(4), 280–288.e1 (2016)

Guha, A., Biswas, A., Grewal, D., Bhowmick, S.: An empirical analysis of the joint effects of shoppers' goals and attribute display on shoppers' evaluations. J. Mark. **82**, 142–156 (2018)

Gunasekaran, A., Patel, C., Tirtiroglu, E.: Performance measures and metrics in a supply chain environment. Int. J. Oper. Prod. Manage. **21**(1/2), 71–87 (2001)

Hansen, R., Sia, S.K.: Hummel's digital transformation toward omnichannel retailing: key lessons learned. MIS Q. Executive **14**(2), 51–66 (2015)

Ho, E., Kowatsch, T., Ilic, A.: The sales velocity effect on retailing. J. Interact. Mark. **28**(4), 237–256 (2014)

Homburg, C., Jozi, D., Kuehnl, C.: Customer experience management: toward implementing an evolving marketing concept. J. Acad. Mark. Sci. **45**, 377–401 (2017)

Ittner, C.D., Larcker, D.F.: Are nonfinancial measures leading indicators of financial performance? An analysis of customer satisfaction. J. Account. Res. **36**(1998), 1–35 (1998)

Ittner, C.D., Larcker, D.F.: Coming up short on nonfinancial performance measurement. Harvard Bus. Rev. **81**, 88–95 (2003)

Kalaignanam, K., Kushwaha, T., Rajavi, K.: How does web personalization create value for online retailers? Lower cash flow volatility or enhanced cash flows. J. Retail. **94**(3), 265–279 (2018)

Karpen, I.O., Bove, L.L., Lukas, B.A., Zyphur, M.J.: Service-dominant orientation: measurement and impact on performance outcomes. J. Retail. **91**(1), 89–108 (2015)

Kashmiri, S., Nicol, C.D., Hsu, L.: Birds of a feather: intra-industry spillover of the target customer data breach and the shielding role of IT, marketing, and CSR. J. Acad. Mark. Sci. **45**, 208–228 (2017)

Kranzbühler, A.M., Kleijnen, M.H.P., Morgan, R.E., Teerling, M.: The multilevel nature of customer experience research: an integrative review and research agenda. Int. J. Manag. Rev. **20**(2), 433–456 (2018)

Kumar, V., Anand, A., Song, H.: Future of retailer profitability: an organizing framework. J. Retail. **93**(1), 96–119 (2017)

Lado, A.A., Dant, R.R., Tekleab, A.G.: Trust-opportunism paradox, relationalism, and performance in interfirm relationships: evidence from the retail industry. Strateg. Manag. J. **29**(4), 401–423 (2008)

Lee, P.K.C., Cheng, T.C.E., Yeung, A.C.L., Lai, K.: An empirical study of transformational leadership, team performance and service quality in retail banks. Omega **39**(6), 690–701 (2011)

Lemon, K.N., Verhoef, P.C.: Understanding customer experience throughout the customer journey. J. Mark. **80**(6), 69–96 (2016)

Lin, Y., Luo, J., Cai, S., Ma, S., Rong, K.: Exploring the service quality in the e-commerce context: a triadic view. Ind. Manag. Data Syst. **116**(3), 388–415 (2016)

Melnyk, S.A., Bititci, U., Platts, K., Tobias, J., Andersen, B.: Is performance measurement and management fit for the future? Manag. Account. Res. **25**(2), 173–186 (2014)

Neely, A., Gregory, M., Platts, K.: Performance measurement system design - a literature review and research agenda. Int. J. Oper. Prod. Manag. **15**(4), 80–116 (1995)

Pickard, T.: KPMG - Global retail trends 2019. Global retail trends 2019 (2018)

Richard, O.C., Stewart, M.M., Mckay, P.F., Sackett, T.W.: The impact of store-unit-community racial diversity congruence on store-unit sales performance. J. Manag. **43**(7), 2386–2403 (2017)

Schaal, K., Hübner, A.: When does cross-space elasticity matter in shelf-space planning? A decision analytics approach. Omega **80**, 135–152 (2018)

Shee, H., Shah Jahan, M., Fairfield, L., Nyoman, P.: The impact of cloud-enabled process integration on supply chain performance and firm sustainability. Supply Chain Manag. Int. J. **23**(6), 500–517 (2018)

Srinivasan, R., Sridhar, S., Narayanan, S., Sihi, D.: Effects of opening and closing stores on chain retailer performance. J. Retail. **89**(2), 126–139 (2013)

Tang, C., Liu, Y., Oh, H., Weitz, B.: Socialization tactics of new retail employees: a pathway to organizational commitment. J. Retail. **90**(1), 62–73 (2014)

Tesco PLC: Welcome to our Annual Report (2018)

Tsai, J.Y., Raghu, T.S., Shao, B.B.M.: Information systems and technology sourcing strategies of e-Retailers for value chain enablement. J. Oper. Manag. **31**(6), 345–362 (2013)

Verhoef, P.C., Kannan, P.K., Inman, J.J.: From multi-channel retailing to omni-channel retailing - introduction to the special issue on multi-channel retailing. J. Retail. **91**(2), 174–181 (2015)

Verhoef, P.C., Lemon, K.N., Parasuraman, A., Roggeveen, A., Tsiros, M., Schlesinger, L.A.: Customer experience creation: determinants, dynamics and management strategies. J. Retail. **85**(1), 31–41 (2009)

Webster, J., Watson, R.T.: Analyzing the past to prepare for the future: writing a literature review. MIS Q. **26**(2), xiii–xxiii (2002)

Wen, N., Lurie, N.H.: The case for compatibility: product attitudes and purchase intentions for upper versus lowercase brand names. J. Retail. **94**(4), 393–407 (2018)

Zentes, J., Morschett, D., Schramm-Klein, H.: Strategic Retail Management - Text and International Cases, 3rd edn. Springer, Wiesbaden (2017)

Artificial Intelligence

A Maturity Model for IT-Related Security Incident Management

Gunnar Wahlgren[1(✉)] and Stewart Kowalski[2]

[1] Department of Computer and System Science, Stockholm University, Stockholm, Sweden
wahlgren@dsv.su.se

[2] Faculty of Computer Science and Media Technology, Gjøvik University College, Norwegian University of Science and Technology, Gjøvik, Norway
stewart.kowalski@ntnu.no

Abstract. The purpose of the study is to validate the ability of a maturity model for measuring escalation capability of IT-related security incident. First, an Escalation Maturity Model (EMM) and a tool were developed to measure the maturity of an organization to escalate IT-related security incidents. An IT tool for self-assessment was used by a representative from three organizations in the Swedish health sector to measure the organization's ability to escalate IT-related security incident. Second, typical security incident scenarios were created. The incident managers from the different organizations were interviewed about their organization's capabilities to deal with these scenarios. Third, a number of independent information security experts, none of whom had seen the results of EMM, ranked how the three different organizations have handled the different scenarios using a measurable scale. Finally, the results of EMM are compared against the measurable result of the interviews to establish the predictive ability of EMM. The findings of the proof of concept study shows that the outcome of EMM and the way in which an organization would handle different incidents correspond well, at least for organizations with low and medium maturity levels.

Keywords: Incident escalation · Incident management · Maturity models · Self-assessment

1 Introduction

Being unable to handle IT-related security incidents can have a devastating effect on an organization. In 2017, the ransomware incident in the UK health sector resulted in that operation being cancelled, ambulances being diverted, and patient records being made unavailable, among other things. When dealing with different incidents, the escalation of the incident to the right individual or groups of individuals is very important because the organization must react appropriately.

To deal with this problem the authors have developed an Escalation Maturity Model (EMM) to measure an organization's capability to escalate IT-related security incident and used this model in several public and private organizations in Sweden. The purpose of this proof of concept study is to evaluate how well the outcome of EMM

W. Abramowicz and R. Corchuelo (Eds.): BIS 2019, LNBIP 353, pp. 203–217, 2019.
https://doi.org/10.1007/978-3-030-20485-3_16

matches how an organization in the Swedish health sector would handle IT-related incidents in practice. The reasons for choosing the health sector for this study were, among other things, the effect that the ransomware incident had in the UK health sector. Furthermore, the health sector is a sector in which one expects that the ability to handle incidents should be established and tested in many organizations.

First, in this proof of concept study, three organizations in the Swedish health sector used the EMM ranking tool in their organizations. Second, several scenarios that include different IT-related security incidents were created. These scenarios were given to the incident managers of these three organizations. The incident manager's descriptions of how their organization would handle the escalation in these scenarios were documented. This documentation was given to 37 independent information security experts. These experts used these descriptions and ranked the escalation maturity level of these three different organizations. Finally, the experts ranking was compared to the EMM tools ranking. It was found that there was a near match between the expert's ranking and the EMM's tool ranking. This indicates the predictive ability of EMM.

2 Background

2.1 Risk Management

The International Organization for Standardization (ISO) [1] has established a standard for information security management. The term IT risk management refers to approaches and methods that lead to cost-effective security solutions. This is done by measuring the security risk to IT systems and then assuring adequate levels of protection. IT security risk management is a continuous process and consists of the following three main steps: (i) risk monitoring, (ii) risk assessment with risk treatment, and (iii) risk communication. The National Institute of Standards and Technology (NIST) has introduced the framework of enterprise-wide risk management using three different levels (tiers) to look at an organization. This multitier concept is described in many of NIST's publications [2–5]. Risk management decisions about IT security are made at all these levels. The three levels are top management (tier 1), middle management (tier 2), and operational staff (tier 3). Top management's decisions are often strategic in nature, while middle management's decisions are tactical. Staff, on the other hand, must deal with real IT security risk incidents and often must react directly to them.

2.2 Escalation

How risk escalation is handled is one of the most important aspects of IT security risk communication. In common language, the term "escalation" is often used in relationship to political and military conflicts [6]. The authors use the term in a slightly different way. Specifically, the term is used when one organizational level seeks assistance or informs a higher level about an issue it cannot handle. Each level must consider when an incident occurs, if the incident would harm the acceptable risk level of the organization. There are three basic alternatives for each level: (i) accept the risk, (ii) mitigate the risk (risk treatment), and (iii) escalate the risk to a level above.

Different reasons can be used to justify escalation. One reason could be budgetary considerations, for example, if implementing new expensive countermeasures is required. Another reason could be that the incident is so serious that it is necessary to get help from a higher organizational level. The organization must of course respond if a crisis occurs and recover from any damage the incident has caused. When an incident does not require immediate action, escalation could mean that new countermeasures need to be installed later if similar incidents are likely to reoccur.

2.3 Incident Management

The number of IT security incidents could be reduced with the help of a sound risk management program. However, some incidents can neither be avoided nor anticipated. Therefore, organizations need incident management to detect an incident quickly, mitigate the impact, and restore services in a trustworthy manner. Various standards and guidelines, such as ISO 2035 [7], ITIL [8], and NIST [9], describe best practices for effective and efficient incident management. Palilingan and Batmetan [10] describe how the ITIL framework is used for incident management in academic information systems.

An IT-related security incident could be, for example, disruption in software and hardware, loss of data, external attacks, or human errors in handling. In the ISO 2035 standard, an information security incident is defined as a single or a series of unwanted or unexpected information security events that have a significant probability of compromising business operations and threatening information security.

2.4 Maturity Models

Nolan [11], in 1973, was the first to present a descriptive stage theory which many consider to be the conceptual genius of maturity models. The stage theory concerns planning, organizing, and controlling activities associated with managing the organization's computer resource. Nolan developed a model with different stages of growth. The capability maturity model (CMM) was first described by Humphrey et al. [12]. They used their maturity models to assess the software engineering capability of contractors and identify five different maturity levels. In the ISO/IEC Technical Report 15504-7, organizational maturity is defined as "an expression of the extent to which an organization consistently implements processes within a defined scope that contributes to the achievement of its business goals (current or projected)" [13]. Pöppelbuß and Röglinger [14] identify three design principles for maturity models: (i) descriptive, (ii) prescriptive, and (iii) comparative. Philips [15] describes how to use a CMM to derive security requirements. ISACA [16] presents how maturity models could be used to recognize which maturity levels different IT security risk management processes are on in an enterprise or organization. Aguiar et al. [17] describes the use of the maturity model in incident management.

3 Research Methods

The overall research aim is to build and evaluate an artifact that organizations can use to measure their capability to handle escalation of IT-related security incidents. A design science approach has been chosen. The main reason this methodology was chosen was that the IT risk landscape is continuously evolving, which creates a systemic need for an organization to have a reusable tool in place to measure its escalation capability to deal with risk.

According to Vaishnavi and Kuechler [18], design science research methodology consists of five steps. In the first step, information about real-world problems is collected. Step two is a tentative design. In the third step, an artifact is developed. The artifact is evaluated in the fourth step, with the help of performance measures. In the last step, the design processes are completed, and conclusions are drawn. The research plans have been divided into three cycles [19]. All steps are not used in all cycles. In the last two cycles only step 3, 4 and 5 are used.

In the first cycle, version 1 of EMM was constructed. Version 1 was evaluated with the help of security specialists from both the private and public sector to generate a second version. In the second cycle, version 2 of EMM was tested on different Swedish organizations. Two tests were made [20]. First, version 2 was tested on two of Sweden's largest banks. Next, version 2 was tested on several other Swedish organizations. In the third cycle, version 3 of EMM was developed and tested, which is described later in this paper.

4 Approach

4.1 The Difference Between the Versions

In version 1 of EMM, only the maturity levels for the different maturity attributes were defined. The main change in version 2 of EMM was that EMM had been completed with a query package for self-assessment that made it possible to determine the different maturity levels manually, using the answers to the various questions. Regarding version 3 of EMM, besides developing a PC-based tool, the number of maturity attributes had increased to eight, and the query package was expanded with a number of questions that mainly concerned privacy issues.

4.2 EMM Version 3

As Fig. 1 shows, the maturity model consists of a matrix with different maturity levels as rows and different maturity attributes as columns. ISACA's [16] maturity attributes have been used as a starting point, but they have been adapted around escalation of IT-related security incidents. The same five maturity levels as Humphrey [12] have been used and, as ISACA, a sixth level, "non-existent," has been added. When the authors selected the different attributes they think are essential for assessing an organization's capability to escalate different types of IT-related security incident, the following approach was used.

Level \ Attribute	A Awareness	B Responsibility	C Reporting	D Policies	E Knowledge	F Procedures	G Means	H Structure
0 Non-existent								
1 Initial								
2 Repeatable								
3 Defined								
4 Managed								
5 Optimized								

Fig. 1. Escalation maturity model.

First, the incident must be detected. If this is possible, the person in charge must be **aware** that it is an IT-related security incident. To be aware, **knowledge** of different incidents is required. It is then necessary that the person in charge knows his/her **responsibility** for further handling the incident. The next step consists of handling the incident, which means that **procedures** must be in place to guide the correct behavior. These procedures must be anchored in a **policy** that the management defines. If the incident is escalated directly, the person in charge must know to whom, that is, pre-defined groups (organizational **structure**) that can handle the incident must exist. If the incident is escalated later, this must be **reported** to the management. **Means** such as appropriate risk analysis methods for analyzing incidents must be available.

The maturity model for escalation capability has six different maturity levels.

0. Non-existent means that different processes are not applied, and there is no need for any kind of measures.
1. Initial means that the need for measures has been identified and initiated, but the processes that are applied are ad hoc and often disorganized.
2. Repeatable means when measures are established and implemented, and the various processes follow a regular pattern.
3. Defined means when measures are defined, documented, and accepted within the organization.
4. Managed means that the processes are monitored and routinely updated.
5. Optimized means that processes are continuously evaluated and improved using various performance and effective measures that are tailored to the organization's goals.

Eight different maturity attributes are identified.

A. Awareness deals with aspects of how aware employees are of various IT-related security incidents.
B. Responsibility deals with allocation of responsibilities within the organization for IT-related security incidents.
C. Reporting deals with the reporting channels and how regular reporting of IT-related security incidents is done.

D. Policies deal with different policies for IT-related security incidents.
E. Knowledge deals with the different skills and knowledge needed to handle IT-related security incidents.
F. Procedures deal with various procedures for handling IT-related security incidents.
G. Means deal with various tools for handling IT-related security incidents.
H. Structure deals with various predefined groups for handling IT-related security incidents.

4.3 Query Package for EMM Version 3

Together with EMM, a query package to support self-assessment was also developed. The query package will be used by the organization to answer the different questions in the package. The answer to each question is "Yes" or "No." When all the questions have been answered, it is possible to determine the maturity level of the different maturity attributes. The number of questions in EMM version 3 is 67.

4.4 IT Tool

Version 3 of EMM has been used to develop a PC-based tool. The tool will be used by an organization in the self-assessment process to enter answers to the questions in the query package. The tool includes a help function to assist the organization when answering the questions. When all the questions are answered, the tool will automatically calculate the maturity level of each attribute. The tool will also suggest what action the organization could take when answering "No" to a question.

5 Proof of Concept Study

EMM is intended to be used and has been used by organizations in all sectors of society. However, this study concentrates on organizations within the Swedish health sector. The Swedish health sector is organized among the state, county councils, and municipalities. The state is responsible for the overall health policy. For this purpose, the state has several different agencies. The county council is responsible for various hospitals and health centers. The municipalities are responsible for the care of the elderly and support for those whose treatment is completed. The study involves three different organizations in the health sector. The reason for the choice was organizations belonging to different levels in the health sector and that have extensive IT operations and that, among other thing, have an appointed incident manager. Organizations 1 and 2 are government agencies, while organization 3 is a hospital run by a county council. The study was conducted in spring 2017.

The study is divided into two parts. In the first part, a representative from the different organizations used EMM together with the IT self-assessment tool to measure the organization's maturity level to handle IT-related security incidents. The representatives belong to the tactical level and work, for example, as information security managers and therefore should be the persons who have the overall knowledge of the different attributes of EMM, which in most cases handles different organizational aspects.

In the second part, an incident manager at the operational level, who of course is the person who normally handles various incidents, was interviewed about how the organization would handle a number of fictional incident scenarios. The results of the interviews were summarized and documented. This summary was given to 37 independent information security experts. These experts used these descriptions and ranked the escalation maturity level of these three different organizations. Finally, the experts ranking was compared to the EMM tools ranking. It was found that there was a near match between the expert's ranking and the EMM's tool ranking. This indicates the predictive ability of EMM.

5.1 Part 1 of the Study: Use of EMM

Figure 2 presents the outcomes of the first part. The figure shows how well the different organizations meet the maturity levels for the different attributes. The maturity level of the different maturity attributes for Organization 1 shows a slightly mixed picture. Only three of the attributes reach the "Defined" level, while all other attributes end up at a lower level. For example, for the attribute "Responsibility," the answers from the IT security manager show that the division of responsibility between different categories of employees is not clear, which is a prerequisite for reaching the "Initial" level. Another example is the maturity attribute "Reporting," which shows that no regular reporting routines to the organization's management occur, which is a prerequisite for reaching the "Repeatable" level.

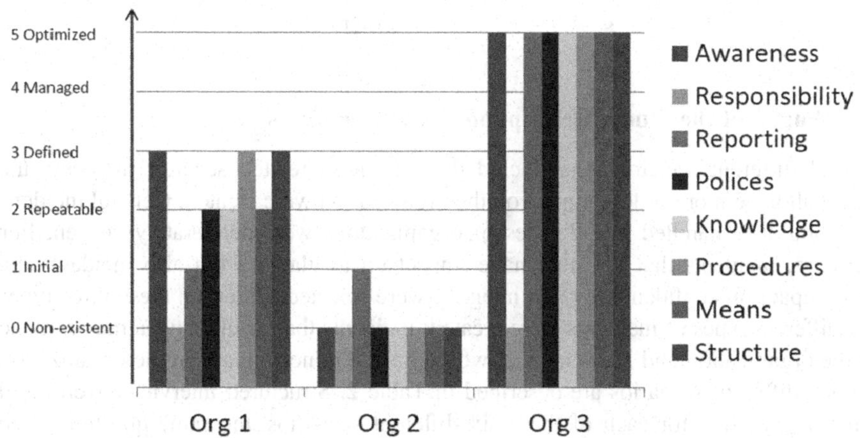

Fig. 2. Maturity levels for the different attributes.

For Organization 2, the responses from information security officers show that for most of the maturity attributes, the organization does not even reach the lowest maturity level "Initial." Only the maturity attribute "Reporting" reaches the level "Initial."

Because the answers to most of the questions are "No," EMM indicates that the organization has a very low level of maturity for handling IT-related security incidents.

For Organization 3, the information security manager answered "Yes" to most of the questions in EMM, which means that the organization meets the highest maturity level "Optimized" for most of the maturity attributes. Only the maturity attribute "Responsibility" shows a deviating value, ending up at the maturity level "Non-existent" because the division of responsibility between different categories of employees is not clear.

For each of the organizations, the average of the different attributes is calculated, where an attribute having the maturity level "Non-existent" has the value 0, while an attribute having the maturity level "Initial" has the value 1, the maturity level "Repeatable" has the value 2, and so on. The result of the calculation shows the average for the different organizations from 0 to 5 in Table 1. Then the result from EMM is converted to a mutual scale and transformed into the different maturity levels as follows.

1. Non-existent to Initial {0, 2} as low maturity level
2. Repeatable to Defined {2, 4} as medium maturity level
3. Managed to Optimized {4, 5} as high maturity level

Table 1. Result from EMM.

Organization	Average	Mutual scale
A	2.0	Medium
B	0.1	Low
C	4.4	High

5.2 Part 2 of the Study: Description of the Interviews

Several different scenarios that included different IT-related security incidents that would all have more or less impact on the organization were created. General incidents that should be handled by all types of organizations were deliberately chosen. Furthermore, incidents that not only have impact on availability but also incidents that have impact on confidentiality and integrity were selected. Based on these three types, two different types of incidents were created: incidents that would have a major impact on the organization and incidents that would have a minor impact on the organization. The six different scenarios are described in Table 2. Structured interviews were used, which means that for each of these six different scenarios, the same questions were asked of the incident manager at the different organizations. The following questions were used.

- How was the incident detected?
- Are the employees aware that an incident occurred?
- In what way do they know that they are responsible for the incident?
- How is the effect of the incident analyzed?

Table 2. Description of the different scenarios.

Impact	Major impact	Minor impact
Availability	**Incident 1.** For the business, vital systems cannot be run due to fire at data providers (internal or external) for long periods of time	**Incident 2.** Overload attacks (denial of service) prevent the running of vital systems for the business for a shorter time
Confidentiality	**Incident 3.** People outside the organization have access to a sensitive database containing personal data, and this has been so for a long unknown time	**Incident 4.** By accidental registration of access permissions, some employees have gained access to a number of data in a database to which they are not entitled
Integrity	**Incident 5.** Data in a database have been corrupted due to a program error, and this has been in progress for a long time	**Incident 6.** Due to a previous interruption of a system, data have been registered manually. When the data are entered into the system later on, some information will be lost

- Do the employees know if the incident should be escalated?
- Do the employees know how to handle the incident?
- Do the employees know how to report the incident?
- Does the organization have predefined groups that handle different incidents?
- Are any resources available if a serious incident occurs?

The incident managers from the different organizations were interviewed about how the organization would handle the various scenarios described in Table 2. The results of the interviews were documented and were verified with each incident manager. First, a general summary is provided of how the organization handles IT-related security incidents. Then how the organization would handle the incident in question is briefly described for each scenario.

Organization A. Only the employees of the IT department seem to be aware of the different types of IT-related security incidents. The employees have received training and know the divisions of responsibility because different roles are defined and documented. Documented procedures are available, such as escalation routines that define how different incidents are handled. An incident manager is appointed, as is a major incident manager who handles major incidents. Incidents will be classified into one of four different categories, according to ITIL, and will be reported. If necessary, incidents will be reported to other organizations. The organization has predefined groups such as a crisis management team that can handle serious incidents with a documented continuity plan. The IT provider has a backup facility. Organization A would handle the incidents in the following way.

- Incident 1 will immediately be detected and reported to the incident manager, who will escalate the incident to the crisis management team and report to other organizations.

- Incident 2 will also be immediately detected and, provided that it is not classified as "major," it will not be escalated and instead will be logged for a possible future action.
- Incident 3 will be detected internally by various functions or by affected persons and, in the worst case, by the media. The incident will be handled by the incident manager, escalated to the crisis management team, and reported to other organizations.
- Incident 4 will be detected afterwards by the system administrator using a monthly report from the system. The escalation of the incident depends on what the employees who have received incorrect access to the information have done with it.
- Incident 5, which may be difficult to detect, will probably be detected by an internal control function. The incident will be handled by the incident manager, escalated to the crisis management team, and reported to other organizations.
- Incident 6 will be detected by a control function. If the corrupted information does not affect patient safety, the incident will not be escalated, but it will lead to a review of various routines.

Organization B. At least those working in the IT department are aware that an incident occurred, and they also know their responsibilities, even if formal roles have not been defined. An incident manager is appointed and, if necessary, incidents are reported to other organizations. No formal analyses of what impact incidents cause are performed. No crisis management team exists, so incidents are escalated to the unit manager, who will contact the next level, if necessary. In a crisis, the organization's top management will handle the incident. The procedures for managing incidents are not documented. At least the IT department knows how to report incidents. The organization is located in different places, so it is possible to move the IT operation. Organization B would handle the incidents in the following way.

- Incident 1 will be detected immediately, primarily by IT operation. The incident will be escalated via the incident manager, but it will probably take some time because defined groups, such as crisis management teams, are missing.
- Incident 2 will also be detected immediately by IT operation. The incident will be logged for future analysis and will possibly lead to some form of action, but it is doubtful that this analysis would be based on a formal risk analysis.
- Incident 3 will be detected, in the worst case by the media, but it can take a long time. The incident will be escalated, but this will also take time because the incident manager must contact the organization's senior management and established communication channels are missing. Nevertheless, after a while, a crisis will be defined.
- Incident 4 depends on what the employees who have received incorrect access to the information have done with it. If the information has not been used, then the incident will not be escalated, but instead will be logged for future analysis and could possibly lead to a review of various routines.
- Incident 5, which may take a long time to be identified, will probably be detected by IT support. The incident will be handled by the IT manager and escalated to the organization's top management, and a crisis will be defined.

- Incident 6 will probably be detected by IT support. The incident will be escalated to the unit manager and eventually will lead to a review of various routines.

Organization C. The hospital has extensive experience in dealing with serious incidents in the health sector, and this also applies to the organization's management of IT-related security incidents. The organization uses various processes for managing IT-related security incidents that are defined in ITIL. In general, the impact of an incident for the organization will be classified into four categories according to ITIL, together with the impact of the incident on availability, confidentiality, and integrity. Incidents will usually be detected by the employees who contact the service desk, which, if necessary, will escalate the incident. An appointed person works as incident manager, and other five people alternate as standbys in this role outside normal working hours. Documented processes such as escalation and reporting routines are available and are updated regularly. The organization has established predefined groups that can handle different types of incidents and report them to other organizations. Furthermore, the organization has backup facilities, and the same applies for the IT provider. Organization C would handle the incidents in the following way.

- Incident 1 will be detected immediately and reported to the incident manager, who will escalate the incident to the crisis management team.
- Incident 2 will also be detected immediately. If it is solved within 30 min, it will not be escalated and instead will be logged for a possible future action.
- Incident 3 will also be escalated. All employees know how to identify an incident and that personal information is sensitive information. The incident will primarily be handled by the information security manager and chief physician.
- Incident 4 will be handled by the information security manager, and the escalation of the incident depends on what the employees who have received incorrect access to the information have done with it.
- Incident 5 will be escalated and primarily handled by the information security manager. If the altered data might affect patient safety, it is likely that all or part of the organization will be switched to manual routines until the corrupted information has been corrected.
- Incident 6. It is not clear the incident will be handled. If the corrupted information affects patient safety, the incident will be treated as a health-care incident.

5.3 Part 2 of the Study: Evaluation of Interviews by Independent Security Experts

At the Swiss CISO Summit in autumn 2018, the authors had the opportunity to evaluate the predictive ability of EMM. 37 independent information security experts, none of whom had seen the results of EMM, ranked the documentation of how the three organizations handle the different scenarios using the following classification.

- **L = Levels 0 and 1:** Low. The organization has limited capability to escalate incidents (for example, no documented procedures, unclear responsibility for each employee, limited awareness and education of employees, no functional roles like an appointed incident manager).

- **M = Levels 2 and 3:** Medium. The organization has some capability to escalate incidents (for example, documented procedures (for escalation, reports to management and other organizations) documented continuity plans, division of responsibility within the organization, defined roles like an appointed incident manager and crisis teams, training plans for employees, awareness among different types of employees).
- **H = Levels 4 and 5:** High. The organization has extensive capability to escalate incidents (for example, continuous updating and improvement of documented procedures, continuity plans, training plans, defined roles, and division of responsibility).

The information security experts who participated in the Swiss CISO summit were divided into five different groups, and each group was placed at a separate table. At each table, one participant acted as table host and another one as rapporteur. The total number of participants at each group/table was around 7. Each participant received a document with a description of the interviews. All participants in a group received the same document, a description of the interviews from either Organization A, B, or C. The distribution of documents among the different groups is shown in Table 3.

Table 3. Consensus result from the different groups

Group	Org A	Org B	Org C	How sure	Nr	Sector	Years CISO	Years RM	Years IM
1	M			VS	8	Private	7	4	2
2		L		VS	7	Private Public	10	6	5
3			M	S	6	Private Public	10	10	10
4	M			S	7	Private Public	12	8	9
5		L		VS	9	Private	12	12	10

The participants first read the description to reach their own opinion. Then each group had a discussion so that a group consensus could be established. The result of the discussion, which took about an hour, was documented by the rapporteur, who also documented how sure each group was about their consensus using the following scale: (i) very sure (VS), (ii) sure (S), (iii) unsure (U), or (iv) very unsure (VU). In addition, the rapporteur documented the following information about the participants in each group.

- Nr. = Number of persons at each table
- Sector = Predominant sector represented at each table, if any
- Years CISO = Average years of experience as CISO/security expert
- Years RM = Average of years of experience in IT security risk management
- Years IM = Average of years of experience in IT security incident management

Finally, the rapporteur presented the results for the other groups. A summary of the result for each group is shown in Table 3. The table shows that both groups 1 and 4 ranked Organization A as medium (M), which means that the organization has some ability to escalate incidents. Group 1 was very sure (VS) about the result, while Group 4 was only sure (S). Furthermore, both groups 2 and 5 ranked Organization B as low (L), which means that the organization has limited ability to escalate incidents. Both groups 2 and 5 were very sure (VS) about the result. Because the five groups participated in other activities at the Swiss CISO Summit during the day, only one group was able to rank Organization C. Group 3 ranked Organization C as medium (M), which means that the organization has some ability to escalate incidents. Group 3 was sure (S) about the result.

Using a measurable scale, when assessing the documentation of the various interviews, is in some way always subjective. However, the fact that several highly experienced security experts have come to a consensus after an in-depth discussion shows that a ranking is possible.

5.4 Comparison Between the Result of EMM and the Result of the Interviews

First a discussion of how different approaches that can be used to evaluate the predictability of EMM. One approach is, of course, to use EMM first and then wait until a real serious incident occurs, which can take a considerable amount of time if it happens at all. Another approach is to use EMM afterwards when a serious incident has occurred. The disadvantage of this approach is that when a serious incident occurs, various measures will be taken that would probably change the outcome of EMM if it is executed afterwards. Another disadvantage with both approaches is that information about serious incidents is often confidential, which makes it difficult for an external party to make an evaluation. This is the background why this proof of concept study has used fictional incidents when the predictability of EMM is evaluated.

Finally, a comparison is made of how well the result from EMM matches the result of the interviews. This comparison shows that Organizations A and B have a clear match between the result of EMM and the result of the interviews. For Organization C, the result does not match so well. According to the ranking of the interview by the independent security experts, the different incidents would be handled less efficiently than the EMM results show. Table 4 illustrates the results of the comparison.

Table 4. Comparison between the use of EMM and the interviews.

Organization	Result from EMM	Result from interviews
A	Medium	Medium
B	Low	Low
C	High	Medium

6 Conclusion and Future Research

Although the number of organizations is limited to three, the conclusion of this proof of concept study shows that the outcome of EMM and the way in which an organization would handle different incidents correspond well, at least for the organizations with low and medium maturity levels. The authors have developed EMM and the tool for self-assessment that has been used in previous tests. The authors' contribution with this proof of concept study is that by comparing the outcome of EMM with how organizations would handle different incidents in practice, the predictability of EMM has increased. This means that organizations with greater certainty can use EMM to measure the organization's capability to handle various IT-related security incidents.

After some modification of the scenarios and the associated questions, the research will continue. New similar studies will be conducted with more material (more organizations) so it will be possible to determine with even greater certainty that the outcome of EMM indicates an organization's escalation capability of IT-related security incidents.

References

1. ISO - International Organization for Standardization: Information technology: information security risk management, ISO/IEC 27005 (2011)
2. NIST - National Institute of Standards and Technology: Guide for applying risk management framework to federal information systems. NIST Special Publication 800-37 Revision 1 (2010)
3. NIST - National Institute of Standards and Technology: Guide for conducting risk assessment. NIST Special Publication 800-30 Revision 1 (2011)
4. NIST - National Institute of Standards and Technology: Managing information security risk. NIST Special Publication 800-39 (2011)
5. NIST - National Institute of Standards and Technology: Information security continuous monitoring (ISCM) for federal information system and organizations. NIST Special Publication 800-137 (2011)
6. Kahn, H.: On Escalation: Metaphors and Scenarios. Praeger, Santa Barbara (1986)
7. ISO - International Organization for Standardization: Information technology – security techniques — information security incident management, ISO/IEC 27035 (2016)
8. Brewster, E., Griffiths, R., Lawes, A., Sansbury, J.: IT Service Management: A Guide for ITIL Foundation Exam Candidates, 2nd edn. BCS, The Chartered Institute for IT (2012)
9. NIST - National Institute of Standard and Technology: Computer Security Incident Handling Guide. NIST Special Publication 800-61 Revision 2 (2012)
10. Palilingan, V., Batmetan, J.: Incident management in academic information system using ITIL framework. In: IOP Conference Series: Materials Science and Engineering, vol. 306 (2018)
11. Nolan, R.: Managing the computer resource: a stage hypothesis. Commun. ACM **16**(7), 399–405 (1973)
12. Humphrey, W., Edwards, R., LaCroix, G., Owens, M., Schulz, H.: A method for assessing the software engineering capability of contractors Technical report, Software Engineering Institute, Carnegie Mellon University (1987)

13. ISO - International Organization for Standardization: Information technology – process assessment; assessment of organizational maturity, ISO/IEC Technical report 15504-7 (2008)
14. Pöppelbuβ, J., Röglinger, M.: What makes a useful maturity model? A framework of general design principles for maturity models and its demonstration in business process management. In: Proceedings of the Nineteenth European Conference on Information Systems - ECIS 2011, Association for Information Systems electronic Library – AISeL (2011)
15. Philips, M.: Using a Capability Maturity Model to Derive Security Requirements. SANS Institute, Bethesda (2003)
16. ISACA: The risk IT framework. Rolling Meadows, IL (2009)
17. Aguiar, J., Pereira, R., Vasconcelos, J., Bianchi, I.: An overlapless incident management maturity model for multi-framework assessment (ITIL, COBIT, CMNI-SVC). Interdisc. J. Inf. Knowl. Manag. **13**, 137–163 (2018)
18. Vaishnavi, V., Kuechler, W.: Design research information systems. http://desrist.org/design-research-in-information-systems. Accessed Jan 2019
19. Wahlgren, G., Kowalski, S.: A maturity model for measuring organizations escalation capability of IT-related security incidents in Sweden. In: Proceedings of the 11th Pre-ICIS Workshop on Information Security and Privacy, Dublin, Association for Information Systems electronic Library - AISeL (2016)
20. Wahlgren, G., Kowalski, S.: IT security risk management model for handling IT-related security incidents: the need for a new escalation approach. In: Maleh, Y. (ed.) Security and Privacy Management, Techniques, and Protocols, pp. 129–151. IGI Global, Hershey (2018)

A Framework to Monitor Machine Learning Systems Using Concept Drift Detection

Xianzhe Zhou, Wally Lo Faro(✉), Xiaoying Zhang, and Ravi Santosh Arvapally

Mastercard, 2200 Mastercard Blvd, O'Fallon, MO 63368, USA
{steven.zhou,wally.lofaro,evelyn.zhang,
ravisantosh.arvapally}@mastercard.com

Abstract. As more and more machine learning based systems are being deployed in industry, monitoring of these systems is needed to ensure they perform in the expected way. In this article we present a framework for such a monitoring system. The proposed system is designed and deployed at Mastercard. This system monitors other machine learning systems that are deployed for use in production. The monitoring system performs concept drift detection by tracking the machine learning system's inputs and outputs independently. Anomaly detection techniques are employed in the system to provide automatic alerts. We also present results that demonstrate the value of the framework. The monitoring system framework and the results are the main contributions in this article.

Keywords: Monitoring system · Concept drift · Machine learning · Anomaly detection · Framework

1 Introduction

The human body receives more sensory inputs than it can cope with so it has evolved to filter out most of them automatically, only focusing on what is anomalous. Enterprises have grown in complexity to a point where an analogy with the human body is apt. The need for autonomous agents monitoring enterprises for anomalies is felt as well. In this article, we look through the lens of concept drift to illustrate such a monitoring system applied to another machine learning based system. The monitored system can be a classifier, such as those presented in the work of Arvapally et al. [1], other machine learning based applications, or any source of data that displays traceable patterns over time. The monitoring system described herein currently operates at Mastercard.

The remainder of this article is organized as follows. A brief review of concept drift and related work is given in Sect. 2. Section 3 introduces framework of the monitoring system. In Sect. 4, the monitoring system is discussed in detail under the design framework along with the results.

W. Abramowicz and R. Corchuelo (Eds.): BIS 2019, LNBIP 353, pp. 218–231, 2019.
https://doi.org/10.1007/978-3-030-20485-3_17

2 Related Work

A machine learning model can malfunction every so often due to production issues or changes in input data. A monitoring system is needed to promptly flag model performance glitches. The essence of the monitoring system is to detect concept drift which is defined as [2]:

$$\exists X : P_{t0}(X, y) \neq P_{t1}(X, y) \quad (1)$$

where P_{t0} denotes the joint distribution between input variable X and output variable y at time t0. There has been a substantial research into concept drift detection methodologies [2,4–7] for various types of drifts [2], demonstrated by Figs. 1 and 2. Because of the slowly changing nature of the underlying data in our application, we are more concerned about sudden drifts than others.

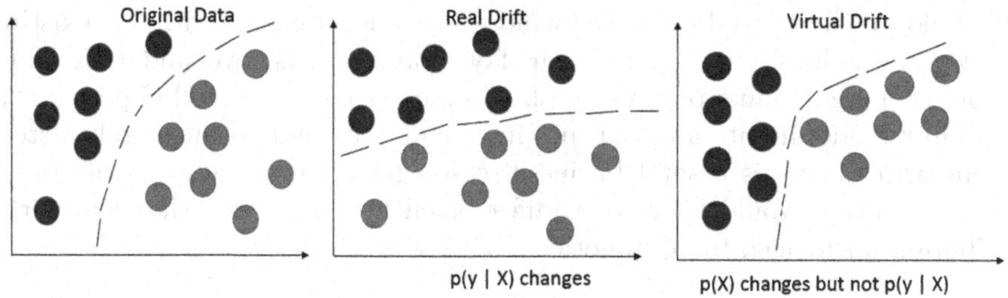

Fig. 1. Types of drift: circles represent instances; different colors represent different classes

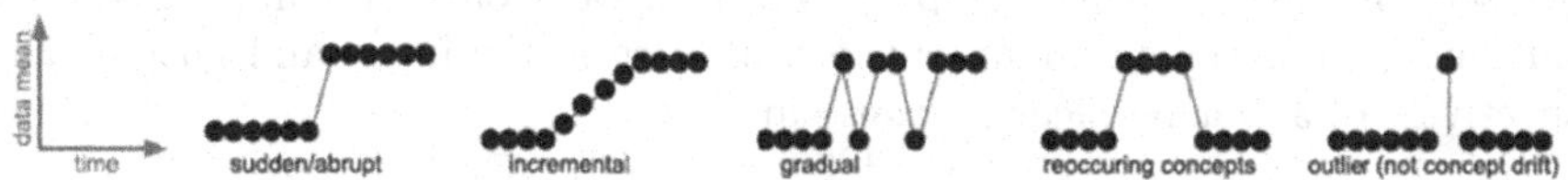

Fig. 2. Patterns of concept change

Moreover, probability chain rule,

$$P(X, y) = P(y|X) * P(X) \quad (2)$$

allows us to monitor a model's inputs $P(X)$ and outputs $P(y|X)$ separately during operation. In other words, we can track changes in P(X, y) by comparing $P_{t0}(X)$ to $P_{t1}(X)$, and $P_{t0}(y|X)$ to $P_{t1}(y|X)$. This division of supervision, or modular design, simplifies business operation.

In recent research, we noticed some researchers developing methods that can detect concept drift when concept drift is labeled [16]. This is based on application of both single classifier as well as ensemble classification techniques [16,17].

These ensemble methods use multiple algorithms for concept drift detection following which a voting procedure will be conducted. These methods also study how to handle when limited labeled data is available. The underlying challenge here is, in most cases concept drift labels are either unavailable or very expensive to come up with. Availability of concept drift labels is very useful in developing accurate models. Our proposed framework is suitable when the concept drift labels are unavailable. In addition, our proposed framework can monitor both supervised and unsupervised learning models.

In addition, our framework does not specifically focus on any particular kind of dataset. Some recent research shows methods to detect concept drift in twitter [18]. Though the underlying nature of the data is streaming, the proposed methods in [18] are more applicable to specific Twitter data rather than general datasets. Some recent studies have shown fuzzy logic based windowing for concept drift detection in gradual drifts [15]. These adaptive time-windowing methods are definitely useful since they will increase the accuracy of the overall methodology. This fuzzy-logic based windowing concept can go with any existing framework. Finally, the modular design of our framework is novel and adds value. For the proposed framework, we applied anomaly detection techniques [8–11] to monitor model inputs and outputs independently, and to ultimately detect concept drift. There is a solid business reason to monitor these distributions. Namely, an event would have to manifest itself in such a way that leaves the distribution unchanged to go unnoticed. This seems unlikely.

3 Framework of the Monitoring System

As we have seen, the probability chain rule allows us to monitor the inputs and outputs of a machine learning system independently. As a result, the proposed framework consists of two components namely the input tracer and the output tracer. Figure 3 presents the proposed framework of the monitoring system. Input and output tracers monitor the anomalies in the input and output data respectively of a machine learning system.

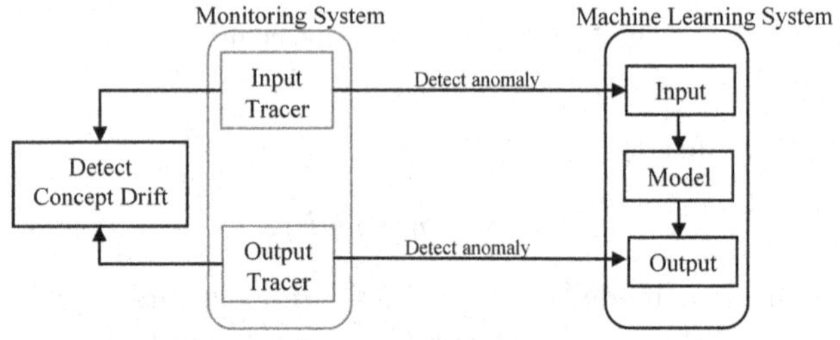

Fig. 3. Monitoring system

Although different techniques are employed in each tracer, both the input tracer and the output tracer follow the same design sequence, inspired by the work of Trotta et al. [3]. Figure 4 presents three steps of the design sequence which is common to both tracers.

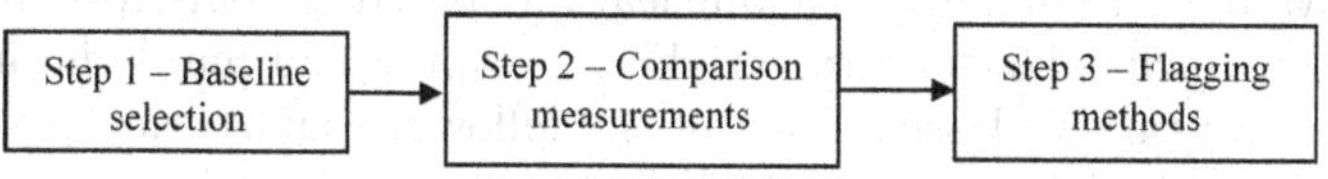

Fig. 4. Design framework

The three steps in the design sequence are explained below.
Step 1 - Detecting concept drift in data is the goal of our proposed framework. One way to detect drift is to track the distributional changes in data, where a baseline must be defined to make comparison. Domain understanding and data observation help in selecting a baseline. One has to consider three different criteria in selecting baselines: which historical values are included in the baseline? Whether to apply a smoothing technique? How frequently should the baseline be updated? Unavailability or inadequate labeled data is commonly observed. This leaves us with the option to leverage historical distributions of the data as our baseline.

Step 2 - After determining the baseline, a comparison measure is needed for benchmarking. Since there are several similarity measures for distributions available, the following criteria should be taken into account:

- Does the quantitative measurement align with the qualitative features of objects being measured?
- Can a threshold be created?
- The computational complexity to assess the scalability

Step 3 - In step 3, a flagging mechanism is designed and developed to automate the alerting effort. The goal of step 3 is to automatically alert on out-of-pattern values generated by step 2. The mechanism consists of two different functions. The first function is developed based on anomaly detection algorithms and the second function is based on heuristic rules provided by the domain experts. Identified anomalies are served as output to the business users to take informed business decisions. This step may involve further processing of the result, for example, providing a ranked list of anomalies.

Both the input and output tracers we are describing follow the three steps in the design sequence described above. Although these three steps are common to the tracers, the data fed into tracers exhibits different characteristics. Input tracer monitors the input data that goes into our machine learning system. While the output tracer monitors the output provided by the machine learning system. Detailed explanations of both input and output tracer are provided in the following sections.

4 The Monitoring System

This section presents the details of the developed monitoring system. The objective of the system is to monitor machine learning based systems. We are referring to a classifier which takes feature variables as inputs and maps the input data to one of several target classes. Each of the following sub-sections explains the business background, the solution, and the results.

4.1 Input Tracer

Business Background. Feature variables (FVs) are category profiles, created for each system account at Mastercard. Table 1 presents a snippet of how FVs are stored in a database. Equivalent to an index, a FV reflects recent activities of an account in a category. FVs are constructed periodically (weekly) for each account, and fed into a classifier. The data quality of FVs is crucial to the success of the classifier. The input tracer is designed to track the data quality of all FVs with the following expectations:

- Effective at alerting data quality incidents
- Scalable to handle large data volume
- Provide guidance for debugging when a data quality incident occurs

Table 1. Feature Variable (FV) examples in a database

Account number	FV	FV values	Creation date
1	FV 22	3.2	01-01-2018
1	FV 22	3.5	01-08-2018
1	FV 5	17.8	01-01-2018
1	FV 5	12.6	01-08-2018
2	FV 22	100.9	01-01-2018
2	FV 5	0	01-01-2018

Solution. The solution is based on the three steps presented in the framework.

Step 1 - Baseline selection: For the input tracer, the baseline is the FV value distribution from the previous week. This baseline is selected because of the following reasons:

- Values of a single FV are observed to have stable distribution over time.
- FV value distribution contains a large number of data points and hence noise can be tolerated.

- FV values can have cyclical fluctuations and hence a long term average is unsuitable.

Step 2 - Comparison Measurements: In order to track distribution changes, weekly FV value distributions are compared to those from the previous week. Multiple metrics are considered by leveraging information theory such as Kullback-Leibler divergence, Jeffrey's divergence and squared Hellinger distance. Despite being successful at signaling distribution changes, KL divergence and the other two measurements can only be computed after distributions are transformed to probability mass functions. The transformation is expensive when a distribution contains a large number of data points. This poses a scalability challenge in production and results in the abandonment of the metrics in this case. Then we turn to moment matching to compare two distributions while addressing the scalability challenge. If moments of two samples are matched, we assume both sample batches come from the same underlying distribution. Here, a moment M at time t is the expectation of a random variable X raised to a power p where p can be any positive integer, given N samples in a sample set:

$$M_{t,p} = \frac{1}{N} * \sum_N X_t^p \tag{3}$$

In this context, the first and second moments (mean and the mean of the squared values) are employed. After first and second moments are calculated for values of a FV over different weeks, week-over-week percentage changes are computed:

$$WOW\%_t = \frac{M_{t,p} - M_{t-1,p}}{M_{t-1,p}} \tag{4}$$

The percentage changes are given different weights based on business needs. In the developed system, the first moment WOW percentage change received 66% weight and the second moment WOW percentage change received 33%. The past 52 weeks of weighted WOW percentage changes are then fed into an anomaly detection model for training. The visualization of moments can help to spot anomalies, and provide clues to unravel a data quality incident. However, an automatic alerting/flagging system is needed, which is discussed below.

Step 3 - Flagging Methods: The objective of this step is to explore ways to provide automated alerts to the business users. For the input tracer different clustering algorithms are considered to enable automated alerts. We explored two different algorithms. DBSCAN, a density based clustering algorithm and K Means clustering algorithm. Application of these two algorithms present a need for post-clustering analysis and computation. This involves splitting and/or combining clusters to identify outlier cluster resulting in additional complexity. After exploration we settled with Isolation Forest algorithm [14] to identify outliers. Isolation Forest is an unsupervised anomaly detection method. Being an ensemble method, Isolation Forest is efficient at performing outlier detection. One reason Isolation Forest appeals is that it is a native binary classifier. Isolation Forest also controls the size of each cluster by user-defined parameters such as contamination rate. Essentially, the algorithm performs anomaly detection using a ranking system. Each data point is ranked based on average path lengths, which is a representation of the likelihood of being an outlier. For more reading on Isolation Forest please see article [12,14].

One drawback of Isolation Forest in this framework is that normal cyclical variations can be misclassified as anomalies. When a data series has very little fluctuation, even a small change, which can be natural noise, makes a difference in the predictive outcome. This drawback is mitigated by employing a rule-based exception handler on the results produced by Isolation Forest. For example, in order to handle seasonal fluctuations, thresholds for anomaly are adjusted for certain periods of time. These rules are defined by domain experts based on operational needs.

Because of the reasons explained above, the alerting system consists of two layers. The first layer is an application of Isolation Forest and the second layer is a rule based exception handler. The second layer rules are applied on the output obtained from Isolation Forest algorithm (layer 1). The output from second layer is offered as results to the business owner. Details about layer 1 and 2 are presented below.

Layer 1: Isolation Forest
Training data: weighted 1st moment WOW% and 2nd moment WOW% of past 52 weeks
Model parameters:
- *Number of estimators*
- *The number of features to train each base estimator*
- *Contamination(proportion of outliers in the data set)*

Layer 2: Rules Layer
Rule based exception handler, used to adjust the predictive result from layer 1

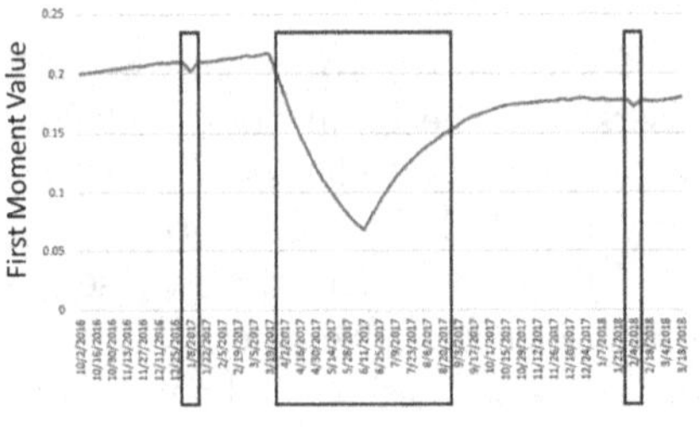

(a) Time Series of First Moment, FV 77

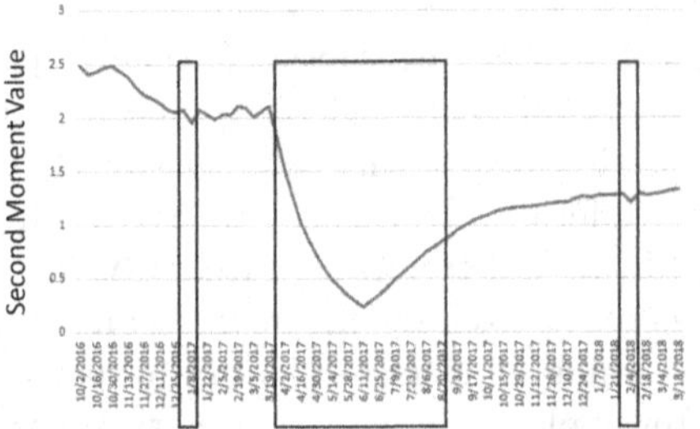

(b) Time Series of Second Moment, FV 77

Fig. 5. Plots present time series of first and second moment of FV 77 (Color figure online)

Results. The above steps were applied on FV 77. Due to the absence of reliable labels, we rely on visual data analysis (see Fig. 5) and domain understanding to judge the effectiveness of the tracer. Figure 6 presents the results produced by the input tracer. The results in Fig. 6 are after the application of layer 1 and layer 2 rules. The input tracer (see Fig. 6) alerts on the following dates: 2017-01-08, 2017-01-15, 2017-03-26 to 2017-07-16, and 2018-02-04. Based on visual analysis, we notice there are three periods (highlighted by yellow boxes in Fig. 5a and b) during which anomalous behavior occurs. The alerts from the input tracer (Fig. 6) are consistent with our visual data analysis. In addition, the business owners are in agreement with the alerts produced by the system. The input tracer has been tested with a variety of FVs. The system is scalable and effective at capturing outliers.

Input Tracer Conclusion. The proposed input tracer is successful at alerting data quality incidents that occur among feature variables. Ranking data points for outliers with Isolation Forest helps business users prioritize and understand the data behind the drift. This provides information to users to investigate and debug when a data quality incident occurs.

4.2 Output Tracer

Business Background. Often it is possible to see the effect of a production incident on a classifier by examining a machine learning model's output. Given the slowly changing nature of the input data, a dramatic shift in the distribution of model outputs signals a potential issue/incident. An incident is an unplanned interruption to an IT service or reduction in the quality of the service impacting customers. A production issue/incident could be a result of one of several reasons. For example, applying the classifier to incorrectly scored accounts or misprocessed feature variables (input data i.e. FVs) may lead to incidents.

The focus of the output tracer is on the outputs produced by a machine learning system. The output tracer is therefore designed to identify anomalous classifier behaviors. By examining the distribution of all classes, the output tracer can alert out-of-pattern class distributions so that swift corrective actions can be taken.

Solution. The key observation and assumption here are similar to the one made by the input tracer: aggregate account activities for each account family should not change dramatically over time. The output from a machine learning system is categorized into one of the target classes (mapped classes). The mapped classes produced by the classifier are expected to have a stable distribution for each account family. This statement is true unless an incident or a technical issue occurs while the system is in use. The output tracer is turned into a problem of monitoring the distribution changes of classes over time. The solution follows the three steps outlined in the framework (Sect. 3).

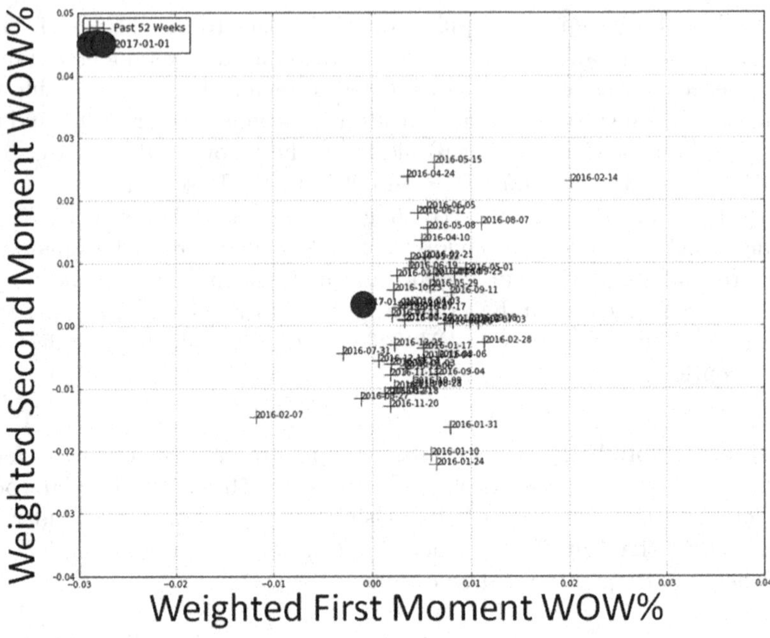

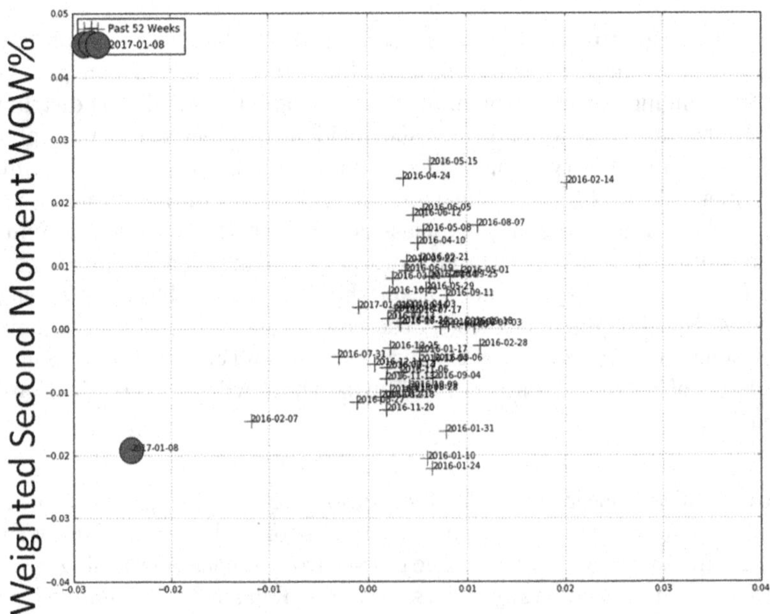

Fig. 6. Weighted First and Second Moment WOW% Scatter Plots, Colored Based on Isolation Forest Outcomes (Best Viewed in Color). *Circle: test data, Crosses: training data (prior 52 weeks), Green: inliers, Red: outliers* (Color figure online)

Step 1 - Baseline selection: The first step to identify anomalous behavior is to define what is normal. Taking both business context and potential seasonality into account, the baseline, or the reference distribution is defined by averaging past ten same day-of-week's class distributions. For example, given Monday 7/16/2018, the reference distribution is calculated by taking the average of class distributions of previous Monday (7/9), the Monday before (7/2) and so on. This granularity was determined by domain experts. In comparison to the input tracer, the baseline of the output tracer involves more computation, by averaging class distributions of the past ten same days-of-week. However, although the computational complexity is high in output tracer, the size of the output data is smaller. Therefore the scalability challenge is easier to handle. A large time window is selected to minimize the impact of noise because the number of data points in a class distribution could be small. On the other hand, the window should not be oversized, so that the latest trend can be reflected. Equal weight is given to each of the ten distributions, and hence no single distribution is dominating the reference distribution. Since data labels are unavailable, abnormally mapped classes are currently included in building the reference distribution. Abnormal class mappings can be removed in future after they are detected and recorded.

Step 2 - Comparison Measurements: Unlike feature variables (inputs), the class data (outputs) has less of a scalability challenge but is more volatile in nature. With those differences in mind, different similarity measurements and anomaly detection methods are evaluated and selected. Kullback-Leibler divergence, also known as relative entropy (RE), is chosen to measure the difference between the reference distribution q(x) and the current distribution p(x). The higher the relative entropy value is, the more skewed current distribution is compared to the reference distribution.

$$D(p||q) = \sum p(x) \ln(\frac{p(x)}{q(x)}) \tag{5}$$

For Eq. (5), p(x) is the current distribution, and q(x) is the reference distribution. The current distribution p(x) is calculated at the end of each day.

Unlike squared Hellinger distance, KL divergence is not restricted between zero and one. A sizable distribution change will be translated to a change of large magnitude in KL divergence, which facilitates the creation of a threshold. KL divergence lacks symmetry but this can be solved by averaging D(p||q) and D(q||p). A small positive number replaces all zero probability densities.

Step 3 - Flagging Methods: Account families that have fewer observations tend to have more volatile class distributions which lead to higher RE values. So, relative entropy alone cannot be used to detect anomaly. As a result, a second factor is introduced to measure the size of an account family. The second factor is computed by taking the natural logarithm of the average number of activities performed under the account family. Visualization of RE values against the size of account family can be used to identify which account families have skewed class distributions. The objective of this step is to automate the alerting system using the two variables mentioned above.

The alerting system comprises of two layers: the first layer is based on DBSCAN. The second layer consists of human-defined rules. These rules are applied on top of the cluster results produced by DBSCAN. Details about layer 1 and 2 are presented below.

Layer 1: DBSCAN
The first factor – the RE values of all account families for one day
The second factor – the "size" of an account family derived from the number of activities under the account family
Model parameters:
- *The maximum distance (eps) for a neighborhood*
- *The minimum number of samples in a neighborhood for core points*

Layer 2: Rules Layer - rule based exception handler, used to adjust the predictive result from layer 1 and to handle cases in which a large number of account families go wrong.

The RE values and account family sizes are fed into a clustering algorithm to detect anomalies. In this case, DBSCAN (Density-based Spatial Clustering for Applications with Noise) [13] algorithm is tested and selected to detect anomalies. According to Scikit Learn, DBSCAN views clusters as high-density areas separated by low density areas [12]. One drawback of the algorithm is that DBSCAN does not necessarily produce two clusters. Based on the density of data points and parameters chosen, the algorithm may produce more than two clusters. In those circumstances, the cluster with the largest number of points is deemed as the "normal" cluster and all remaining clusters are considered anomalies. The parameters (thresholds) of DBSCAN are selected from both mathematical and business perspectives.

Other anomaly detection algorithms were also tested, such as isolation forest. However, the fact that isolation forest always selects a fixed percentage of anomalies does not apply to the condition here in the output tracer. The non-linear nature of the data (as shown in Fig. 7) also negatively impacts the performance of isolation forest.

Layer 2 is applied on the results produced by the layer 1. This will handle cases in which anomalies occur in large number of account families. Rules can also be added to accommodate special situations. For example, a rule restricting the size of an account family can be applied here to remove cases where the sample size is too small to have business values. The alerts from layer 2 are served as output to business users.

Results. Figure 7a presents relative entropy values against the natural logarithm of account family sizes, which are derived from the output of a machine learning system. The plot clearly shows a dense cluster with a few scattered points (see Fig. 7a). This phenomenon aligns with the common assumption of anomaly detection: anomalies are the minority group that behave substantially

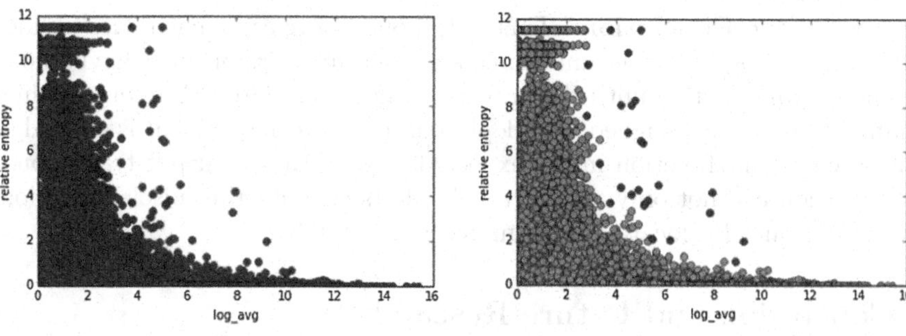

(a) Relative Entropy Values versus Natural Logarithm of Account Family Sizes

(b) Anomalies versus Normal Points (Best View in Color)

Fig. 7. Plots present relative entropy and account family size before and after DBSCAN algorithm applied (Color figure online)

differently from the majority. Figure 7b presents a plot of data with labels produced by the DBSCAN algorithm. The figure shows two different clusters (red and blue). Data points colored blue are outliers captured by the DBSCAN algorithm and red points are inliers.

The results are then verified by comparing baseline distribution with current distribution. Points from Fig. 7b are randomly selected for verification. The current distribution and reference distribution are plotted side by side to check if the detected anomalies indeed display any anomalous behaviors. Figure 8a is an example of a normal case (red points in Fig. 7b). Figure 8b is an example of the distribution of a detected anomaly (blue points in Fig. 7b). From Fig. 8b, we can clearly notice how current distribution is significantly different from its reference distribution.

Since there is no labelled data, results are evaluated with feedback from the business. False positives are tuned to be as low as possible to avoid too many false alarms, whereas in the meantime incident information is being collected to evaluate false negatives.

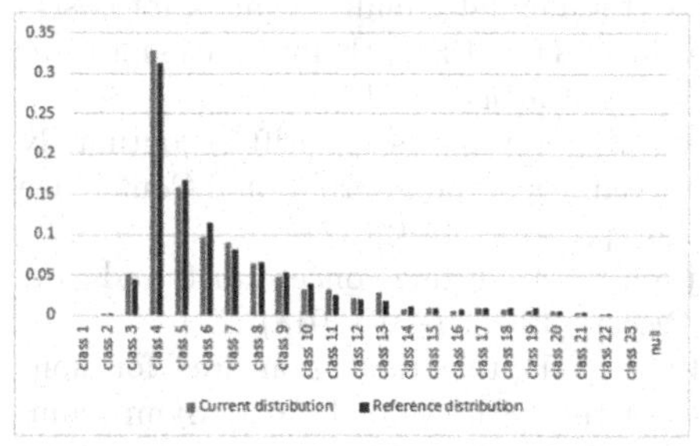

(a) Distribution of a normal point

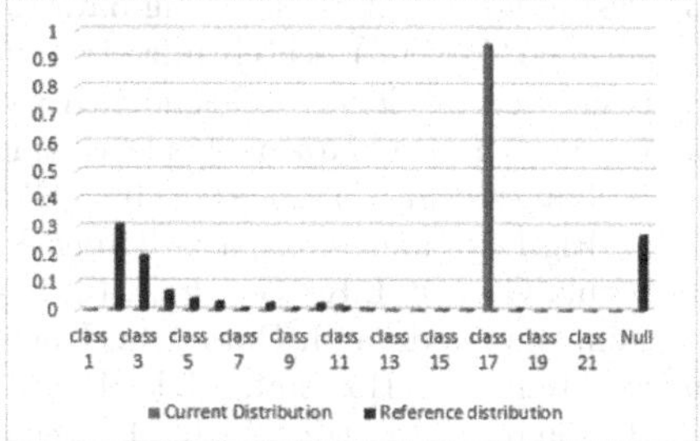

(b) Distribution of a detected anomaly

Fig. 8. Current distributions versus reference distributions (Color figure online)

Output Tracer Conclusion. Timely response to operational issues or incidents is critical because relevant teams are often unaware of misbehaviors of a machine learning system until the impact becomes very large. As more machine learning systems are designed and deployed, a monitoring system is needed to ensure the systems function in the expected way. With the output tracer, business operation can not only receive timely alerts on potential misclassifications of activities, but also locate the anomalies immediately.

5 Conclusion and Future Research

In this article, we presented methods to detect and to alert when concept drift occurs. The input and output tracer monitor the data that goes into a classifier and the classes produced by the classifier respectively. And the modular design of the developed system ensures the two models work independently. The two cases supporting the proposed methods demonstrates success in alerting drifts. Receipt of these alerts allows users to study the data and monitor the system in detail, and this clearly has a huge benefit.

In the future, we would like to continue research in two different directions. On one hand, we will continue experimentation to study the relationship between properties of different data sets and drifts, especially when data changes are unstable. On the other hand, we would like to study how these alerts can assist in developing online machine learning algorithms. Exploring drift and anomaly detection with real time data streams would be interesting as well. In conclusion, concept drift has a huge role to play in resolving challenges in the area of machine learning model deployment and model maintenance.

References

1. Arvapally, R.S., Hicsasmaz, H., Lo Faro, W.: Artificial intelligence applied to challenges in the fields of operations and customer support. In: 2017 IEEE International Conference on Big Data (IEEE Big Data), Boston, pp. 3562–3569 (2017)
2. Gama, J., Zliobaite, L., Bifet, A., Pechenizkiy, M., Bouchachia, A.: A survey on concept drift adaptation. J. ACM Comput. Surv. (CSUR) **46**, 44:1–44:37 (2014)
3. Pastorello, G., et al.: Hunting data rouges at scale: data quality control for observational data in research infrastructures. In: 2017 IEEE 13th International Conference on e-Science (e-Science), Auckland, pp. 446–447 (2017)
4. Gamage, S., Premaratne, U.: Detecting and adapting to concept drift in continually evolving stochastic processes. In: ACM Proceedings of the International Conference on Big Data and Internet of Thing, London, pp. 109–114 (2017)
5. Webb, G., Lee, L.K., Goethals, B., Petitjean, F.: Analyzing concept drift and shift from sample data. J. Data Min. Knowl. Discov. **32**(5), 1–21 (2018)
6. Gholipur, A., Hosseini, M.J., Beigy, H.: An adaptive regression tree for non-stationary data streams. In: Proceedings of the 28th Annual ACM Symposium on Applied Computing (ACM), Coimbra, pp. 815–817 (2013)
7. Jadhav, A., Deshpande, L.: An efficient approach to detect concept drifts in data streams. In: IEEE 7th International Advance Computing Conference (IEEE IACC), Hyderabad, pp. 28–32 (2017)

8. Chandola, V., Banerjee, A., Kumar, V.: Anomaly detection: a survey. ACM Comput. Surv. (CSUR) **41**(3), 15 (2009)
9. Ding, M., Tian, H.: PCA-based network traffic anomaly detection. J. Tsinghua Sci. Technol. **21**(5), 500–509 (2016)
10. Zhang, L., Veitch, D., Kotagiri, R.: The role of KL divergence in anomaly detection. In: Proceedings of the ACM SIGMETRICS Joint International Conference on Measurement and Modeling of Computer Systems, San Jose, pp. 123–124 (2011)
11. Laptev, N., Amizadeh, S., Flint, I.: Generic and scalable framework for automated time-series anomaly detection. In: Proceedings of the 21th ACM SIGKDD International Conference on Knowledge Discovery and Data Mining (ACM), Sydney, pp. 1939–1947 (2015)
12. Pedregosa, F., et al.: Scikit-learn: machine learning in Python. J. Mach. Learn. Res. **12**, 2825–2830 (2011)
13. Ester, M., Kriegel, H.P., Sander, J., Xu, X.: A density based algorithm for discovering clusters in large spatial databases with noise. In: Proceedings of the Second International Conference on Knowledge Discovery and Data Mining, Portland, pp. 226–231 (1996)
14. Liu, F.T., Ting, K.M., Zhou, Z.H.: Isolation based anomaly detection. ACM Trans. Knowl. Discov. Data (TKDD) **6**(1), 3 (2012)
15. Liu, A., Zhang, G., Lu, J.: Fuzzy time windowing for gradual concept drift adaptation. In: IEEE International Conference on Fuzzy Systems (FUZZ-IEEE), Naples, pp. 1–6 (2017)
16. Geng, Y., Zhang, J.: An ensemble classifier algorithm for mining data streams based on concept drift. In: 10th International Symposium on Computational Intelligence and Design (ISCID), Hangzhou, pp. 227–230 (2017)
17. Hu, H., Kantardzic, M.M., Lyu, L.: Detecting different types of concept drifts with ensemble framework. In: 17th IEEE International Conference on Machine Learning and Applications (ICMLA), Orlando, pp. 344–350 (2018)
18. Senaratne, H., Broring, A., Schreck, T., Lehle, D.: Moving on Twitter: using episodic hotspot and drift analysis to detect and characterise spatial trajectories: In: 7th ACM SIGSPATIAL International Workshop on Location - Based Social Networks (LBSN), Dallas (2014)

Determining Optimal Multi-layer Perceptron Structure Using Linear Regression

Mohamed Lafif Tej(✉) and Stefan Holban

Faculty of Automation and Computers, Politehnica University of Timisoara, Timisoara, Romania
afiftej@gmail.com, stefan.holban@cs.upt.ro

Abstract. This paper presents a novel method to determine the optimal Multi-layer Perceptron structure using Linear Regression. Starting from clustering the dataset used to train a neural network it is possible to define Multiple Linear Regression models to determine the architecture of a neural network. This method work unsupervised unlike other methods and more flexible with different datasets types. The proposed method adapt to the complexity of training datasets to provide the best results regardless of the size and type of dataset. Clustering algorithm used to impose a specific analysis of data used to train the network such us determining the distance measure, normalization and clustering technique suitable with the type of training dataset used.

Keywords: Multi-layer Perceptron · Linear regression · Clustering methods · Pattern recognition · Artificial neural network

1 Introduction

Determining the structure of Multi-layer Perceptron is a critical issue in the design of a Neural Network [1]. Until now, there is no general equation to define the structure of Multi-layer Perceptron, which can deal with different kind of problems to be resolved by the neural network. Each problem needs a particular structure that responds to his requirements. Methods currently used do not rely on the complexity of the problem must be solved by the Multi-layer Perceptron. Most currently used methods are very limited, time-consuming and supervised [2] such us Growing and Pruning Algorithms, Exhaustive Search, Evolutionary Algorithms and so on. In this paper, a novel method to determine the optimal Multi-layer Perceptron structure using Linear Regression will be introduced. The idea is to group the dataset used to train the Multi-layer Perceptron using conventional methods of pattern recognition [3, 4] according to specific criteria until we get a set of useful parameters, which will be used in the design of Multi-layer Perceptron structure. The results obtained from clustering the dataset used to train the network are used as independent variables to define a linear regression models [5] used to determine the Multi-layer Perceptron structure. The equation defined by the linear regression used to minimize the distance between a fitted line and all the data points. The regression model aims to achieve maximum accuracy in determining the number of hidden layers and the number of neurons in these layers.

W. Abramowicz and R. Corchuelo (Eds.): BIS 2019, LNBIP 353, pp. 232–246, 2019.
https://doi.org/10.1007/978-3-030-20485-3_18

2 Related Work

The design of the structure of a neural network is an extremely active area of research and does not yet have any definitive guiding theoretical principles. The currently used methods are very limited and time-consuming such as Growing and Pruning algorithms [6], exhaustive search, and evolutionary algorithms [7]. Here are some widely spread methods for determining the number of hidden neurons.

Many researchers use numerous thumb rules such as the number of hidden neurons should be between the size of the input and output layers. The number of hidden neurons should be: (number of inputs + outputs) * (2/3). The number of hidden neurons should be less than twice the number of input layer neurons [8]. These rules provide a starting point but do not achieve the best architecture only after a number of tests based on trial and error. Trial and Error approach does not yield good results except by accident, sometimes called exhaustive search [9]. Exhaustive Search approach makes searching through all possible topologies and then select the one with the least generalization error. The disadvantage of this method is time-consuming.

The Growing neural network algorithm was initially proposed by Vinod et al. [10]. Growing Algorithms method makes searching through all possible topologies and then select the one with the least generalization error. Search in this method stops if the generalization error does not have remarkable change, unlike exhaustive search.

Pruning Algorithms method tries to train an oversized network, and then determines the relative importance of weights by analyzing them. This method prunes the weights with the least importance and then repeats the task. The disadvantage of this method is that the analysis of weights is time-consuming.

In this paper, we proposed a method to determine the structure of a Multi-layer Perceptron based on the complexity level of the considered problem making it more flexible with different datasets types than classical methods. In addition, this method makes the design of Multi-layer Perceptron unsupervised.

3 Multiple Linear Regression Method Used

Following a set of criteria in the analysis of clusters obtained through hierarchical clustering of the dataset used to train the neural network, which results a number of parameters can be useful to define a linear regression model to determine the structure of Multi-layer Perceptron [11]. Parameters obtained from clustering will be evaluated using statistical hypothesis testing [12] to be able to identify whether it exists dependencies between these parameters and the number of hidden layers and the number of hidden neurons. The parameters selected through this evaluation used as independent variables of the regression models [13].

Figure 1 presents a framework of the regression model. The model shows how to use results obtained from clustering the training dataset to determine the regression model used to generate the optimal number of hidden layers and the number of neurons in these layers.

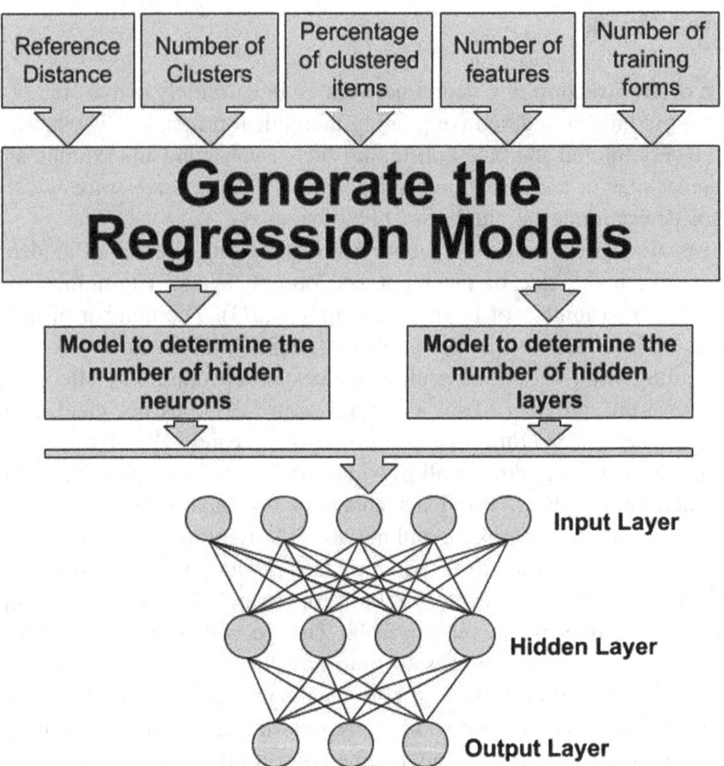

Fig. 1. The framework of the regression model

3.1 Regression Analysis

Regression analysis is a statistical technique to predict a quantitative relationship between a dependent variable and a set of independent variables [14]. The defined regression equation depends on the assumption concerning the relationship between the dependent variable and the independent variables [15]. The linear regression equation seeking to minimize the errors to fit the data points to a straight regression line representing the equation. Using information obtained by observations or measurements, the equation is defined. The indicator of multiple determination coefficient R^2 is required to determine the relationship between the Independent variable and the dependent variables. R^2 expresses the variation of the dependent variable affected by the variation of independent variables. The indicator of multiple determination coefficient is essential special if supported by other statistical indicators [16].

The mathematically Multiple Linear Regression model having this form:

$$f = X \to Y$$

$$f(X) = w_0 + \sum\nolimits_{j=1}^{n} w_j x_j \tag{1}$$

The regression models consist of unknown parameter w, the dependent variable Y and the independent variable X.

3.2 Statistical Hypothesis Testing

The statistical hypothesis testing [17] are used to examine parameters obtained through hierarchical clustering of training dataset to select a number of parameters to determine the regression model. The hypothesis testing used to prove that the regression model is significant. Depending on the null hypothesis H_0, which assume no significant relationship between the independent variables X and the dependent variable Y.

H_0: There is no relationship between the clustering results and the structure of Multi-layer Perceptron
Ha: There is a relationship between the clustering results and the structure of Multi-layer Perceptron

The probability coefficients of independent variables (P-value) have a value of less than 0.05 based on parameters proposed to be independent variables of the regression model.

F-Test analysis [18] used for the analysis of variance will be taken as an evidence to prove that the structure of Multi-layer Perceptron depends on the selected factors.

3.3 The Independent Factors Selected

Based on statistical hypothesis testing and F-Test a set of factors are proposed to determine the regression equation in addition to that it has been proven that there is a link between all factors in models. Moreover, relatively small positive and negative correlations exist [19]. The selected factors prove the effectiveness and efficiency of the proposed model through the Multiple Coefficient of Determination [20] and the Multiple Correlation Coefficient [21] where they obtain results close to 1.

The proposed factors obtained by clustering the training dataset will be used as independent variables to determine the regression models:

- The number of obtained cluster
- The percentage of grouped items
- The reference distance
- The number of training forms
- The number of features in the input

Moreover, the quality measure of the network structure was considered as an independent factor. The quality measure factor takes into account the configuration and interconnection layers [22].

The proposed regression models consist of two models the first model used to determine the number of hidden layers and the second model used to determine the number of hidden neurons.

Regression Model to Determine the Number of Hidden Layers. A set of factors prove the ability to influence on the dependent variable y, which represent the number of hidden layers of a Multi-layer Perceptron. Using the statistical hypothesis testing mentioned previously and experimental results it is turned out that the dependent variable y depending on changes of the following independent variables.

- X_1: The number of obtained cluster multiplied by the reference distance
- X_2: The reference distance
- X_3: The percentage of grouped items
- X_4: The quality measure of the network structure

The independent variables X_1, X_2, and X_3 obtained through clustering of the training dataset. In addition to that, a quality measure of the network structure X_4 is taken as independent factors. The quality measure factor depends on the reference distance and the structure of Multi-layer Perceptron.

A Multiple Linear Regression model representing the equation to determine the number of hidden layers of the Multi-layer Perceptron have the following mathematical form:

$$y = a_0 + \sum_{j=1}^{4} a_j x_j \tag{2}$$

The dependent variable y represents the number of hidden layers and a_0, a_1, a_2, a_3, and a_4 present the constants used to predict the dependent variable y. a_0 is the intercept parameter and a_1, a_2, a_3, and a_4 are the slope parameters.

Based on the percentage of contribution of each independent variable in the regression model and the absolute values of partial correlation coefficients let us concluded that dependent variable y is influenced by several factors. Among these factors is Reference Distance, which has an important influence on the number of hidden layers.

Regression Model to Determine the Number of Hidden Neurons. The regression model used to calculate the number of hidden neurons will be determined using a set of factors selected in accordance with the above considerations from the results obtained through clustering of the training dataset. The number of hidden neurons depending on changes in the following independent variables.

- X_1: The number of features in the input
- X_2: The number of obtained cluster
- X_3: The reference distance
- X_4: The quality measure of the network structure

The independent variable X_1 represents the number of features of the training dataset. X_2 and X_3 present the obtained number of cluster and the reference distance respectively obtained using clustering of the training dataset.

A Multiple Linear Regression model representing the equation to determine the number of hidden neurons of the Multi-layer Perceptron have the following mathematical form:

$$y = a_0 + \sum_{j=1}^{4} a_j x_j \quad (3)$$

The dependent variable y represents the number of hidden neurons. The obtained number of hidden neurons will be evenly distributed to the hidden layers if the number of hidden layers exceeds one layer. Therefore, each hidden layer contains a number of neurons equal to others.

The independent factors influence on the number of hidden neurons with varying levels. The factor that has the highest influence being the Reference Distance.

The number of hidden neurons obtained using regression method will be divided equally by the number of hidden layers.

3.4 Clustering of the Training Dataset

The proposed regression method depends mainly on the results obtained from clustering of the training dataset. The most convenient clustering algorithm for the proposed method is Agglomerative Hierarchical Clustering algorithm [23]. Each cluster obtained through Agglomerative Hierarchical Clustering seeks to ensure the highest similarity of objects within the cluster and at the same time the highest dissimilarity between clusters [24]. Clusters obtained using Agglomerative Hierarchical Clustering can contain several sub-clusters then there will be a hierarchical clustering. The hierarchical clustering is a set of nested clusters that build a cluster tree (Dendrogram) to represent objects. The root of the tree represents the cluster, which group all other clusters and objects. In some cases, the leaves of the tree represent clusters of one objects. The Agglomerative Hierarchical Clustering algorithm [25–29] consider each object as a single cluster and then try to join the closest clusters until obtaining only one single cluster. The optimal number of clusters is determined by making a cut of all segment with a length greater than a predefined value [30]. This reference value (Reference Distance) is chosen according to specific criteria.

The value of Reference Distance, which is appropriate to obtain the optimal number of clusters, must attain a set of criteria. Implementation of the following criteria can make the number of obtained clusters useful for the proposed regression method to determine the structure of Multi-layer Perceptron.

The first criterion requires grouping at least ninety percent of the items of the training dataset. The ninety percent of items grouped considered sufficient where the result could cover the entire training dataset.

The second criterion requires that the number of clusters should be taken as few as possible in order to minimize the size of the network with reason that the increase of the number of clusters causes an increase in the number of hidden layers and the number of hidden neurons using the proposed method. With a few numbers of hidden layers and neurons the complexity of Multi-layer Perceptron reduce [31].

The third criterion requires a Reference Distance value in which any increase on it does not affect the number of obtained clusters [32–34]. While taking into account the condition, which should be avoided such as the very short value of Reference Distance for which each leaf of the tree, represents a cluster of one object or a relatively large value of Reference Distance for which grouping all objects in one cluster.

The fourth criterion requires the right selection of distance metrics (such as Manhattan and Euclidean) and linkage methods (single, complete, and average linkage) appropriate to the clustering algorithm and the type of training dataset [35]. A good choice of distance metrics increases the accuracy of the proposed method.

By following these criteria, the results of clustering of training dataset can generate a set of parameters useful to construct the regression models used for determining the optimal structure of a Multi-layer Perceptron.

4 Experimental Results

A number of experimental tests will be conducted to prove the effectiveness of the proposed method. The training dataset will be trained using different Multi-layer Perceptron structure then compared to results of the proposed method.

In this paper, the Waveform Database Generator Version 1 Dataset used to prove the validity of the proposed method. Waveform dataset consists of 21 attributes and 5000 instances. Dataset classes are generated from a combination of two of three "base" waves.

4.1 Experimental Results Obtained from Clustering of the Training Dataset

Agglomerative Hierarchical clustering is used to cluster the Waveform dataset. The number of clusters varies based on the value of Reference Distance. According to the criteria described above the perfect Reference Distance value is 8.

Figures 2 and 3 below presents the number of clusters and the percentage of objects clustered and the corresponding values of Reference Distance.

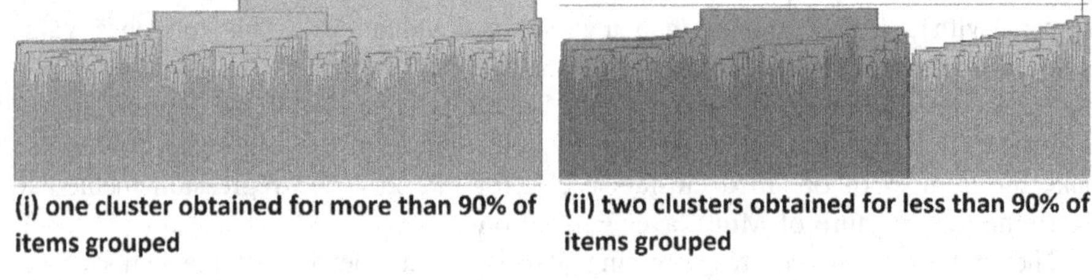

(i) one cluster obtained for more than 90% of items grouped
(ii) two clusters obtained for less than 90% of items grouped

Fig. 2. Clusters obtained based on the percentage of items grouped

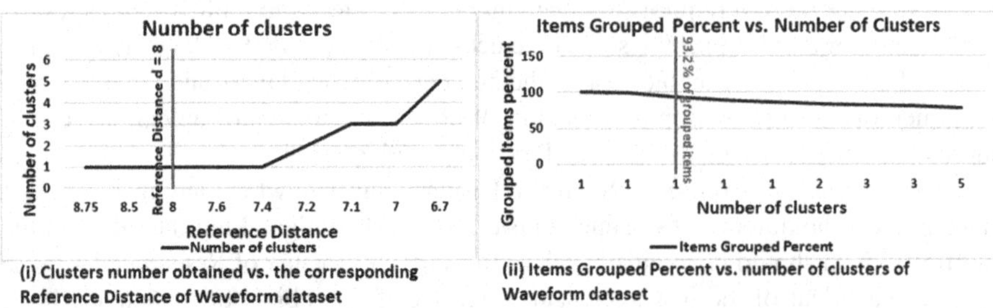

Fig. 3. Clusters number obtained vs. the corresponding reference distance of waveform dataset and items grouped percent vs. number of clusters of waveform dataset

The appropriate parameters selected in this case study is Normalization type "standard", clusters distance "Average Link" and "Manhattan" distance.

Figure 2 presents the obtained Dendrogram and the corresponding number of clusters obtained according to the reference distance values. Based on criteria listed above the optimal number of clusters is one cluster.

According to the criteria described above, we conclude that the ideal number of clusters is one cluster with 93.2% of items clustered for a value of Reference Distance equal to 8. The results obtained can be useful for determining the regression models used to construct the optimal structure of Multi-layer Perceptron for training the Waveform dataset.

4.2 Calculating the Number of Hidden Layers Using Regression Model

Based on the clustering of Waveform dataset the independent factors used to determine the Eq. (2) for calculating the number of hidden layers has the following values: $X_1 = 1 \times 8$, $X_2 = 8$, $X_3 = 93.2$, $X_4 = 98.38$.

According to the criteria described above, the selected value of Reference Distance is 8 for a percentage of grouped items more than 90% and the corresponding number of clusters is one, therefore, the value of X_1 will be 8×1 equal to 8. X_2 represents the value of Reference Distance therefore, $X_2 = 8$. X_3 represents the percentage of grouped items therefore $X_3 = 93.2$ as we see in Fig. 3. For X_4 which represents the quality measure of the network structure, it was determined by creating a structure based on the number of hidden layers equal to the obtained number of clusters corresponding to the selected Reference Distance and for the number of neurons is determined using the Formula (9). Based on that the quality measure of the network structure $X_4 = 98.38$ corresponding to the selected Reference Distance.

Using the above values in Eq. (2) will result in $y = 1$.

$Y = 1$, this concludes that the optimal number of hidden layers using the proposed method is equal to one layer.

The values of the Multiple Determination Coefficient R^2 obtained is close to 1. R^2 expresses the level of variation of the number of hidden layers affected by the variation of selected independent variable X_1, X_2, X_3, and X_4. It proves the validation of the proposed model and the successful choice of independent factors.

4.3 Calculating the Number of Hidden Neurons Using Regression Model

The implementation of Eq. (3) to calculate the number of neurons in the hidden layers use the following values of independent factors obtained from the clustering of Waveform dataset.

Based on the clustering of Waveform dataset the independent factor used to determine the Eq. (3) for calculating the number of hidden layers has the following values.

$X_1 = 21$, $X_2 = 1$, $X_3 = 8$, $X_4 = 98.38$

Using the above values in Eq. (3) will result in $y = 74$.

$Y = 74$, this concludes that the optimal number of hidden neurons using the proposed method is equal to 74 hidden neurons.

R^2 obtained is close to 1. R^2 expresses the level of variation of the number of hidden neurons affected by the variation of selected independent variable X_1, X_2, X_3 and X_4. It proves the validation of the proposed model and the successful choice of independent factors.

4.4 Comparison of the Proposed Method with Classical Methods

To validate the results obtained using the proposed method a comparison with widely spread methods are conducted. The proposed regression method will be compared with the classical methods so that we can prove the validity of the proposed method. The following classical formulas will be used in this comparison:

In – number of input neurons
Out – number of output neurons
Hidden – number of hidden neurons
Training – number of training forms

$$Hidden = 1/2(In + Out) \tag{4}$$

$$Hidden = SQRT\,(1/2\,(In + Out)) \tag{5}$$

$$Hidden = (In + Out) * 2/3 \tag{6}$$

$$Hidden = Training\,/10\,(In + Out) \tag{7}$$

$$Hidden = (Training - Out)\,/\,In + Out + 1 \tag{8}$$

$$Hidden = 1/2\,(In + Out) + SQRT\,(Training) \tag{9}$$

Formula (10): The number of hidden neurons should be between the size of the input layer and the size of the output layer. Formula (11): The number of hidden neurons should be less than twice the size of the input layer. A set of datasets used such as Waveform Database Generator dataset, Image Segmentation dataset, Glass identification dataset, Landsat dataset, Sonar dataset, ECG dataset, QRS dataset, P-wave dataset and T-wave datasets. Table 1 Presents specifications of datasets used.

Table 1. Specifications of neural networks

Dataset	Sonar	ECG	P-wave	QRS	T-wave	Landsat	Glass	Segmentation	Waveform
Input neurons	60	6	2	2	2	4	9	19	21
Output neurons	2	16	16	16	16	7	7	7	3
Training items	208	452	452	452	452	6435	214	2310	5000

Table 2 presents the number of hidden neurons using the classical method:

Table 2. Number of hidden neurons using the classical method

Dataset	Formula (4)	Formula (5)	Formula (6)	Formula (7)	Formula (8)	Formula (9)	Formula (10)	Formula (11)
Sonar	31	5.75	41	0	6	45	2 < x < 60	x < 120
ECG	11	3.32	14	2	89	32.26	6 < x < 16	x < 12
P-wave	9	3	12	2	235	30.26	2 < x < 16	x < 4
QRS	9	3	12	2	235	30.26	2 < x < 16	x < 4
T-wave	9	3	12	2	235	30.26	2 < x < 16	x < 4
Landsat	5	2.24	7	58	1615	85.22	4 < x < 7	x < 8
Glass	8	2.83	10	1	31	22.63	9 < x < 7	x < 18
Segmentation	13	3.61	17	8	129	62	19 < x < 7	x < 38
Waveform	12	3.46	16	20	241	82.71	21 < x < 3	x < 42

Table 3 presents the number of hidden neurons using the proposed method:

Table 3. Number hidden neurons using the proposed method

Dataset	Neurons	Layers
Sonar	35	2
ECG	20	1
P-wave	10	2
QRS	21	1
T-wave	20	3
Landsat	24	2
Glass	207	2
Segmentation	75	1
Waveform	74	1

Comparison Based on the Training Time Using Classical Methods vs. the Proposed Method. A comparison of the training time using classical methods vs. the proposed method results that the training time of the proposed method does not have the best training time for all datasets but for ECG, P-wave and QRS perform well. For example, the training time of ECG is 0.75 s using the proposed method and the best classical method record is equal to 1.47 s. The failure of the proposed method with some datasets to obtain the best training time is because of the number of neurons selected. The training time depends mainly on the number of neurons in the network and the size of dataset regardless of the used formula. Since the number of neurons is selected based on the complexity of the problem, therefore, the training time is affected by the complexity of the problem using the proposed method.

Comparison Based on the Percentage of Accuracy Using Classical Methods vs. the Proposed Method. Table 4 below shows a comparison of the results in terms of percentage of classification accuracy [36] using classical methods vs. the proposed method.

Table 4. Comparison of the percentage of accuracy using classical methods vs. the proposed method

Dataset	Formula (4)	Formula (5)	Formula (6)	Formula (7)	Formula (8)	Formula (9)	Formula (10)	Formula (11)	proposed method
Sonar	81.25	80.76	80.76	74.5192	81.73	81.25	81.7308	81.25	82.2115
ECG	57.9646	59.292	59.292	60.8407	59.292	59.292	57.9646	59.5133	60.8407
P-wave	53.7611	53.9823	53.9823	53.9823	53.9823	54.2035	53.9823	53.9823	54.2035
QRS	59.0708	58.8496	58.8496	59.5133	59.9558	58.8496	59.5133	59.5133	60.177
T-wave	53.9823	54.2035	53.9823	54.2035	56.1947	56.6372	56.6372	54.2035	56.6372
Landsat	76.7535	50.3006	80.5611	76.7535	82.5651	82.3647	84.8703	85.3213	85.6595
Glass	85.0467	72.8972	82.7103	61.6822	97.6636	94.3925	85.0467	82.7103	98.5981
Segmentation	97.5325	95.2381	97.9221	95.8442	97.6623	98.8312	95.7576	98.8312	99.1342
Waveform	95.74	89.28	97.32	97.86	98.22	98.4	96.82	97.86	98.6

As observed from Table 4 and Fig. 4, the proposed method has the best percentage of accuracy for most datasets. The classical methods sometimes get good results but it depends on the database. For example, formula (7) has a good percentage of accuracy for ECG dataset and the lowest percentage accuracy for sonar and Landsat datasets. Formula (8) obtained the highest percentage of accuracy compared to other classical methods for Glass dataset while getting the lowest percentage for ECG dataset. Formula (9) perform well with datasets Glass, Segmentation and waveform but for other dataset have a medium percentage of accuracy.

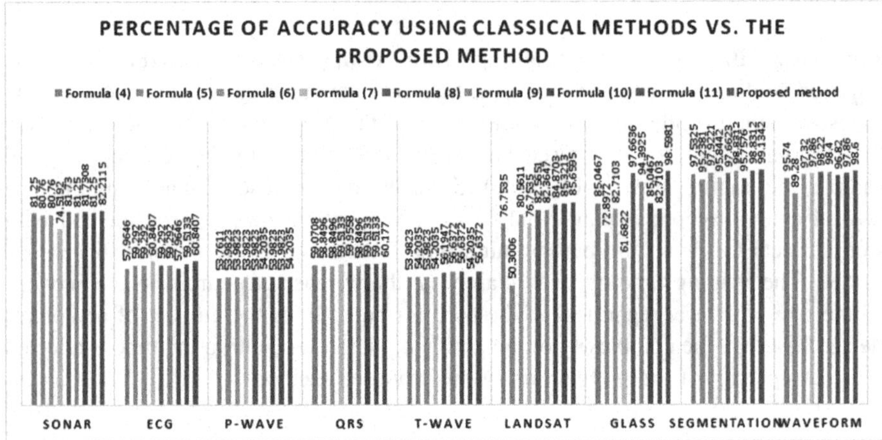

Fig. 4. Comparison of the percentage of accuracy using classical methods vs. the proposed method

Comparison Based on the Error/Epoch Using Classical Methods vs. the Proposed Method. Table 5 presents a comparison of the error/epoch [37] using classical methods vs. the proposed method:

Table 5. Comparison of the error/epoch using classical methods vs. the proposed method

Dataset	Formula (4)	Formula (5)	Formula (6)	Formula (7)	Formula (8)	Formula (9)	Formula (10)	Formula (11)	proposed method
Sonar	0.014593	0.035628	0.013602	0.046503	0.013424	0.014582	0.00485	0.01448	0.00487
ECG	0.031626	0.034692	0.031409	0.035248	0.031574	0.030976	0.03186	0.03314	0.03507
P-wave	0.042172	0.042182	0.042169	0.042182	0.042229	0.042180	0.04216	0.04218	0.04209
QRS	0.035101	0.035733	0.035060	0.036108	0.035463	0.035133	0.03524	0.03610	0.03548
T-wave	0.041593	0.042493	0.041322	0.042407	0.039277	0.039284	0.03913	0.04240	0.03913
Landsat	0.046556	0.087793	0.037348	0.046556	0.028518	0.028037	0.02892	0.03111	0.02671
Glass	0.036890	0.061292	0.036047	0.084371	0.003342	0.010134	0.03689	0.03604	0.00267
Segmenta-	0.005022	0.010025	0.004892	0.008103	0.003333	0.002705	0.00941	0.00365	0.00239
Waveform	0.026669	0.052205	0.017591	0.014369	0.011738	0.010616	0.02026	0.01436	0.00934

As observed in Table 5 and Fig. 5, the proposed method has the lowest values of error/epoch for most datasets. The classical methods sometimes get good results with formulas (8) (9) (10) these results somewhat acceptable compared to other classical methods. Formulas (5) and (7) have the highest values of error/epoch.

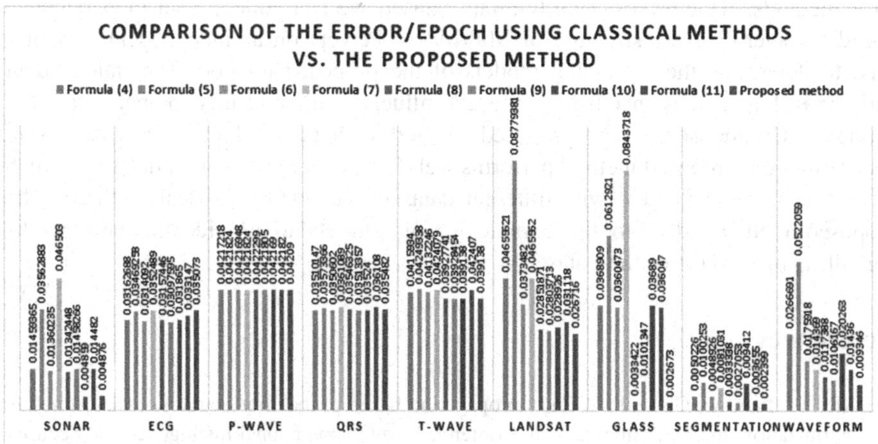

Fig. 5. Comparison of the error/epoch using classical methods vs. the proposed method

Comparison of Classical Methods vs. the Proposed Method Conclusion. The proposed method get the best percentage accuracy for most datasets, unlike the classical methods. The classical formulas (4) (5) (6) (10) (11) perform well with small datasets which have a few training items. Whereas Formulas (7) (8) (9) perform well with large datasets because they take into consideration the number of training items. Formula (9) is better than (7) and (8) which mean the SQRT of training items have a positive effect on the results. Formulas (4) (5) (6) depend mainly on the number of input and output neurons making it effective for a small datasets while do not perform well with large datasets which have complex problems to solve. The training time depends mainly on the number of neurons in the network and the size of dataset regardless the used formula. The results of error/epoch obtained is almost similar to the result of the percentage of accuracy. Comparison of the proposed method with classical methods leads us to deduce that the proposed method performs well for the different type of datasets, which mean that the proposed method is more flexible with different datasets types than classical methods. The proposed method adapt to the complexity of datasets to provide the best results regardless of the size of the dataset. In some cases, the dataset is chosen with a size more than required, which leads to bad results using classical methods but this problem is avoided by using the proposed method since it focuses on the complexity of the problem to be solved regardless the size of the dataset.

5 Conclusion

It is noticeable that Pattern Recognition plays a significant role in the determination of the optimal structure of Multi-layer Perceptron using the proposed method. The proposed method makes the design of Multi-layer Perceptron unsupervised and helps to dispense with the need for designer experience and the waste of time using trial and error methods. By clustering the training dataset, we can collect a set of parameters useful to determine the structure of Multi-layer Perceptron as independent variables used to determine the regression models of the proposed method. The independent variable Reference Distance has the highest influence on the results compared to other variables. Comparison of the proposed method with classical methods leads us to deduce that the proposed method performs well for the different type of datasets, which mean that is more flexible with different datasets types than classical methods. The proposed method adapt to the complexity of datasets to provide the best results regardless of the size of the dataset.

References

1. Xie, Y., Fan, X., Chen, J.: Affinity propagation-based probability neural network structure optimization. In: Tenth International Conference on Computational Intelligence and Security (CIS), pp. 85–89. IEEE, November 2014. https://doi.org/10.1109/cis.2014.156
2. Thomas, A.J., Petridis, M., Walters, S.D., Gheytassi, S.M., Morgan, R.E.: On predicting the optimal number of hidden nodes. In: International Conference on Computational Science and Computational Intelligence (CSCI), pp. 565–570. IEEE, December 2015. https://doi.org/10.1109/csci.2015.33

3. Bishop, C.: Pattern Recognition and Machine Learning. Springer, New York (2006). ISBN 978-1-4939-3843-8
4. Pan, H., Liang, D., Tang, J., Wang, N., Li, W.: Shape recognition and retrieval based on edit distance and dynamic programming. Tsinghua Sci. Technol. **14**(6), 739–745 (2009). https://doi.org/10.1016/S1007-0214(09)70144-0
5. Amiri, S.S., Mottahedi, M., Asadi, S.: Using multiple regression analysis to develop energy consumption indicators for commercial buildings in the US. Energy Build. **109**, 209–216 (2015). https://doi.org/10.1016/j.enbuild.2015.09.073
6. Dora, S., Sundaram, S., Sundararajan, N.: A two stage learning algorithm for a growing-pruning spiking neural network for pattern classification problems. In: International Joint Conference on Neural Networks (IJCNN), pp. 1–7. IEEE, July 2015. https://doi.org/10.1109/ijcnn.2015.7280592
7. Sheela, K.G., Deepa, S.N.: Review on methods to fix number of hidden neurons in neural networks. Math. Prob. Eng. (2013). http://dx.doi.org/10.1155/2013/425740
8. Berry, M.J., Linoff, G.: Data Mining Techniques: For Marketing, Sales, and Customer Support. Wiley, New York (1997). ISBN 0471179809
9. Esfe, M.H., et al.: Thermal conductivity of Cu/TiO2–water/EG hybrid nanofluid: experimental data and modeling using artificial neural network and correlation. Int. Commun. Heat Mass Transfer **66**, 100–104 (2015). https://doi.org/10.1016/j.icheatmasstransfer.2015.05.014
10. Vinod, V.V., Ghose, S.: Growing nonuniform feedforward networks for continuous mappings. Neurocomputing **10**(1), 55–69 (1996). https://doi.org/10.1016/0925-2312(95)00024-0
11. Faraway, J.J.: Extending the Linear Model with R: Generalized Linear, Mixed Effects and Nonparametric Regression Models, vol. 124. CRC Press, Boca Raton (2016)
12. Dangeti, P.: Statistics for Machine Learning. Packt Publishing Ltd, Birmingham (2017)
13. Brown, S.H.: Multiple linear regression analysis: a matrix approach with MATLAB. Alabama J. Math. **34**, 1–3 (2009)
14. Austin, P.C., Steyerberg, E.W.: The number of subjects per variable required in linear regression analyses. J. Clin. Epidemiol. **68**(6), 627–636 (2015). https://doi.org/10.1016/j.jclinepi.2014.12.014
15. Sasaki, T., Kinoshita, K., Kishida, S., Hirata, Y., Yamada, S.: Effect of number of input layer units on performance of neural network systems for detection of abnormal areas from X-ray images of chest. In: IEEE 5th International Conference on Cybernetics and Intelligent Systems (CIS), pp. 374–379. IEEE, September 2011. https://doi.org/10.1109/iccis.2011.6070358
16. Naseem, I., Togneri, R., Bennamoun, M.: Linear regression for face recognition. IEEE Trans. Pattern Anal. Mach. Intell. **32**(11), 2106–2112 (2010). https://doi.org/10.1109/TPAMI.2010.128
17. Pozo, F., Vidal, Y.: Wind turbine fault detection through principal component analysis and statistical hypothesis testing. Energies **9**(1), 3 (2015). https://doi.org/10.3390/en9010003
18. Cohen, P., West, S.G., Aiken, L.S.: Applied Multiple Regression/Correlation Analysis for the Behavioral Sciences. Psychology Press, New York (2014). ISBN 9781135468255
19. Wang, W., Morrison, T.A., Geller, J.A., Yoon, R.S., Macaulay, W.: Predicting short-term outcome of primary total hip arthroplasty: a prospective multivariate regression analysis of 12 independent factors. J. Arthroplasty **25**(6), 858–864 (2010). https://doi.org/10.1016/j.arth.2009.06.011

20. Ghaedi, M., Reza Rahimi, M., Ghaedi, A.M., Tyagi, I., Agarwal, S., Gupta, V.K.: Application of least squares support vector regression and linear multiple regression for modeling removal of methyl orange onto tin oxide nanoparticles loaded on activated carbon and activated carbon prepared from Pistacia atlantica wood. J. Colloid Interface Sci. **461**, 425–434 (2016). https://doi.org/10.1016/j.jcis.2015.09.024
21. Chatterjee, S., Hadi, A.S.: Regression Analysis by Example. Wiley, New York (2015)
22. Krizhevsky, A., Sutskever, I., Hinton, G.E.: Parallelizing neural networks during training. U. S. Patent 9,811,775, Google Inc. (2017)
23. Bouguettaya, A., Yu, Q., Liu, X., Zhou, X., Song, A.: Efficient agglomerative hierarchical clustering. Expert Syst. Appl. **42**(5), 2785–2797 (2015). https://doi.org/10.1016/j.eswa.2014.09.054
24. Ng, M.K., Li, M.J., Huang, J.Z., He, Z.: On the impact of dissimilarity measure in k-modes clustering algorithm. IEEE Trans. Pattern Anal. Mach. Intell. (3), 503–507 (2007). http://doi.ieeecomputersociety.org/10.1109/TPAMI.2007.53
25. Karypis, G., Han, E.H., Kumar, V.: Chameleon: hierarchical clustering using dynamic modeling. Computer **32**(8), 68–75 (1999). https://doi.org/10.1109/2.781637
26. Murtagh, F., Contreras, P.: Algorithms for hierarchical clustering: an overview, II. Wiley Interdisc. Rev. Data Min. Knowl. Discov. **7**(6), e1219 (2017). https://doi.org/10.1002/widm.1219
27. Dalbouh, H.A., Norwawi, N.M.: Improvement on agglomerative hierarchical clustering algorithm based on tree data structure with bidirectional approach. In: Third International Conference on Intelligent Systems, Modelling and Simulation (ISMS), pp. 25–30. IEEE, February 2012. https://doi.org/10.1109/isms.2012.13
28. Aggarwal, C.C., Reddy, C.K. (eds.): Data Clustering: Algorithms and Applications. CRC Press, Boca Raton (2013). ISBN 1466558210, 9781466558212
29. Gath, I., Geva, A.B.: Unsupervised optimal fuzzy clustering. IEEE Trans. Pattern Anal. Mach. Intell. **11**(7), 773–780 (1989). https://doi.org/10.1109/34.192473
30. Langfelder, P., Zhang, B., Horvath, S.: Defining clusters from a hierarchical cluster tree: the dynamic tree cut package for R. Bioinformatics **24**(5), 719–720 (2007). https://doi.org/10.1093/bioinformatics/btm563
31. Zhao, Z., Xu, S., Kang, B.H., Kabir, M.M.J., Liu, Y., Wasinger, R.: Investigation and improvement of multi-layer perceptron neural networks for credit scoring. Expert Syst. Appl. **42**(7), 3508–3516 (2015). https://doi.org/10.1016/j.eswa.2014.12.006
32. Raghuvanshi, A.S., Tiwari, S., Tripathi, R., Kishor, N.: Optimal number of clusters in wireless sensor networks: an FCM approach. In: International Conference on Computer and Communication Technology (ICCCT), pp. 817–823. IEEE, September 2010. https://doi.org/10.1109/iccct.2010.5640391
33. Wang, L.C., Wang, C.W., Liu, C.M.: Optimal number of clusters in dense wireless sensor networks: a cross-layer approach. IEEE Trans. Veh. Technol. **58**(2), 966–976 (2009). https://doi.org/10.1109/TVT.2008.928637
34. Liu, X., Croft, W.B.: Experiments on retrieval of optimal clusters. Technical report IR-478, Center for Intelligent Information Retrieval (CIIR), University of Massachusetts (2006)
35. Kumar, V., Chhabra, J.K., Kumar, D.: Performance evaluation of distance metrics in the clustering algorithms. INFOCOMP **13**(1), 38–52 (2014)
36. Piczak, K.J.: Environmental sound classification with convolutional neural networks. In: IEEE 25th International Workshop on Machine Learning for Signal Processing (MLSP), pp. 1–6. IEEE, September 2015. https://doi.org/10.1109/mlsp.2015.7324337
37. Lillicrap, T.P., Cownden, D., Tweed, D.B., Akerman, C.J.: Random synaptic feedback weights support error backpropagation for deep learning. Nature Commun. **7**, 13276 (2016). https://doi.org/10.1038/ncomms13276

An Effective Machine Learning Framework for Data Elements Extraction from the Literature of Anxiety Outcome Measures to Build Systematic Review

Shubhaditya Goswami[1], Sukanya Pal[1], Simon Goldsworthy[2,3], and Tanmay Basu[1(✉)]

[1] Ramakrishna Mission Vivekananda Educational and Research Institute, Belur Math, West Bengal, India
gogol.goswami@gmail.com, sukanyapal19@gmail.com, welcometanmay@gmail.com

[2] Taunton and Somerset NHS Foundation Trust, Beacon Centre, Taunton, Somerset, UK
Simon.Goldsworthy@tst.nhs.uk

[3] University of the West of England, Bristol, UK

Abstract. The process of developing systematic reviews is a well established method of collecting evidence from publications, where it follows a predefined and explicit protocol design to promote rigour, transparency and repeatability. The process is manual and involves lot of time and needs expertise. The aim of this work is to build an effective framework using machine learning techniques to partially automate the process of systematic literature review by extracting required data elements of anxiety outcome measures. A framework is thus proposed that initially builds a training corpus by extracting different data elements related to anxiety outcome measures from relevant publications. The publications are retrieved from Medline, EMBASE, CINAHL, AHMED and Pyscinfo following a given set of rules defined by a research group in the United Kingdom reviewing comfort interventions in health care. Subsequently, the method trains a machine learning classifier using this training corpus to extract the desired data elements from new publications. The experiments are conducted on 48 publications containing anxiety outcome measures with an aim to automatically extract the sentences stating the mean and standard deviation of the measures of outcomes of different types of interventions to lessen anxiety. The experimental results show that the recall and precision of the proposed method using random forest classifier are respectively 100% and 83%, which indicates that the method is able to extract all required data elements.

Keywords: Information extraction · NLP · Health informatics · Systematic review · Text mining · Machine learning

---

S. Goswami and S. Pal have equal contribution in this work.

W. Abramowicz and R. Corchuelo (Eds.): BIS 2019, LNBIP 353, pp. 247–258, 2019.
https://doi.org/10.1007/978-3-030-20485-3_19

1 Introduction

A systematic review is an article that summarizes information about a clinical question or topic from different health care publications, where the topic is discussed [1]. The synopsis of different data elements for a particular topic collected from a referred article is generally included in systematic reviews [1]. Some examples of data elements are population of an intervention, inclusion criteria for testing the effect of a drug etc. The clinical researchers manually extract these data elements from the publications and build a systematic review over the years [2]. Despite their widely acknowledged usefulness, the data extraction phase of the systematic review process is time-consuming [1,2]. A research collaboration was initiated by a research Radiation TherapisT (RTT) at Taunton and Somerset NHS Foundation Trust in UK to identify and characterize effective comfort interventions used in health care practices to comfort a patient undergoing a clinical procedure [3]. A systematic review was planned to summarize and appraise published evidence of the effective comfort interventions used in health care disciplines.

The objective of this work is to develop an effective machine learning framework to identify data elements of interest from the relevant literature of anxiety outcome measures in the clinical trials of comfort intervention. There have been many research works on the effectiveness of certain interventions for reducing the anxiety levels of patients undergoing operative treatment. An intervention is a combination of strategies designed to promote behavioural changes or improve health condition of individuals. The population is divided into two groups - intervention and control group. The group receiving the intervention is called the intervention group. The control group is defined as the group that does not receive the intervention and is then used as a benchmark to measure how the other tested participants are responding. For comfort interventions, the anxiety levels are measured on the basis of the scores provided by the participants of the intervention and control group for different anxiety related questions in a questionnaire provided to them by the experts, both at the beginning and end of the treatment. The mean and standard deviation values of the anxiety scores for different interventions mentioned in the related publications are thus the required data elements in this work that have to be extracted in order to create a systematic review.

The main challenge in this kind of data is that the required data elements do not appear in fixed patterns in the articles. The elements can be found either in tables or in plain text. Moreover, the data elements may occur in different contexts in an article. It is difficult to find the required data elements by searching the free texts and hence a tool is required to accomplish this task.

A machine learning framework is thus proposed here to identify data elements related to the interventions of anxiety. The work has been carried out in collaboration with the RTT, who collected and annotated the data. The proposed framework consists of two phases. In the first phase, it builds a training corpus by extracting the sentences containing the means and standard deviations of different anxiety outcome measures from the publications collected over

Medline, EMBASE, CINAHL, AHMED and Pyscinfo following a given set of rules defined by the RTT. Therefore, in the second stage, a machine learning classifier is trained using the training corpus to extract the desired data elements from new publications. The bag of words are used as features and the conventional tf-idf weighting scheme is followed to create the term document matrix [4]. The experiments are conducted on 48 publications related to anxiety outcome measures. The training corpus is developed using 40 files randomly chosen from the entire collection. It contains two classes - anxiety and non-anxiety. The anxiety class comprises of the sentences that contain the required data elements of anxiety outcome measures. The rest of the sentences that does not contain any term related to anxiety belong to the non-anxiety category. The aim is to automatically extract the mean and standard deviation of scores of different types of interventions of anxiety from the rest 8 publications, which is considered as test set. The required data elements of these 8 publications are manually annotated by the RTT to evaluate the performance of the proposed framework. Three different classifiers viz., Support Vector Machine (SVM), Random Forest and Logistic Regression have been explored to accomplish this task. Random forest classifier outperforms the other classifiers in terms of precision, recall and f-measure on the test set. The experimental results show that the proposed framework using random forest classifier is able to identify the sentences containing the required data elements with 100% recall and 83% precision on the test set. Consequently it can be concluded that the proposed framework will be effective for data extraction from the literature of anxiety outcome measures.

The paper is organized as follows. The related works to this study are described in Sect. 2. Section 3 explains the proposed framework. The experimental evaluation is presented in Sect. 4. The merits and limitations of the proposed method is discussed in Sect. 5. Finally, we conclude with the scopes of future works in Sect. 6.

2 Related Works

There are a lot of research works to create machine learning frameworks that extract information from clinical text, most of which focus on named entity recognition (NER) or mapping to medical concepts. NER is the task of identifying named entities e.g., patient names, drug names, hospital names, clinical procedure, protein names etc. in text [5]. The NER methods are broadly divided into two types - traditional machine learning models like SVM, conditional random fields (CRF) and deep learning models e.g., recurrent neural networks using word embeddings [5]. Informatics for Integrating Biology and the Bedside (i2b2), a US National Center for Biomedical Computing have organized different workshops on Natural Language Processing (NLP) for automatic de-identification of protected health information (PHI) from relevant electronic health records [6–10]. There are 18 types of PHI as defined by US Health Insurance Portability and Accountability Act and these workshops released individual corpora with manually annotated PHIs as gold standards for performance evaluation.

Thus the systems developed over these i2b2 workshops are focused on training machine learning or deep learning models by identifying potential features using the gold standards. Gobbel et al. created a token order specific naive Bayes classifier called RapTAT that maps free text phrases to pre-defined concepts over Systematized Nomenclature of Medicine - Clinical Terms (SNOMED CT) [11]. There have been some research works for identifying medical named entities in languages other than English e.g., in Chinese [12], French [13] and Swedish [14]. Nevertheless these studies mostly address the issue of identifying regular named entities over clinical records.

Basu et al. [1] developed a method to extract relevant data elements related to congestive heart failure. They have used different systematic reviews over PubMed related to congestive heart failure to build the training corpus. Marshal et al. developed a web based system, named RobotReviewer that uses machine learning and NLP for finding the Risk of Bias (RoB) of how a particular clinical study was performed [15]. It generates a report summarizing the key information from the randomized controlled trials, which includes, e.g., details concerning trial participants, interventions, and reliability in relation to RoB only [15]. Specifically, this system uses the Cochrane RoB Checklist[1] to assess RoB in randomized controlled trials which includes six areas, namely, random assignment generation, allocation concealment, blinding of participants and personnel, blinding of outcome assessment, incomplete outcome data, and selective reporting. Each area is given either "low", "high" or "unclear" risk of bias [16]. It may be noted that the information regarding the outcome measures related to interventions of anxiety can not be identified by RobotReviewer as examined by the RTT.

There are a growing number of research works to identify depression or mental illness over social media [17]. Choudhury et al. developed a crowd-sourcing strategy of collecting ground truth data on depression from Twitter, and devised a variety of measures such as language, emotion, style and user engagement to train a SVM classifier to identify depression over social media postings [18]. Shen et al. proposed a multimodal depressive dictionary learning method to detect depressed users in Twitter [19]. It may be noted that these studies deal with the methods to build training corpora specific to some particular social media e.g., Twitter. However, there is hardly any study that discuss the issue of identifying anxiety outcome measures from relevant publications.

3 Proposed Framework

A supervised machine learning framework is introduced in this paper to identify the sentences containing the required data elements related to anxiety outcome measures. The framework consists of two major steps as described below.

[1] https://training.cochrane.org/handbook.

3.1 Building Training Corpus

The data used for this work include 48 articles in PDF format related to anxiety outcome measures. The PDFs are converted to free text using Fitz[2], a library in Python. The resulting free texts from the PDF documents consist of a lot of special characters, which are wrongly converted. For example, '=' symbols in the PDF documents are converted to '1/4' in the free texts. In such cases, all the '1/4' symbols in the free text documents are replaced by '=' using an appropriate regular expression. Similarly, if the special character '�' comes between two digits in the free text then it is replaced by a decimal point, otherwise, it is removed from the text. Some texts do not have space between the delimiter of a sentence and the first character of the next sentence. In that case, a space is added there. Subsequently, the sentences are extracted following the delimiters. Different such rules have been applied to clean the free text documents.

The training corpus is developed by randomly choosing 40 out of 48 text documents. The aim is to build two classes - anxiety and non anxiety. In principle, the anxiety class should contain the sentences that have the required data elements related to anxiety outcome measures and the sentences that do not contain any data elements or terms related to anxiety should construct the non-anxiety class. The following rules are used to build the training corpus.

- Certain keywords related to anxiety outcome measures are used to identify a sentence that belong to the anxiety class. This set of keywords are mentioned in Algorithm 1. The keywords have been suggested by the RTT.
- The sentences that do not contain any such keyword or the term 'anxiety' are assigned to the non-anxiety class.
- There are some sentences that contain either any keyword or the term 'anxiety', but do not contain the required anxiety outcome measures. These sentences are discarded from the training corpus.

Initially, the sentences of each of these 40 text documents are extracted. Subsequently, a sentence is either labelled as anxiety or non-anxiety class or ignored according to the above rules. The required data elements may exist in different tables in the PDFs. A table is identified as a sentence from the text document. A rule is applied that if the word "TABLE" is followed by at least 2 newlines in the same sentence then that sentence is considered as a table. Note that, only certain rows of a table can contain the required data and the rest of the rows may either belong to the non-anxiety class or ignored. Thus the training corpus is developed. The detailed steps to build the training corpus is described in Algorithm 1.

3.2 Identification of Desired Data Elements

The stopwords are discarded from the texts using the standard English stop word list in NLTK[3]. All the remaining terms in the sentences are lemmatized

[2] https://pypi.org/project/PyMuPDF/1.9.2/.
[3] https://www.nltk.org.

Algorithm 1. Proposed Framework to Build the Training Corpus

Input : 1) A given set of free text documents
2) *Keywords* ← {stai, s-stai, staiy-1, state-trait, bai beck anxiety inventory, state anxiety, trait anxiety, anxiety level}
Steps:

```
for each document do
    extract sentences from a document following regular delimiters
    for each sentence in a document do
        convert each character to lower case
        if sentence is a table following the given rules then
            break table into rows
            for each row in table do
                if it contains any of the given Keywords then
                    anxiety_class ← row
                else if it has no mention of 'anxiety' then
                    non-anxiety_class ← row
                end if
            end for
        else
            if the sentence contains the given Keywords and digits and the word
              'SD' then
                anxiety_class ← sentence
            else if the word 'anxiety' does not exist then
                non-anxiety_class ← sentence
            end if
        end if
    end for
end for
return anxiety_class, non-anxiety_class
```

to map the grammatical variations of the same word to their root word. The documents contain lots of digits, which are removed, since they are not good features to train the classifier in this framework.

The text documents are generally represented by the bag of words (BOW) model [4]. In this model, each document in a corpus is generally represented by a vector, whose length is equal to the number of unique terms, also known as vocabulary. Let us denote the number of documents of the corpus and the number of terms of the vocabulary by n and m respectively. Number of times the ith term t_i occurs in the jth document is denoted by tf_{ij}, $i = 1, 2, ..., m$; $j = 1, 2, ..., n$. Document frequency df_i is the number of documents in which a particular term appears. Inverse document frequency determines how frequently a term occurs in a corpus and it is defined as $idf_i = log(\frac{n}{df_i})$. The weight of the ith term in the jth document, denoted by w_{ij}, is determined by combining the term frequency with the inverse document frequency as follows:

$$w_{ij} = tf_{ij} \times idf_i = tf_{ij} \times log(\frac{n}{df_i}), \quad \forall i = 1, ..., m \text{ and } \forall j = 1, ..., n$$

This term weighting scheme is known as *tf-idf* weighting. The documents can be efficiently represented using the vector space model in most of the text mining algorithms [4]. In this model each document d_j is considered to be a vector $\mathbf{d_j}$, where the ith component of the vector is w_{ij}, i.e., $\mathbf{d_j} = (w_{1j}, w_{2j}, ..., w_{nj})$. This *tf-idf* weighting scheme is used to represent the document vectors in the proposed framework. The term-document matrix of the training corpus following the tf-idf weighting scheme is used to train a state of the art classifier. Therefore the classifier identifies the class labels of the sentences in the individual documents of the test set.

4 Experimental Evaluation

4.1 Experimental Settings

The performance of random forest, SVM and logistic regression classifiers have been explored to classify the sentences of a document in the test set. It may be noted that these classifiers are widely used for text classification. The parameters of these classifiers are tuned using 10-fold cross validation on the training set. Therefore the best set of parameters are used to classify the sentences of the documents in the test set. The sentences are either categorized to anxiety class or to the non-anxiety class. The classifiers are implemented using scikit-learn[4], a machine learning tool in Python [20].

4.2 Evaluation Criteria

The performance of the proposed framework is evaluated by using the standard precision, recall and f-measure [4]. The precision and recall for two class classification problem can be computed as

$$\text{Precision} = \frac{\text{TP}}{\text{TP} + \text{FP}}$$
$$\text{Recall} = \frac{\text{TP}}{\text{TP} + \text{FN}}$$

Here TP stands for *true positive* and it counts the number of sentences correctly predicted to the anxiety class. FP stands for *false positive* and it counts the number of sentences that actually belong to the non-anxiety class, but predicted as anxiety. FN stands for *false negative* and it counts the number of sentences that actually belong to the anxiety class, but predicted as non-anxiety. TN stands for *true negative* and it counts the number of sentences correctly predicted to the non-anxiety class. The f-measure combines recall and precision with an equal weight in the following form:

$$\text{F-measure} = \frac{2 \times \text{recall} \times \text{precision}}{\text{recall} + \text{precision}}$$

[4] http://www.scikit-learn.org.

The closer the values of precision and recall, the higher is the f-measure [21]. F-measure becomes 1 when the values of precision and recall are 1 and it becomes 0 when precision is 0, or recall is 0, or both are 0. Thus f-measure lies between 0 and 1. A high f-measure value is desirable for good classification.

4.3 Analysis and Results

The training and test corpora respectively contain 7343 sentences and 1855 sentences. The training corpus has 43 sentences for anxiety class and 7300 sentences for non-anxiety class. The bag of words model has the vocabulary size of 5789, which is reduced to 5098 by stopwords removal and lemmatization. The performance of random forest, support vector machine and logistic regression classifiers to classify the sentences of the 8 documents of the test set are reported in Table 1. Table 1 shows that random forest classifier is able to find all the true positive sentences for each document in the test set and consequently there is no false negative sentence for each document. On the other hand, the performance of SVM in terms of number of true negative and false positive sentences is better than random forest and logistic regression classifiers for most of the documents in the test set. However, the objective of this work is to identify sentences that contain the required data elements for anxiety outcome measures. Note that here it is more important for a classifier to reduce the number of false negative sentences than false positive sentences, since false negative sentences are containing the required data elements. Hence it can be concluded from Table 1 that random forest classifier outperforms SVM and logistic regression. An example of false positive sentence as identified by all three classifiers is as follows.

Table 1. Performance of different classifiers

File	True positive			False negative			False positive			True negative		
	RF[a]	SVM[b]	LR[c]	RF	SVM	LR	RF	SVM	LR	RF	SVM	LR
File 1	**4**	3	2	**0**	1	2	34	**21**	21	196	**209**	209
File 2	**1**	1	1	**0**	0	0	2	**2**	2	163	**163**	163
File 3	**5**	3	4	**0**	2	1	14	**9**	9	257	**262**	262
File 4	**3**	3	3	**0**	0	0	20	**5**	6	159	**174**	173
File 5	**11**	5	5	**0**	6	6	54	32	**30**	293	315	**317**
File 6	**4**	4	4	**0**	0	0	**20**	21	24	**139**	138	135
File 7	**6**	4	5	**0**	2	1	70	**40**	42	106	**136**	134
File 8	**9**	5	5	**0**	4	4	49	**23**	32	236	**262**	253

[a]RF: Random Forest; [b]SVM: Support Vector Machine; [c]LR: Logistic Regression classifiers.

The results from the linear regression analysis indicated that lavender-sandalwood aromatherapy statistically significantly reduced anxiety (p = .032) compared with placebo use [22].

Table 2. Performance of different classifiers after refinement

File	True positive			False negative			False positive			True negative		
	RF[a]	SVM[b]	LR[c]	RF	SVM	LR	RF	SVM	LR	RF	SVM	LR
File 1	**4**	3	2	**0**	1	2	0	0	0	196	**209**	209
File 2	**1**	1	1	**0**	0	0	0	0	0	163	**163**	163
File 3	**5**	3	4	**0**	2	1	2	2	2	257	**262**	262
File 4	**3**	3	3	**0**	0	0	1	1	1	159	**174**	173
File 5	**11**	5	5	**0**	6	6	3	3	3	293	315	**317**
File 6	**4**	4	4	**0**	0	0	0	0	0	**139**	138	135
File 7	**6**	4	5	**0**	2	1	4	4	4	106	**136**	134
File 8	**9**	5	5	**0**	4	4	3	3	3	236	**262**	253

[a]RF: Random Forest; [b]SVM: Support Vector Machine; [c]LR: Logistic Regression classifiers.

Table 3. Precision, Recall, F-measure of different classifiers after refinement

Classifier	Precision	Recall	F-Measure
Random Forest	**0.83**	**1**	**0.9**
Support Vector Machine	0.76	0.76	0.75
Logistic Regression	0.78	0.79	0.77

This sentence is classified to the anxiety class as it has more similarity with the sentences of the anxiety class than that of non-anxiety class using the bag of words model with tf-idf weighting scheme. To reduce the number of such false positive sentences identified by individual classifiers, a rule is used on the sentences that are classified to the anxiety class. The rule is to remove all the sentences from the anxiety class that do not contain the term 'SD', where 'SD' stands for standard deviation. The performance of the classifiers after applying this rule are reported in Table 2. It can be observed from Table 2 that the false positive sentences identified by the individual classifiers are significantly reduced. The precision, recall and f-measure of the individual classifiers are reported in Table 3. Table 3 shows that random forest classifier outperforms the other classifiers in terms of precision, recall and f-measure. The effectiveness of the proposed framework for extraction of data elements of anxiety outcome measures can be observed from these results.

5 Discussion

It may be noted that no clear system is currently available that automatically extracts outcome measures of comfort interventions to populate the results of a systematic literature review paper. It has been mentioned in Sect. 2 that Basu et al. developed a method to extract relevant data elements related to congestive heart failure [1]. They use different systematic reviews over PubMed related to congestive heart failure to build the training corpus. They introduced a sentence matching technique to retrieve the sentences from the full text articles that correspond to the manually extracted data elements mentioned in the relevant systematic reviews of congestive heart failure. Therefore they have trained a SVM classifier using the bag of words model to extract data elements of interest from new publications. The proposed method for building training corpus is different than the system developed by Basu et al. [1] as it does not require any existing systematic reviews in this regard. It only needs some keywords as identified by the RTT to extract relevant sentences to build the training corpus. In the second stage the proposed method performs different text classification algorithms e.g., SVM, random forest using the bag of words model to identify the data elements of interest. Eventually the random forest classifier is found to be significant for the proposed study rather than SVM, which was used by Basu et al. [1]. The same has been discussed in Sect. 4.3.

The proposed framework has shown promising results in extracting the data elements of anxiety outcome measures, but it is not without limitations. It does not populate the results directly into a systematic review paper and also does not make any judgments about the results. Note that this system works on the free texts and it can not read data from the texts inside a figures or charts. Furthermore, some specific rules have been used in the proposed system to clean the free text documents and subsequently the training corpus is developed. These rules may have to be changed as per the requirements for the outcome measures of other diseases.

Although further work is required, the proposed system has a lot of potential. Currently it can support clinical researchers by ensuring the quality of data extraction and the developments will ensure timely release of results from systematic reviews. There is a potential that the latest evidence is released sooner and subsequently could benefit the outcomes of patients.

6 Conclusions

A method to automatically build training corpus followed by a sentence classification framework is proposed to identify required data elements for anxiety outcome measures from relevant publications. The sentence classification framework is developed using conventional bag of words features and state of the art classifiers. The value and validity of the framework is observed from the empirical analysis. However, the random forest classifier identifies all sentences containing required data elements, but it extracts a large number of false positive sentences. Deep learning methods using different types of word embeddings e.g., word2vec, Glove can be explored in future to improve the performance. Future development of the current system will also include a direct feed into a systematic review paper and make judgments on the extracted data in the form of a narrative synthesis through a linguistic training corpus that could be used by clinical researchers. Moreover, the effectiveness of the proposed framework may be tested for extracting required data elements of outcome measures from comfort interventions of various other diseases.

References

1. Basu, T., et al.: A novel framework to expedite systematic reviews by automatically building information extraction training corpora. arXiv preprint arXiv:1606.06424 (2016)
2. Jonnalagadda, S.R., Goyal, P., Huffman, M.D.: Automating data extraction in systematic reviews: a systematic review. Syst. Rev. **4**(1), 78 (2015)
3. Goldsworthy, S.D., Tuke, K., Latour, J.M.: A focus group consultation round exploring patient experiences of comfort during radiotherapy for head and neck cancer. J. Radiother. Pract. **15**(2), 143–149 (2016)
4. Basu, T., Murthy, C.: A supervised term selection technique for effective text categorization. Int. J. Mach. Learn. Cybern. **7**(5), 877–892 (2016)
5. Yadav, V., Bethard, S.: A survey on recent advances in named entity recognition from deep learning models. In: Proceedings of the International Conference on Computational Linguistics (COLING), pp. 2145–2158 (2018)
6. Uzuner, Ö., Luo, Y., Szolovits, P.: Evaluating the state-of-the-art in automatic de-identification. J. Am. Med. Inform. Assoc. **14**(5), 550–563 (2007)
7. Uzuner, Ö., Solti, I., Cadag, E.: Extracting medication information from clinical text. J. Am. Med. Inform. Assoc. **17**(5), 514–518 (2010)
8. Halgrim, S.R., Xia, F., Solti, I., Cadag, E., Uzuner, Ö.: A cascade of classifiers for extracting medication information from discharge summaries. J. Biomed. Semant. **2**(3), S2 (2011)
9. Stubbs, A., Kotfila, C., Uzuner, Ö.: Automated systems for the de-identification of longitudinal clinical narratives: overview of 2014 i2b2/UTHealth shared task Track 1. J. Biomed. Inform. **58**, S11–S19 (2015)
10. Stubbs, A., Filannino, M., Uzuner, Ö.: De-identification of psychiatric intake records: overview of 2016 CEGS N-GRID shared tasks Track 1. J. Biomed. Inform. **75**, S4–S18 (2017)
11. Gobbel, G.T., et al.: Development and evaluation of raptat: a machine learning system for concept mapping of phrases from medical narratives. J. Biomed. Inform. **48**, 54–65 (2014)
12. Zhang, B., Lu, M., Fang, Y.: A feature-enhanced entity recognition method for Chinese electronic medical records. In: 2018 9th International Conference on Information Technology in Medicine and Education (ITME), pp. 9–14. IEEE (2018)
13. Goeuriot, L., et al.: Overview of the CLEF eHealth evaluation lab 2015. In: Mothe, J., et al. (eds.) CLEF 2015. LNCS, vol. 9283, pp. 429–443. Springer, Cham (2015). https://doi.org/10.1007/978-3-319-24027-5_44
14. Dalianis, H., Velupillai, S.: De-identifying swedish clinical text-refinement of a gold standard and experiments with conditional random fields. J. Biomed. Semant. **1**(1), 6 (2010)
15. Marshall, I.J., Kuiper, J., Banner, E., Wallace, B.C.: Automating biomedical evidence synthesis: RobotReviewer. In: Proceedings of the Conference. Association for Computational Linguistics. Meeting, vol. 2017, p. 7. NIH Public Access (2017)
16. Higgins, J.P.T., Green, S.: Cochrane Handbook for Systematic Reviews of Interventions, 5th edn. Cochrane Collaboration, London (2011)

17. Guntuku, S.C., Yaden, D.B., Kern, M.L., Ungar, L.H., Eichstaedt, J.C.: Detecting depression and mental illness on social media: an integrative review. Curr. Opin. Behav. Sci. **18**, 43–49 (2017)
18. De Choudhury, M., Counts, S., Horvitz, E.: Social media as a measurement tool of depression in populations. In: Proceedings of the Annual ACM Web Science Conference, pp. 47–56 (2013)
19. Shen, G., et al.: Depression detection via harvesting social media: a multimodal dictionary learning solution. In: Proceedings of the Twenty-Sixth International Joint Conference on Artificial Intelligence (IJCAI-17), pp. 3838–3844 (2017)
20. Pedregosa, F., et al.: Scikit-learn: machine learning in Python. J. Mach.Learn. Res. **12**, 2825–2830 (2011)
21. Basu, T., Murthy, C.A.: A feature selection method for improved document classification. In: Zhou, S., Zhang, S., Karypis, G. (eds.) ADMA 2012. LNCS (LNAI), vol. 7713, pp. 296–305. Springer, Heidelberg (2012). https://doi.org/10.1007/978-3-642-35527-1_25
22. Trambert, R., Kowalski, M.O., Wu, B., Mehta, N., Friedman, P.: A randomized controlled trial provides evidence to support aromatherapy to minimize anxiety in women undergoing breast biopsy. Worldviews Evid.-Based Nurs. **14**(5), 394–402 (2017)

A Model for Inebriation Recognition in Humans Using Computer Vision

Zibusiso Bhango and Dustin van der Haar(✉)

Academy of Computer Science and Software Engineering,
University of Johannesburg, APK Campus,
Cnr University Road and Kingsway Avenue, Johannesburg 2006, South Africa
zbhango@gmail.com, dvanderhaar@uj.ac.za

Abstract. The cost of substance use regarding lives lost, medical and psychiatric morbidity and social disruptions by far surpasses the economic costs. Alcohol abuse and dependence has been a social issue in need of addressing for centuries now. Methods exist that attempt to solve this problem by recognizing inebriation in humans. These methods include the use of blood tests, breathalyzers, urine tests, ECGs and wearables devices. Although effective, these methods are very inconvenient for the user, and the required equipment is expensive. We propose a method that provides a faster and convenient way to recognize inebriation. Our method uses Viola-Jones-based face-detection for the region of interest. The face images become input to a Convolutional Neural Network (CNN) which attempts to classify inebriation. In order to test our model's performance against other methods, we implemented Local Binary Patterns (LBP) for feature extraction, and Support Vector Machines (SVM), Gaussian Naive Bayes (GNB) and k-Nearest Neighbor (kNN) classifiers. Our model had an accuracy rate of 84.31% and easily outperformed the other methods.

Keywords: Computer vision · Convolutional Neural Networks · Machine learning · Inebriation recognition · Support Vector Machines · k-Nearest Neighbor · Naive Bayes

1 Introduction

Substance abuse has been a social issue in need of addressing for centuries now [8]. The human costs of substance use problems regarding lives lost, medical and psychiatric morbidity and social disruptions by far surpasses the economic costs [1]. Substances intercept and alter the messages going to the nervous system, resulting in altered perception. Usually, alcohol induces euphoria, relaxation or hyperventilation, thereby changing the mood of the drinker considerably. The euphoria is the feeling the user is after, and the user will continue consuming alcohol to keep getting the same effect. However, tolerance builds up swiftly; increased doses are required to satisfy the same level of effects, leading to dependence. When unattended, this can lead to an accidental fatal alcohol poisoning.

W. Abramowicz and R. Corchuelo (Eds.): BIS 2019, LNBIP 353, pp. 259–270, 2019.
https://doi.org/10.1007/978-3-030-20485-3_20

Depending on the type of alcohol and its effectiveness, it is often difficult to identify alcohol drinkers, especially in their early stages. Physical ways to detect alcohol abuse include rapid heart rate, high blood pressure, poor muscle coordination, total mental confusion and dilated pupils, excessive sweating, among others [11]. However, as alcoholics continue drinking alcohol, many behavioural traits become more apparent. These include constant depression, introversion, lack of cleanliness and personal hygiene, valuable possessions going missing, intellectual ineptitude and lack of problem-solving skills [2].

More Americans die from substance overdose than they do in car accidents [3]. In order to tackle this social issue, there is a need for novel methods to gain more insight and combat abuse and addiction. The most common way of detecting alcohol abuse is by using a breathalyser. This method, although useful, is quite invasive and requires participation from the user. There are also legal implications that come with this approach. Although law enforcement officers have the right to breathalyse people, private citizens do not necessarily share that right. Due to the sensitive information captured by breathalysers, there are also ethical issues connected with it.

Due to these issues, there is a need for an alternative way to recognise inebriation. There's a need for a system that is faster, more convenient and generally accepted by people. The side effects of alcohol consumption make it possible to recognise drunkenness in an image or video. A biometrics system that recognises inebriation using computer vision will save time and effort, and is generally accepted by people. It will bring about real-time inebriation recognition faster, without effort from the user.

The rest of the paper aims to outline the problem at hand and compare methods used to recognize inebriation and provide the best-performing ones. The next sections are divided as follows. In problem background, we outline the problem at hand, which is how to detect inebriation in humans using computer vision. In related Work, we describe tried methods in the literature that tackle similar problems to ours. In the experimental setup, we describe our proposed approach in detail and explain the methods that we will use to preprocess, extract and classify features as either inebriated or sober. In results, we provide results on our methods and compare them against each other and others found in the literature. In conclusion, we provide our findings from the research and future work.

2 Problem Background

The human body handles adversity well, such as dealing with the injection of toxins and poisons. The human consumption of alcohol affects physical and cognitive functions and has legal consequences such as drunk driving and underage drinking. The average human body eliminates 12 g of alcohol per hour [4]. Blood-Alcohol Concentration (BAC) is the most common metric used to measure the amount of alcohol in the human body at a given time, expressed in grams of alcohol per litre of body fluid. In countries such as Romania, the Czech Republic and Hungary, it is illegal to drive with any alcohol content in your system.

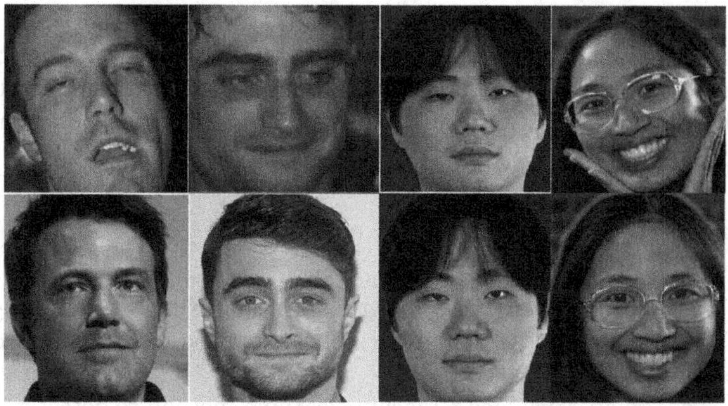

Fig. 1. Some samples from the database used for data sampling to test our model. The top row consists of drunk individuals, while the bottom row consists of sober individuals.

For China, Estonia, Poland and Sweden, among others, it is illegal to drive with a 0.02% BAC. In most Western European countries such as France, Germany and Greece, you're illegally driving under the influence if you have 0.05% BAC. The USA, New Zealand and the UK have a more lenient BAC of 0.08%. A BAC of 4 g/L is likely to result in a coma while a BAC of 4.5–5.0 g is likely to result in death [4] (Fig. 1).

Globally, alcohol consumption leads to approximately 3.3 million deaths each year [5]. Excessive alcohol use is the third leading lifestyle-related cause of death in the United States [6]. It results in physical harm, mental malfunction and is responsible for 1 in 10 deaths among adults aged 20–64 years in the United States annually [7]. Drunk driving endangers the intoxicated driver as well as other sober drivers on the road. Despite these facts, binge drinking (which is defined as 4 or more drinks for women on a single occasion and 5 or more drinks for men on a single occasion) is still on the rise [7].

Some of the effects of alcohol include lower inhibitions, lower caution, loss of fine motor coordination and inability to do complex tasks or general problem-solving. Alcohol consumption also results in slurred speech, weakened balance, slow reaction times and staggering walk or inability to walk. Glossy appearance to eyes, blurry and double vision, loss of memory, heavy sweating, slower pupil response, slowed heart rate and breathing and reduced blood pressure can also result from drinking alcohol [11]. In some instances, nausea, vomiting or loss of consciousness can also occur.

Existing methods to detect alcohol consumption in people involve urine and saliva testing and using a breathalyzer. Expensive equipment is needed such as ones used to capture heart biosignals, infrared cameras or breathalyzers [8]. These methods are useful but very invasive. We believe there is a better way of recognizing inebriation, a way that is just as efficient but non-invasive and less inconvenient for both the subject and the one doing the inebriation testing.

When an individual is drunk, their appearance, the way they talk, walk or behave changes drastically, and it's possible to differentiate that using computer vision. Eye gaze, face pose and facial expression changes and these features can be used to differentiate between an image or video of a drunk person and that of a sober person.

3 Related Work

Since our method focuses on recognizing inebriation using computer vision, we looked at existing methods in the literature that tackled inebriation recognition and computer vision.

Aiello and Agu [7] developed a machine learning method to detect a drinker's Blood Alcohol Content (BAC) from their gait by classifying accelerometer and gyroscope sensor data collected from the drinker's smartphone. Alcohol-sensitive physical attributes such as weight, height and gender were taken into account when classifying. They used 34 intoxicated individuals (14 males and 20 females) for data sampling, and generated time and frequency domain features such as sway (gyroscope) and cadence (accelerometer). They used sensor-impairment goggles on the subjects to simulate the effects of alcohol on the body. Using this kind of special equipment is expensive and simulating intoxication effects limits the model's generalizability. They managed to implement feature normalization to account for differences in walking styles and automatic outlier elimination to reduce the effects of accidental falls. Their inebriation classifier had an accuracy of 72.66%.

Yadav and Dhall [8] proposed a new dataset called DIF (Dataset of Intoxicated Faces) containing RGB face videos of drunk and sober people obtained from online sources. 80 video samples were used, 30 being sober individuals and 50 being inebriated individuals. They analyzed the face videos to extract features related to eye gaze, face pose, and facial expressions. They implemented a convolutional neural network for feature extraction and a recurrent neural network to model the evolution of these multimodal facial features. The experiment showed that the eye gaze and facial features are discriminative for their dataset. They achieved 75.54% classification accuracy on the DIF dataset, thereby showing that face videos can be effectively used to detect drunkenness in humans.

Tseng and Jan [9] developed a unified deep learning network architecture that uses both semantic segmentation and object detection to detect people, cars, and roads simultaneously. They did this by creating a simulated environment in the Unity engine, which they used as a dataset. The simulated environment contained people, cars, roads, grass and the sky. They used the Single Short Multibox Detector (SSD), which enabled the network to detect objects of different sizes, making the predictions size-invariant and more accurate. Their proposed network performed end-to-end prediction well on the tested dataset, achieving 99.46% accuracy, although there are no details on the dataset used.

Al-Theiabat and Aljarrah [10] developed a motion analysis system which analyses tackle scenes in soccer games. They developed a computer vision system to detect if a soccer player was intentionally falling to earn their team a free kick or penalty kick. The tackle scenes go through five stages of processing: identification of the falling player, extraction of tracking points, motion tracking, features extraction and scene classification. They tracked using Kanade-Lucas-Tomasi optical flow with the aid of pyramid levels and forward-backward error algorithm. They used 25 samples; 12 being actual fouls and 13 being dives. They executed classification using Weka software with Naive Bayes tree (NB tree) classifier. Their system had an 84% classification accuracy.

Computer vision research has been pursued for many years, but little work has been done on recognizing inebriation. Most research in the literature on substance abuse uses private datasets and expensive equipment such as sensors, which makes it difficult to measure the performance of algorithms used. We propose a system that will use a dataset made up of publicly available aggregated face images of inebriated and sober individuals gathered on the internet, and inexpensive equipment and focus on algorithms to improve on classifying inebriated and sober individuals.

4 Experimental Setup

4.1 Methodology

In our approach, we use the existing literature on computer vision such as [16] and inebriation recognition such as [8] to derive a model that uses computer vision to detect inebriation among individuals. We will use a secondary dataset to train our model and test its viability objectively. A prototype will be designed to test our model, and performance metrics used to test the performance of the prototype and its ability to detect inebriation in people are Accuracy, Precision, recall, f1 score, True Positive Rate (TPR), False Positive Rate (FPR) and Equal Error Rate (EER).

4.2 Data Sampling

In order to test the performance of our method, we used RGB face images consisting of inebriated and sober individuals collated by the authors. The dataset is made up of publicly available aggregated face images of inebriated and sober individuals gathered on the Internet.

Our dataset consists of 153 inebriated individuals and 101 sober individuals, both males and females. Images consist of people of various age groups. Since no datasets of inebriated and sober individuals exist, we are testing our classifier in the wild. We took images of reported inebriated individuals on the Internet, such as celebrities, and used their sober images and others to create our benchmark (Fig. 1).

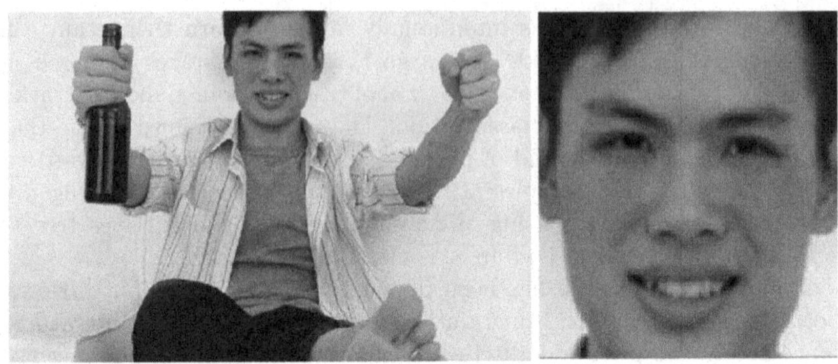

Fig. 2. Input image before preprocessing (left). Resulting ROI image after implementing the Viola-Jones face detection algorithm [15] (right).

5 Model

5.1 Preprocessing

Our model uses RGB images containing people as input. The first step in our classification process is extracting the Region of Interest (ROI), which is the face. The input image is converted to a grayscale format to remove noise. We use the Viola-Jones object detection algorithm [15] on the grayscaled image to get our ROI (as depicted in Fig. 2). The algorithm has four features: Haar feature selection, creating an integral image, AdaBoost training and cascading classifiers. Haar features contain what's common among people's faces, such as the eye region is darker than the cheeks and the nasal region being lighter than the eye region. Integral images, which allow integrals for the Haar extractors to be calculated by only adding four numbers, are used to improve efficiency. Face detection takes place inside a detection window. Adaboost training is used to train the algorithm. We test every window for face images using Haar-like features. We then use the windows containing the minimum error rates as the windows containing our face images. After Adaboost training, we use cascading classifiers to find the windows containing face images for classification. Cascading classifiers are split into classes, with each class containing fewer features to check than the next one. Each window is tested for face images using these classes. Only those images passing the test in each class are sent to other classes for further testing. If an image passes the final class, we have positively identified the face image.

5.2 Convolutional Neural Network

Convolutional Neural Networks (CNN's) have been widely used in computer vision for image or video recognition. CNN's, like any other neural network, are made up of neurons with learnable weight and bias. Each neuron receives several

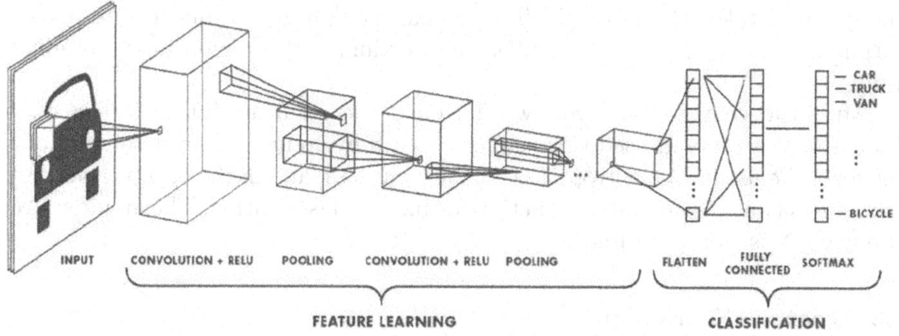

Fig. 3. Convolutional Neural Network architecture, taken from [12].

inputs, takes a weighted sum over them, passes it through an activation function and outputs the result to another neuron. The entire network has a loss function, with the primary goal being to lower the loss function output and converge to a solution.

CNN's are made up of four layers: convolution, ReLU, pooling and the fully connected layer. This is what separates it from the other neural networks. In the convolution operation, we take a filter of a specific size and slide it over our image to get the dot product between the filter chunks of our image, resulting in a feature map. The feature map's pixels will be altered, and its size will also change. The convolution operation captures the local dependencies in the original image. After every convolution operation, a ReLU (Rectifier Linear Unit) is used. ReLU is an elementwise operation that replaces all negative values in the feature map with zero. Its purpose is to introduce non-linearity. The ReLU equation is as follows:

$$f(x) = x^{+} = \max(0, x) \tag{1}$$

where x is the pixel in the feature map, pooling is used to reduce the dimensionality of each feature map while retaining the most essential information. This makes the feature maps dimension smaller and more manageable. We can perform convolution, ReLU and pooling operations multiple times. We then flatten our resulting feature maps and use them as input to a fully connected layer of the Neural Network.

In our implementation, we implemented a LeNet convolutional neural network because of its simplicity and efficiency. A 3×3 filter was used for convolution operation, and 32 filters in all were used. All face images went through preprocessing to detect faces and were turned back to RGB color images. The shape of the input image is (64, 64, 3). We chose a smaller filter size because a larger one can overlook the crucial features and miss them.

We used Max-pooling because it extracts most crucial information better than average pooling. For each region on the image represented by a filter, max-pooling takes the most significant pixel of that region and create a new output matrix where each element is the maximum of a region in the original input.

After the convolution layer, we flattened the resulting image into a one-dimensional array and used this as input to a fully connected layer to train our network. We used cross-entropy to calculate our loss, and in the output layer, we used softmax as the activation function for binary classification. The architecture of our CNN is shown on Fig. 3.

5.3 Feature Extraction

In our alternative pipeline to compare against CNN, we used Local Binary Patterns (LBP's) for feature extraction. The LBP algorithm is rooted in 2D texture analysis. It works on the idea of summarising a local structure in an image by comparing each pixel to all of its neighbours. Each pixel is taken as a centre, each of the eight neighbourhoods is compared against the centre; a pixel with a higher value is converted to 1, and 0 if it's smaller. With 8 surrounding pixels, you end up with 2^8 possible combinations, commonly known as Local Binary Patterns (Fig. 4).

Fig. 4. Grayscaled image (left). Resulting LBP image (right).

After getting the LBP codes, a histogram is then generated from the resulting image. This histogram becomes our feature space, which is used for classifying images.

5.4 Classification

For classification, we used three classifiers: Support Vector Machines (SVM), Gaussian Naive Bayes (GNB) and k-Nearest Neighbor (kNN). SVM is a supervised machine learning algorithm used for classification and regression problems. It uses the kernel trick to transform your data, then based on those transformations, it finds the optimal boundary between the classes. SVMs work efficiently in high dimensional spaces and use a subset of training points in the decision function (commonly known as support vectors), which is memory efficient. We used the Radial Basis Function (RBF) for training our model.

The Naive Bayes is a supervised learning algorithm which applies Bayes' theorem with the "naive" assumption of conditional independence between every pair of features given the class variable's value [13]. The classifier aggregates information using conditional probability and assumes independence among features. It is based on finding functions describing the probability of belonging to a specific class given features. Naive Bayes classifiers are extremely fast compared to other classification algorithms, and they do well to alleviate the curse of dimensionality. However, Naive Bayes classifiers are known to suffer from a weak assumption, making them bad estimators [13].

kNN is a learning algorithm which is most popular for classification purposes [14]. The idea behind kNN is to find a predefined number of training samples closest in distance to the new input, and predict the new input's class from these. Distance can be any metric measurement, although Euclidean distance is the most commonly.

6 Results

In order to measure the performance of our model, we calculated the accuracy, precision, recall, f1-score, equal error rate (EER) and Receiver Operating Characteristic (ROC) curve for each pipeline, including CNN. The ROC curve used to plot the True Positive Rate (TPR) versus the False Positive Rate (FPR) to measure the performance of the system when classifying inebriation. The Area Under the Curve determines whether the system performs well or not. The EER is when the FPR and the FNR are equal. Our model accuracy and model loss are shown in Figs. 5 and 6, respectively.

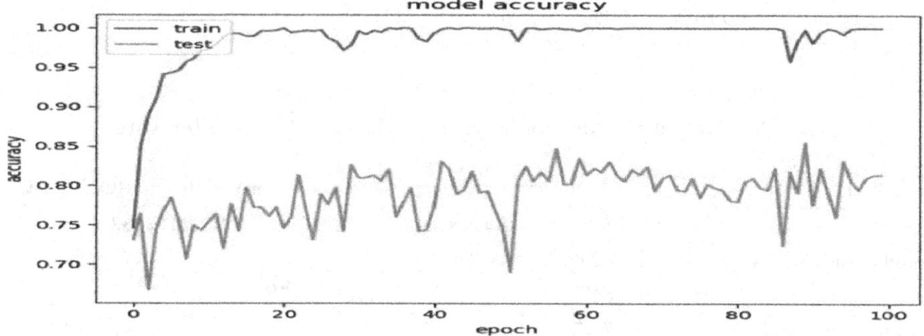

Fig. 5. CNN model accuracy curve during training and testing.

Our Convolutional Neural Network model had an accuracy of 84.31%. Our precision was 84.38%, the recall was 71.05%, f1-score of 77.14% and we achieved

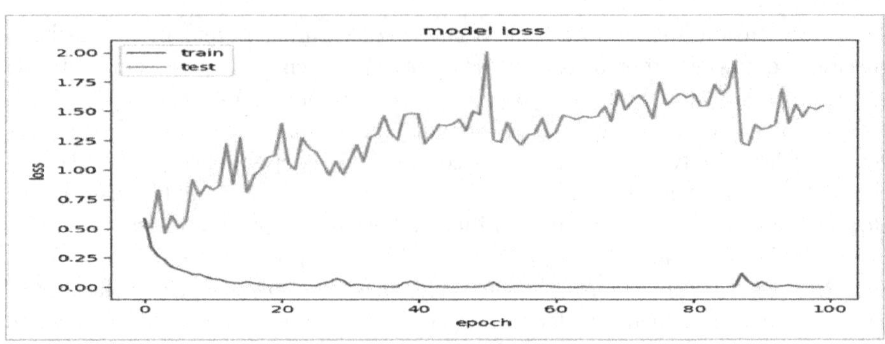

Fig. 6. CNN model loss curve during training and testing.

an EER of 22.2%. As shown in Table 1, our model outperforms the other pipelines (Gaussian Naive Bayes, k-Nearest Neighbor and Support Vector Machines), each with Local Binary Patterns used as features.

Table 2 shows the performance of our model against similar systems in the literature. We achieved higher accuracy and precision. Al-Theiabat and Aljarrah [10] achieved a higher recall and f1-score on a much smaller dataset consisting of 25 samples. Our recall and F1-score is very competitive, and we had a better EER than the rest.

Table 1. Comparing CNN with other classifiers

Method	Accuracy	Precision	Recall	F1-Score	EER	ROC Area
CNN	**84.31%**	**84.38%**	**71.05%**	**77.14%**	**22.21%**	**83%**
LBP-GNB	62.75%	50%	52.63%	51.28%	39.06%	63%
LBP-kNN	63.73%	51.35%	50%	50.67%	39.53%	64%
LBP-SVM	66.67%	60%	31.58%	41.38%	37.5%	66%

Table 2. Comparing our model with similar classifiers in literature.

Method	Accuracy	Precision	Recall	F1-Score	EER	ROC Area
CNN	**84.31%**	**84.38%**	71.05%	77.14%	**22.21%**	83%
Aiello and Agu [7]	72.66%	72.3%	72.7%	72.1%	-	**89.2%**
Al-Theiabat and Aljarrah [10]	84%	76.47%	**100%**	**86.7%**	-	-
Yadav and Dhall [8]	75.54%	-	76%	-	-	-

This proves that our model is feasible, and computer vision can be used to recognize inebriation using convolutional neural networks. We chose Local Binary Patterns because of how effective they are in texture analysis and their potential in differentiating facial appearances of inebriated individuals from those sober. However, Local Binary Patterns did not provide a distinct feature space for machine learning algorithms such as SVM, GNB and kNN to classify inebriation efficiently.

7 Conclusion

Substance abuse has taken many lives. It alters perception, and when unhandled, can lead to dependence. Its effects are rapid heart rate, high blood pressure, poor muscle coordination, total mental confusion, dilated pupils and excessive sweating, amongst others. One approach to combatting this is through inebriation detection.

Current methods of detecting inebriation, such as urine tests, blood tests, and breathing tests, using breathalyzers, using ECG to capture heart signals, fitness devices and wearable devices are effective but very inconvenient to the users. The equipment used is also costly.

In this paper, we proposed a model that uses computer vision to recognize inebriation. Only a camera is needed to achieve inebriation recognition, and there is no user participation required, such as breathing into a breathalyzer. Our model achieved excellent results, with an accuracy rate of 84.31%. We achieved results superior to the alternative implementations that use Local Binary Patterns with varying classifiers. We then compared our model to other similar models in literature and our model has higher accuracy.

Implementing a computer vision-based inebriation recognition system will make it easier and faster to detect inebriation, thereby increasing its application in other real-life problem domains. These domains include transportation, where we test drivers/pilots for alcohol use before embarking trips. In medicine, medical practitioners can be tested before diagnosing patients or performing surgeries. In social places, we can monitor customers in bars against excessive drinking. Such a system will reduce accidents and deaths, and can save people from alcohol dependence.

There is not much research done on using computer vision to detect inebriation or substance abuse and addiction. We believe our method is worthy of further research. Using deep neural networks such as Recurrent Neural Networks (RNN) to detect inebriation in videos can potentially improve on our model considerably. There is currently very little publicly available datasets of inebriated and sober individuals to test models with, and this provides uncertainty on whether a model is accurate or not. Developing a dataset that can be used to test algorithms will certainly improve inebriation recognition. Our model's performance gives us the optimism that computer vision can indeed be used to recognize inebriation.

References

1. O'Connor, P.G., Samet, J.H.: Substance Abuse. J. Gen. Intern. Med. **17**, 398–399 (2002)
2. NIDA: Drugs, Brains, and Behavior: The Science of Addiction, 1 July 2014. https://www.drugabuse.gov/publications/drugs-brains-behavior-science-addiction/addiction-health. Accessed 29 Apr 2018
3. Fan, Y., Zhang, Y., Ye, Y., Li, X., Zheng, W.: Social media for opioid addiction epidemiology: automatic detection of opioid addicts from twitter and case studies. In: Proceedings of the 2017 ACM on Conference on Information and Knowledge Management, Singapore (2017)
4. Mirielli, E., Webster, L.: Modeling alcohol absorption and elimination from the human body: a case study in software development: nifty assignment. J. Comput. Sci. Coll. **30**, 110–112 (2015)
5. Toroghi, M.K., Cluett, W.R., Mahadevan, R.: Multiscale metabolic modeling approach for predicting blood alcohol concentration. IEEE Life Sci. Lett. **2**, 59–62 (2016)
6. Arnold, Z., LaRose, D., Agu, E.: Smartphone inference of alcohol consumption levels from gait. In: 2015 International Conference on Healthcare Informatics, pp. 417–426 (2015)
7. Aiello, C., Agu, E.: Investigating postural sway features, normalization and personalization in detecting blood alcohol levels of smartphone users. In: 2016 IEEE Wireless Health (WH), pp. 1–8 (2016)
8. Yadav, D.P., Dhall, A.: DIF: dataset of intoxicated faces for drunk person identification. ArXiv e-prints (2018)
9. Tseng, Y.H., Jan, S.S.: Combination of computer vision detection and segmentation for autonomous driving. In: 2018 IEEE/ION Position, Location and Navigation Symposium (PLANS), pp. 1047–1052 (2018)
10. Al-Theiabat, H., Aljarrah, I.: A computer vision system to detect diving cases in soccer. In: 2018 4th International Conference on Advanced Technologies for Signal and Image Processing (ATSIP), pp. 1–6 (2018)
11. Solutions Recovery: Physical Impact of Alcohol Abuse. https://www.solutions-recovery.com/alcohol-treatment/physical-impact/. Accessed 5 Jan 2019
12. Prabhu: Neural network with many convolutional layers (2018)
13. Jahromi, A.H., Taheri, M.: A non-parametric mixture of Gaussian naive Bayes classifiers based on local independent features. In: 2017 Artificial Intelligence and Signal Processing Conference (AISP), pp. 209–212 (2017)
14. Okfalisa, Gazalba, I., Mustakim, Reza, N.G.I.: Comparative analysis of k-nearest neighbor and modified k-nearest neighbor algorithm for data classification. In: 2017 2nd International conferences on Information Technology, Information Systems and Electrical Engineering (ICITISEE), pp. 294–298 (2017)
15. Viola, P., Jones, M.J.: Robust real-time face detection. Int. J. Comput. Vis. **57**, 137–154 (2004)
16. LeCun, Y., Haffner, P., Bottou, L., Bengio, Y.: Object recognition with gradient-based learning. Shape, Contour and Grouping in Computer Vision. LNCS, vol. 1681, pp. 319–345. Springer, Heidelberg (1999). https://doi.org/10.1007/3-540-46805-6_19

Interaction of Information Content and Frequency as Predictors of Verbs' Lengths

Michael Richter(✉), Yuki Kyogoku, and Max Kölbl

Natural Language Processing Group, Universität Leipzig, Leipzig, Germany
richter@informatik.uni-leipzig.de,
kyogokull@gmail.com, max.w.koelbl@gmail.com

Abstract. The topic of this paper is the interaction of Average Information Content (IC) and frequency of aspect-coded verbs in Linear Mixed Effect Models as predictors of the verbs' lengths. For 30 languages in focus, it came to light that IC and frequency do not have a simultaneous, positive impact on the length of verb forms: the effect of the IC is high, when the effect of frequency is low and vice versa. This is an indication of *Uniform Information Density* [13–16]. Additionally, the predictors IC and frequency yield high correlations between predicted and actual verbs' lengths.

Keywords: Information Content · Frequency · Linear mixed models · Economy in interactions · Aspect

1 Introduction

This paper presents a study on the interactions of Average Information Content (IC), frequency, and lengths of verbs, coded by aspect, in 30 natural languages taken from Universal Dependency-treebanks [1] (the set of languages is given in Table 2 in the appendix). We aim to find out, whether verb lengths are contingent on their entropy and frequency and starting from this, we pose the research question: whether and how do the predictors IC [2, 3] and FREQUENCY interact in a linear model when predicting the lengths of aspectual verb forms? Our starting point is the distinction between two aspect types, namely *perfective* and *imperfective*. When marked with perfective aspect, a verb expresses a completed event and when marked with imperfective aspect, it expresses a non-completed, unbounded event (see Sect. 3.1 below).

FREQUENCY is the classical Zipfian predictor and expresses the idea that "The magnitude of words tend, on the whole, to stand in an inverse […] relationship to the number of occurrences"[4].

The IC of a sign can be interpreted as the amount of surprise, when encountering that sign, in our case specific verb form, given a context. If the answer to the question "Is this the verb form I expected in this context?" turns out to be "Yes" then the occurrence of that verb form is not surprising and the IC is low. But if the answer is: "Not really" or even "Definitely not", then there is a surprisal effect and consequently, the IC of that verb form is higher. IC represents the intuition that longer words should carry more information than shorter words, otherwise the greater length would be not

W. Abramowicz and R. Corchuelo (Eds.): BIS 2019, LNBIP 353, pp. 271–282, 2019.
https://doi.org/10.1007/978-3-030-20485-3_21

economical. The method we apply is a linear regression analysis employing a *Linear Mixed-Effect Model* (LMM; [5]).

Our aim is motivated in general by observations of coding asymmetries in grammars (e.g. [6–8]) and the FORM-FREQUENCY CORRESPONDENCE principle [9] that more frequent linguistic forms tend to be shorter than rare ones. The need to explain this principle is motivated in particular by findings of [3] and [10] who both found that IC is a stronger predictor of word lengths than frequency. In [3], IC is estimated from unstructured n-grams in the context of targets, whereas [10] utilised for the estimation of IC from contexts verb dependencies and observed strong interactions between the predictors. In a previous study [11], which focused on lengths of aspect-coded verbs in Russian utilising verb dependents as contexts, the significance of the predictor IC could be conformed and, in addition, significant interactions between IC and frequency came to light: the predictors do not have a simultaneous impact on length. The conclusion was that this could be due to an economy principle of the interaction of IC, frequency, and word lengths [11].

In the present study, we aim to challenge these results by considering a larger set of languages, contexts and number of verbs. We estimate the IC (i) from unstructured contexts, i.e. uni-, bi- and trigrams to the left of the target verbs and (ii) from dependency frames of verbs.

Interactions of Average Information Content, frequency, and lengths of linguistic signs in natural language have hardly been considered in research so far. For formal languages, *Shannon's Source Coding Theorem* [12] which states the principle of coding economy. Based on this, the specific research question of this study is whether economic principles can also be proven in natural languages within the interactions between IC, frequency and verbs' lengths. In addition, we relate interaction of information in natural language to the *Uniform Density Principle* (UID, [13–16]) which says that in language, information is uniformly distributed in an *information continuum* [17] and that for the sake of processability of messages, information must not exceed the capacity of communication channels, that is, in order to ensure an even flow of information, peaks and troughs of information should be avoided. UID can be used as feature for classifiers in applications of Artificial Intelligence such as data mining/information extraction, machine translation of texts, text generation, parsing [17, 18] and automatic summarization of texts.

2 Related Work

IC is a variant of conditional entropy and is based on conditional probabilities (the estimate is given in (1) below). Language models which utilise conditional probabilities, are employed in parsing tasks [19]. [19–21] argue that those models have cognitive plausibility, since humans tend to make predictions from contexts, when they parse natural language. [19] uses the concept of *surprisal*, that is, the negative log-probability of words, in order to describe the effect of information based on conditional probabilities of words, that is, words in contexts on (human) sentence processors.

To our best knowledge, there are not many studies on the correlation between IC and word lengths. [3] show for ten Indo-European languages that IC, calculated on the basis of syntactic contexts (bigrams, trigrams, and 4-grams), is a better predictor of word length than frequency. According to the authors, the effect of frequency is largely due to its correlation with IC. [3] ascribe the attested correlation of word length and information content to the principle of uniform information density, which says roughly that the information rate of communication over time is kept as constant as possible. They point out that IC is known to influence the amount of time, which speakers take to pronounce a word, and conclude by suggesting Zipf's law in the following way: "the most communicatively efficient code for meanings is one that shortens the most predictable words – not the most frequent words" [3]. More recently, [10] has investigated, whether the length of words can be predicted by IC, when it is estimated from syntactic dependents rather than from unstructured contexts of target words for Arabic, Chinese, English, Finnish, German, Hindi, Persian, Russian and Spanish. [10] confirms the hypothesis that words with a higher IC tend to be longer. For Russian, [11] found strong impacts both of IC, estimated from syntactic dependents, and of frequency on lengths of aspectually marked verbs.

3 Linguistic Theory

3.1 Aspect

Aspect expresses temporal structures of verbs, verb phrases and sentences and can be connected to time in tenses. However, whereas time places an event on a timeline relative to a given reference point, aspect defines the (temporal) perspective taken on an event [22]. The two major aspect categories are *perfective* and *imperfective*. Whereas a perfective form expresses reference to a single, completed, particular event, an imperfective form is devoid of any semantic commitment to whether or not the event is single, completed, or particular. These features allow imperfectives to refer to single, unbounded events that are only partially realised at the reference time, to refer to multiple events, and to refer to non-particular (generic) events. It is generally assumed that aspect can be expressed both lexically and grammatically (see [7]). While lexical aspect, also known as *Aktionsart*, is taken to be an inherent property of verb meaning (regardless of any specific grammatical realisation), grammatical aspect pertains to the way aspect is encoded by means of grammar (typically morphosyntactically), which can vary depending on how a speaker decides to construe an event. Consider *I love vegetables* and *I am loving vegetables*: the lexical aspect of the English verb *love* can be defined as a 'state' (see [23] for a definition of actional classes). Accordingly, in the present it is usually expressed grammatically by the simple present verb form, whose function is, among other things, to express the imperfective aspect. Notwithstanding, speakers are free to conceptualise the same event differently and employ, for example, the present progressive. In this case, the event is still in the present, but it is presented as an activity rather than a state. The use of the present progressive is arguably rather exceptional with 'state' verbs, but it is not impossible. The relationship between lexical aspect and grammatical aspect is a notoriously very complex one, both because it is not always easy or even possible to decide the lexical aspect class of a given verb lemma by abstracting from specific instantiations in speech/text and because grammatical tenses such as the present or the present progressive in English are complex structures expressing many bits of information conjointly, whose interpretation can also be strongly dependent on syntactic context. By determining the default lexical aspect for each given verb lemma in a language, the following hypothesis can be tested: given a verb lemma, those word forms which express the lexical default aspect are, on average, shorter than those expressing any other 'non-default' aspect category. Returning to the 'love'-example from above, the default form *love* is shorter than non-default *loving*.

In fact, this prediction is a particular instantiation of the hypothesis, that higher frequency words are more predictable than lower frequency words. Predictable items need less salient formal representation, and consequently high frequency items are expected to be expressed by shorter forms. This (binary) opposition is known in the linguistic literature as 'coding asymmetry' and has so far been investigated mostly qualitatively in (theoretical and typological) linguistics [8, 9]. In our study, the 'aspect coding asymmetry'-hypothesis restricts the notion of coding asymmetry to aspect encoding.

3.2 Coding Asymmetries

Telic dynamic verbs have perfective (and generally past-tense) meaning when used without any overt marker, while *atelic* verbs have imperfective meaning, when used without an overt marker. *Telic* verbs tend to occur in the perfective aspect and atelic verbs tend to occur in the imperfective aspect. *Telicity* refers to processes which are completed, while *atelicity* refers to uncompleted ones, in other words, ongoing processes. Thus, a telic verb encoded with perfective aspect is the expected case, the default case, while it would be a surprisal and the non-default case if it were used with the imperfective aspect. Analogously, an atelic verb encoded with imperfective aspect would be the expected case, that is, default, while a perfective coding would be a surprisal, that is, non-default. Default forms tend to occur more frequently and are zero coded across languages, while non-default forms tend to be coded overtly by special imperfective or perfective markers, respectively, as illustrated for the language Inuktitut in Fig. 1 (cf. [24]):

	Atelic	Telic
Imperfective	Ø coding (e.g. *pisuk-* 'is walking')	overt coding (e.g. *ani-liq-* 'is going out')
Perfective	overt coding (e.g. *pinasuk-jariiq-* 'finished working')	Ø coding (e.g. *ani-Ø* 'went out')

Fig. 1. Coding asymmetries in Inuktitut

The grey shaded fields contain the expected, the defaults, and thus shorter forms, while the non-highlighted fields contain the surprisals, the non-defaults, and thus the longer verb forms.

3.3 Default Coding

In order to render the data cross linguistically uniform, we reduced oppositions within each treebank to the binary imperfective-perfective distinction, which can be found in all languages displaying morphological aspect. We subsumed the habitual and progressive aspects under the imperfective and the resultative aspect under the perfective aspect. Verb forms in the prospective aspect have been ignored, since its value is not clear with respect to the imperfective and perfective opposition.

A default or non-default aspect was determined for each verb. For all verbs of a given language, minima and maxima were determined with respect to the number of occurrences with perfective and imperfective aspect. By comparisons of mean values (t-tests), it was checked whether minima and maxima originate from different distributions. If minima and maxima differ significantly, the more common aspect form of a verb i.e. the maximum, was interpreted as its default aspect.

3.4 Resources and Method

As data resource, we utilised UD treebanks (versions 2.1/2.3, [1]), which comprise 102 treebanks and 60 languages. In the appendix, we list the set of exploited UD-corpora with the respective number of verb lemmas and verb forms.

We employed an LMM ([5], see (3) below) with the fixed effects IC and FREQUENCY respectively. A successful application of that model to all the UD treebanks, where aspect is encoded as a verb morphological feature, would provide evidence for the existence of an aspect-related coding asymmetry. In the UD corpus, only 45 treebanks annotate the morphological feature *Aspect*. We took FREQUENCY as the negative logarithm of the probabilities p of word forms, that is to say, Shannon Information (SI) as given in (1):

$$FREQUENCY = SI = -\,log_2(p) \tag{1}$$

The estimation of IC as average information content is given in (2) [4, 5]:

$$IC = E(-log_2(P(W = w|C = ci))) \tag{2}$$

IC is the expectation value of the negative log of conditional probability of a verb w (marked with imperfective or with perfective aspect) given its contexts c_i. The LLM we used is given in (3):

$$y = \beta_0 + \beta_1 FREQUENCY + \beta_2 IC + \beta_3 FREQUENCY \cdot IC + W + \varepsilon \tag{3}$$

In (3), y-values are lengths of the verb forms, whereby lengths are the number of characters of a word. In contrast to previous studies, we took the lengths of individual verb forms and not the average lengths of the verb lemmas. The model comprises two predictors, i.e. fixed effects, FREQUENCY and IC, the interaction between the two fixed effects and in addition, a random effect W. ε is an error term. We experimented with three different random effects: IC, frequency and default-value. Random effects define grouping variables with an associated intercept. The high predictive power of

LMM results from the possibility to fix intercepts from random effects during regression. This makes it possible to test value differences in the dependent variable for group-dependent systematics.

4 Results

4.1 Correlations Between Predicted and Actual Verbs' Lengths

Experiments with two different types of contexts were carried out in order to estimate the IC: (i) uni, bi- and trigrams and (ii) verb dependencies, as given in Table 1 (see http://universaldependencies.org/guidelines.html):

Table 1. The set of verb dependencies, used for the estimation of IC.

Dependency label	Function in the sentence	Example (dependencies in boldface)
nsubj	nominal subject	**they** laugh
csubj	clausal subject	**that they are coming**, is beautiful
nsubj:pass	nominal passive subject	**gold** has been found
csubj:pass	clausal passive subject	**when it will happen**, will be told
obj	object	they found **gold**
iobj	indirect object	she gave **Jack** an apple
advmod	adverbial modifier	that is **very** delicious
advcl	adverbial clause modifier	**during the storm**, it was cold
obl	oblique nominal	she passes **him** the salt
ccomp	clausal complement	you see **what they want**

Utilising uni-, bi- and trigrams as contexts for the estimation of IC, the predictors IC and FREQUENCY and their interaction were highly significant ($p < .01$) for all languages. We took R^2 as correlations between predicted verb lengths and actual verb lengths. For models with IC as random effects, it was differentiated, whether IC was estimated from uni, bi- or trigrams. A recurring pattern in the majority of languages is that correlations with the IC estimated from unigrams are higher than correlations with IC estimated from bigrams, whose IC in turn performs better than IC estimated from trigrams. This means that the target verbs convey more information to unigrams in context than to bi- and trigrams respectively. This pattern can also be seen in the models with the random effects FREQUENCY and DEFAULT (Note that only the unigram-based results of random effects FREQUENCY and DEFAULT are shown in Fig. 2 of the correlations below). Very high correlations are yielded with models which utilise IC from unigrams, and, in addition, employ (i) IC and (ii) FREQUENCY as random effects. Surprisingly, models with DEFAULT as a random effect produce only moderate correlations. Amongst the 30 languages, Old Church Slavonic ($\rho = .98$) and Russian ($\rho = .96$) yielded the highest correlations between predicted and actual words' lengths.

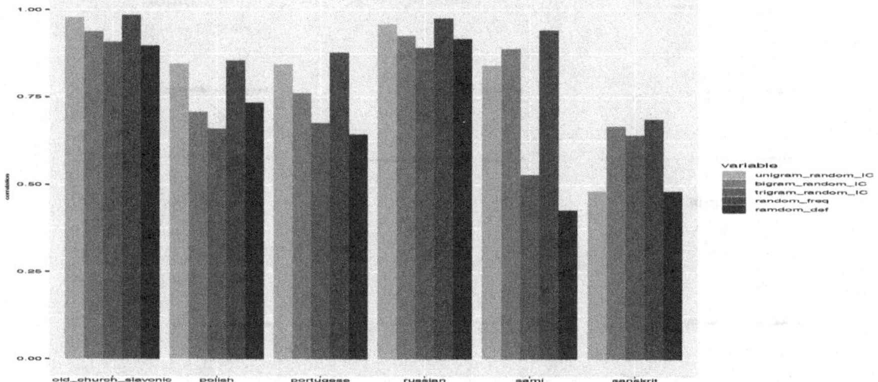

Fig. 2. Correlations between the predicted and actual verb lengths in Old Church Slavonic, Polish, Portuguese, Russian, Sami and Sanskrit. (Color figure online)

In contrast, in some languages such as Buryat, Hungarian, Marathi, and Sanskrit, we observe lower correlations with all models and in addition, deviations from the systematics of descending correlations from unigrams and bi- to trigrams. This may be due to the small size of these corpora. A general finding is that, when uni-, bi- and trigrams are used as contexts, IC and FREQUENCY are both significant predictors of length. These findings are consistent with the observations on [5] and [10]. As illustrative examples, in Fig. 2 are given: (i) the correlations within the high performing languages Old Church Slavonic, Polish, Portuguese, Russian, Sami and (ii) the recurring pattern that IC estimated from unigrams outperforms IC estimated from bi- and trigrams (see above) (the blue and the velvet bars give correlations from unigrams). The deviating results of Sanskrit are an example of the bias of a small corpus size (see above):

Utilising verb dependencies as contexts, the best results were achieved using lengths of lemmas as random effect. When utilising dependency frames for estimation of IC, we observed for much less languages both a significant impact of the predictors IC and FREQUENCY and, in addition, a significant interaction of IC and FREQUENCY ($p < .05$). For Czech, Polish and Latin, the correlations between predicted and actual verbs lengths were Czech: $\rho = .73$, *Polish*: $\rho = .78$ and *Latin*: $\rho = .56$.

4.2 Interactions Between IC and Frequency

With the IC extracted from uni-, bi- and trigrams, an almost identical interaction-pattern can be observed in all languages in focus, illustrated in Fig. 3 for six typologically diverse existing and extinct languages, that is Basque, Czech, Gothic, Hindi, Russian and Latin with the IC estimated from unigram-contexts (random effects: IC and FREQUENCY).

The regression lines in the diagrams on the bottom left are almost diagonals in all languages, which expresses a strong positive impact of the predictor IC on verb length. Subsequently, the regression line flattens out, indicating a weakening of the impact. The final state, i.e. when the impact of FREQUENCY is at its highest, is either a horizontal regression line, indicating a lack of influence of IC, or even a slope from left

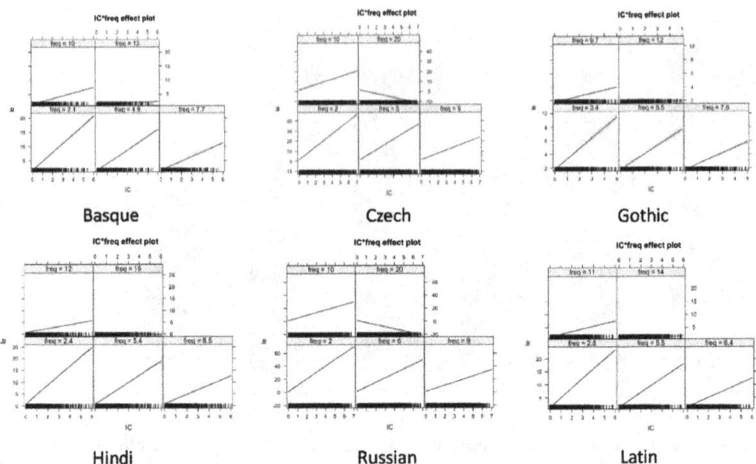

Fig. 3. Interactions between IC estimated from n-grams and FREQUENCY in Basque, Czech, Gothic, Hindi, Russian and Latin.

to right which is the case in Czech and Russian. This shows a negative impact, i.e. high IC values correspond to shorter verbs. It holds that the effect of IC weakens when the effect of FREQUENCY becomes stronger: with frequent verb forms and thus with weak impact of FREQUENCY, the influence of IC is strong; with rare verb forms which cause a strong surprisal-effect and have high values of the predictor FREQUENCY, IC has no or even an opposite effect. In case of a negative correlation between IC and verb lengths, the effect of FREQUENCY is strong enough to compensate for that negative influence of the IC on lengths: IC and FREQUENCY do have a simultaneous positive influence on the length of verb forms which confirms the findings in our previous study [11].

Utilising dependency frames as contexts, we get significant interactions between FREQUENCY and IC only in Czech, Polish and Latin, as illustrated in Fig. 4. As with IC estimated from n-grams, the strong positive impact of the predictor IC decreases while simultaneously the impact of FREQUENCY increases until the regression line is flattened: the impact of IC is annihilated:

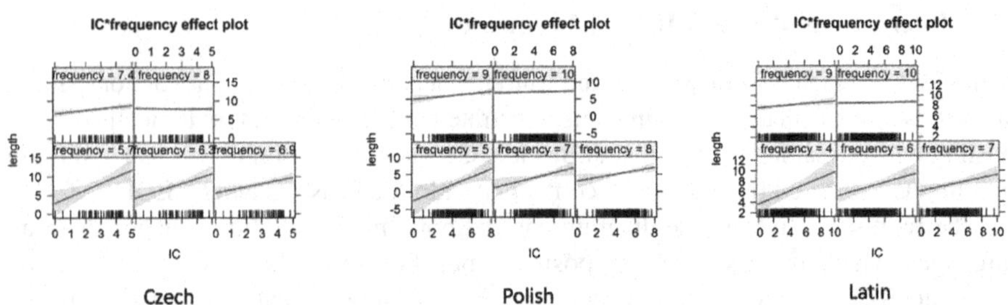

Fig. 4. Interactions between IC estimated from dependency frames and FREQUENCY in Czech, Polish and Latin.

5 Conclusion

Our study reveals that lengths of aspect-coded verbs are contingent on IC and frequency: IC and FREQUENCY (i.e. Shannon's local information content) together form a model with high predictive power. However, the strength of this correspondence depends to a high degree on the contexts which are utilised for the estimation of the IC. Whereas with IC from n-grams and with IC and FREQUENCY as random effects, high correlations were found between predicted and actual verb lengths in most of the languages studied and strong interactions were found in all languages in focus, dependency frames for the estimation of IC were performing much poorer: we found high and moderate correlations for Czech, Polish and Latin in combination with strong interactions when using dependency frames for the estimation of IC. The finding is, that there is a systematics in the transmission of information from target words to contexts: unigrams receive from target words more information than bigrams and these again more information than trigrams. The question whether there exists an interaction between IC and frequency, that is, information content in context and Shannon's local information content can be answered in the affirmative for all languages in focus when using n-grams for the estimation of IC, however for fewer languages, when using dependencies. We draw the conclusion that within our corpora, dependency frames, unlike n-grams, in many languages, do not produce high surprisal effects. One possible explanation is a lack of variation within the dependency frames. In addition, an interesting finding is, that the model which utilised DEFAULT as random effect is outperformed by models using IC and FREQUENCY as random effects.

Within the interaction of IC and FREQUENCY, the effect of an economic principle comes to light: a non-simultaneous impact of the two fixed effects IC and FREQUENCY. Both effects are significant, however to the degree that one predictor weakens, the other predictor becomes stronger. The economy consists in the fact that to predict lengths, one predictor suffices, although both predictors have a significant impact on lengths. Rare verb forms have a high local information content, i.e. FREQUENCY, but do not tend to convey high amounts of information to their contexts, i.e. IC, since in this case FREQUENCY is sufficient for predicting verb lengths. And in contrast, IC is strong with frequent verbs with low FREQUENCY-values. This economic principle seems to be effective in the interaction of form/coding, IC and frequency in the natural languages which we focused on in this study. The economy relates to Shannon's Source Coding Theorem [12], which captures codings in n-ary alphabets and does not concern spontaneous linguistic behavior of humans, but rather a deeper layer of language and has presumably developed during the development of languages: IC and FREQUENCY both correspond to length/coding but only one effect at a time t correlates with length. We relate the interaction of FREQUENCY (Shannon Information) and IC to the UID-principle, since the total amount of information which both features convey is a continuum: in other words, the situation does not occur that both features provide simultaneously either high or low amounts of information at the same time, rather the amount of information is balanced and thus, as stated above, an even flow of information is ensured. This is an essential precondition of incremental sentence comprehension in applications such as extraction of information from information systems, machine translation of texts and incremental parsing [14, 21, 25]: the most probable meanings of signs/words and the most probable syntactic functions are assigned on the basis of *Uniform Information Density* which is defined as the mean of variance of information within sentences [26]. The further disclosing of these complex relationships that is, interactions of information theory-based features and, in addition, their application in machine learning will be a topic of further research.

Acknowledgments. This work was funded by the Deutsche Forschungsgemeinschaft (DFG, German Research Foundation) – project number: 357550571.

Appendix

Table 2. The UD corpora in focus, the respective sums of lemmas and verb forms.

Corpus	$\sum$ lemmas	$\sum$ word forms
Ancient Greek (grc-ud-test.conllu)	1117	2519
Arabic (ar-ud-test.conllu)	588	1064
Basque (eu-ud-test.conllu)	555	1544
Belarusian (be-ud-test.conllu)	100	117
Bulgarian (bg-ud-test.conllu)	796	1247
Buryat (bxr-ud-test.conllu)	391	991
Chinese (zh-ud-test.conllu)	889	889
Czech (cs-ud-test.conllu)	2442	6663
Gothic (got-ud-test.conllu)	502	1242
Greek (el-ud-test.conllu)	383	736
Hindi (hi-ud-test.conllu)	231	620
Hungarian (hu-ud-test.conllu)	490	721
Italian (it-ud-test.conllu)	437	700
Kazakh (kk-ud-test.conllu)	403	1036
Kurmanji (kmr-ud-test.conllu)	169	520
Latin (la_proiel-ud-test.conllu)	676	1732
Latvian (lv-ud-test.conllu)	970	1607
Marathi (mr-ud-test.conllu)	28	51
North Sami (sme-ud-test.conllu)	476	1016
Old Church Slavonic (cu-ud-test.conllu)	518	1402
Polish (pl-ud-test.conllu)	748	1116
Portuguese (pt-ud-test.conllu)	408	699
Russian (ru_syntagrus-ud-test.conllu)	2324	7217
Sanskrit (sa-ud-test.conllu)	138	214
Slovak (sk-ud-test.conllu)	573	899
Slovenian (sl-ud-test.conllu)	686	1108
Spanish (es-ud-test.conllu)	553	876
Turkish (tr-ud-test.conllu)	397	1509
Ukrainian (uk-ud-test.conllu)	773	1117
Urdu (ur-ud-test.conllu)	93	243

References

1. Nivre, J., Agić, Z., Ahrenberg, L., et al.: Universal dependencies 2.0–CoNLL 2017 shared task development and test data. LINDAT/CLARIN digital library at the Institute of Formal and Applied Linguistics (FAL) (2017)
2. Cohen Priva, U.: Using information content to predict phone deletion. In: Proceedings of the 27th West Coast Conference on Formal Linguistics, pp. 90–98. Cascadilla Proceedings Project (2008)
3. Piantadosi, S.T., Tily, H., Gibson, E.: Word lengths are optimized for efficient communication. PNAS **108**(9), 3526–3529 (2011)
4. Zipf, G.: The Psycho-Biology of Language. Houghton Mifflin, Boston (1935)
5. Bates, D., Mächler, M., Bolker, B., Walker, S.: Fitting linear mixed-effects models using lme4. J. Stat. Softw. **67**(1), 1–48 (2015)
6. Greenberg, J.H.: Language Universals With Special Reference to Feature Hierarchies. Mouton, The Hague (1966)
7. Croft, W.: Verbs: Aspect and Causal Structure. Oxford University Press, Oxford (2012)
8. Haspelmath, M.: Creating economical patterns in language change. In: Good, J. (ed.) Semantics and Contextual Expressions, pp. 185–214. Oxford University Press (2008)
9. Haspelmath, M., Calude, A., Spagnol, M., Narrog, H., Bamyaci, E.: Coding causal noncausal verb alternations: a form–frequency correspondence explanation. J. Linguist. **50** (3), 587–625 (2014)
10. Levchina, N.: Communicative efficiency and syntactic predictability: a crosslinguistic study based on the universal dependency corpora. In: Proceedings of the NoDaLiDa 2017 Workshop on Universal Dependencies, (UDW 2017) (2017)
11. Celano, G.A, Richter, M., Voll, R., Heyer, G.: Aspect coding asymmetries of verbs: the case of Russian. In: Barbaresi, A., Biber, H., Neubarth, F., Osswald, R. (eds.) KONVENS 2018. Proceedings of the 14th Conference on Natural Language Processing, pp. 34–39 (2018)
12. Shannon, C.E., Weaver, W.: A mathematical theory of communication. Bell Syst. Tech. J. **27**, 623–656 (1948)
13. Jaeger, T.F.: Redundancy and reduction: Speakers manage syntactic information density. Cogn. Psychol. **61**(1), 23–62 (2010). https://doi.org/10.1016/j.cogpsych.2010.02.002
14. Levy, R., Jaeger T.F.: Speakers optimize information density through syntactic reduction. In: Proceedings of the 20th Conference on Neural Information Processing Systems (NIPS) (2007)
15. Aylett, M., Turk, A.: The smooth signal redundancy hypothesis: a functional explanation for relationships between redundancy, prosodic prominence, and duration in spontaneous speech. Lang. Speech **47**(1), 31–56 (2004)
16. Genzel, D., Charniak E.: Entropy rate constancy in text. In: Proceedings of ACL, pp. 199–206 (2002)
17. Crocker, M.W., Demberg, V., Teich, E.: Information density and linguistic encoding (IDeaL). KI Künstliche Intelligenz **30**(1), 77–81 (2016)
18. Agrawal, A., Agarwal, S., Husain, S.: Role of expectation and working memory constraints in Hindi comprehension: an eyetracking corpus analysis. J. Eye Mov. Res. **10**(2), 1–15 (2017)
19. Hale, J.: A probabilistic earley parser as a psycholinguistic model. In: Proceedings of NAACL (2001)
20. Altmann, G., Kamide, Y.: Incremental interpretation at verbs: restricting the domain of subsequent reference. Cognition **73**(3), 247–264 (1999)
21. Levy, R.: Expectation–based syntactic comprehension. Cognition **106**(3), 1126–1177 (2008)

22. Villupillai, V.: Zero Coding in Tense-Aspect Systems of Creole Languages. John Benjamins, Amsterdam (2012)
23. Vendler, Z.: Linguistics in Philosophy. Cornell University Press, Ithaca (1967)
24. Bohnemeyer, J., Swift, M.: Event realization and default aspect. Linguist. Philos. **27**(3), 263–296 (2004)
25. Levy, R.: Memory and surprisal in human sentence comprehension. In: van Gompel, R.P.G. (ed.) Sentence Processing, pp. 78–114. Psychology Press, Hove (2013)
26. Collins, M.X.: Information density and dependency length as complementary cognitive models. J. Psycholinguist. Res. **43**(5), 651–681 (2014)

Automated Prediction of Relevant Key Performance Indicators for Organizations

Ünal Aksu[1,2(✉)], Dennis M. M. Schunselaar[2], and Hajo A. Reijers[1,2]

[1] Utrecht University, Utrecht, The Netherlands
{u.aksu,h.a.reijers}@uu.nl
[2] Vrije Universiteit Amsterdam, Amsterdam, The Netherlands
d.m.m.schunselaar@vu.nl

Abstract. Organizations utilize Key Performance Indicators (KPIs) to monitor whether they attain their goals. For this, software vendors offer predefined KPIs in their enterprise software. However, the predefined KPIs will not be relevant for all organizations due to the varying needs of them. Therefore, software vendors spend significant efforts on offering relevant KPIs. That relevance determination process is time-consuming and costly. We show that the relevance of KPIs may be tied to the specific properties of organizations, e.g., domain and size. In this context, we present our novel approach for the automated prediction of which KPIs are relevant for organizations. We implemented our approach and evaluated its prediction quality in an industrial setting.

Keywords: Key Performance Indicators · Prediction · Relevance

1 Introduction

Organizations measure the performance of their business processes to determine whether they attain their goals. As a means for that, Key Performance Indicators (KPIs) are used [20]. *Average duration of product delivery* is a KPI that organizations use to monitor their product delivery processes. By tracking this KPI, organizations can predict how much staff must be assigned to their product delivery processes to keep the duration of a product delivery below a certain threshold, e.g., on average 3 days.

To support organizations in process performance measurement, software vendors offer predefined KPIs in their software products. With this, they aim to provide the maximal set of KPIs that may be relevant for most organizations. However, predefined KPIs will not work successfully in all organizations because they want relevant KPIs aligned to their specific goals [20]. For example, *taken*

Supported by the NWO AMUSE project (628.006.001): a collaboration between Vrije Universiteit Amsterdam, Utrecht University, and AFAS Software in the Netherlands. This work is a result of the AMUSE project. See amuse-project.org for more information.

W. Abramowicz and R. Corchuelo (Eds.): BIS 2019, LNBIP 353, pp. 283–299, 2019.
https://doi.org/10.1007/978-3-030-20485-3_22

leave per day is a relevant KPI for a production organization, whereas it may not be relevant for a university, which has similar number of employees. For this reason, software vendors include Business Intelligence (BI) functionality into their software products and let organizations develop custom KPIs. Although organizations may do this, custom development of KPIs still requires a significant effort both from software vendors and organizations [20].

Numerous studies have been conducted for determining relevant KPIs for organizations [3,4,14,15,28]. In these studies, relevant KPIs are either defined from scratch or selected from a set of KPIs (e.g., a KPI library) for each organization. Moreover, in these studies, the identified reasons that make certain KPIs relevant for one organization are not usually reusable at determining the KPIs for another. Therefore, for current approaches tailoring KPIs is a manual endeavor that needs to be repeated for each organization, and requires a significant effort both from software vendors and organizations.

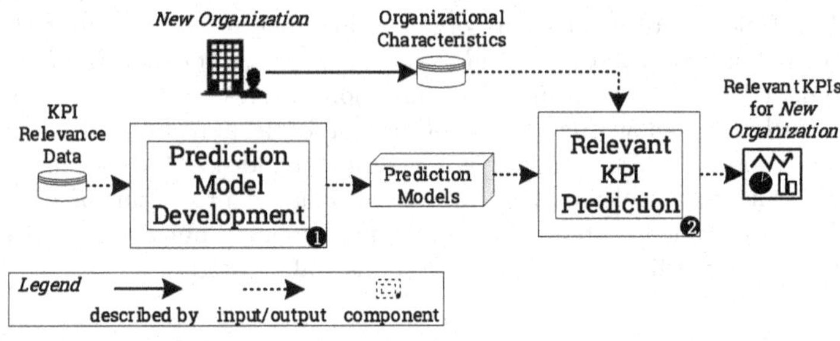

Fig. 1. Our approach for predicting relevant KPIs for organizations

Within this paper, we propose a novel approach for the automated prediction of relevant KPIs for organizations. The approach takes a set of prediction models aimed at predicting the relevance of KPIs and the characteristics of a new organization, e.g., domain, location, and number of employees. By checking the given organizational characteristics of that organization against the relevance factors of KPIs encoded in the prediction models, the approach predicts which KPIs are relevant to the organization (see ❷ in Fig. 1). To determine the relevance factors of KPIs and develop prediction models (see ❶ in Fig. 1), the approach uses the known relevance values of a set of KPIs for a number of organizations, which needs to be given in the form of a specific input, KPI Relevance Data. By means of the automatically determined relevance factors of KPIs, we automate the prediction of relevant KPIs for organizations, which is manually repeated for every organization in current approaches. Thus, our approach sets itself apart from the state of the art. We evaluate the prediction quality of our approach by applying it in a real-life setting at a Dutch ERP software vendor. In this context, we discuss the results that we obtained.

In Sect. 2, we present our approach aimed at the automated prediction of relevant KPIs for organizations. The details of the implementation of the approach are given in Sect. 3. Afterwards, in Sect. 4, we evaluate the prediction quality of our approach by applying it in a real-life, industrial setting and present the results obtained in the application. Section 5 is devoted to the discussions on the implications of the obtained results. In Sect. 6, we provide an overview of related work on providing relevant KPIs to organizations. Finally, we present our conclusions and potential directions for future work in Sect. 7.

2 Approach

In this section, we explain the details of our approach on the automated prediction of relevant KPIs for organizations. As introduced in Sect. 1, there are two tasks: predicting relevant KPIs and developing prediction models. They are taken care of by the components Prediction Model Development and Relevant KPI Prediction. The former uses prediction models and the organizational characteristics of a new organization as inputs; the latter uses KPI Relevance Data as the only input. For the sake of simplicity, first, the definitions of organizational characteristics, prediction models, and KPI Relevance Data are listed below. Then, the details of each component is given.

Definition 1 ***Organizational Characteristics*** *contain the values of a set of characteristics (e.g., domain, location, and number of employees) by which organizations can be characterized. For example, Organization o1 = {domain = Retail ∧ location = Amsterdam ∧ numberOfEmployees= [10–19] ∧ doesExport = Yes ∧ industryClassification-MainGroup = 47 ∧ industryClassification-SubGroup = 8109 }.*

Definition 2 ***Prediction Models*** *are aimed at predicting the relevance of KPIs. Each prediction model encodes a KPI, the factors that are the determinants to what extent the KPI will be relevant for organizations, and a prediction modeling technique, which outperforms predicting the relevance value of the KPI for organizations using those relevant factors.*

Definition 3 ***KPI Relevance Data*** *is a 2-tuple: (1) the relevance values of a set of KPIs for a number of organizations where that KPI set is considered as the comprehensive set from which a sub-set will be selected and (2) the key characteristics of these organizations with their values, i.e., Organizational Characteristics. For example, in Fig. 3, an excerpt from a sample KPI Relevance Data is depicted.*

In our approach, the relevance value of a KPI can be a numeric value from a scale, namely KPI Relevance Scale. As the KPI Relevance Scale, we use a five-points Likert-type scale: [1, 5], where a higher value denotes a higher relevance. The reason for using a five-points scale is that it has been recommended by many researchers [5, 10, 27] as the optimal number of relevance categories.

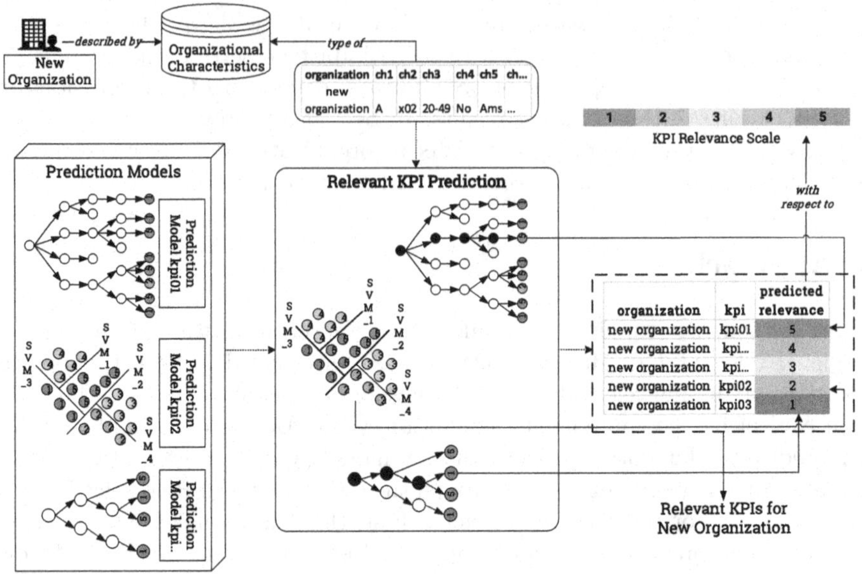

Fig. 2. Predicting relevant KPIs for a new organization

Relevant KPI Prediction: To predict which KPIs are relevant for a new organization, two inputs are required: the organizational characteristics of that organization and prediction models. The prediction modeling technique encoded in each prediction model is executed with the given organizational characteristics. Thus, a predicted relevance value will be obtained for the KPI. In the output, the obtained relevance values are sorted from highest to lowest. Afterwards, the KPIs that have the highest predicted relevance value, a value of 5 in the KPI Relevance Scale used in our approach (see in Fig. 2), are marked as the set of relevant KPIs for the new organization. However, this marking is flexible and one can say that a value of either 4 or 5 may be presented as the set of relevant KPIs for the new organization. For this, to what extent a KPI is used for making decisions about the related business process in an organization may be a reason.

Prediction Model Development: This component takes KPI Relevance Data as input. For each KPI in the input, an analysis task is performed to determine what organizational characteristics are the determinants of the relevance value of a KPI for organizations. The reason for performing the task per KPI is that relevance factors may vary from one KPI to another. For example, "number of employees" may be the only factor that makes a KPI relevant for organizations, whereas for those organizations the relevance of another KPI may be dependent on both "number of employees" and "organization type", e.g., whether it is a non-profit organization.

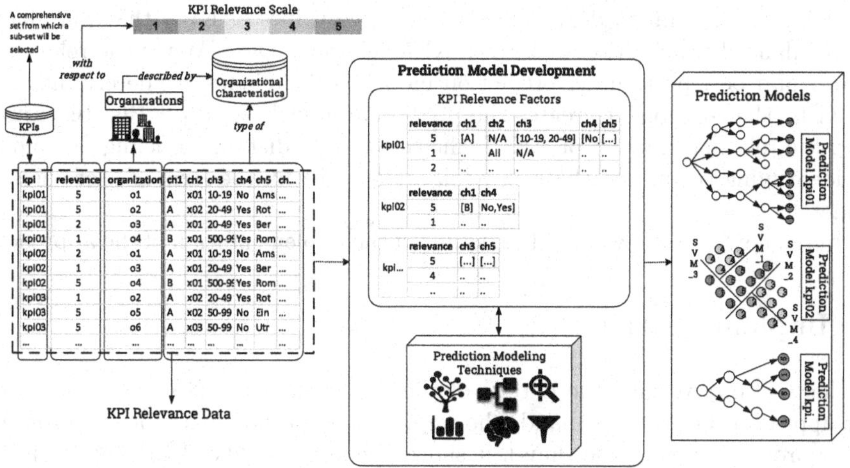

Fig. 3. Creation of the prediction models for predicting relevant KPIs

Since the organizational characteristics in the given KPI Relevance Data are raw data, they need to be transformed into features to better represent the underlying patterns in the given KPI Relevance Data. That transformation is done by the encoders within this component. More specifically, a one-hot encoder is used for each feature. Then, a feature subset is selected for each prediction modeling technique employed within the component–employed techniques are listed in the implementation documentation of the approach[1]. This feature subset selection helps to make sense of the features for prediction modeling techniques. To keep the best performing features in subsets, the worst performing feature at each iteration is eliminated, and then the dependencies between features are uncorrelated by a dimensionality reduction.

Afterwards, the component trains and tests each prediction modeling technique to find out the best performing prediction modeling technique for each KPI. The reason for that is a prediction modeling technique may not outperform for predicting the relevance values of all KPIs since the relevance values in a given KPI Relevance Data may not be the same for all KPIs. Moreover, each prediction modeling technique has its own noise handling mechanism. For example, while Random Forest may be the best for an imbalanced set of relevance values, other prediction modeling techniques, e.g., Ada Boost may perform poorly. While training a prediction modeling technique, the component chooses a set of appropriate hyperparameters to discover the parameters that may result in more accurate predictions. To do so, the component uses a cross-validated grid-search algorithm. As a result of the train and test, the component identifies the best performing prediction modeling technique at finding the relevance factors of each KPI. For this, the balanced accuracy metric [9,25] is used.

[1] The implementation of our Automated Relevant KPI Determination Approach is available at http://amuse-project.org/software/.

By doing so, we aim to deal with the relevance values of KPIs that may have an imbalanced distribution in a given KPI Relevance Data. When the relevance factors and best performing prediction modeling techniques are determined for all KPIs, the component creates the prediction models for the KPIs. In particular, the relevance factors of a KPI, the selected prediction modeling technique for identifying them, and the KPI itself are encoded in the form of a prediction model.

In the next section, we give the details of the implementation of the approach.

3 Implementation

In this section, we give the details of the implementation (see Footnote 1) of our approach. On the one hand, the implementation is a constructive proof of the approach. On the other hand, it shows the applicability of the approach. In Fig. 4, the technical details of the implementation are depicted.

As explained, to predict relevant KPIs, the approach requires prediction models as input. This is taken care of the Prediction Model Development component within the approach. It takes KPI Relevance Data as input and develops prediction models. To accurately capture the knowledge in the given KPI Relevance Data, as shown in Fig. 4, a nested (two-level) stratified cross validation is used: (1) for all KPIs and (2) per KPI. More specifically, both model development and testing will be carried out n-times, which is specified in each stratified cross validation block, using a different sample dataset of the given KPI Relevance Data. By doing so, we aim to develop the prediction models that both capture the patterns in the given KPI Relevance Data, but also generalizes well to unseen organizational characteristics of new organizations.

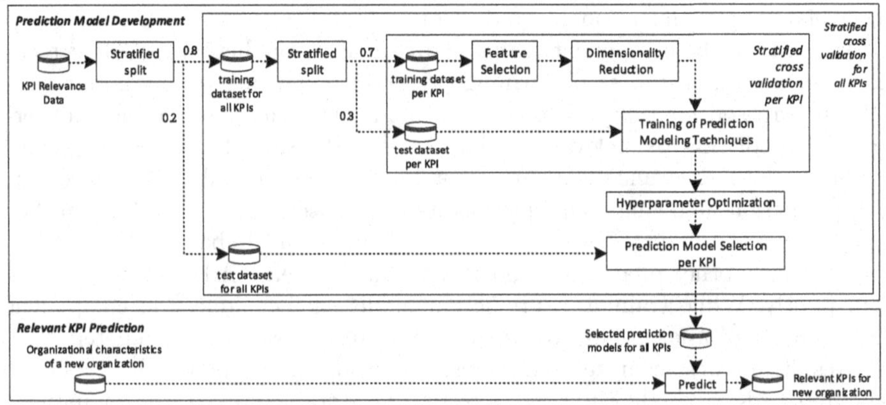

Fig. 4. Technical details of the implementation of our approach

Within the *stratified cross validation block for all KPIs*, prediction model development and testing for all KPIs will be done in 5-folds. This means that in each fold, an 80%/20% stratified split [18] is done to divide the given input into training and test datasets. By doing so, the approach can develop prediction models and test them using the sample datasets that are preserving the percentage of the data points for each KPI and class (i.e., relevance values). As a result, the approach creates two different dataset in each fold: training dataset for all KPIs and test dataset for all KPIs. Similarly, within the *stratified cross validation block per KPI*, the approach does the prediction model development and testing for each KPI in 5-folds. A training dataset per KPI and a test dataset per KPI will be generated in each fold. The aforementioned prediction modeling techniques will be trained and tested using these datasets. In order to avoid over-fitting to the training data, a 70%/30% split is preferred [12].

The approach utilizes feature selection and dimensionality reduction [19] to select the organizational characteristics in the training dataset per KPI that contribute most to the prediction variable (the relevance value of a KPI). Then, the approach applies feature scaling to have standardized range of the values of the selected organizational characteristics. By doing so, we aim to make that each selected organizational characteristic may equally influence the prediction variable.

Afterwards, the approach trains each prediction modeling technique contained in it. Meanwhile, the approach tunes the parameters of each prediction modeling technique to determine the parameter set with which each trained prediction modeling technique performs best. When all prediction modeling techniques are trained, the approach tests them using test dataset per KPI mentioned above. Using the balanced accuracy metric, the approach selects the best performing prediction model for a KPI, from the set of the prediction models that are created in all folds of the *cross-validation block per KPI* and tested. When the best performing prediction model for each KPI is selected, then the approach completes the prediction model generation process. In other words, the selected prediction models are ready for predicting relevant KPIs for organizations.

To achieve a high quality at predicting relevant KPIs for organizations, while developing prediction models within the approach, we use 3 types of meta-algorithms: stacking, boosting, and bagging [29]. In stacking, a meta-technique tries to learn the best combination of the prediction models of the primary prediction modeling techniques, which are combined as a stack. In boosting, the same prediction modeling technique is applied in a chain to learn and fix the prediction errors of prior prediction models developed in the chain. Different sub-samples of the training dataset are taken and multiple prediction models are generated in bagging. Then, these models are aggregated to form a final prediction model, which has a better accuracy value.

Since most properties of organizations are categorical data types and the scale we used for relevance values (KPI Relevance Scale) has multiple points, it is required to support multi-class classification [12] prediction modeling techniques from the machine learning discipline within our approach. Accordingly, our approach employs the prediction modeling techniques that are listed in the implementation documentation of the approach (see Footnote 1). However, our approach is flexible to support continuous (numeric) data types. This can be indicated in the configuration where the approach learns the organizational characteristics contained in a given KPI Relevance Data. Moreover, our approach is extensible to support regression prediction modeling techniques in the case that one may want to predict a decimal value for the relevance value of KPIs instead of a numeric value from a KPI Relevance Scale.

In addition, to obtain better predictive performance, the approach reduces the problem of multi-class classification to multiple binary classification problems while developing prediction models. In this regard, we apply the following strategies: one-vs-rest and one-vs-one [12]. The former involves training a single prediction modeling technique per class, whereas in the latter a particular prediction modeling technique is trained for each different pair of classes.

In the following section, we describe how we evaluate our approach in a practical use of its implementation.

4 Evaluation

In this section, we demonstrate the use of the proposed approach in an industrial setting and evaluate how accurately it predicts relevant KPIs for organizations. In this regard, in the following subsections, we describe how we develop prediction models and use them at predicting relevant KPIs for organizations in a case study.

4.1 Data Collection

In order to develop prediction models, the approach requires KPI Relevance Data. However, although software vendors usually know the key characteristics of the organizations that they deliver their software products to, they are typically not aware of the relevance of the KPIs that they offer to those organizations. Therefore, we investigated whether we could identify a proxy for this type of data. KPI Usage Logs are typical data sources in which software vendors typically keep track of how KPIs are being used by organizations. In general, software vendors either record these logs using the software product in which they offer KPIs for organizations or using a third-party BI tool (e.g., Qlik Sense and Microsoft Power BI), which they use as a means for enabling organizations to develop custom KPIs. KPI Usage Logs is a data source from which one can obtain information on the usage of KPIs. For example, how many times a particular KPI is used in an organization, when that KPI is used, and how much time has spent using the KPI can be obtained from KPI Usage Logs. The obtained information on the usage of KPIs can be seen as the interest of organizations in KPIs. Moreover, one can interpret the interest of organizations in KPIs as the relevance values of KPIs for those organizations. Thus, as a primary proxy for known relevance values of KPIs for organizations, KPI Usage Logs are determined.

A Dutch ERP software vendor, the case study company, records KPI Usage Logs for the KPIs that the company offers to its customers. In the company, we had a training session on the KPIs, which are offered to its customers within its ERP software product. In particular, we examined the KPIs in the Human Resource Management (HRM) area. The reason for that is that human resources form a key asset in any organization. As such, the availability of the employees in an organization is essential in performing its business processes to attain its goals. However, due to various reasons, employees may not be always available, for example, sickness, injury, maternity leave, or vacations. Absence and leave are the two sub areas in HRM that concern the unavailability of employees in organizations. The former deals with the unexpected reasons of unavailability of employees. The latter focuses on the unavailability of employees resulting from statutory rights as granted by labour laws. In this regard, together with two experts who manage the KPIs in the company, we selected 13 KPIs from the absence and 6 KPIs from the leave sub area. While selecting the KPIs, our main consideration was the wide usage of the KPIs by organizations to get sufficient data points such that our approach can predict relevant KPIs for a new organization accurately. The selected 19 KPIs are commonly used by more than 2000 client organizations of the software vendor. Afterwards, the experts defined a set of metrics for transforming the usage of the KPIs into the relevance values of them. Since the defined metrics require a minimum of one year usage of the KPIs, the relevance values of the selected 19 KPIs are obtained for approximately 1100 organizations, which use those KPIs at least for a year.

In addition to the obtained relevance values, the characteristics of those 1100 organizations were required to create KPI Relevance Data and develop the prediction models for the selected KPIs. To determine which characteristics and their values for those organizations are available in the company, we arranged three meetings with various experts. The first meeting was with the following experts: a director–the CIO (Chief Information Officer) of the company–who has knowledge on scoping the KPIs offered to the organizations and a senior product developer who is an expert on designing and developing KPIs. These experts explained the characteristics of the organizations that they consider while scoping and developing the KPIs offered to the organizations. A marketing manager participated in the second meeting and described the characteristics of the organizations that often request adjustments for the offered KPIs. In the last meeting, together with a product manager, we analyzed the data about the organizations to identify what characteristics and their values are available within the company. As a result, we selected the characteristics shown in Table 1. All these characteristics are categorical and the values of them were available for 750 out of the aforeselected 1100 organizations. Moreover, the names of these characteristics are translated from their original definitions in Dutch.

Table 1. Selected characteristics of organizations

Characteristic	Explanation with some example values
Legal form - Main group	The main group of the legal form of an organization, e.g., with or without legal entity
Legal form	The legal form of an organization, e.g. private limited company, foundation or association
Non-profit	A non-profit organization uses the money it earns to help people. However, a profit organization invests the money it earns on developing new products or services to sell them and make more money
Industry classification - Main group	The main group to which the organization is assigned by the Chamber of Commerce within the Netherlands. For example, construction is the main group to which general civil and utility construction organizations are assigned. Transportation and storage is another example for main group
Industry classification - Sub group	The sub group to which the organization is assigned by the Chamber of Commerce within the Netherlands. Construction of residential buildings and construction of railways are two example sub groups of the construction main group. Similarly, passenger transport and freight transport are two example sub groups of the transport and storage main group
Province	The province where the organization is registered
Number of - Employee range	The range of the total number of employees in the organization. For example, 10–19, 20–49, and 50–99
Import	Whether the organization does import
Export	Whether the organization does export
Has subsidiary organizations	Whether the organization has subsidiary organizations

By combining the obtained relevance values of the selected KPIs 19 with the organizational characteristics of the aforementioned 750 organizations, we created the KPI Relevance Data for our evaluation. This means that the required input for developing the prediction models for the KPIs is ready. Accordingly, the approach created 19 prediction models for predicting the relevance of the selected 19 KPIs.

4.2 Applying the Automated Relevant KPI Determination Approach

We predicted relevant KPIs for a set of organizations using the developed prediction models. Since the developed prediction models are for the KPIs in the absence and leave sub areas, the set of the organizations for which relevant KPIs are predicted are accordingly selected. In particular, the organizations are selected from the client organizations of the case study company that are not purchased and not use to the selected 19 KPIs, but use the related functionalities, i.e., absence and leave within the ERP software product of the company. As a result, we selected 261 organizations and predicted relevant KPIs for them.

To determine the prediction accuracy of our approach in the case study, we collaborated with the CIO of the company and a senior product manager in the case study company. The reasons for collaborating with these two experts is that these experts have extensive knowledge both on the organizations for which relevant KPIs are predicted and on the expected relevance values of the selected KPIs for those organizations. Then, we calculated the prediction accuracy of our approach by comparing the predicted relevance values of the KPIs for the organizations with the expected relevance values of the KPIs for these organizations, which are provided by the aforementioned two experts. In Fig. 5, the prediction accuracy of our approach in the case study is depicted.

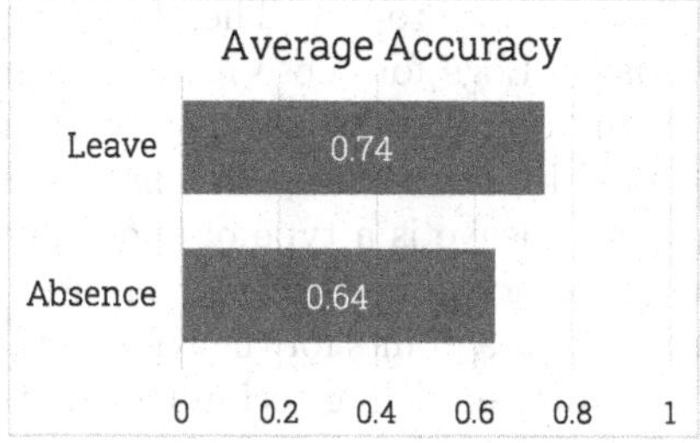

Fig. 5. Prediction accuracy of our approach in the case study

As depicted in Fig. 5, in the case study, our approach achieves a 74% balanced accuracy at predicting the relevance of 6 KPIs in the leave sub area for 261 organizations. Similarly, a 64% balanced accuracy is achieved at predicting the relevance of 13 KPIs in the absence sub area for the same organizations. The weighted average of the prediction quality values for these two sub areas will show the prediction quality of our approach for the HRM area, which is 67%. In the following section, we discuss the implications of those results.

5 Discussion

To consider a certain prediction quality as good, one should look into the context of the application where that quality is measured [8,23,26]. In this regard, we discuss the prediction accuracy of our approach. Since no other study has so far focused on the automated prediction of relevant KPIs for organizations, there is no exact reference to compare the prediction quality of our approach with. However, we think that the prediction quality of our approach shown in Fig. 5 is reasonable [26].

As mentioned in Sect. 3, the problem that our approach tries to solve is a multi-class classification. Having a balanced distribution of each class has a significant effect on prediction quality in multi-class classification. However, this was not the case for the KPIs that we used in the case study. Notably for the lower relevance values (e.g., 1 and 2 with respect to the used KPI Relevance Scale), there were fewer data points than for the higher relevance values (e.g., 4 and 5 with respect to the used KPI Relevance Scale). As a result, the approach was inclined to predict higher relevance values for some KPIs, which are expected to have lower relevance values by the experts in the case study company. This indicates that the prediction quality of the approach has negatively affected due to lacking data points.

One of the possible reasons for a lower prediction quality value in the absence sub area is that there were fewer data points in the used KPI Relevance Data for each KPI in the absence sub area than for each KPI in the leave sub area. In particular, there was a limited variety of small organizations in the known relevance values of the KPIs in the absence sub area. This was because in small organizations, the management of absence-related data is ad-hoc, i.e., absence-related data may not be stored day-to-day. Therefore, there was missing usage information in the KPI Usage Logs for those KPIs for small organizations. By contrast, leave operations in organizations are mostly recorded day-to-day since there are regulations defined by law to keep the data related these leave operations up-to-date. In addition, leave is a type of operation that can be planned ahead, whereas absence has a more unpredictable nature.

As a result of having fewer data points for the KPIs in the absence sub area, as shown in Table 2, the prediction modeling techniques that use linear separation method for input data outperformed at predicting the relevance values of the KPIs in this sub area. However, tree/forest based prediction modeling techniques were majority in the leave sub area since there were more data points for the KPIs this sub area.

Table 2. Outperformed predicting modeling techniques for the KPIs in the HRM area

	Logistic regression	Support Vector Machines (SVMs)	Decision tree	Random forest	Stacked (Decision Tree & SVMs)
Absence	7	3	2	1	0
Leave	0	0	2	2	2

We also had the idea to apply our approach in the finance area. Using the KPI Usage Logs of 109 KPIs in the finance area, we created KPI Relevance Data.

Then, we analyzed this data before applying the approach on it. We found out that for more than 60% of the KPIs, there are fewer data points for 3 out of 5 known relevance values. We decided against actually using the approach although the data was not good enough, i.e., containing fewer data points. Unfortunately, the approach performed worse than predicting the relevance of the KPIs in the HRM area. We examined the failing predictions for the KPIs in the finance area with the two experts together with whom we determined the prediction accuracy of the approach in the HRM case–a director and a senior product developer. These experts pointed out that the expected relevance of the KPIs in the finance area are mostly dependent on various financial characteristics of organizations such as debt, revenue, payment periods of both the customers and suppliers of these organizations, and how the products and services are sold by these organizations. Although our approach is extensible to new organizational characteristic, however; unfortunately, these organizational characteristics are not available in the case study company since these are mostly sensitive data about organizations.

Software vendors that focus on automatically predicting relevant KPIs for their customers and operate various domains can apply our approach. However, if these software vendors may want to predict relevant KPIs for their customers using a different set of KPIs and organizational characteristics than we demonstrated in the case study, they need to provide their KPI Relevance Data to our approach and develop prediction models using the approach. Then, these software vendors can predict relevant KPIs for their customers by executing the approach with the developed prediction models.

6 Related Work

Due to the high interest in both academia and business, there is a broad literature in the field of organizational performance measurement. Notably researchers proposed various approaches dealing with determining relevant KPIs for organizations since KPIs are widely used as a means for measuring the performance of organizations. Within these approaches, creating relevant KPIs afresh for any organization or choosing KPIs from a reference set of KPIs (e.g., a KPI library) as the relevant set for a particular organization are the two common ways of determining relevant KPIs. In this section, we list some of the works, which cover the following question that we are interested in: how are relevant KPIs determined for organizations?

Much work has been conducted on defining relevant KPIs from scratch for organizations in various domains. Granberg and Munoz develop KPIs for airport managing organizations [11]. Similarly, to monitor the performance of airports, a set of KPIs are proposed in [7]. Kaganski et al. [15] describe the development of KPIs for small and medium-sized enterprises (SMEs). While a set of KPIs for the organizations that have highly diverse product families are defined in [24], Elliot et al. [6] specify a set of KPIs for a large pediatric healthcare organization. Since the development of KPIs in the aforementioned works is from scratch and manual, in each work, it is required to have an intensive technical knowledge of the organization to which relevant KPIs are determined. Thus, a significant effort is required to obtain that knowledge.

Apart from the aforementioned works, del Río-Ortega et al. present a meta-model [4] as a basis for working with KPIs. Using the language proposed as part of the meta-model one can model KPIs within the process models of the processes in an organization. Then, the values of the modeled KPIs can be derived from the execution logs of the process models. However, this still requires each organization to determine relevant KPIs for itself and model them using the proposed meta-model. Therefore, this will require a significant effort of each organization.

In some studies [13,16,21,22], researchers focus on selecting a subset from a set of KPIs to determine the relevant set of KPIs for organizations. Within that selection process, researchers mostly consider the sector of an organization or a set of business processes in an organization. However, due to the varying needs of organizations, a KPI subset that is selected as the relevant set for one organization may not be relevant for all other organizations, which are in the same sector or perform similar business processes with that organization. Therefore, that KPI subset selection process needs to be repeated for many organizations. To deal with that, Analytic Network Process (ANP) is utilized [3,14,17,28]. In particular, certain characteristics of KPIs such as reliability, comparability, and understandability are taken into account to determine the priorities of a set of existing KPIs in organizations. This is mostly done together with specific experts in organizations. Then, the KPIs that have the highest priorities are selected as the relevant KPIs for organizations. However, on the one hand, since the considered characteristics of KPIs are subjective to experts, the priority of a KPI may vary from one organization to another. On the other hand, ANP is a time-consuming and complex multi-criteria decision-making method, and therefore requires a significant effort from organizations.

7 Conclusion and Future Work

In this paper, we presented a novel approach aimed at the automated prediction of relevant KPIs for organizations. A set of prediction models aimed at predicting the relevance of KPIs and the organizational characteristics of a new organization are the required inputs by the approach. The approach determines which of the KPIs that are encoded in the prediction models are relevant for that new organization using the relevance factors of the KPIs. To identify these factors automatically and develop prediction models, the approach employs prediction modeling techniques and applies them on the known relevance values of KPIs for organizations, which should be given in the form of a specific input, KPI Relevance Data.

To show the accuracy of our approach, we implemented it and demonstrated in a case study at a Dutch ERP software vendor. Within the case study, together with experts in the company, we selected 19 KPIs from the HRM area that are offered to organizations by the company in its ERP software product. The known relevance values of the selected KPIs were not available in the company. Therefore, we identified KPI Usage Logs as a proxy for known relevance values of KPIs and subsequently we created KPI Relevance Data and developed the prediction models for the selected 19 KPIs together with the experts in the company. Afterwards, the relevance values of the KPIs were predicted for 261 organizations, which are new to those KPIs. Finally, we evaluated the prediction quality of the approach by comparing the predicted relevance values of the KPIs against the expected relevance values of those KPIs, which are provided by two experts in the company. The prediction quality of the approach was of sufficient quality to show the practical usage of the approach. As a result, we automate the selection of relevant KPIs for every organization. For current approaches, this is a manual endeavor that needs to be repeated for every single organization. Thus, we believe that our approach lowers the efforts of software vendors for determining relevant KPIs for their client organizations or the efforts of these organizations doing this themselves.

In future work, we want to extend our approach for determining relevant KPIs for different roles in organizations since the relevance of a KPI might vary from one role to another in organizations. For example, there may be a significant difference in the relevance value of a KPI on daily stock changes between a CEO and for a warehouse employee. Furthermore, sales, purchasing, and logistics are the areas to which we envision extending our approach since their commonality among organizations in addition to the less sensitivity of data for organizations in these areas in comparison to other areas, e.g., finance and accounting. Besides, we plan to develop a decision graph aimed at identifying which visualization best suits for particular KPIs. Thus, engaging dashboards comprising relevant KPIs can be built automatically. Moreover, the approach in this paper and the approach that we presented in [2] are part of our Cross-Organizational Process Mining Framework, which we introduced in [1], and will be together incorporated into the framework. With this, we aim to provide recommendations for organizations using the benchmarks that are developed utilizing relevant KPIs.

References

1. Aksu, Ü., Schunselaar, D.M.M., Reijers, H.A.: A cross-organizational process mining framework for obtaining insights from software products: accurate comparison challenges. In: 18th IEEE Conference on Business Informatics (2016)
2. Aksu, Ü., Schunselaar, D.M.M., Reijers, H.A.: An approach for automatically deriving key performance indicators from ontological enterprise models. In: 7th International Symposium on Data-driven Process Discovery and Analysis (2017)
3. Carlucci, D.: Evaluating and selecting key performance indicators: an ANP-based model. Measuring Bus. Excellence **14**(2), 66–76 (2010)
4. del-Río-Ortega, A., Resinas, M., Cabanillas, C., Cortés, A.R.: On the definition and design-time analysis of process performance indicators. Inf. Syst. **38**(4), 470–490 (2013)
5. Eisenberg, M., Hu, X.: Dichotomous relevance judgments and the evaluation of information systems. Proc. Am. Soc. Inform. Sci. **24**, 66–69 (1987)

6. Elliot, C., Mcullagh, C., Brydon, M., Zwi, K.: Developing key performance indicators for a tertiary children's hospital network. Aust. Health Rev. **42**(5), 491–500 (2018)
7. Eshtaiwi, M., Badi, I., Abdulshahed, A., Erkan, T.E.: Determination of key performance indicators for measuring airport success: a case study in Libya. J. Air Transp. Manag. **68**, 28–34 (2018)
8. Frost, J.: Making predictions with regression analysis, March 2019. https://statisticsbyjim.com/regression/predictions-regression
9. García, V., Mollineda, R.A., Sánchez, J.S.: Index of balanced accuracy: a performance measure for skewed class distributions. In: Araujo, H., Mendonça, A.M., Pinho, A.J., Torres, M.I. (eds.) IbPRIA 2009. LNCS, vol. 5524, pp. 441–448. Springer, Heidelberg (2009). https://doi.org/10.1007/978-3-642-02172-5_57
10. Gluck, M.: Exploring the relationship between user satisfaction and relevance in information systems. Inf. Process. Manage. **32**(1), 89–104 (1996)
11. Granberg, T.A., Munoz, A.O.: Developing key performance indicators for airports. In: 3rd ENRI International Workshop on ATM/CNS (2013)
12. Gravetter, F.J., Wallnau, L.B.: Essentials of statistics for the behavioral sciences (2013)
13. Ioan, B., Nestian, A.S., Tita, S.M.: Relevance of key performance indicators (KPIs) in a hospital performance management model. J. Eastern Europe Res. Bus. Econ. **2012**, 15 (2012)
14. Kachitvichyanukul, V., Luong, H., Pitakaso, R.: A hybrid MCDM approach to KPI selection of the coordination problems of production and sales departments-an empirical study of iron and steel industry of China and Taiwan. In: 13th Asia Pacific Industrial Engineering and Management Systems Conference (2012)
15. Kaganski, S., Paavel, M., Lavin, J.: Selecting key performance indicators with support of enterprise analyse model. In: 9th International DAAAM Baltic Conference "Industrial Engineering" (2014)
16. Kaganski, S., Snatkin, A., Paavel, M., Karjust, K.: Selecting the right KPIs for smes production with the support of PMS and PLM. Int. J. Res. Soc. Sci. **1**(3), 69–76 (2013)
17. Kucukaltan, B., Irani, Z., Aktas, E.: A decision support model for identification and prioritization of key performance indicators in the logistics industry. Comput. Hum. Behav. **65**, 346–358 (2016)
18. Parsons, V.L.: Stratified Sampling (2017)
19. Padmaja, D.L., Vishnuvardhan, B.: Comparative study of feature subset selection methods for dimensionality reduction on scientific data. In: IEEE 6th International Conference on Advanced Computing (2016)
20. Parmenter, D.: Key Performance Indicators: Developing, Implementing, and Using Winning KPIs. Wiley, New York (2015)
21. Peral, J., Maté, A., Marco, M.: Application of data mining techniques to identify relevant key performance indicators. Comput. Stand. Interfaces **54**, 76–85 (2017)
22. Pinna, C., Demartini, M., Tonelli, F., Terzi, S.: How soft drink supply chains drive sustainability: key performance indicators (KPIs) identification. In: 51st CIRP Conference on Manufacturing Systems (2018)
23. Saitta, S.: What is a good classification accuracy in data mining? March 2019. http://www.dataminingblog.com/what-is-a-good-classification-accuracy-in-data-mining
24. Schmidt, M., Schwöbel, J., Lienkamp, M., et al.: Developing key performance indicators for variant management of complex product families (2018)

25. Sokolova, M., Lapalme, G.: A systematic analysis of performance measures for classification tasks. Inf. Process. Manage. **45**(4), 427–437 (2009)
26. Steyerberg, E.W., et al.: Assessing the performance of prediction models: a framework for some traditional and novel measures. Epidemiology **21**(1), 128 (2010)
27. Tang, R., Shaw Jr., W.M., Vevea, J.L.: Towards the identification of the optimal number of relevance categories. J. Am. Soc. Inform. Sci. **50**(3), 254–264 (1999)
28. Van Horenbeek, A., Pintelon, L.: Development of a maintenance performance measurement framework-using the analytic network process (ANP) for maintenance performance indicator selection. Omega **42**(1), 33–46 (2014)
29. Zhou, Z.H.: Ensemble Methods: Foundations and Algorithms. Chapman and Hall/CRC, Boca Raton (2012)

Genetic Programming over Spark for Higgs Boson Classification

Hmida Hmida[1,2(✉)], Sana Ben Hamida[2], Amel Borgi[1], and Marta Rukoz[2]

[1] Faculté des Sciences de Tunis, LR11ES14 LIPAH, Université de Tunis El Manar, 2092 Tunis, Tunisia
hhmida@gmail.com

[2] Université Paris Dauphine, PSL Research University, CNRS, UMR[7243], LAMSADE, 75016 Paris, France

Abstract. With the growing number of available databases having a very large number of records, existing knowledge discovery tools need to be adapted to this shift and new tools need to be created. Genetic Programming (GP) has been proven as an efficient algorithm in particular for classification problems. Notwithstanding, GP is impaired with its computing cost that is more acute with large datasets. This paper, presents how an existing GP implementation (DEAP) can be adapted by distributing evaluations on a Spark cluster. Then, an additional sampling step is applied to fit tiny clusters. Experiments are accomplished on Higgs Boson classification with different settings. They show the benefits of using Spark as parallelization technology for GP.

Keywords: Genetic Programming · Machine learning · Spark · Large dataset · Higgs Boson classification

1 Introduction

Digital transformation that we witness in organizations and companies have generated a huge volume of data. High storage capacities facilitated this phenomenon and provided organizations with their own data lakes[1].

To discover hidden knowledge in this data, many Artificial Intelligence (AI) tools and techniques have been used such as Neural Networks, Decision Trees, etc. Evolutionary Algorithms (EA), and in particular Genetic Programming (GP) [14], are candidate solutions since they have shown satisfying results for a wide range of problems (classification, time series prediction, etc.). However, GP suffers from an overwhelming computational cost. This becomes more noticeable when the handled problem has a very large dataset as input. In fact, the evaluation step (see Fig. 1) is the Achilles heel of GP. It is at the origin of the increasing computational time. Therefore, any solution that applies GP to solve a large scale problem, has to focus on reducing the evaluation cost.

[1] 'A data lake is a collection of storage instances of various data assets additional to the originating data sources.' (Source: Gartner).

W. Abramowicz and R. Corchuelo (Eds.): BIS 2019, LNBIP 353, pp. 300–312, 2019.
https://doi.org/10.1007/978-3-030-20485-3_23

Summarily, mitigating this cost could be achieved by either parallelizing evaluations or reducing their number by means of dataset sampling, hardware acceleration or distributed computing. Hadoop MapReduce[2] and Apache Spark[3] are Big Data tools that implement a new programming model over a distributed data storage architecture. They are *de facto* tools for data intensive applications. However, according to our knowledge, neither Spark libraries, such as Spark built-in library MLlib, nor existing *ad hoc* solutions do provide an implementation of GP adapted to Spark, which hinders its use in Big Data frameworks. Moreover, recent machine learning problems more often than not need very high computing power and resources in order to make a solution in a reasonable time. The challenge of running GP on Spark becomes more defying when only Small clusters are available.

This work outlines how we ported an existing GP implementation to Spark context in order to take the most of its proven potential. We apply this solution to the Higgs Boson classification problem (see Sect. 2.4) and study the effect of varying some GP parameters on learning performance and time. Additionally, we include a sampling algorithm to GP and test it in the same environment configuration used with the whole dataset. We discuss the contribution of this sampling method to the learning process and the effect of varying number of generations, population size and sample size on training time and classifier performance.

The remainder of this paper is organized as follows. Section 2 gives an introduction to GP basics. It presents Spark and recent works that combine these two concepts. Section 3, exposes how we adapted an existing GP implementation to comply with Spark environment. Then, we show why and how a sampling phase is added to GP. Details about experimental design followed by the obtained results are discussed respectively in Sects. 4 and 5. Finally, we end with some concluding remarks and perspectives.

2 Background and Related Works

2.1 Genetic Programming (GP)

In the standard GP, the population is composed of tree-based individuals very close to Lisp programs. It performs the common steps of any EA that are:

1. Randomly create a population of individuals where tree nodes are taken from a given function and terminal sets. Then evaluate their fitness value by executing each program tree against the training set.
2. According to a fixed probability, individuals are crossed and/or mutated to create new offspring individuals.
3. New solutions are evaluated and a new population is made up by selecting best individuals from parents and offspring according to their fitnesses.
4. Loop step 2 and 3 until a stop criteria is met.

[2] https://hadoop.apache.org.
[3] https://spark.apache.org.

The evaluation step is at the origin of the increase of the GP computing time. In fact, it executes all programs (individuals) as many times as the size of the training set. In a classification problem, an individual is a program tree that represents a classification model. Figure 1 illustrates the evaluation of an individual against the training set. Its phenotype is $(if(IN3 > 0.3, IN3, IN0+IN1) < 0.6)$ and is represented by the tree in the same figure. It depicts a single iteration in the evaluation of a single individual within the population. The output is translated in a class prediction and is compared to the given value from the training set. The fitness value is computed, after looping on the whole training set, according to the fitness function adopted (error rate, true positive rate, etc.).

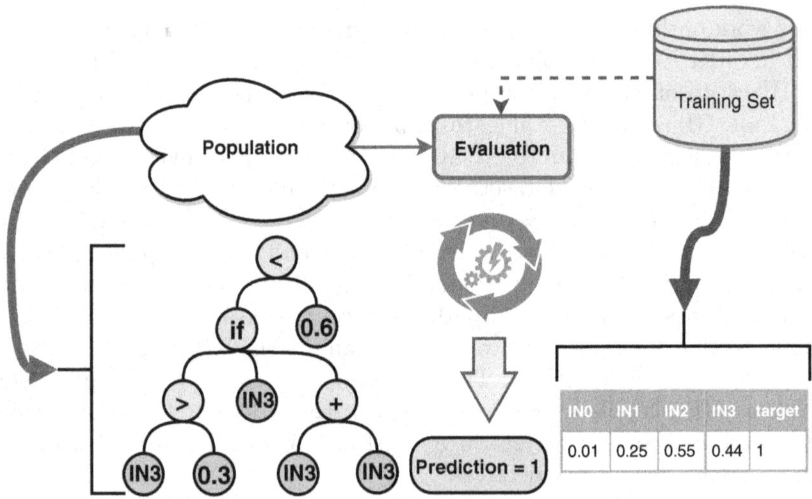

Fig. 1. GP Evaluation.

2.2 Spark and MapReduce

MapReduce is a parallel programming model introduced by Dean et al. in [5] and made popular with its implementation Apache Hadoop. The main idea of this model is that moving computation is cheaper than moving data. On a cluster, data is distributed over nodes and processing instructions are distributed and performed by nodes on local data. This phase is known as the Map phase. Finally, at the Reduce phase, some aggregation is performed to produce the final output. Hadoop Distributed Files System (HDFS) ensures data partitioning and replication for MapReduce. However, it needs many serialization operations and disk accesses. In addition to that, Hadoop MapReduce does not support iterative algorithms which is the case for EA.

Apache Spark is one of many frameworks intended to neutralize the limitations of MapReduce while keeping its scalability, data locality and fault tolerance. It is up to 100 times faster than MapReduce owing to in-memory computing. The keystone of Spark is the Resilient Distributed Datasets (RDD) [19]. An RDD is a typed cacheable and immutable parallel data structure. Operations on RDDs are of two types: transformations (*map. filter, join, ...*) and actions (*count, reduce, reduceByKey, ...*). Spark DAGScheduler, computes an optimized Directed Acyclic Graph exploiting lazy evaluation and data locality. Spark is compatible with different cluster managers (Built-in Standalone, Hadoop YARN, Mesos and Kubernetes) [13].

2.3 Previous Works

Recently, several works deal with parallelizing EA in distributed environments. Qi et al. [17] and Padaruru et al. [15] apply Genetic Algorithms (GA) on software test suite generation where input data is a binary code. Both give a Spark based implementation that parallelizes fitness evaluation and genetic operators. In Chávez et al. [4], the well-known EA library ECJ is modified in order to use MapReduce for fitness evaluations. This new tool is tested using GA to resolve a face recognition problem over around 50 MB of data. Only time measure was considered in this work. In Peralta et al. [16], the MapReduce model is applied to implement a GA that preprocesses big datasets (up to 67 millions instances). It applies an evolutionary feature selection, in a map phase, to create a vector per mapper on top of disjoint subsets from the original dataset. The reducer aggregates the previously created vectors. Funika et al. [7] implement an 'Evaluation Service' that can be solicited through a REST API. It is not an implementation of a specific EA algorithm but rather an outsourcing of the evaluation. This service can be used by any algorithm that requires the evaluation of an expression over a given dataset. They tested this service for 3 different expressions on datasets varying from 1 MB to 1024 MB. Al-Madi et al. [1] present a full GP implementation based on MapReduce. It relies on creating a mapper for each individual by storing the population on HDFS. Each mapper, calculates the fitness value for the involved individual. Then, the reducer collects the population and performs selection, crossover and mutation to create the new population. The focus is on increasing population size (until 50000) and only with small datasets classification problems.

Our proposal is inspired by all these works and mostly by Chávez et al. [4] where the authors have integrated the MapReduce model to the ECJ library in order to run EA on a hadoop cluster. Their goal was to evolve very large populations (up to 3 million individuals), which was achieved by using the checkpointing feature and serializing individuals in an HDFS file. This work is focused on population and did not study the use of massive datasets. Our proposal is rather concerned with training dataset. We both give a transformation of an existing tool, but our solution uses a different underlying infrastructure which is Spark engine and handles a big dataset to assess the advantages of distributing GP evaluation. Another difference is the EA algorithm applied. While Chávez used GA, we use standard tree-based GP. Finally, we extend this solution with a sampling technique.

2.4 HIGGS Dataset

A Higgs or Z Boson is a heavy state of matter resulting from a small fraction of the proton collisions at the Large Hadron Collider [3].

From the machine learning point of view, the problem can be formally cast into a binary classification problem. The task is to classify events as a signal (event of interest) or a background. Baldi et al. [2] published for benchmarking machine-learning classification algorithms a big dataset of simulated Higgs Bosons that contains 11 million simulated collision events [10].

In this work, we propose to handle the whole dataset, which is a big challenge when using EA. Table 1 summarizes the main characteristics of the dataset.

Table 1. Higgs dataset composition.

Total of events	11 millions
Number of attributes	28 real-valued (21 low-level and 7 high-level variables)
Percentage of signals	53%
Training set size	10.5 millions events
Test set size	500 K events

3 Porting DEAP to Spark

DEAP (Distributed Evolutionary Algorithms in Python) [6] is presented as a rapid prototyping and testing framework that supports parallelization and multiprocessing. It implements several EA: GA, Evolution Strategies (ES) and GP. We decided to use this framework for the following reasons: (1) it is a Python package which is one of the 3 languages supported by Spark, (2) it implements standard GP with tree based representation and (3) it is distributed ready. The third point means that DEAP is structured in a way that facilitates distribution of computing tasks. DEAP is not natively compliant with Spark and do not use any of its parallel data structures (RDDs, DataFrames or DataSets).

To adapt DEAP for a parallel computing engine, the usual method is to replace the map method, of the Toolbox class, by a code that calls the desired parallelized operation. Unfortunately, Spark has a constraint that prohibits nesting RDDs. To evaluate an individual, we need to access our dataset stored in an RDD. Consequently, this solution is not feasible. The following paragraph shows how we transformed DEAP to benefit from Spark RDDs.

3.1 Implementation Model

From recent works described in Sect. 2.3, two main scenarios for distributing any EA, and in particular GP, can be laid out:

1. Distribute the whole GP process: each mapper is an independent run that uses local data. It is very close to the co-evolution scheme or *island* model. By the end of this scenario, an aggregation is required to obtain the final solution. This aggregation is run on the driver program. For example, aggregation can be made through a voting using the best individual of each population in a classification problem. Another way is to take the best individual after a test on the whole training set. This solution needs more resources on the cluster.
2. Distribute population: this is suitable when a single population is evolved. Parallelization can be reached by partitioning the population using RDDs. Therefore, individuals are distributed on nodes, and each node processes the local individuals in spite of the total population. To proceed the following phase of the GP (see Sect. 2.1), the output of the previous phase is collected as a new RDD. It is composed of the modified individuals (RDDs are immutable).

The evaluation represents more than 80% of the total time cost in EA [7,9]. We suggest to focus on evaluation which can be easily distributed on Spark cluster even with limited resources seeing that we do not need independent populations. Besides, for machine learning problems involving big training sets, data must be parallelized and then we cannot parallelize population. The fitness function considered in this problem is to maximize correct predictions. To adapt fitness computing to Spark, a first alternative is to replace the default evaluation function used in DEAP. Map operation on *TrainingRDD* does not use the individual but a function (func) representing the Genetic Program. This alternative makes as many reduce operations as the size of population. This generates an important cost even for small populations. A key rule in Spark optimization is to reduce the number of action operations with regard to transformations. Thus, we altered DEAP evaluation so it maps all the population on the *TrainingRDD*. This diminishes action calls to one call per generation. Figure 2 gives the global flowchart of the modified DEAP evaluation algorithm. First, training set is transformed into an RDD (*TrainingRDD*) from a file stored on HDFS and is cached (Fig. 3, line 3). Then, at each generation, the population is evaluated against the training set by mapping their functions on *TrainingRDD* partitioned over worker nodes (Fig. 3, line 9). To get fitness values, a reduce operation is performed (Fig. 3, line 12). After, offspring can be generated on the driver node by applying mutation and crossover. The offspring replaces the old population and program loops until maximum generation number is reached.

3.2 Data Sampling

GP is a costly algorithm and this is intensified with big datasets like HIGGS. For clusters with a small number of nodes, running GP for such problems remains of high-cost. Furthermore, in these datasets, redundancy is inescapable. Sampling is a very suitable technique to deal with this situation. Additionally, depending on the underlying algorithm, sampling may counter overfitting, enhance learning quality and allow large population or more generations per run.

For these reasons, we investigated the use of sampling with the previous implementation. In this work, we started by a simple sampling method which is Random Subset Selection (RSS) [8]. RSS combines simplicity with efficiency. Spark makes available two operations: *sample* and *takeSample*. They produce samples with an oscillating size around the specified target. For efficiency, we used *sample* which is an RDD transformation. *takeSample* is an action and cannot be optimized by Spark DAG Scheduler. The training sample is renewed at each generation before evaluating the population.

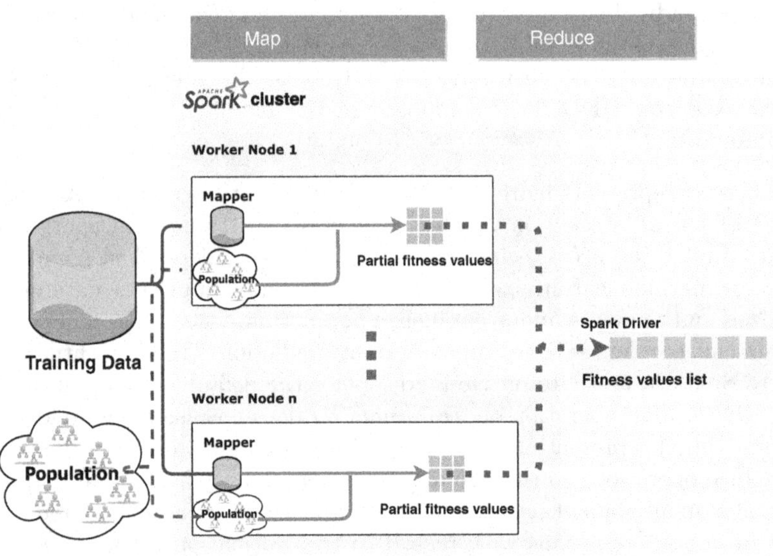

Fig. 2. Flowchart of the modified evaluation.

The commented code snippet in Fig. 3 outlines the steps we made to adapt the DEAP standard GP.

4 Experimental Settings

4.1 Framework

Software Framework. The details of used software are as follows: Spark version: 2.1.0, Hadoop version: 2.9.1, Resource Manager: YARN, Operating System: SMP Debian 4.9.130 and DEAP version: 1.2.2.

Spark Cluster. We used a tiny cluster composed of 4 worker nodes. Each node has a 16 core Intel Core processor at 2.397 GHZ, 45 GB of RAM and 1 TB of HDFS storage space.

In the following experiments, Spark application is submitted to the cluster via spark-submit script. We used the same YARN directives that are optimized for the cluster size and the used dataset accordingly to guidelines in [12] for all the GP runs in order to neutralize the effect of this configuration on results.

On this cluster, 4 Spark executors are deployed per node. The number of available nodes does not allow us further investigation of hardware effect on GP performance.

4.2 GP Settings

General Settings. Based on few runs, we set parameter values (Table 2) for GP. The process of tuning GP settings is beyond the scope of this work and have not been thoroughly studied.

```
from pyspark import SparkContext
sc = SparkContext(appName="DEAPSPARK")
TrainingRDD = sc.textFile("training.csv").cache() #parallelize training set
initGP()
while(generation<maxGeneration): # GP loop
    trainingSubset = TraininigRDD.sample().cache()
    # serialize population and map it on training subset
    popFunctions = [toolbox.compile(ind) for ind in population]
    fitnessRDD = TrainingRDD.map(lambda line:\
        [getPrediction(func,line) for func in popFunctions])
    # compute final fitness using reduce
    fitnessValues =\
        fitnessRDD.reduce(lambda v1,v2:list(map(operator.add,v1,v2)))
    updatePopulationFitnesses(population, fitnessValues)
    # Select the next generation individuals
    offspring = select(population)
    # Apply genetic operators
    offspring = evolve(offspring, crossoverProb, mutationProb)
    population[:] = offspring
```

Fig. 3. Modified DEAP GP loop.

Terminal and Function Sets. The terminal set includes 28 features of the benchmark Higgs dataset with a random constant. The function set includes basic arithmetic, comparison and logical operators reaching 11 functions (Table 3).

Table 2. GP settings.

Parameter	Value
Initialization	Ramped half and half
Tournament size	4
Tree limit	17
Crossover probability	0.9
Mutation probability	0.04
Generations and population size	61 different combinations

Table 3. GP terminal and function sets.

Function (node) set	
Arithmetic operators	+, −, ∗, /
Comparison operators	<, >, =
Logic operators	AND, OR, NOT
Other	IF (IF THEN ELSE)
Terminal set	
Higgs Features	28
Random Constants	1
Boolean values	True, False

5 Results and Discussion

Since the main objective is to tackle computation cost of GP in supervised learning problems, the first recorded measure is learning time. It comprises the elapsed time from the initialization of the first population until the last generation. It does not include the time for evaluation against the test set.

Reducing time must not be at the expense of learning quality. Then, by the end of each run, the best individual based on the fitness function is evaluated on the test dataset. Results are recorded in a confusion matrix from which accuracy, True Positive Rate (TPR) and False Positive Rate (FPR) are calculated. The objective function used with GP is the classification accuracy:

$$Accuracy = \frac{True\ Positives + True\ Negatives}{Total\ patterns} \tag{1}$$

We tried 61 different configurations obtained by varying population size, generation number and sample size. Each configuration is run 11 times. Then we compute the average learning time and the overall best individual performance

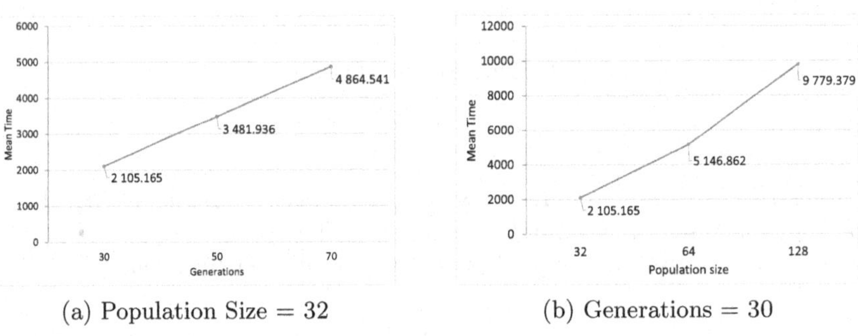

(a) Population Size = 32 (b) Generations = 30

Fig. 4. FSS Mean Time.

metrics. By these experiments, we intend to trace the speed gain in learning time and how it reacts to population size and generations number changing.

The results are reported in Fig. 4 in which 3 population sizes are tested (32, 64 and 128 with 30 generations) and 3 generations numbers (30, 50 and 70 with 32 individuals per population) with full dataset (FSS).

It is noticeable that parallelizing GP on Spark, facilitates its use for solving large classification problem like Higgs Boson classification. With only 4 nodes, a GP run takes on average 4864.541 s in 70 generation with 32 individuals and 9779.379 s in 30 generations with 128 individuals. A serial execution of GP on a single node takes more than 20 s to evaluate one individual against the total dataset which could take more than 44800 s for 70 generations with 32 individuals. Parallelizing evaluations under Spark achieves a speedup over nine times. This can be boosted by deploying more nodes on the cluster. Otherwise, learning time increases linearly with respect to the population size and generations number when using the same Spark settings.

Then, to allow using large populations and more generations without adding nodes, we injected RSS to the modified GP. We used the same settings but more population size values (from 32 to 8192) and generation numbers (from 30 to 1500) with a sample size fixed at 10000 instances. Results are exposed in Fig. 5. The mean learning time for 1500 generations evolving 128 individuals is 3116.744 s and for 30 generations with 8192 individuals is 763.9 s. The two curves have the same pace. This is owing to the fact that evaluation is the predominant phase in GP. Also, with low values of population size and generations number, time curve is almost flat or with slight slope. It means that time is affected mainly by Spark scheduling delays. This is tightly related to the cluster use tuning. This remains valid for target sample size (Table 4).

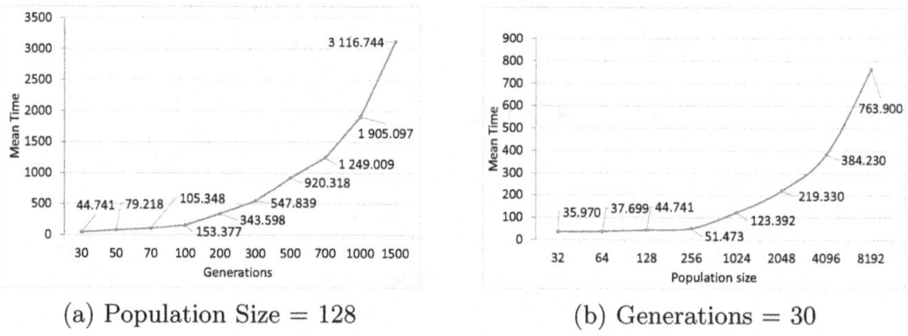

(a) Population Size = 128

(b) Generations = 30

Fig. 5. RSS Mean Time.

We tried the values: 1000, 5000, 10000, 50000 and 100000 instances per sample. While population size and generations number are set to 128 and 300.

To weigh the differences between using the whole dataset and a sampled subset for training, we juxtapose the results of full dataset (FSS) with those obtained by RSS with the same number of generations and population size in Fig. 6.

Table 4. RSS sample size effect on time.

Sample size	1000	5000	10000	50000	100000
Mean time (S)	384.748	411.198	547.839	971.921	1684.747

Also, we keep an eye on learning performance in Fig. 7. Undoubtedly, RSS outperforms GP without sampling for the 5 different target sizes. As regards learning performance, in terms of accuracy, experiments using RSS are less efficient with low number of generations or population size. It surpasses using the full dataset with large populations or more generations. On the one hand, in Fig. 7(a), it's for more than 50 generations that all RSS variants have best accuracy. On the other hand, Fig. 7(b) shows that RSS outperforms the use of the entire dataset only with a population size of 128. It is important to notice that we do not focus on enhancing learning performance and do not use any enhancing technique (e.g. feature engineering). Nevertheless, the best accuracy obtained is 66.93% with RSS (population: 128, generations: 1500, sample: 10000) in 3190.02 s. The best result in [18] is 60.76% realized with logistic regression.

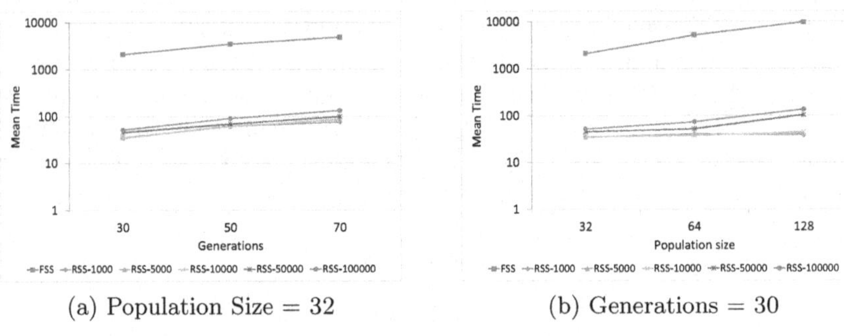

(a) Population Size = 32 (b) Generations = 30

Fig. 6. RSS vs FSS mean learning time.

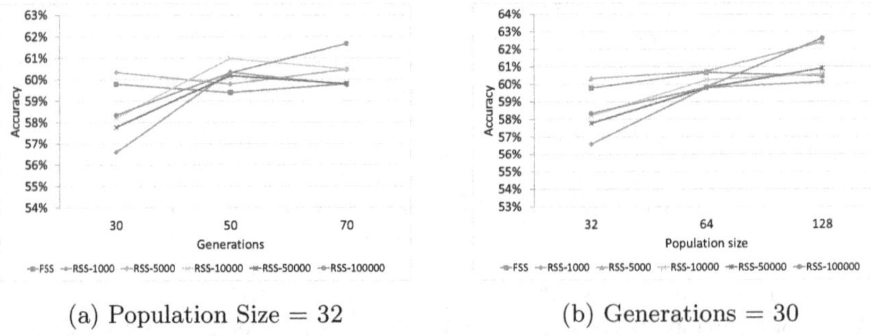

(a) Population Size = 32 (b) Generations = 30

Fig. 7. RSS vs FSS Best individual accuracy.

6 Conclusion

We presented, in this paper, details about reshaping DEAP library by parallelizing evaluation on Spark cluster. We obtained encouraging results that proclaim Spark as an efficient environment and is suitable for distributing GP evaluations. Then, we integrate a simple sampling technique that preserves learning performance while providing the possibility to probe GP with large populations or for a high number of generations. We studied experimentally the effect of varying 3 parameters: population size, generations number and sample size.

This work provides a Spark compliant GP implementation without the need to code it from scratch. Thus, it can be used in resolving different machine learning problems. Although it has been successfully tested on a small cluster, the size of the underlying cluster on the overall performance has to be investigated.

A logical extension is to study the impact of different Spark configurations (number of nodes, RDD partitioning, partition sizes, etc.). This will help to find the most suitable execution settings. A second path is to check the feasibility of adding other sampling techniques and in particular active sampling that prove to be advantageous in the context of machine learning. Hierarchical sampling, that we used in [11], is a promising candidate.

References

1. Al-Madi, N., Ludwig, S.A.: Scaling genetic programming for data classification using mapreduce methodology. In: Fifth World Congress on Nature and Biologically Inspired Computing, NaBIC 2013, 12–14 August 2013, pp. 132–139. IEEE (2013)
2. Baldi, P., Sadowski, P., Whiteson, D.: Searching for exotic particles in high-energy physics with deep learning. Nature Commun. **5** (2014)
3. Baldi, P., Sadowski, P., Whiteson, D.: Enhanced higgs boson to $\tau+$ $\tau-$ search with deep learning. Phys. Rev. Lett. **114**(11), 111–801 (2015)
4. Chávez, F., et al.: ECJ+HADOOP: an easy way to deploy massive runs of evolutionary algorithms. In: Squillero, G., Burelli, P. (eds.) EvoApplications 2016. LNCS, vol. 9598, pp. 91–106. Springer, Cham (2016). https://doi.org/10.1007/978-3-319-31153-1_7
5. Dean, J., Ghemawat, S.: MapReduce: simplified data processing on large clusters. In: Brewer, E.A., Chen, P. (eds.) 6th Symposium on Operating System Design and Implementation (OSDI 2004), San Francisco, California, USA, 6–8 December 2004, pp. 137–150. USENIX Association (2004)
6. Fortin, F.A., De Rainville, F.M., Gardner, M.A., Parizeau, M., Gagné, C.: DEAP: evolutionary algorithms made easy. J. Mach. Learn. Res. **13**, 2171–2175 (2012)
7. Funika, W., Koperek, P.: Scaling evolutionary programming with the use of apache spark. Comput. Sci. (AGH) **17**(1), 69–82 (2016)
8. Gathercole, C., Ross, P.: Dynamic training subset selection for supervised learning in Genetic Programming. In: Davidor, Y., Schwefel, H.-P., Männer, R. (eds.) PPSN 1994. LNCS, vol. 866, pp. 312–321. Springer, Heidelberg (1994). https://doi.org/10.1007/3-540-58484-6_275

9. Giráldez, R., Díaz-Díaz, N., Nepomuceno, I., Aguilar-Ruiz, J.S.: An approach to reduce the cost of evaluation in evolutionary learning. In: Cabestany, J., Prieto, A., Sandoval, F. (eds.) IWANN 2005. LNCS, vol. 3512, pp. 804–811. Springer, Heidelberg (2005). https://doi.org/10.1007/11494669_98
10. Higgs Dataset: http://archive.ics.uci.edu/ml/datasets/HIGGS
11. Hmida, H., Hamida, S.B., Borgi, A., Rukoz, M.: Scale genetic programming for large data sets: case of higgs bosons classification. Procedia Comput. Sci. **126**, 302–311 (2018). The 22nd International Conference, KES-201
12. Karau, H., Warren, R.: High Performance Spark, 1st edn. O'Reilly, Sebastopol (2017)
13. Kienzler, R.: Mastering Apache Spark 2.x. Packt Publishing, Birmingham (2017)
14. Koza, J.R.: Genetic Programming: On the Programming of Computers by Means of Natural Selection. MIT Press, Cambridge (1992)
15. Paduraru, C., Melemciuc, M., Stefanescu, A.: A distributed implementation using apache spark of a genetic algorithm applied to test data generation. In: Companion Material Proceedings of Genetic and Evolutionary Computation Conference, 15–19 July 2017, pp. 1857–1863. ACM (2017)
16. Peralta, D., del Río, S., Ramírez-Gallego, S., Triguero, I., Benitez, J.M., Herrera, F.: Evolutionary feature selection for big data classification: a MapReduce approach. Math. Probl. Eng. **2015**, 11 (2015)
17. Qi, R., Wang, Z., Li, S.: A parallel genetic algorithm based on spark for pairwise test suite generation. J. Comput. Sci. Technol. **31**(2), 417–427 (2016)
18. Shashidhara, B.M., Jain, S., Rao, V.D., Patil, N., Raghavendra, G.S.: Evaluation of machine learning frameworks on bank marketing and Higgs datasets. In: 2nd International Conference on Advances in Computing and Communication Engineering, pp. 551–555 (2015)
19. Zaharia, M., et al.: Resilient distributed datasets: a fault-tolerant abstraction for in-memory cluster computing. In: Proceedings of the 9th USENIX Symposium on Networked Systems Design and Implementation, NSDI 2012, 25–27 April 2012, pp. 15–28. USENIX Association (2012)

Genetic Algorithms for the Picker Routing Problem in Multi-block Warehouses

Jose Alejandro Cano[1(✉)], Alexander Alberto Correa-Espinal[2], Rodrigo Andrés Gómez-Montoya[3], and Pablo Cortés[4]

[1] Universidad de Medellín, Carrera 87 # 30-65, Medellín, Colombia
jacano@udem.edu.co

[2] Universidad Nacional de Colombia, Carrera 80 # 65-223, Medellín, Colombia

[3] ESACS – Escuela Superior en Administración de Cadena de Suministro, Calle 4 # 18-55, Medellín, Colombia

[4] Universidad de Sevilla, Camino de los Descubrimientos s/n, 41092 Sevilla, Spain

Abstract. This article presents a genetic algorithm (GA) to solve the picker routing problem in multiple-block warehouses in order to minimize the traveled distance. The GA uses survival, crossover, immigration, and mutation operators, and is complemented by a local search heuristic. The genetic algorithm provides average distance savings of 13.9% when compared with s-shape strategy, and distance savings of 23.3% when compared with the GA with the aisle-by-aisle policy. We concluded that the GA performs better as the number of blocks increases, and as the percentage of picking locations to visit decreases.

Keywords: Order picking · Picker routing · Genetic algorithm · Artificial intelligence · Multi-block warehouse · Warehouse management

1 Introduction

Order picking is the most repetitive and most labor-intensive operation in warehouses management, being the transport the activity involving more than 50% of the time and cost required for processing orders [1]. Thus, it is necessary to reduce travel times taking into account that they do not add value and represent cost and movement savings within warehouses and distribution centers [2].

In response, the picker routing problem plans a tour to retrieve items of customer orders, starting and ending at the depot, visiting several storage locations and minimizing travel distance and travel time [3]. Traditionally this problem has been addressed for single-block warehouses [4], however, it is necessary to consider multi-block warehouses which are common in modern warehouses and distribution centers, providing a significant shorter routing distance [5]. Therefore, it is required to provide new solution approaches to the picker routing problem considering realistic warehouse environments, facilitating an efficient integration with the order batching problems and batch sequencing problem.

Currently, the solution approaches for the picker routing problem in multi-block warehouses are related to routing strategies as the s-shape strategy, largest gap strategy,

W. Abramowicz and R. Corchuelo (Eds.): BIS 2019, LNBIP 353, pp. 313–322, 2019.
https://doi.org/10.1007/978-3-030-20485-3_24

aisle-by-aisle strategy [6]; heuristics such as block-aisle1 and block-aisle2 [7], branch-and-bound procedures [6], savings and nearest neighbor heuristics [8]; and graph optimization algorithms [9]. However, routing strategies and heuristics are not easy to understand and memorize pickers in multi-block layouts, and they provide non-optimal solutions for travel distances [10]. On the other hand, optimal solutions usually require high computational time, which is not feasible for the operations planning in warehouse environments. Hence, the use of metaheuristics predominates for solving realistic picker routing problems, improving the solutions coding to work with large problems [11], finding near-optimal and good-quality solutions through the exploration of the solution space [12].

2 The Picker Routing Problem in Multi-block Warehouses

The picker routing problem aims to minimize the traveled distance and can be modeled as a Steiner TSP for multi parallel-aisle layouts where some nodes can be visited more than once and other nodes do not have to be visited [8]. Even, the STSP can be formulated as a classical TSP by calculating the minimum distances between each pair of picking locations [13, 14].

In a 2D warehouse, the Manhattan distance d_{ij} between two picking positions i and j belonging to the same block is calculated using Eq. 1, considering (0, F) as the coordinate of the lower left corner of the warehouse and (0, B) as the coordinate of the upper left corner of the warehouse.

$$d_{ij} = \begin{cases} |x_i - x_j| + |y_i - y_j| \text{ if } i \text{ and } j \text{ belong to the same aisle} \\ |x_i - x_j| + min\{|y_i - B| + |B - y_j|, |y_i - F| + |F - y_j|\} \text{ otherwise} \end{cases} \quad for\, 1 \leq i \neq j \leq L \quad (1)$$

For multi-block warehouses, where $h \in H$ and H is the set of blocks in a warehouse configuration, B_g represents the back y-coordinate for the block h, and F_h represents the front y-coordinate for the block g. Therefore, two picking positions i and j are represented by $i = (x_{\mathrm{ih}}, y_{\mathrm{ih}})$ and $j = (x_{\mathrm{jh}}, y_{\mathrm{jh}})$ identifying the block for each picking position. Figure 1 shows the configuration of a 3-block warehouse, where (0, F_1) and (0, B_1) are the front and back y-coordinates for block 1, (0, F_2) and (0, B_2) are the front and back y-coordinates for block 2, and (0, F_3) and (0, B_3) are the front and back y-coordinates for block 3.

Thus, if i belongs to block 3 and j belongs to block 1 then $i = (x_{\mathrm{i3}}, y_{\mathrm{i3}})$ and $j = (x_{\mathrm{j1}}, y_{\mathrm{j1}})$, and the minimum distance between i and j (d_{ij}) is calculated using Eq. 2. Note that the back and front y-coordinate correspond to the block of the starting storage position i.

$$d_{ij} = |x_{i2} - x_{j1}| + \min\{|y_{i2} - B_3| + |B_3 - y_{j1}|, |y_{i2} - F_3| + |F_3 - y_{j1}|\} \; for\, 1 \leq i \neq j \leq L \quad (2)$$

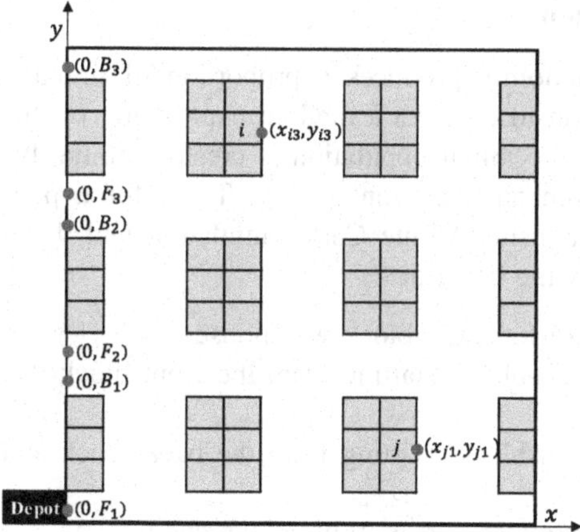

Fig. 1. A warehouse layout with three blocks.

Based on Eqs. 1 and 2, the minimum distances between each pair of picking locations in multiple-block warehouses can be computed in matrices. Additionally, we consider the following assumptions for the picker routing problem addressed in this study:

- All tours start and end at the Depot.
- Each item is stored in one and only one storage position.
- The distances between two storage locations are rectilinear (Manhattan).
- Only horizontal movements are considered (low-level picker-to-part systems).
- Pickers are able to traverse the aisles in both directions.
- The horizontal distance within picking aisles is not considered.
- The customer orders to be retrieved in a tour were previously validated by an order batching process, ensuring that the capacity of the orders cannot exceed the capacity of the picking vehicle.

3 Genetic Algorithms for the Picker Routing Problem

The design of the genetic algorithm to solve the picking routing problem is as follows:

3.1 Encoding Chromosome

The encoding process on single picker routing problem is responsible for representing in each gene a picking position, and the size of a chromosome is determined by the number of picking positions to visit [15]. If a picker must visit G unique picking positions, each chromosome will have G genes. The sequence of the genes represents the order picker's visiting order. This solution approach is applicable to a variety of warehouse layouts and is independent of the length, alignment, and number of picking aisles.

3.2 Initial Population

The proposed GA randomly produces a proportion of initial population based on Monte Carlo simulation to perform a random exploration of the solution space, and another proportion of the initial population is created arbitrarily to improve the efficiency of the algorithm and solution quality. Thus, for a population size *N*, *N*–9 individuals are created using Monte Carlo simulation and 9 individuals are created according to the following heuristics:

- The s-shape strategy for multi-block warehouses.
- The s-shape for each block, starting from the front block and ending at the back block.
- The s-shape for each block, starting from the back block and ending at the front block.
- The aisle-by-aisle strategy for multi-block warehouses.
- Visit each aisle from the front to the bottom of the warehouse.
- A sequence of picking locations in ascending order according to their number.
- A sequence of picking locations in descending order according to their number.
- A sequence of picking locations in ascending order according to their number and a local search process based en SWAP movements improves the fitness value.
- A sequence of picking locations in descending order according to their number and a local search process based en SWAP movements improves the fitness value.

3.3 Fitness Function

Based on Eqs. 1 and 2, the fitness function measures the sum of distances between the picking positions following the sequence in which the picking positions appear on the chromosome. As shown in Eq. 3, the fitness function includes the distance from the depot to the first picking position and the distance from the last picking position. In Eq. 3, *gen(g)* represents the number of the picking location assigned to the gen *n* in a chromosome.

$$Fitness = d_{0,gen(1)} + \sum_{g=1}^{G-1} d_{gen(g),gen(g+1)} + d_{gen(G),0} \tag{3}$$

3.4 Survival Operator

In each generation, a percentage of the chromosomes best fitness value is selected ensuring the survival of individuals with high-quality genetic information in each generation. The number of top chromosomes in each generation is determined by the survival probability *Ps*.

3.5 Selection Method

The linear ranking method is used to select the parent chromosomes for the crossover operator, assigning more probability to individuals with the best fitness, and less probability to chromosomes with worst fitness. Then, a roulette wheel selection provides random numbers to choose each pair of parents for the crossover operator.

3.6 Crossover Operator

The GA uses a two-point crossover method selecting two crossing points randomly on each parent. The genes of the segment bounded by the two crossing points of each parent are exchanged with each other (See Fig. 2). This method can provide non-feasible offspring by repeating picking positions and disappearing some picking positions in a chromosome. Such situations are dealt with by applying a correction mechanism that eliminates repeated genes from the new segment in each offspring. As shown in Fig. 2, the picking positions with the smallest number that have not been assigned are chosen and assigned to the first available gene, and repeat this procedure until completing each chromosome. The crossover probability *Pc* determines the number of chromosomes created in each generation by the crossover operator.

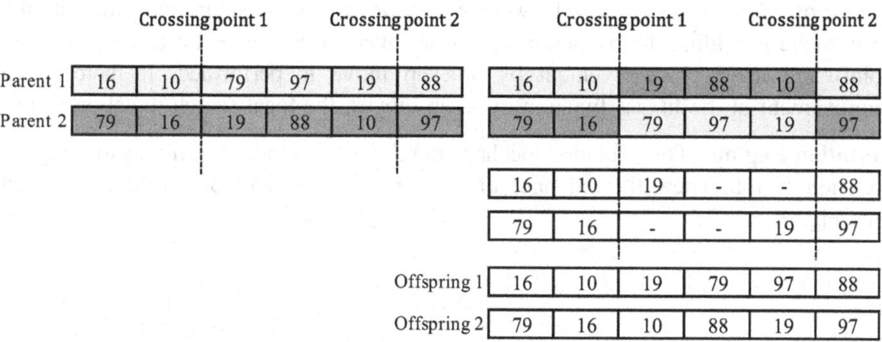

Fig. 2. Crossover operator and correction mechanism.

3.7 Immigration Operator

In each generation, new individuals are created using Monte Carlo simulation to prevent the genetic algorithm from local optima during the evolutionary process. The number of chromosomes introduced in each generation is determined by the migration probability *Pi*.

3.8 Mutation Operator

The mutation operator is used to increase the genetic diversity of the individuals in a new generation, and the number of chromosomes to mutate in each generation is determined by the mutation probability *Pm*. Three mutation methods can be used (SWAP Mutation, Local Search Mutation, and Mutation Region), and one of them is chosen randomly for each chromosome to mutate.

SWAP Mutation. This is the swapping mutation method (exchange mutation) where two randomly chosen genes are exchanged through a SWAP movement as shown in Fig. 3.

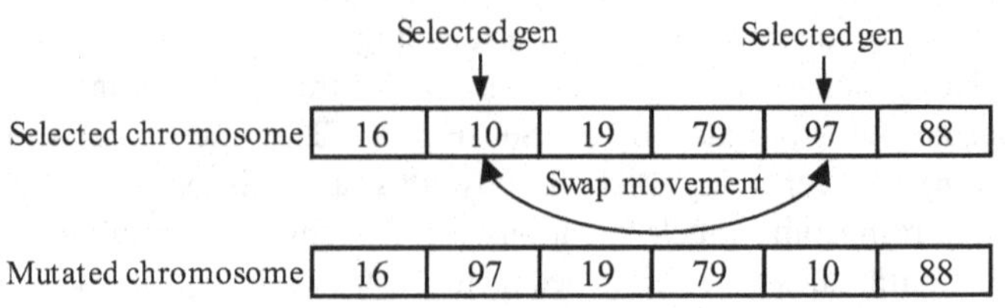

Fig. 3. SWAP mutation.

Local Search Mutation. A neighborhood structure based on SWAP movements is used to improve the solutions by interchanging two genes in a chromosome. A sequence of swap moves is made between each pair of genes in a chromosome, and the swap move providing the best improvement is taken as the best solution (new starting point) and another improvement by a swap move is performed. If there is no improvement in the fitness function by swap moves, the local search mutation stops.

Mutation Region. This mutation mechanism selects two random positions to delimit a mutation region. Then, the sequence of the genes in the mutation region is reversed (See Fig. 4).

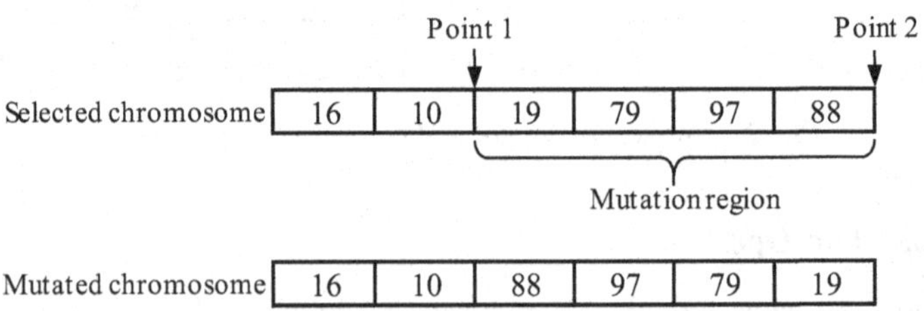

Fig. 4. Mutation region mechanism.

3.9 Stopping Criterion

The genetic algorithm stops when the number of desired iterations is satisfied, or when a number of consecutive iterations are executed without obtaining an improvement in the global fitness.

3.10 Local Search Procedure

After the algorithm has gone through all generations, it concludes with a local search procedure to refine the best solution found. In the local search phase, a neighborhood structure based on SWAP movements is used to improve the solutions by interchanging two genes in a chromosome. This local search phase uses the same procedure of the Local Search Mutation.

4 Experiments

The GA control parameters for solving the picker routing problem are tabulated in Table 1. Some control parameters such as the population size, generations, and consecutive iterations without improvement in the best-known solution depend on the size of the picker routing problem, i.e., the number of picking positions to visit in a tour.

Table 1. Control parameters for the GA.

Control parameter	Value
Population size	20 + G/2
Crossover probability (*Pc*)	75%
Mutation probability (*Pm*)	20%
Immigration probability (*Pi*)	10%
Survival probability (*Ps*)	15%
Generations	$40 + \lceil G/3 \rceil$
Consecutive iterations without improvement in the best-known solution	$\lceil G/4 \rceil$

We tested the GA performance in 54 scenarios (2 × 3 × 3 × 3), which result from the combination of the experimental parameters shown in Table 2. Each scenario is replicated 10 times by varying the number of items in each order, and therefore varying the picking positions to visit in each tour, obtaining 540 experimental instances. The number of picking positions to visit in each tour results from selecting the unique picking locations from the orders that must be retrieved.

Table 2. Experimental parameters for the GA.

Control parameter	Value
Blocks	2, 3
Aisles	5, 10, 15
Positions per aisle side	5, 10, 15
Number of customer orders	3, 5, 7
Number of items in each order	Uniform [1, 5]

The GA was developed using Visual Basic, on a 3.10 GHz speed processor with 4.0 GB RAM.

5 Results and Discussion

The GA performance was compared with the S-shape routing policy and the aisle-by-aisle routing policy [6] with an adaptation that forces traversing the sub-aisles located at the front or back of the warehouse. The comparison of the GA and the benchmark is measured using Eq. (4), where D_{GA} and $D_{BENCHMARK}$ respectively represent the traveled distance provided by the GA and the benchmarks.

$$\%\Delta = \frac{D_{GA} - D_{BENCHMARK}}{D_{BENCHMARK}} \times 100 \quad (4)$$

Table 3 summarizes the comparative results between GA and benchmarks highlighting that GA provides average savings of 13.9% over s-shape strategy and provides savings up to 25.9% when considering 3 blocks, 15 aisles, 10 storage positions per side, and 7 customer orders. Similarly, GA provides average savings of 23.3% over aisle-by-aisle strategy and provides savings up to 37.4% when considering 3 blocks, 15 aisles, 15 storage positions per side, and 7 customer orders.

Table 3. Distance savings for the GA.

Distance savings	GA vs s-shape	GA vs aisle-by-aisle
Maximum	25.9%	37.4%
Average	13.9%	23.3%
Minimum	2.6%	8.4%

On the other hand, the minimum savings of GA compared with S-shape aisle-by-aisle are provided when considering 2 blocks, 5 aisles, 5 storage positions per side, and 7 customer orders. Figure 5 shows that as the size of the warehouse and the number of blocks increases, the GA performs better than the routing strategies, providing shorter traveled distances.

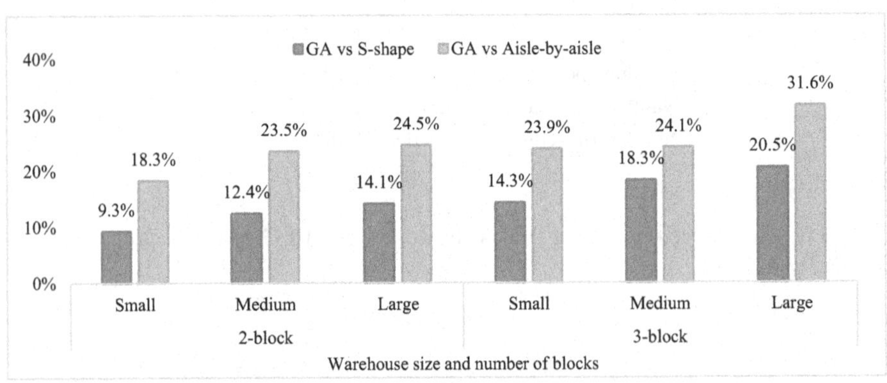

Fig. 5. Average distance savings of GA by warehouse size and number of blocks.

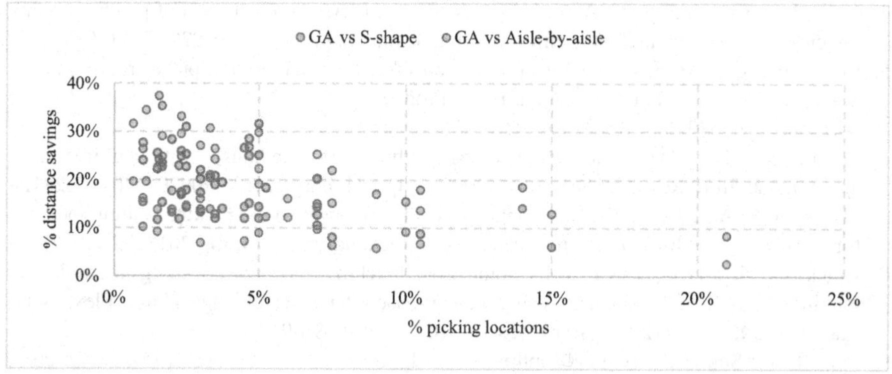

Fig. 6. Distance savings of GA compared to s-shape and aisle-by-aisle routing policies.

Likewise, Fig. 6 shows a trend in which, as the percentage of picking locations to visit in a warehouse decreases, the GA performance increases with respect to benchmarks.

Additionally, for all evaluated instances, the GA provides a computation time that varies between 9.4 and 236.6 ms. Therefore, the results demonstrate that the proposed GA provides high-quality solutions in short computing times facilitating its integration in joint order batching, sequencing and picker routing problems that require the execution of picking routes multiple times for each batch [16].

6 Conclusion

In this article, we proposed a GA considering survival, crossover, immigration, and mutation operators, and the results of the algorithm are refined with a local search procedure. Experiments show that the proposed GA provides high-quality solutions in short computing times. The efficiency of picker routing operations is greatly raised with the GA when the number of blocks and warehouse size increase, and when the percentage of picking locations decreases. Due to the efficiency of the GA, we recommend to use this algorithm in joint order picking problems requiring performing picking routes several times.

References

1. Tompkins, J.A., White, J.A., Bozer, Y.A., Tanchoco, J.M.A.: Facilities Planning. Wiley, New Jersey (2010)
2. Koch, S., Wäscher, G.: A grouping genetic algorithm for the order batching problem in distribution warehouses. J. Bus. Econ. **86**, 131–153 (2016). https://doi.org/10.1007/s11573-015-0789-x
3. Hsieh, L.-F., Huang, Y.-C.: New batch construction heuristics to optimise the performance of order picking systems. Int. J. Prod. Econ. **131**, 618–630 (2011). https://doi.org/10.1016/j.ijpe.2011.02.006

4. Cano, J.A., Correa-Espinal, A.A., Gómez-Montoya, R.A.: An evaluation of picking routing policies to improve warehouse efficiency. Int. J. Ind. Eng. Manag. **8**, 229–238 (2017)
5. Li, J., Huang, R., Dai, J.B.: Joint optimisation of order batching and picker routing in the online retailer's warehouse in China. Int. J. Prod. Res. **55**, 447–461 (2017). https://doi.org/10.1080/00207543.2016.1187313
6. Roodbergen, K.J., De Koster, R.: Routing methods for warehouses with multiple cross aisles. Int. J. Prod. Res. **39**, 1865–1883 (2001). https://doi.org/10.1080/00207540110028128
7. Shouman, M.A., Khater, M., Boushaala, A.: Comparisons of order picking routing methods for warehouses with multiple cross aisles. AEJ Alexandria Eng. J. **46**, 261–272 (2007)
8. Kulak, O., Sahin, Y., Taner, M.E.: Joint order batching and picker routing in single and multiple-cross-aisle warehouses using cluster-based tabu search algorithms. Flex. Serv. Manuf. J. **24**, 52–80 (2012). https://doi.org/10.1007/s10696-011-9101-8
9. Jang, H.Y., Sun, J.U.: A graph optimization algorithm for warehouses with middle cross aisles. Appl. Mech. Mater. **145**, 354–358 (2012). https://doi.org/10.4028/www.scientific.net/AMM.145.354
10. Scholz, A., Wäscher, G.: Order batching and picker routing in manual order picking systems: the benefits of integrated routing. Cent. Eur. J. Oper. Res. **25**, 491–520 (2017). https://doi.org/10.1007/s10100-017-0467-x
11. Tsai, C.-Y., Liou, J.J.H., Huang, T.-M.: Using a multiple-GA method to solve the batch picking problem: considering travel distance and order due time. Int. J. Prod. Res. **46**, 6533–6555 (2008). https://doi.org/10.1080/00207540701441947
12. De Santis, R., Montanari, R., Vignali, G., Bottani, E.: An adapted ant colony optimization algorithm for the minimization of the travel distance of pickers in manual warehouses. Eur. J. Oper. Res. **267**, 120–137 (2018). https://doi.org/10.1016/j.ejor.2017.11.017
13. Lu, W., McFarlane, D., Giannikas, V., Zhang, Q.: An algorithm for dynamic order-picking in warehouse operations. Eur. J. Oper. Res. **248**, 107–122 (2016). https://doi.org/10.1016/j.ejor.2015.06.074
14. Cano, J.A., Correa-Espinal, A.A, Gómez-Montoya, R.A.: Mathematical programming modeling for joint order batching, sequencing and picker routing problems in manual order picking systems. J. King Saud Univ. Eng. Sci. (2019). https://doi.org/10.1016/j.jksues.2019.02.004
15. Cano, J.A., Correa-Espinal, A.A., Gómez-Montoya, R.A.: Solving the order batching problem in warehouses using genetic algorithms. Inf. Tecnológica. **29**, 235–244 (2018). https://doi.org/10.4067/S0718-07642018000600235
16. Cano, J.A., Correa-Espinal, A.A., Gómez-Montoya, R.A.: A review of research trends in order batching, sequencing and picker routing problems. Espacios. **39**, 3 (2018)

ICT Project Management

Subject-Orientation as a Means for Business Information System Design – A Theoretical Analysis and Summary

Matthes Elstermann(✉) and Jivka Ovtcharova

Institute for Information Management in Engineering, KIT,
Kriegsstraße 77, 76133 Karlsruhe, Germany
matthes.elstermann@kit.edu

Abstract. (Business) Information systems become more and more complex due to an increase in the volume of data, but also due to more and more interconnected elements that all need to be orchestrated to perform as a uniform system. Correctly understanding and describing (business) processes, is one of the cornerstone foundations in the creation of almost all information systems. While the systems themselves have become more complex, and the means to program them have evolved over the last decades, the means to analyze and communicate about the processes they execute have stagnated on a simplistic level from the 1960s. Over the last 15 years, there has been work done in the development of concepts and tools on the topic of subject-orientation and subject-oriented (business) process modeling and management (S-BPM) that is different from earlier, classical process description approaches. This paper analyzes and argues about the shortcomings and discrepancies of those classical approaches and argues how subject-orientation may be an improvement when employed as a means in the design and development of information systems.

Keywords: Subject-orientation · S-BPM · Business process modeling · Process analysis · Information system design

1 Introduction

1.1 Motivation

This paper is a theoretical analysis of the description paradigm of subject-orientation in the context of the design and creation of business information systems. This work is a summary of several years of working with and analyzing the concepts and tools of the paradigm. It argues why it may be a preferable choice over existing concepts and their shortcomings when it comes to information system conceptualization and design. Our goal here is to provide the reader with the profound understanding of what subject-orientation is and what the fundamental differences between the classical process description concept and the subject-oriented paradigm are. Thereby we want to provide a basis for discussions that go beyond the assumption that subject-orientation is simply just another form of describing processes with arrows and boxes, but rather a

W. Abramowicz and R. Corchuelo (Eds.): BIS 2019, LNBIP 353, pp. 325–336, 2019.
https://doi.org/10.1007/978-3-030-20485-3_25

fundamentally different thinking structure and consequently a tool that should be considered in the context of business information system genesis.

2 Relevance: Processes and Business Information Systems

Describing business process is an essential aspect of conceiving business information systems. Unsurprisingly, every business information system will support the processes of a business. Understanding the according goals and supposed tasks that lead to the fulfillment of those goals therefore is crucial.

The artifacts containing the available information on these tasks and goals may have different forms: They may be called, e.g., user stories, requirements, or specifications and range from short natural language texts to formal models. However, they will in some way always contain more or less precise descriptions about the (business) processes to be supported. These descriptions will then be transformed into an information system that can execute or support those processes (Figs. 1 and 2).

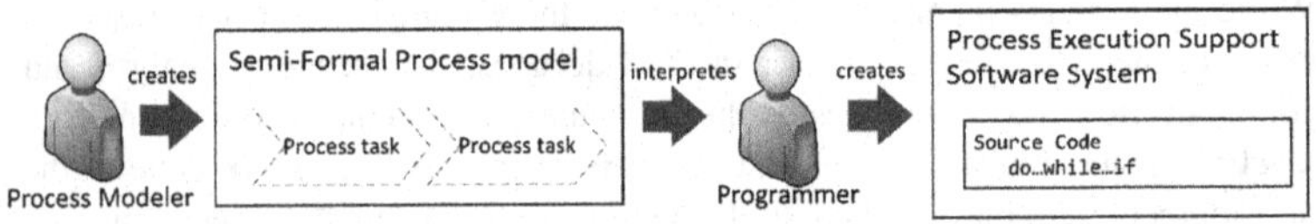

Fig. 1. Indirect execution of process models in an information system (error-prone)

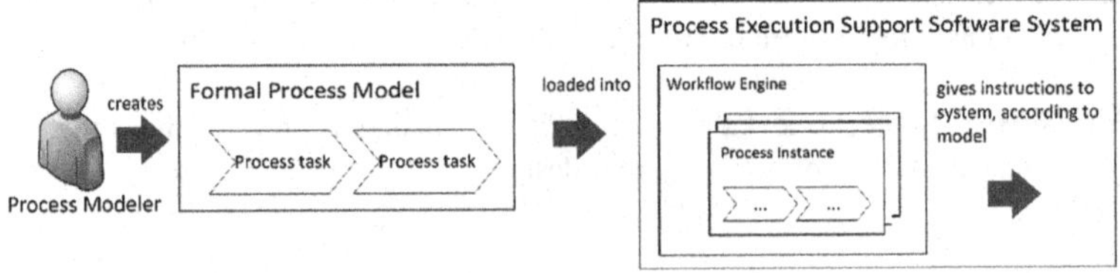

Fig. 2. Direct execution of process models – execution via a workflow engine

3 Classical Process Thinking

'Process' is an abstract concept, representing the idea that certain consecutive or parallel, observable events, actions of actors, or states of objects are related in some logical-causal and/or time-dependent way. Thus, the concept of '*process*' gives us the ability to think and communicate about 'a process' (process thinking).

When tasked with the question of what a process is or how to describe it, what often is stated by people is something along the lines of Fig. 3: the input-task-output model.

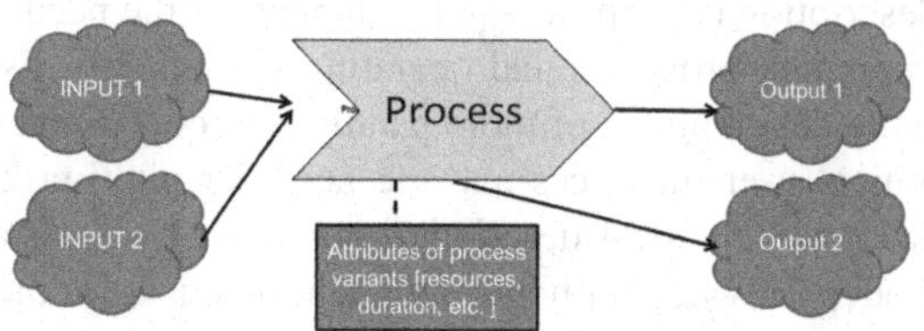

Fig. 3. Classical Process Concept with multiple inputs, outputs, and attributes

This input-task-output model comes with simple notions of being able to chain process steps after on another and notions of branching and looping. Furthermore, there is a single **abstraction mechanism**: the sub-process idea that allows summarizing details in a super- process task in a hierarchical tree structure.

The concept can be seen as the basis and can be found in all existing standard methods and notations used to describe business processes, be it DIN Flow Charts [1], UML Activity Diagrams [2], the, in business-context widely used, Event-Driven Process Chains (EPC) [3], or the currently modeling standard that is the Business Process Model and Notation (BPMN) [4]. The input-task-output concept is also well established in engineering domains where it is being taught to young engineers as can be seen in elementary teaching material such as [5] or [6].

It works well for "simple" linear processes where especially the boundary conditions and rules are well understood and agreed on. E.g., a production process in a factory can be naturally structured hierarchically (factory → production lines → individual machines). Furthermore, there all inputs and outputs are on the same abstraction level – namely none at all: they are all within the physical domain. If abstraction was necessary all inputs and outputs fit well together, making the need to break the simple and well-mannered tree structure of the according process description very unlikely.

For consideration in the domain of business information systems, another analogy can be made: the input-task-output concept is the central concept behind procedural programming languages such as C, PASCAL, or (partly) BASIC. All these languages are Turing complete [7]. Assuming the same for process modeling languages, it is theoretically possible to describe any computational process with them! But if every conceivable (computable) process can be described this way, why need another paradigm?

3.1 Indicators for Shortcomings

The principle of Turing completeness as well as many years of application prove that in principle there is no reason to have anything beyond the classical description concept. Still there are problems, but they are subtle and come mostly to the surface when the need arises to combine several simple individual process descriptions into larger, more complex models. There are a few indicators for those problems with the classical procedural concept and its practical limits:

The most prominent is the existence of another programming paradigm: Object-Orientation [8]. Its adoption, from a scientific concept born in the 1960s, to the most widely known programming paradigm [9, 10], allows to suspect that the procedural input-task-output description concept was not sufficient for the needs and requirements of ever increasing complex programs and operating systems [9]. Especially the challenge to coordinate the development of more parallel and concurrent activities with the complexity of graphical user interfaces and the need for multi-tasking increased the renunciation rate from the classical description approach by the start of the 1990s'. While theoretically adept, it was simply not sufficient practical for the task of being the fundament for a single monolithic but coherent process description (code base) that supposedly was created by a growing group of heterogeneous people over time[1]. For the domain of business process management the articulation of these problem is even more subtle. One example may be the existing differentiation between the concept of *'workflow'* vs. the concept of *'process'* that was encountered not in literature but in interviews with practitioners. The former was used to refer to formal process description that where explicitly created for workflow engines. *'Processes'* on the other hand was used for descriptions that supposedly could not be expressed as "*mere workflows*" and subsequently were modeled for most parts in Power Point[2]. We can only suppose that this is due to tediousness of modeling high abstraction process formally – leading to the well-known phenomenon of *'process tapestries'*[3]. The consequence of this type of informal modeling on the other hand are process descriptions that bend or even brake the semantics and principle modeling structure in the attempt to express complex circumstances within the simplistic confines of a linear model and not having the time or capacities to model precise. We have encountered such an example in a research project concerned with strategic product planning. As depicted in Fig. 4, the model, provided by a project partner, supposedly depicted a linear sequence of tasks organized in a tree structure. However, a detailed analysis showed that the described tasked were only visually lined up and synchronized: Neither was there a singular aspect that could actually run through all process steps, nor was there a coherent time frame or timeline.

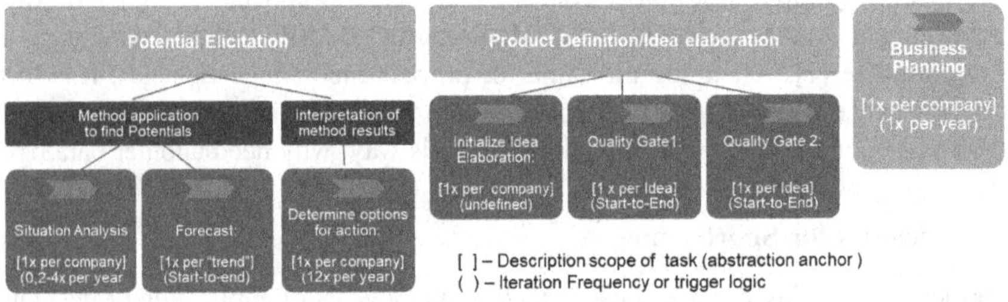

Fig. 4. Exemplary process model structure with consistently changing scopes and incompatible temporal dependencies.

[1] Further discussions of the advantages and disadvantages of the procedural vs. object-oriented programming paradigm should be common knowledge in the domain of business information systems. Otherwise [10] or any Google search should provide an in-depth.

[2] Supposedly, the most widely used process-modeling tool in existence.

[3] In programming, the equivalent would be the concept of *"spaghetti code"*.

Truthfully, that is only anecdotal evidence, but we have encounter those on multiple occasions and the main observation remains: for business process modeling, in principle, only the classical procedural description paradigm is used.

This finding let to the question whether the paradigm of subject-orientation and subject-oriented business process management (S-BPM[4]) are indeed different and potentially advantageous in this regard.

4 A Short Introduction to Subject-Orientation

4.1 Basic Definitions

Subject-Orientation is a modeling or description paradigm for processes. It is derived from the structure of natural languages and requires the explicit and continuous consideration of active entities as the conceptual center of description. Active entities (subjects) and passive elements (objects) must always be distinguished and activities or tasks can only be described in the context of a subject. The interaction between subjects is of particular importance and must explicitly be described as an exchange of information that cannot be omitted.

As a research discipline, the paradigm is based on the works of Albert Fleischmann [11]. The fundamental concept, however, traces its root into research of parallel running system and actor based concepts. Furthermore, the principle idea (but not the explicit consideration) can be found in many areas of IT, e.g., in the advice for programming saying that code or threads (active entities) should ideally be separated from data and data storage (passive data objects)[5].

4.2 Pass

While in principle it is possible to model subject-oriented with other modeling languages, currently, only the Parallel Activity Specification Schema (PASS), conceived by Fleischman as an integral part of S-BPM can be considered a fully subject-oriented, graphical process-modeling notation [11, 12].

In accordance with the subject-oriented principle, any process model must first include a description of the involved active units or roles – the subjects – and the messages they may exchange. This information is contained in a so called Subject Interaction Diagram (SID). **Temporal or causal information** about when or in what order the messages are being exchanged is not contained in an SID.

The flow of activities are described in individual Subject Behavior Diagrams (SBD) that exist for each subject appearing in the SID. Tasks or blocks in the SBDs are called **states**. A subject always is said to *be in* exactly one state at a time. Being "*in a*

[4] A process management discipline that is oriented towards and heavily incorporates the Subject-Orientated modeling paradigm.

[5] This, however, is only an advice, since, e.g., the object-oriented programming language Java does not have the formal means to express the distinction of threads classes and data classes and mixing is possible.

state" implies that the activity associated with the corresponding state is being executed. **Do States** (yellow) denote that a subject is performing an activity, task, or function that does not require input from other subjects. **Send** and **Receive States** denote the interaction with other subjects. They are exited via corresponding **Send** and **Receive Transitions** that denote the condition that must be fulfilled in order to leave the states. For Receive Transitions that is the active reception of a message from a defined sender. For Send Transitions, the exit condition is the completed transmission of the denoted message to the indicated recipient. It is important to note that forks (multiple outgoing transitions) are always XOR splits in SBDs.

One and only one state in an SBD may be denoted as Initial State (▶). Equally, a state may be denoted to be an end or final State (■), but an SBD may have more than one final state. A process is considered finished when every subject has reached a state denoted as End State. If not all subjects have done so, an End State may be left again, either because of a subject's own decisions in case the end state is a Do-State or when a message reactivates the subject if the final state is a receive state (Fig. 5).

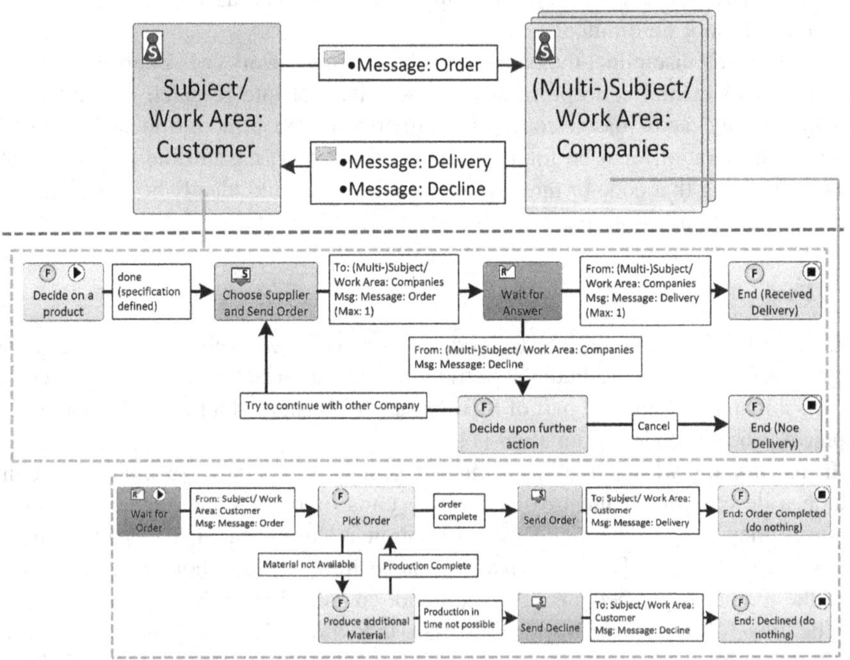

Fig. 5. Example process modeled in PASS with Subject Interaction Diagram (SID - top) and individual Subject Behavior Diagrams (SBD – lower two graphs)

5 Discussion of Subject-Orientation

5.1 Non-ideal Aspects of Subject-Orientation

Uncommon Modeling Concept: Foremost, it must be mentioned that subject-orientation and its concepts are not widespread or at least uncommon. While the principle idea of subject or active entity does exist in every business process modeling approach in one form or another – e.g., swim-lanes or organization units in EPC and BPMN – those approaches are not oriented towards the subject and do require that information to be included. The consequence is that especially people trained in or used to classical description approaches tend to misunderstand or be confused by PASS models upon first encounter[6]. E.g., typically it is attempted to interpret a temporal flow in an SID. In addition, the strict separation of actions and interaction (send and receive), while being logical, is unusual and requires readers first to understand the notation.

More Complexity for Small Processes Models: Further interpretation difficulties may stem from an increase of perceived complexity. Especially for small processes, subject-oriented PASS diagrams will be larger than comparable classical conceived models. Mostly for two reasons:

With PASS, the model will always be split into SID and SBDs and, in consequence, have multiple model parts where classical approaches may only have a single graph[7].

Equally, the requirement to explicitly model communication increases the number of model elements per-se and an increase of elements and text in any graphical model makes the model harder to comprehend simply by containing more information.

Extra Effort Necessary for Linear Processes Without Interactions: The explicit modeling of communication does not only increase the model size, but also the effort for the modeling activity itself: Having to define a message, as well as its sending and reception, and finding a descriptor for that message that is agreed on by involved stakeholders and process natives is tedious. It does result in more precise and executable models, but not being able to "sweep" the communication aspect "under the carpet" during modeling may be a factor for rejection.

Complex to Model Trilateral Communication: An especially obvious example of the increase of effort required by PASS is caused by the fundamental modeling concept that only allows the description of bilateral communication. Models with trilateral communication or even more involved parties may get rather complex. E.g., a negotiation process where three parties need to reach a common agreement.

---

[6] Teaching observation: people without prior formal process modeling experience seem to have less problems adopting to the SID/SBD structure of PASS in contrast to formally schooled process models that futilely try to apply the classical linear modeling structures also to subject-oriented models.

[7] Which is the actual problem, as real-life process rarely tend to not fit on one slide.

No Official Technical ISO or OMG Standard (Yet): In contrast to older notations like BPMN or EPC there is no formal description standard accepted by a global standardization organization such as the International Standard Organization (ISO) or the Object Management Group (OMG) yet. This situation does not limit the effectiveness or logic of the paradigm or PASS, but it may foster reservations against the application within organizations that try to apply only established methods and technologies that can be used independently from a single tool vendor. However, an according official standard for PASS has been created within the Subject-Oriented community and is planned to be proposed to the OMG [13].

5.2 The Advantages of Subject-Orientation/ PASS

Powerful Yet Compact: While being powerful expression-wise, at the same time PASS is very simple. It consists of merely five core symbols accompanied by the according connectors. As the thesis of [14] has shown, that is enough to cover all of [15] 's workflow patterns that are deemed necessary for a notation to be considered a full-fledged process modeling language. This is in stark contrast to e.g. BPMN's currently featured 116+ symbols of which, according to [16], only a fraction is ever used by a single modeler.

Modeling Linear and Cyclic Concepts Simultaneously: The expressive power of PASS can best be seen in its ability to have linear/start-finish-logic concepts (e.g., projects) together with iterative consideration (e.g., yearly fiscal reporting or strategic planning), formally correct, within the same coherent process model. As far as known, PASS is the only process language with that ability.

Multiple Abstraction Mechanisms: Where necessary, the classical task/sub-task abstraction mechanisms for individual activities as well as the object-related mechanisms like inheritance or aggregation can be applied in PASS[8].
With the concept of subject, however, another and especially powerful mechanism is available. Subjects cannot only be used as containers to describe behaviors within a process. In the form of interfaces, subjects may serve as placeholders to refer to other process models. This can be used to either describe follow-up processes or refer to a sub-process-entity with more details on a specific matter [17].

Another powerful tool is the so-called Multi-Subject mechanism. It allows to intuitively, yet formally, define the possible creation of multiple sub-processes instances within the context of one process. This kind of expressiveness can formally only be matched by the advanced concept of "Colored Petri-Nets" [18] that graphical-wise are not nearly as intuitive to model and to comprehend as the declaration of a multi-subject.

Formal Process Modeling Language: At the same time, PASS is a precise and formal process modeling language with a well-defined interpreter concept for workflow

[8] Obviously, object-oriented abstraction concepts like inheritance (is-a) are used for passive data-objects. In PASS, these are the messages and according business objects transferred by them.

engines [19]. Therefore, PASS fulfills the base requirement of being usable for creating actual machine-readable and therefore executable process models. This is a significant advantage, and much in contrast to most other process description means.

Simple Mapping of Process Tasks and Users in Information Systems: The concept of subject has another practical use when working with actual workflow engines as part of business information systems. For such a system, it is usually necessary to define which users have the rights to execute parts/steps of a process. With classical description concept, that kind of assignment is usually done during modeling when each process step is individually assigned to a user or role.

With subject-oriented models, the matching element it is not the atomic task, but rather the subject. This allows separating such concerns from modeling.

Aligned With Human Information Gathering and Thinking Structures: The subject-oriented paradigm is closely aligned to the structure of natural languages and the fundamental concepts of information exchange between human beings. The pattern of having to state the subject first (in the SID) corresponds to the natural structure of human languages and the way humans are used receive information[9]. Due to its closeness to natural languages and the rather non-abstract structure, the only requirement for learning subject-orientation is simply a solid competence in natural languages.

Natural Context Separation: PASS itself does not only follow the general structure of natural languages, the subject as a model-organizing element also allows separating or splitting larger models intuitively. This concept is called Natural Context Separation:

Models of large and complex processes will always need to be separated into several parts (split up or abstracted) that are easier to grasp individually. With conventional process models, this separation will be done arbitrarily and up to the choice of the individual modeler. With complex interwoven process models, containing multiple loops and interaction, splits are hard to set if done at all (*process tapestries*).

With PASS, a split-up automatically occurs through the modeling of several subjects and separation of concerns between the SID and SBD. Consequently, when processes are complex enough to require multiple models to comprehend them, PASS is providing a mechanism for such splits according to structures humans are familiar with.

Explicit Modeling of Communication: The requirement to explicitly model interactions may be seen as a drawback, however, at the same time it is a positive aspect for subject-oriented process models. Usually, it is the communication or points of interaction that are the neuralgic element of a process. Explicitly identifying and naming messages and their transmission increases the chance for better process understanding and better process models:

[9] More than half of the world's languages have a subject-object-verb (SOV) structure. Among them Turkish, Japanese, or Latin. Roughly 30% have the subject-verb-object (SVO) structure e.g. the English or German languages [23].

First, because what may be considered relevant or irrelevant is very subjective. A communication act considered "too simple" or "common knowledge" by experienced process participants may be rather crucial for new employees or for external developers.

Secondly, the requirement to explicitly model and name interactions beckons asking questions that are more precise. E.g., naming a message exchange merely "message" is very likely to lead to the question "what kind of message" by a reviewer or stakeholder. Being required to think about appropriate descriptors (names) for messages also is helpful for self-reflection upon one's work within a process.

Thirdly, the concept increases the chances for uncovering inconsistencies in process and terminology early on during process modeling activities. E.g., consider a situation in an organization where a department A refers to a message as "request" while department B refers the same message as "order". In PASS, this situation will easily be identified, and it can be clarified whether the explicitly stated "order" and "request" are indeed the same message or two different aspects. However, if the different names only appear in accompanying textual descriptions, people working with that model may overlook it and face problems due to misunderstandings and confusion[10].

Facilitates Decentralized, Parallel Process Exploration and Modeling: While PASS fosters well-structured, detailed, and easy to comprehend process models, its principle structure also is practical for the task of model creation itself:

When any process-modeling activity starts, modelers usually need to gain a complete understanding of the process that is to be modeled. Especially boundary conditions and limitations need to be discovered, understood, described, and documented (AS-IS). This activity can be called "process exploration" and usually implies interviewing the involved process natives. This task is, by the nature of human communication, error-prone and likely to be done iteratively and separately with different interviews partners. PASS' structure enables the modeling of the different narratives separately, possibly by different modelers in parallel. The individual SBDs and their corresponding SIDs can afterward easily be matched to one another by comparing the communication. A perfect match is not required initially. Furthermore, inevitable changes due to new information gained during exploration or verification can very easily be incorporated into a PASS model as they may affect only the communication between two subjects and do not require a complete remodeling of the overall process concept and it does not require the direct participation of all process natives at the same time.

Overall, this increases the chances that the process modeling activities will be faster, therefore require fewer resources, and/or result in models that have better fidelity.

Ideal for Training and Teaching: Resulting from the previously mentioned aspects, the last but not least, favorable aspect of subject-oriented models is their ability to function out-of-the-box very well as training and teaching material for new process participants that may be required to learn and understand their part in a process or

[10] E.g., a programmer may consider "request" and "order" as two different data items that need to be implemented into a business information system, possibly costing multiple hours of unnecessary development work before found to be the same. Alternatively, worse, if not found, causing confusion and misunderstandings when a system goes live.

interaction with a new business information system. This is due to the separation of concerns that is gained by having first and SID and then individual SBDs. Possible trainees required to understand their role in a given process do not need to understand the full process model or system. They only need to be concerned with the, single SBD for the area of activity they will be responsible for. This reduces the necessary initial information load on a trainee or reader, without the need to create additional process models or model excerpts for training purposes.

6 Conclusion and Outlook

In this paper, we presented an in-depth theoretical analysis of the process description means of subject-orientation and how it compares with the classical process description paradigm. The paper should have provided an inside into what subject orientation is and what possible, drawbacks and strong points are.

Subject-orientation is by no means a silver bullet solving all problems when encountering the task of business process analysis and system engineering. Also, it is clear, that the possible advantage of using subject-orientation over a more classical process description means is hard to measure and indefinitely smaller than the advantage of using any means of process description over using none at all. The paper also did not consider the practical aspects of available software tools to, e.g., actually engage in subject-oriented process modeling with PASS. An overview of a variety of available tools for S-BPM can found in [21].

Nevertheless, the positive aspects of subject-orientation outweigh the non-favorable aspects and from the arguments made here, it should be reasonable to deduct that the paradigm and subject-oriented business process modeling may be a viable if not necessary option when approaching the design of (business) information systems. This is especially true in a complex world where classical, linear thought structures are not sufficient anymore.

Furthermore, when business information systems are supposed to be created for the support of human beings, instead of the other way around, it becomes of utmost importance to give those humans the means to express themselves, their needs, and their values while tasking them with as little complexity as possible. Subject-orientation as far as we can see, is the best solution for that. We have successfully used it and will continue to use subject-orientation for the analysis and conceptual creation of information systems and the business process they should support – and we can recommend it to others as well.

References

1. DIN 66001 (1966). https://web.archive.org/web/20150502020126/http://www.eah-jena.de/~kleine/history/software/DIN66001-1966.pdf. (Zitat vom 19 Feb 2017)
2. Object Management Group. UML Superstructure, v2.1.1. The Object Management Group, 07 February 2005. http://www.omg.org/cgi-bin/doc?formal/2007-02-05

3. Scheer, A.W.: ARIS - vom Geschäftsprozeß zum Anwendungssystem. Springer, Heidelberg (2002). https://doi.org/10.1007/978-3-642-56300-3
4. Object Management Group. Business Process Model and Notation™ (BPMN™) - Version 2.0 (2011). http://www.omg.org/spec/BPMN/2.0/. (Zitat vom 20 Feb 2017)
5. Jakoby, W.: Projektmanagement für Ingenieure. Springer, Wiesbaden (2015). https://doi.org/10.1007/978-3-658-02608-0
6. Walter, U.: Scrip zur Vorlesung: Systems Engineering (2012). https://campus.tum.de/tumonline/LV_TX.wbDisplayTerminDoc?pTerminDocNr=9255. (Zitat vom 20 May 2016)
7. Herken, R.: The Universal Turing Machine: A Half-Century Survey. Springer, Wien (1995)
8. Parbel, M.: Programmiersprachen 2017: Vielfalt ist gefragt 13 Oct 2017 https://www.heise.de/developer/meldung/Programmiersprachen-2017-Vielfalt-ist-gefragt-3861018.html
9. Buchwald, H.: The power of 'As-Is' processes. In: Buchwald, H., Fleischmann, A., Seese, D., Stary, C. (eds.) S-BPM ONE 2009. CCIS, vol. 85, pp. 13–23. Springer, Heidelberg (2010). https://doi.org/10.1007/978-3-642-15915-2_2
10. Meyer, B.: Object-oriented Software Construction, 2nd edn. Prentice Hall PTR, Upper Saddle River (1997)
11. Fleischmann, A.: Distributed Systems: Software Design and Implementation. Springer, Heidelberg (1994). https://doi.org/10.1007/978-3-642-78612-9. ISBN: 978-3-642-78614-3
12. Fleischmann, A., et al.: Subjektorientiertes Prozessmanagement: Mitarbeiter einbinden, Motivation und Prozessakzeptanz steigern. Hanser, München (2011)
13. Elstermann, M., Krenn, F.: The semantic exchange standard for subject-oriented process models. In: S-BPM ONE 2018, Linz, Austria. ACM (2018)
14. Tölle, N., Graef, N.: Evaluation, Mapping und quantitative Reduktion von Workflow Patterns (Control-Flow) 15 May 2009. http://www.aifb.kit.edu/web/Thema3493
15. van der Aalst, W.M.P., et al.: Workflow patterns. Distrib. Parallel Databases **14**, 5–51 (2003)
16. Muehlen, Mz, Recker, J.: How much language is enough? theoretical and practical use of the business process modeling notation. In: Bellahsène, Z., Léonard, M. (eds.) CAiSE 2008. LNCS, vol. 5074, pp. 465–479. Springer, Heidelberg (2008). https://doi.org/10.1007/978-3-540-69534-9_35
17. Elstermann, M., Fleischmann, A.: Modeling complex process systems with subject-oriented means. In: S-BPM ONE 2019, Sevilla, Spain. ACM (2019)
18. Jensen, K., Kristensen, L.M.: Coloured Petri Nets: Modelling and Validation of Concurrent Systems. Springer, Heidelberg (2009). https://doi.org/10.1007/b95112
19. Börger, E.: A Subject-Oriented Interpreter Model for S-BPM (2012). http://www.di.unipi.it/~boerger/Papers/Bpmn/SbpmBookAppendix.pdf
20. Fleischmann, A., et al.: An Overview to S-BPM Oriented Tool Suites. ACM, Darmstadt (2017)
21. Fleischmann, A., Schmidt, W., Stary, C., Obermeier, S., Börger, E.: Subject-Oriented Business Process Management. Springer, Heidelberg (2012). https://doi.org/10.1007/978-3-642-32392-8
22. Dryer, M. S.: Chapter Order of Subject, Object and Verb. The World Atlas of Language Structures Online (2017). http://wals.info/chapter/81. (Zitat vom 06 Feb 2017)
23. Schmidt, W., Fleischmann, A., Gilbert, O.: Subjektorientiertes Geschäftsprozessmanagement. HMD Praxis der Wirtschaftsinformatik **46**, 52–62 (2009)

Quality of Research Information in RIS Databases: A Multidimensional Approach

Otmane Azeroual[1,2,3(✉)], Gunter Saake[2], Mohammad Abuosba[3], and Joachim Schöpfel[4]

[1] German Center for Higher Education Research and Science Studies (DZHW), Schützenstraße 6a, 10117 Berlin, Germany
Azeroual@dzhw.eu

[2] Otto-von-Guericke-University Magdeburg, Universitätsplatz 2, 39106 Magdeburg, Germany

[3] University of Applied Sciences (HTW) Berlin, Wilhelminenhofstraße 75 A, 12459 Berlin, Germany

[4] GERiiCO-Labor, University of Lille, 59650 Villeneuve-d'Ascq, France

Abstract. For the permanent establishment and use of a RIS in universities and academic institutions, it is absolutely necessary to ensure the quality of the research information, so that the stakeholders of the science system can make an adequate and reliable basis for decision-making. However, to assess and improve data quality in RIS, it must be possible to measure them and effectively distinguish between valid and invalid research information. Because research information is very diverse and occurs in a variety of formats and contexts, it is often difficult to define what data quality is. In the context of this present paper, the data quality of RIS or rather their influence on user acceptance will be examined as well as objective quality dimensions (correctness, completeness, consistency and timeliness) to identify possible data quality deficits in RIS. Based on a quantitative survey of RIS users, a reliable and valid framework for the four relevant quality dimensions will be developed in the context of RIS to allow for the enhancement of research information driven decision support.

Keywords: Research information systems (RIS) · Research information · Utility · System acceptance · Data quality dimensions · Data quality measurement · Data quality improvement · Reliability · Validity · Structural equation modeling

1 Introduction

For the operation of a research information system (RIS) as a central source of information in academic institutions, the quality of the research information and the reliability of derived statements is of central importance. RIS is a database and tool of research administration that specifically supports the management and provision of research information and its activities (such as affiliation of persons to institutions, publications, research projects, patents, etc.). The peculiarity of RIS is to understand, manage, evaluate and further develop the portfolio of scientific research activities of academic institutions.

W. Abramowicz and R. Corchuelo (Eds.): BIS 2019, LNBIP 353, pp. 337–349, 2019.
https://doi.org/10.1007/978-3-030-20485-3_26

In addition, RIS provides them with a sound basis for decision-making and reporting, in which the research information from different heterogeneous sources (e.g. human resources, financial budgets, libraries, etc.) are brought together. The reason for this is not at least the intention to merge the collected research information into a homogeneous amount, to bring it into logical context and to be able to evaluate and present research-relevant decisions consequently. Since research information serves the interests of various data users (e.g., academic institutions, funding bodies, companies, etc.), the reports should be generated from a high-quality RIS. If this research information is incorrect, incomplete or inconsistent, this may have significant implications for institutions.

To have valid and valuable results, it is indispensable to define quality dimensions for data management, measuring, achieving, maintaining and ensuring the highest quality of research information, in addition to the application of methods and techniques (e.g., data profiling and data cleansing) [1, 2].

Data quality dimensions help to structure the research information in RIS and make the success measurable for the decision maker [3]. They provide a way to measure and manage data quality and information [13]. As discussed in the various studies [7, 9, 10, 12], there are a diversity of data quality problems in definition and measurement that are essential to ensuring high data quality [19]. Without quality control, data quality will progressively decrease [5].

The paper firstly examines the quality of RIS and its impact on user acceptance, and then proposes a framework as a structural equation model (SEM) to support quality measurement in RIS. With this model, it is possible to find out to what extent the investigated data quality dimensions have an influence on the improvement of the research information in RIS.

Research on this topic so far, by euroCRIS or the German DINI AG FIS, often stressed the general importance of data quality. Our paper tries to add more detailed insight, based on the four quality dimensions (correctness, completeness, consistency and timeliness) and their relationship to the process of improvement in the RIS. To estimate the reliability and validity of the data quality dimensions for the improvement process in the RIS, results of a quantitative online survey by the "QuestionPro" software (between February 2018 and September 2018) with universities and academic institutions from Germany and other European countries are presented. More information about the survey is provided in [4].

The paper tries to answer the following questions:

- Which aspects are important for describing the data quality in RIS?
- Which data quality dimensions are important for RIS to check and measure research information?
- What data quality problems will be exchanged during collection, integration and storage of research information in RIS?
- How to detect data quality problems in RIS?
- At which point of data processing does a data quality check by the RIS take place?
- Which methods and techniques are used to improve and increase data quality in RIS?
- How high is the data quality in RIS?

Factor analysis and Cronbach alpha test are used to assess the consistency, reliability and validity of the results [11, 14, 16].

The paper has four sections: (1) the introduction to the topic and methodology; (2) the concept of data quality in the context of RIS and the user acceptance based on data quality in RIS; (3) presentation of results; (4) a framework for measuring and improving the quality of research information in RIS. Finally, the paper ends with a conclusion of the most important results and an outlook.

2 State-of-the-Art Data Quality

The increase in research information and its sources presents universities and academic institutions with difficult challenges, furthermore data quality is becoming more important. The term data quality is defined in various ways both in the literature and by experts. Wand and Wang [17] conclude that data quality issues occur with inconsistencies between the view of the information system and the view of the real world. The occurring deviations can be determined based on data quality dimensions such as completeness, correctness and consistency. English [8] differentiates between the quality of the data definition, the architecture, the data values and the data presentation. Wang and Strong [18] evaluate in an empirical study of general data quality dimensions across four categories. Data quality was therefore determined contextually or based on the data values (inner data quality). Furthermore, the data quality must satisfy what the user and system demands.

From these different approaches to the topic of data quality can be defined in the context of RIS as *fitness for use* and describes the suitability of the data objects for users in a particular context [18], they must be correct, complete, consistent and current.

Four data quality dimensions, as defined by Wand and Wang [17] will be explained, which are considered relevant in the context of data quality in RIS.

- *Correctness:* The research information is consistent in content and form with the data definition. Correct research information contains the contentwise correct information in the predefined formats of the attributes.
- *Completeness:* On the one hand, the criterion refers to the completeness of the research information in the transmission of data between the different systems. A record is complete if no data has been lost during the transformation from System A to System B. On the other hand, a record is complete if all necessary values are included.
- *Consistency:* The consistency of the research information refers to the correctness of the stored data in the sense of a consistent and complete representation of the relevant aspects of reality.
- *Timeliness:* The research information is current if it reflects the actual property of the described object in a timely manner. The research information is not outdated.

In addition to data security, ease of use and other variables, data quality is one of the main conditions of user acceptance of RIS. This is primarily about trust - trust in the system, in its provider and in its administration. A system that does not reliably identify or correct data problems, or that itself is a source of data quality defects, can (and will) not be trusted. Perceived quality problems affect the subjective performance expectations of the system. Data quality problems or poor data quality can have different causes. In order to improve data quality, the cause must be known. Because only if the cause is remedied, a lasting improvement of the data quality can be achieved. However, in the case of the RIS, poor data quality is all the more problematic in terms of strategic and sometimes highly sensitive information and decision-making aids, such as personal or financial data. The perceived data quality has a direct impact on the expected benefit and thus, indirectly, on the intended and real use of the system. User acceptance is not only a matter of ergonomics and system quality, but also of organization, communication and legal protection. In this sense, data quality is a necessary but not sufficient condition for user acceptance. But one can also ask the question differently: What incentives does the system and its organizational environment provide for the scientists involved? What "facilitating conditions" are created by science management to support acceptance by scientific staff?

3 Results

This section presents the research results of the quantitative study. The survey was addressed to 240 German universities and research institutions and 30 European universities. A total of 51 German universities and research institutions and 17 European universities responded. According to the survey, the responding institutions implemented their RIS 2 to 8 years ago.

The survey's main objective was to assess the management of data quality, i.e. processes to define, measure, and improve data quality. The first question identifies main aspects of the concept of "data quality". According to the respondents, most important is the quality of data provision (overall RIS), as well as the quality of the data content and the quality of the definitions (see Fig. 1).

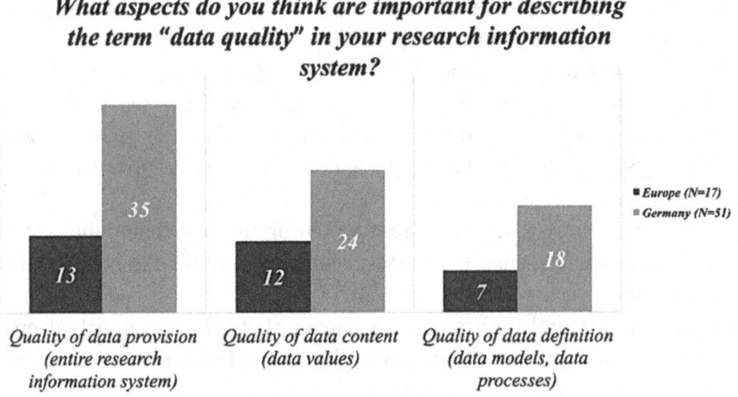

Fig. 1. Aspects describing the concept of data quality in the context of RIS ($N = 68$).

Data quality can differ between the data definition and architectural quality, the quality of the data values as well as the quality of the data presentation and groups these into [8]:

- The quality of the data standards (guidelines that support a consistent, accurate, clear and understandable data definition).
- The quality of the data definitions (semantic aspects and business rules).
- The quality of the information system architecture (general design of data models and databases in terms of reuse, stability and flexibility).

Former research defined data quality with some universal dimensions. The survey tries to assess which of these dimensions are of particular importance to institutions for examining and measuring data quality in RIS.

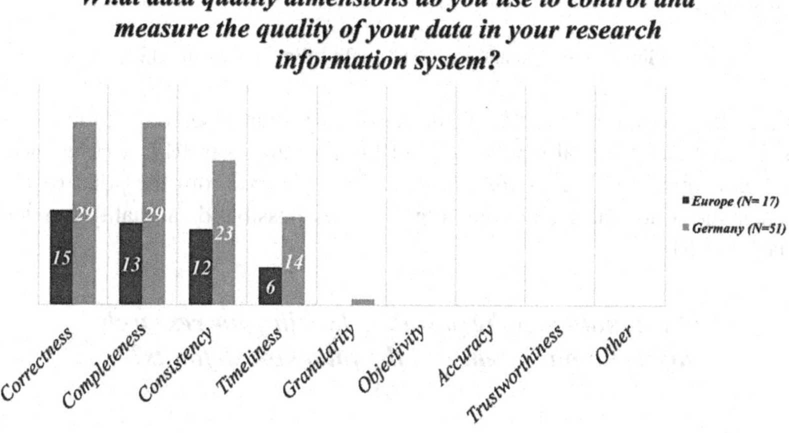

Fig. 2. Data quality dimensions for RIS (N = *68*).

The survey reveals that the respondents evaluate the correctness, completeness, consistency and timeliness of the research information as most important (see Fig. 2). To monitor data quality, these dimensions should be objectively measurable, automatically collected which requires querying the data sources to have values for processing. For larger data sources, good sampling and extrapolation techniques should be used. Automatic assessments should be conducted as often as possible, and simple procedures should be used to not burden RIS unnecessarily. Therewith, e.g. the correctness, completeness and consistency of the research information is verified or at least well assessed. Research information that meets 80% (good) to 100% (very good) of the correctness, completeness, and consistency of data represents a precise reflection of real-world system states to information system states and can be used to justify about data quality [17]. Because such reasoning can be made to improve data quality [17]. In addition, the respective degree of fulfillment of the requirements can be determined by the data user. Figure 3 presents a model for classifying data quality dimensions in the context of RIS.

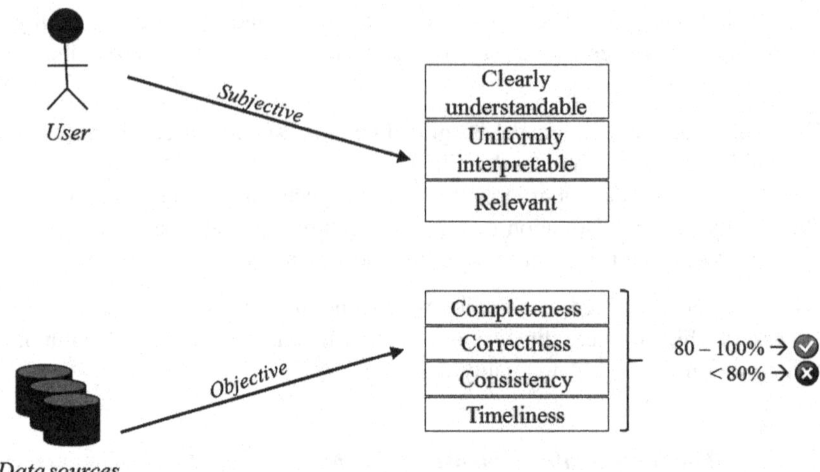

Fig. 3. Classification of data quality dimensions in RIS.

During the collection and storage phase of important research information from various internal and external data sources of the institutions in RIS, a large variety of data problems arise which must be processed by the RIS. From the point of view of universities and academic institutions, Fig. 4 shows possible data quality problems of data quality in RIS.

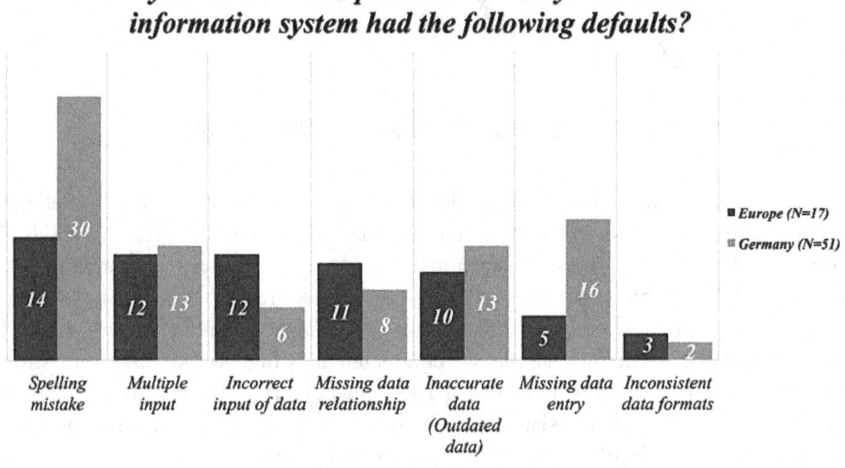

Fig. 4. Data quality problems in RIS (*N* = *68*).

Poor data quality leads to wrong decisions, employee dissatisfaction and rising costs. In order to be able to recognize errors at an early stage and treat them efficiently, the following questions must be answered in institutions:

- Will the data quality in RIS get worse or better?

- Which source system causes the most/least data quality problems in RIS?
- Can patterns or trends be recognized by the data quality check in RIS?

Data quality problems are continuously detected in institutions by plausibility checks. The analysis of quality problems in RIS are illustrated in Fig. 5.

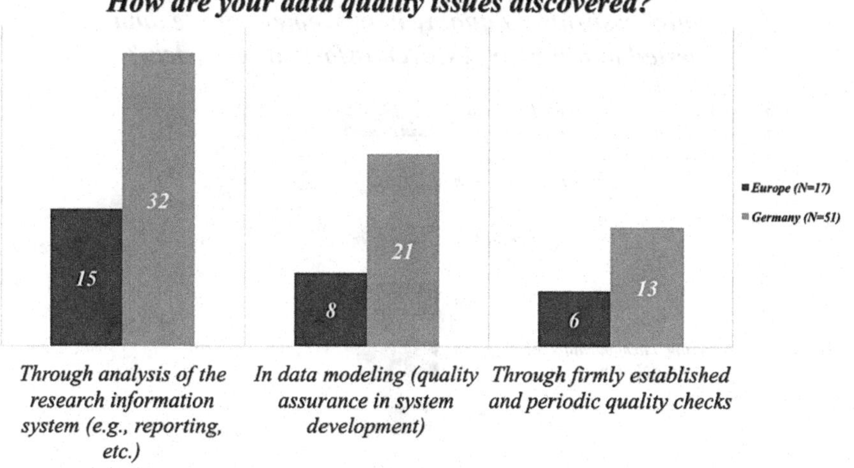

Fig. 5. Data quality checks in RIS ($N = 68$).

The quality checks performed by RIS on institutions take place most during the data processing in the data storage in the RIS as well as the import of internal and external data sources and data presentation (see Fig. 6).

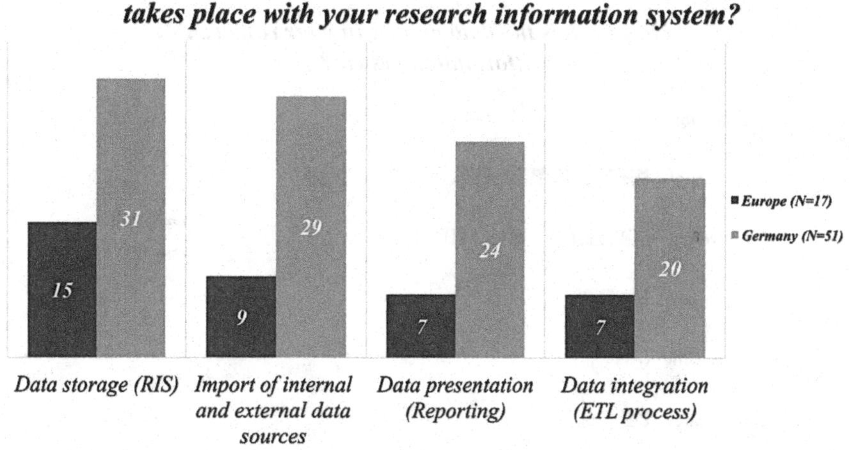

Fig. 6. Data processing with the quality checks by RIS ($N = 68$).

Which techniques, methods and measures are used to improve the RIS data quality? The majority of the respondents use data cleansing methods, while pro-active approaches and data profiling rank second. Re-active approaches and ETL processes seem rather rare. Figure 7 shows the results.

With which techniques, methods and measures is the quality assurance / quality improvement of the data carried out in your research information system?

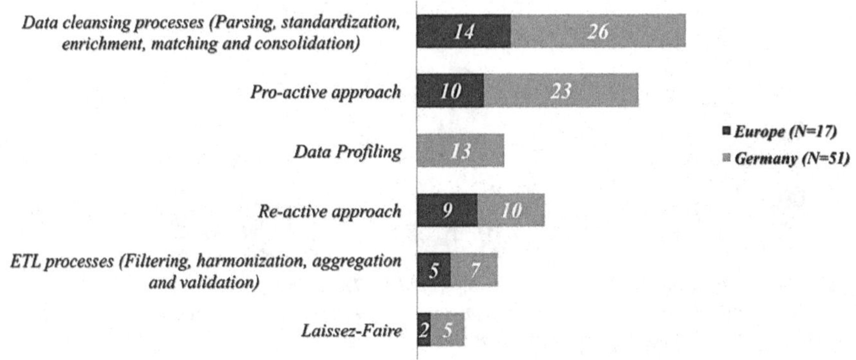

Fig. 7. Techniques, methods and measures to improve data quality in RIS (*N* = *68*).

Many respondents attach great importance to data quality in RIS (see Fig. 8). High quality contributes to the fact that working with the RIS is perceived by the users as pleasant and easy. High and reliable data quality creates trust in RIS. Users not only work more efficiently and more powerfully, but also more securely, which in turn increases user acceptance. High data quality adds value and provides benefits to universities and research institutions which will further increase user acceptance.

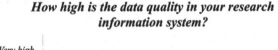

How high is the data quality in your research information system?

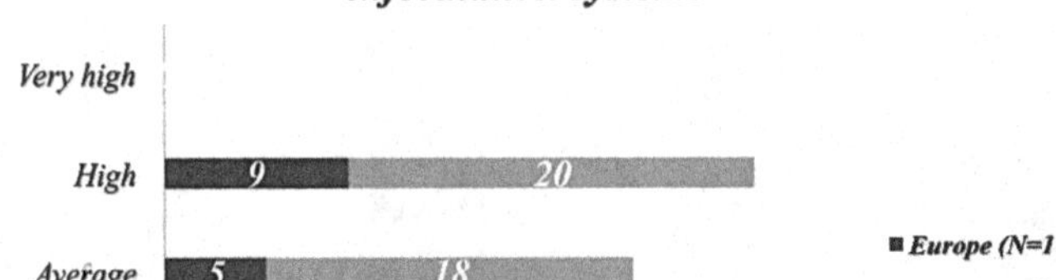

Fig. 8. Degree of data quality of RIS in German and European universities and academic institutions (*N* = *68*).

4 Supporting Framework for Research Information Quality Dimensions

To make a statement about the dependency relationship of the data quality dimensions for the improvement process in the RIS, it is necessary to consider important dimensions to each other. For such a consideration and estimation of reliability and validity based on the results of the quantitative survey, a flexible framework will be used as a structural equation model (SEM) [14]. The data quality in RIS is measured by the variables correctness, completeness, consistency and timeliness. The framework is shown in Fig. 9.

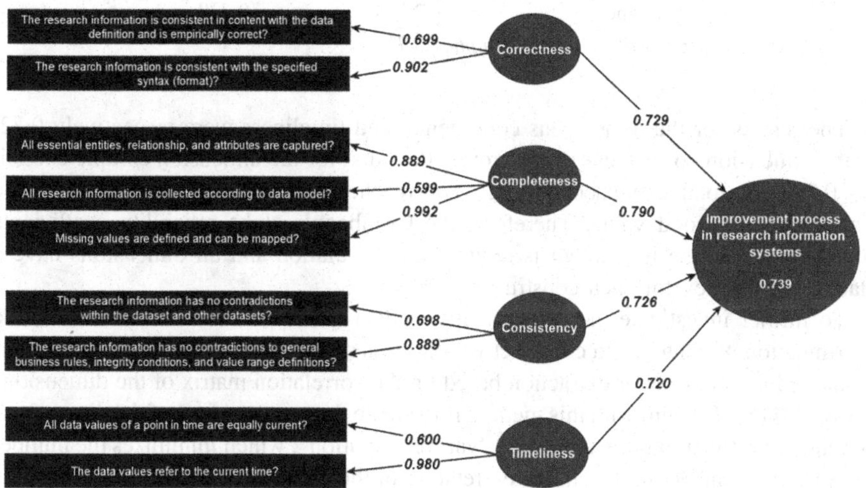

Fig. 9. Framework as a structural equation model for data quality dimensions in RIS.

The framework allows the measurement of the observable variables of data quality dimensions. They represent the latent variables of the constructor. Each latent variable is operationalized by directly observable or ascertainable variables. In this framework, a reflective model is used because the latent variables affect the respective indicators. The number next to the arrow describes the relationship between the latent variable and the corresponding indicator. This number is to be interpreted as a factor load and indicates how strong the reliability and validity is to the latent variable. The developed framework can support institutions at every step from design, execution, analysis, and improvement to assessing and overcoming the data quality issues of a RIS.

To evaluate the reliability and validity of the scales for data quality dimensions and design factors for the improvement process in the RIS, a statistical analysis with R on the survey data was applied to assess Cronbach's alpha analysis and principal component analysis (PCA). Ensuring that respondents can accurately answer questions about data quality dimensions has been limited to the universities and academic institutions that have been using RIS for a long time. For the survey, a Likert scale was used, with four possible answers from "very important" to "unimportant".

Cronbach's alpha determines the reliability of the dimensions and its value is between 0 and 1, with values less than 0.5 considered "unacceptable" and higher than 0.65 "good" [6]. Table 1 calculates and displays the reliability of the coefficient for each individual dimension.

Table 1. Result of Cronbach's alpha for data quality dimensions.

Dimensions	Number of items	Alpha
Correctness	2	0.729
Completeness	3	0.790
Consistency	2	0.726
Timeliness	2	0.720
Overall Cronbach's alpha	0.739	

The results for the dimensions consistency and timeliness were respectively 0.72, for the dimension correctness it was around 0.73 and for the dimension completeness it was 0.79. The total Cronbach's alpha value reaches a value of 0.74 and is therefore considered as a good value. Therefore, the Cronbach's alpha reliability coefficients values show that the instrument is reliable for calculation and all dimensions have a relative consistency for each construct.

To further investigate the relationships of data quality dimensions or factors, the determination of content and construct validity was made using PCA. This is a "method of data reduction or factor extraction based on the correlation matrix of the dimensions involved" [15]. The aim with this method is to create the measurement of the dimensions with the orthogonal rotation technique "varimax rotation", which minimizes the number of connections and simplifies the interpretation of the dimensions [11]. To calculate the factors, a coefficient greater than 0.5 was chosen to make the factor matrix more reliable, with the eigenvalue (variance) greater than 1 and Kaiser-Meyer-Olkin (KMO) greater than 0.5 for measuring the adequacy of the sample [15]. The correlation of the factor values is called loadings and these explain the relationship between dimensions and factor. Using the factor loading, one can see which dimensions are highly correlated with which factor and which dimensions can be assigned to that factor [15]. Table 2 below shows the results and calculation of PCA for the validity of the data quality dimensions.

Table 2. Result of PCA for data quality dimensions.

Dimensions	Number of items	Factor loading	Eigenvalues	% of variance
Correctness	CorrQ1	0.699	1.73	19.34
	CorrQ2	0.902		
Completeness	CompQ1	0.889	5.01	51.84
	CompQ2	0.599		
	CompQ3	0.992		
Consistency	ConsQ1	0.698	1.33	15.75
	ConsQ2	0.889		
Timeliness	TimeQ1	0.600	1.01	13.07
	TimeQ2	0.980		

The KMO-value for items of the dimension correctness was 1.20, for the completeness 1.64, for the consistency 0.86 and for the timeliness the KMO-value was 0.80. Thus, the KMO values were above 0.5 for all four dimensions, indicating that the sample size was adequate and that there were enough items for each factor. All factors of the tested dimensions had a factor load of more than 0.5, which means that all items can be loaded with the same factor. For the first two factors, correctness and completeness, the extracted variance was 19.34% and for others 51.84%. The eigenvalue for both factors is thus greater than 1. However, for consistency and timeliness, the eigenvalue is also greater than 1, with an extracted variance of 15.75% and 13.07%. Thus, for all factors, the items are compared to related dimensions and can be grouped into one factor.

As an overall assessment of the framework, it can be summarized that the data quality dimensions positively influence the improvement process in the RIS. The analysis results prove the good reliability and validity of the nine data quality items in RIS. The result of the PCA shows that the items were highly valid in the construct and demonstrate good statistical properties for testing the developed framework. For academic institutions that have problems integrating different information systems or external data sources, it is advisable to consider these four reliable and valid dimensions to optimize data processing processes and ensure data quality in the RIS.

5 Conclusion

An institution needs research information to monitor and evaluate its research activities and make strategic decisions about different application and usage scenarios. For a holistic view of the research activities and their results, the introduction of a RIS is therefore essential. It is equally essential that such a system provide the required information in a secured quality. In order to make the best possible decisions in academic institutions, they must be based on research information that has to meet high requirements. The right research information must exist and be available in the right place at the right time.

Decisions made on the basis of bad or inadequate information due to poor data quality may not be optimal. Poor data quality in RIS poses a challenge in terms of time and cost to institutions, which should not be underestimated. Especially when integrating research information from heterogeneous systems into the RIS, it may happen that the data formats of the fields in the source system do not match. It is possible that the source data is in the wrong format or in the wrong range of values. To overcome this challenge, the source systems must be measured, adjusted and controlled so that these constellations can no longer occur.

The term data quality in the context of RIS refers primarily to the first aspect, especially the correctness, completeness and consistency of data and the information derived from it. Implicitly, the second aspect is where the timeliness of the information is to be considered, because outdated information is generally no longer correct in dynamic environments, such as in academic institution. Through the analysis of the survey results and the developed framework, the most important dimensions for the improvement process in RIS could be identified, which are crucial for the measurement of data quality in the RIS. The measurement of these four dimensions can be done with each RIS (see [3] for further details on the data measurement in RIS). The concept presented in the paper offers an appropriate way to measure and improve the processing of data quality in RIS.

The quantitative analysis of this paper has shown that data quality is a critical success factor in user acceptance. To ensure the sustainable use of such a system, it requires the greatest possible user acceptance on the part of the science management, the system administrators and the scientists themselves. User acceptance is based on trust in data quality, which requires continuous quality management. Data quality should therefore be treated as a high priority business process, not only to guarantee and enhance the added value of the information produced, but also to ensure the confidence or user acceptance in universities and academic institutions with a RIS, which in turn the responsible use of such systems is indispensable and at the same time, in the sense of positive feedback, can contribute to the quality of all data. Further work is needed for a better understanding of the relationships between data quality, satisfaction, acceptance and perceived usefulness of RIS, with a larger sample including in-house developments and second generation systems.

References

1. Azeroual, O., Saake, G., Abuosba, M.: Data quality measures and data cleansing for research information systems. J. Digital Inf. Manage. **16**(1), 12–21 (2018)
2. Azeroual, O., Saake, G., Schallehn, E.: Analyzing data quality issues in research information systems via data profiling. Int. J. Inf. Manage. **41**, 50–56 (2018)
3. Azeroual, O., Saake, G., Wastl, J.: Data measurement in research information systems: metrics for the evaluation of data quality. Scientometrics **115**(3), 1271–1290 (2018)
4. Azeroual, O., Schöpfel, J.: Quality issues of CRIS data: an exploratory investigation with universities from twelve countries. Publications **7**(1), 1–18 (2019)
5. Batini, C., Cappiello, C., Francalanci, C., Maurino, A.: Methodologies for data quality assessment and improvement. ACM Comput. Surv. **41**(3), 1–52 (2009)
6. Bovee, M., Srivastava, R.P., Mak, B.: A conceptual framework and belief-function approach to assessing overall information quality. Int. J. Intell. Syst. **18**(1), 51–74 (2003)
7. Engemann, K.: Measuring data quality for ongoing improvement: a data quality assessment framework. Benchmarking Int. J. **21**(3), 481–482 (2014)
8. English, L.P.: Improving Data Warehouse and Business Information Quality: Methods for Reducing Costs and Increasing Profits. Wiley, New York (1999)
9. Ge, M., Helfert, M.: A review of information quality research - develop a research agenda. In: Proceedings of the 12th International Conference on Information Quality, MIT, Cambridge, MA, USA, November 9–11, January 2007 (2007)
10. Heinrich, B., Kaiser, M., Heinrich, B.: How to measure data quality? A metric-based approach. In: Twenty Eighth International Conference on Information Systems, Montreal, pp. 101–122, December 2007 (2007)
11. Jolliffe, L.T., Cadima, J.: Principal component analysis: a review and recent developments. Phil. Trans. A Math. Phys. Eng. Soc. **374**(2065), 20150202 (2016)
12. Madnick, S.E., Wang, R.Y., Lee, Y.W., Zhu, H.: Overview and Framework for data and information quality research. J. Data Inf. Qual. (JDIQ) **1**(1), 1–22 (2009)

13. McGilvray, D.: Executing Data Quality Projects: Ten Steps to Quality Data and Trusted Information. Morgan Kaufmann, Boston (2008)
14. Miller, M.B.: Coefficient alpha: a basic introduction from the perspectives of classical test theory and structural equation modeling. Struct. Equ. Model. Multi. J. **2**(3), 255–273 (1995)
15. Panahy, P.H.S., Sidi, F., Affendey, L.S., Jabar, M.A.: A framework to construct data quality dimensions relationships. Indian J. Sci. Technol. **6**(5), 4422–4431 (2013)
16. Schmitt, N.: Uses and abuses of coefficient alpha. Psychol. Assess. **8**, 350–353 (1996)
17. Wand, Y., Wang, R.Y.: Anchoring data quality dimensions in ontological foundations. Commun. ACM **39**(11), 86–95 (1996)
18. Wang, R.Y., Strong, D.M.: Beyond accuracy: what data quality means to data consumers. J. Manage. Inf. Syst. **12**(4), 5–33 (1996)
19. Wang, R.Y., Ziad, M., Lee, Y.W.: Data Quality, vol. 23. Springer, New York (2002)

Modifying Consent Procedures to Collect Better Data: The Case of Stress-Monitoring Wearables in the Workplace

Stéphanie Gauttier(✉)

University of Twente, Drienerlolaan 5, 7522NB Enschede, The Netherlands
s.e.j.gauttier@utwente.nl

Abstract. Smart wearables can be used in the workplace for organisations to monitor and decrease the stress levels of their employees so they can work better. Such technologies record personal data, which employees might not want to share. The GDPR makes it compulsory to get employees' consent in such a scenario, but is seen as asking a yes/no question. We show that implementing this consent procedure is not enough to protect employees and make them adopt devices. Based on interviews, we argue that more control must be given to employees on which data is collected and why through an ongoing engagement and consent procedure. It could lead to higher technology adoption rates and data quality.

Keywords: Consent · Digital organisations · Smart wearables · Ethics · Technology acceptance · Human enhancement

1 Introduction

Microchips implants to control the environment [9], piercing like implants to get new senses[1], wearable devices to manage our stress [15], some have started to propose the use of cyborg-like technology to individuals and organisations. While individual use of such technologies can be seen as an application of the right to morphological freedom [20], their use under the impulse of an employer is more problematic as employees might not have a choice but to change themselves for the purpose of better work performance. Indeed, these technologies are meant to increase the physical, cognitive, and psychological abilities of the individuals so that they reach higher levels of happiness than confined in the natural limits of their bodies [14]. They have only an indirect impact on work as they are targeting generic abilities and designed to benefit to the individual, at work and after work when doing multiple different tasks. It is because I perceive directions better that I perform better in orientation tasks, it is because I have a microchip that I open door without movement and am faster in carrying elements. Completing tasks faster, better, in new ways is seen as intelligence augmentation [6] and is a promise of increased productivity in the workplace.

[1] https://www.theguardian.com/technology/2017/jan/06/first-humans-sense-where-north-is-cyborg-gadget.

W. Abramowicz and R. Corchuelo (Eds.): BIS 2019, LNBIP 353, pp. 350–360, 2019.
https://doi.org/10.1007/978-3-030-20485-3_27

Notwithstanding the possibility offered by enhancement technologies to create value for business, and it being a market in itself, the use of enhancement technologies at work raises both moral and legal issues. These technologies can record personal data. In such a case, issues related to the ownership of the data and to the respect of the employees' privacy arise.

Since 2018, in Europe it is compulsory to ask for specific and explicit consent when recording personal data as per the General Data Protection Regulation (GDPR). Personal data is interpreted as being anything which relates to or allows to identify an individual. This leads to a series of yes or no question in research protocols or before users can use a new technology or service. It allows individuals to decide how their personal data is handled and so to ensure their privacy is respected as much they can.

In many ways, one could argue that users of human enhancement technologies proposed by their organisation should be given the same right to give consent and that their interests would be protected by it. Companies themselves insist on having given a choice to employees and on that use is voluntary [9]. However, there are several challenges attached to gaining meaningful specific consent when considering the use of human enhancement technologies in the workplace. Firstly, these technologies are considered as emerging: the consequences of their use are not revealed yet, and not all of them can be anticipated. This means it can be difficult to obtain informed consent. Secondly, the use of these technologies in the workplace, a social and political environment, raises issues in terms of (perceived) pressure to consent. The interests of the employees might not be protected by simply asking for consent. Privacy has been shown to have an impact on technology acceptance in other domains [23], suggesting that individuals might consider ethical issues when choosing to use a technology. Implementing a consent procedure that reassures potential users about what can be done with their personal data is required. How can the limits of consent be overcome so as to guarantee a protection of the interests of the employees? Can it also improve the quality of the data gathered?

The thesis outlined in this paper is that to preserve the autonomy of participants, it is necessary to go beyond the formality of seeking consent. Rather, one needs to increase the control that the participants have over the course of the experiment on the data collected and its use.

Firstly, we analyse the concept of consent and show that seeking consent is meant to protect autonomy. Then, we present the case of an ongoing research project where hospitals are to ask their staff to use wearable technologies to monitor physical indicators of stress where the different approaches to consent prove to be ineffective in protecting the participants' autonomy. Thirdly, we introduce how displacing the focus from consent to control is more satisfactory in ensuring the participants' autonomy efficiently, in spite of its costs.

2 Seeking Consent as a Procedure to Protect Autonomy

The concept of consent is most discussed in the sphere of bioethics, from which we will draw in this section. Indeed, historically, the notion of consent, i.e. that an individual agrees to take part in a (medical) procedure, appeared during the Nuremberg trials and was set out in the 1947 Nuremberg code as a way to protect individual's autonomy. Seeking consent is needed in order to make sure that individuals are not subject to procedures they do approve. Going further, Beauchamp and Childress [3] define personal autonomy as when '*The autonomous individual acts freely in accordance with a self-chosen plan, analogous to the way an independent government manages its territories and establishes its policies*' (pp 99–100). Protecting the personal autonomy of individuals requires giving them the opportunity to evaluate how a proposed course of action fits with their own desired trajectory. In this paper, we will consider only the case when individuals with capacity to give consent, i.e. adults with the needed cognitive abilities, are asked for consent or dissent. This means the individuals we consider are able to set their plans and evaluate options.

To achieve a form of consent which does protect the personal autonomy of such individuals, it is necessary to add two conditions: the consent must be properly informed and freely given (1964 Helsinki Declaration). This means that seeking consent is more than asking a yes or no question but is a process throughout which individuals get information on the purposes, procedures, of the research they are taking part in. This information must not be deceitful and must be provided in a language intelligible to the individuals whose consent is sought. Furthermore, this process must be implemented so that there are no threats, coercion, or persuasion to agree. While in the light of history some may read the requirement for no pressure and violence as the absence of physical coercion, this concept must be understood more broadly as there can be a form of psychological pressure to agree, depending on social pressure. Individuals are not isolated when they make their decision to consent or dissent and this social context, with its emotional and embodied aspects, needs to be considered. Since the 70s, there is a turn towards such a form of relational autonomy. The place of individuals or groups who have a responsibility in engaging with the individuals to inform their decision is to be considered. Protecting their autonomy, in this relational view, means that paternalistic approaches where an organisation would decide a priori what is the course of action to follow, are prohibited as each individual has a pre-established plan and can reject propositions which divert from the plan. Surrogate decisions and paternalism are to be avoided here and a principle of non-domination is to be followed. The absence of coercion, persuasion or threats also means that individuals must not face only bad alternatives if they dissent. Consent allows a sense of personal integrity as it enables the individual to follow their plan for themselves and their bodies. Consent is related to the concept of self-ownership [12], implying that we have ownership over our bodies and selves, and perhaps our data.

While asking for consent is an additional procedure, it can reinforce trust. The relationship between trust and consent is seen as positive [16] as it means that one can trust that the researcher will respect the terms of consent and the trust put in him by the individual, so no abuses will occur.

Consent can take different forms. It can be broad, or specific. Specific forms of consent are to be sought in current legislation (GDPR). This means that individuals are asked for consent on each aspect of the procedure or data collection which involves their personal data. They can object to certain tools (being recorded, with voice being personal data), while still giving consent to be part of a research project on a given topic. This means that more flexibility is required from the side of the researcher or of any organisation collecting personal data. Consent can also be sought at multiple times in order to verify that the individual still agrees to continuing the procedure.

3 Seeking Consent Is not Enough to Protect Autonomy: The Case of Stress-Monitoring Wearables in the Workplace

The literature shows that such conceptions of consent do not allow to protect autonomy and self-ownership in the case of the use of wearables at work, even though seeking consent as mentioned in the law is seen as a solution to avoid ethical issues in practice.

3.1 Wearables at Work: A New Take on Consent Is Needed

We investigate how hospitals consider these stress-monitoring wearables and how nurses could be asked to use them. The Information System literature looking at technology acceptance in hospitals does not consider the use of wearables for physicians, but the use of tools used to cure patients or wearables for patients. Given the fundamental difference between the types and aims of the technologies explored in these cases and the one we are investigating, we do not base our argument on this previous research. There is little evidence of the impact of smart wearable technologies [7]. Extant research in the domain of consumer wearables has also put the emphasis on privacy and data security [1]. Literature on the use of wearables at work emphasizes the risks of surveillance [17, 24].

The extant literature on wearables at work shows that gaining meaningful consent is difficult. While the literature does not tackle consent directly, we look at consenting to participation is as a form of technology acceptance. In the acceptance of wearable, perceived benefits outweigh perceived risks, so that users might be incentivized to surrender their data and privacy for a greater benefit [25]. When thinking of the workplace, benefits might be tangible and take the form of securing a job or a monetary compensation. Privacy trade-offs and impacts on health however, might not be recognized by employees early on but rather through long term use. This means that consent cannot be meaningfully given at the beginning of an experiment or when a technology is just introduced as the individuals do not realize what they are consenting to.

This is made even more significant considering the effectual approach followed by the project, which is often seen in entrepreneurship and innovation and characterizes cases where organisations try to put a means to an end. It implies that the organisation must make decisions looking at what it can afford, but also at the risks individuals are ready to take [21]. Effectual reasoning is related to overtrust in the project in organisations. In our case, we look at a situation where the use of technology is suggested, but the organisation still has to make sense of how to make the technology useful for its own purposes and to think about how to implement it an efficient and ethical manner so that individuals use it. This means that the organisation has to reflect on how to appropriate the technology along the way and how employees can use this technology. It makes it difficult to ask for specific consent in the first place, as the possibilities coming from the use of the technologies are still to be discovered. Options can be given to individuals, but they risk being meaningless as they have not been tested beforehand.

Finally, users appropriate a system through time [5, 13, 19] and learn how to use it, so that their attitudes can change overtime: they can pause their usage or segment it [10]. As a reaction to automation, individuals can also misuse, disuse or abuse the device purposefully [4, 18]. Seeking consent at only one moment does not address the actual behaviors of users who change opinions over time.

Consent could be offered at several moments so that individuals have the possibility to withdraw. However, this is not the optimal solution for several reasons. Firstly, a withdrawal might intervene after a problematic situation has occurred and trust decreased. Secondly, there might be felt-pressure not to withdraw due to social pressure as described above. Thirdly, this would mean a total non-use of the system, while solutions to adapt uses could be found so that the investment in the information system can still bring some returns (even though not at the scale at which it was thought at first).

Turning towards the ethics literature on consent for the use of mobile health devices highlights that translating the notion of consent, born in the medical and bioethical realms, to medical technologies requires adjustments which imply more extensive procedures and less authority for individual laypersons [2]. The technology we intend to use deals with medical data (stress and mental health, but also as it captures data on heartbeat, blood pressure, skin conductivity), and is of mobile nature, suggesting that the extant literature on the limits of consent apply and, given the sensitivity of the data, more extensive procedures to inform and collect consent are needed. Because the use of the technology is not vital in our case, laypeople can have more control in deciding what technology to use and when. Doing otherwise, especially in a hierarchical setting like in a company would act paternalistically. This can be revisited in different contexts, given the expertise and ability to assess the technology of individuals.

An additional difficulty is added by the workplace setting. Indeed, the workplace also refers to a setting where autonomy is socially-embedded, and where the decision not to participate in the use of a technology can be meaningful and stigmatizing. For instance, [9] shows that employees proposed the use of microchip implants refer to the need to protect their image and not to be laggards when they are interrogated about their attitude towards the microchips. There needs to be a process where users have the possibility to influence the implementation plan and make decisions, rather than having to withdraw from the process totally.

We have shown that regular consent procedure might not be enough to protect the autonomy of employees. To gain meaningful consent, procedures coming from bioethics need to be adapted: the focus should be displaced from personal autonomy to relational autonomy. These procedures also need to take into account the hierarchical relationship at play and employees should be given control over the use of the technology. In practice, not so much thought is given to the informed consent procedure, even if managers are aware of the sensitivity of the data.

3.2 An Illustrated Case: The Role of Consent in the Implementation of Wearables for Stress-Monitoring Purposes

The first step of this research consisted in exploratory interviews with 10 managers of a hospital in Italy and 5 managers of a rehabilitation center in the Netherlands. These 'managers' span from administrative staff to scientific and medical staff. The interviewees were asked what the sources of stress in their organisations are, what form of data would be useful to fight stress, how it could be used for management, what could be consequences for employees, and how they would get employees to wear and use the device. For the purpose of this conference paper, we performed a preliminary analysis of the transcripts looking at (1) how sensitive is the data to be collected perceived to be and (2) what are the ethical concerns identified by the managers, including how to get consent.

Interviewees proceeded by themselves to an assessment of the potential benefits and harms of introducing stress-monitoring wearables for the organisation (see Table 1). It implicitly mentioned the need for fairness in the representation of stress through the data (the word fairness was not used but the concept was described). These potential harms underline that the data to be collected could span outside of working hours and that the organisation would get a database from which the health status of individuals could be inferred. Rules on what needs to be inferred to protect employees (detect the premises of a heart attack) or what should not be known by the hospital are difficult to establish. Rules on how to handle the needs for reorganization are also needed. Involving the employees in shaping the policies around what is meant to be done with their data could be a way to avoid backlash.

Table 1. Potential benefits and harms of the stress-monitoring wearables

Potential Benefits	Potential Harms
Less sick leaves, less burn outs	More complaints which are difficult to handle, demands to change wards
Better team management	Unfairness of the data (measuring stress objectively might not be possible), making it difficult to use
Less errors	Difficulty to explain the organisation will never be stressless – how to divide stressful times (night shifts) fairly
Better communication	Difficulty to separate stress coming from personal situations and stressing coming from work
Less stress	Difficulty in deciding how the data could be analysed for maximum usefulness
Possibility to prove stress	Inferring elements about employees' health
Possibility to use the device to regain control over work conditions	

This is not to say that all potential harms were listed. Indeed, they were focusing on what could happen to the organisation rather than the individual. They were also adopting a consequentialist perspective, without consideration of other approaches to assessing the technology (deontological, virtue-based approaches for instance). Besides, it can be difficult to forecast risks. Recent literature suggests that new frameworks are required in order to proceed to the ethical assessment of technologies for cognitive enhancement [8], and that approaches going beyond the traditional use of checklist-based technology assessments are required as they do not allow to account for the new issues that can arise with emerging technologies [11]. It is difficult to predict what can occur once data is aggregated from several users or an intermediary service or party appears to process the data. There is a need for a dynamic procedure, where users can shape the extent of consent as they are in the experiment and discover potential issues.

However, there is an awareness that stress comes from work, but also from personal situations at home, and so measuring stress levels might be an inquiry in the personal sphere of the individual. Without the device, stressed nurses can be talking to their Head and decide to divulge a problem, which is not the same as the organisation wondering about the data collected and, potentially, scheduling an appointment to discuss someone's stress. The origin of the conversation and the dynamic of how one can decide to retain or share information is different.

Interviewees are also aware that stress is related to personality as it is subjectively felt: some individuals are stressed in a situation A when others are not, perhaps because they have developed better coping mechanisms. Monitoring stress levels, be it for research within an organisation or as a part of the regular functioning of the organisation, is not a trivial undertaking as it allows to record data on the personal experience that one has at work and how this experience is dealt with. It is recording deeply-personal data.

Furthermore, the wearable aims at, ultimately, being able to manage stress. The impact of the device is on the employees, not only on a task (as would be for a regular work tool). For an organisation to decide unilaterally that its employees need compulsory support in this domain could be seen as a paternalistic decision. Besides, the recording of personal data requires asking for consent[2].

Enthusiasm about the device is expressed by employees only when thinking about how it can be used to show to managers how one works best, and the thought that these might not be followed up on raises skepticism on the device. Having individuals defining the purposes of the data collection seem rather important.

Even if these elements seem to point at the need to ask for thorough consent, the notion of consent was mentioned just a few times and only to be rapidly dismissed. For instance, managers from different sections explain that "*As you as you ask for consent, it's okay, no problem*", and another one says that "*If you ask for consent, it's legal, the rest is a moral problem but legally we are fine*". It was assumed by participants that asking for consent and staying in the legal limits for hours of monitoring were the only elements they had to comply with to avoid legal and ethical issues. Consent was discussed broadly, even though this does not mean that specific consent was not considered. Rather, the project being at an initial phase, the participants could not yet dwell on details and discuss specific areas where consent would be needed.

The interviews show that for the device to bring the desired outcome, which is reducing stress, devices have to be worn by teams and monitor stress throughout time, possibly at work and outside of work. It gives to the organisation data that can be used

[2] This holds to be true regardless of the format taken by the device: it could be embedded in uniforms, without making the collection of data active by default.

to infer elements about the health of employees, and not only to potentially manage them better. This makes the use of the device sensitive. In order to protect the autonomy of the employees and ensure that they consent to a use of the data that will not trigger harmful consequences for them, we need to find ways to give employees control over what data they want to contribute and to what aim. What is at stake is to avoid the exploitation of employees by organisations. Other fields, such as biobanking, have pushed towards engagement and stakeholder participation for consent [22] for similar reasons. We propose in the next section some reflections on what the ongoing engagement can concern.

4 From Consent to Control

4.1 More Control for More Autonomy

One way to overcome the issues identified in the literature and through the interviews is to give control back to the employees over the usage of the device and of the data that is being collected through the design of flexible data collection plans. Before going further in explaining why this might be beneficial, a few examples illustrating what is meant here by control and flexibility must be given:

- The users decide what hypotheses they want to check with the data that will be collected: they might want to measure their stress levels before and after certain events, in specific team configurations, which they know could be useful in order to obtain useful and meaningful data, i.e. data that can be used to inform workflow management;
- The users have control over the duration of use of the device during the day;
- The users have control over the choice of moments when they want to use the device, and so can stop their use when their experience of the device gets in the way of their work and priorities;
- The users have control over who sees their data;
- The procedure is ongoing and the person responsible for the technology implementation checks at regular intervals how the technology is used by surveying individuals.

This approach requires an ongoing engagement of the (potential) users with the purposes of the data collection. In this way, they can evaluate what are the purposes to be pursued with the data collection in order to protect or increase their autonomy, even though the data collected might be seen as a way of controlling employees[3]. The process, as it is ongoing, allows the users to reflect on the unintended consequences of the data collection and the data aggregation so that they can adapt their use. This can prevent rejecting the system, which comes at cost when it occurs have an organisation has invested in an information system.

[3] This paradox between autonomy and control has been described in the literature (Gilbert and Sutherland, 2013).

For the organisation, giving control to the users (or here research participants) is rather scary. Indeed, it implies the risk of having data that is not easily comparable or statistically significant. Synergies and insights that would make sense at scale might not be revealed by scattered patterns of use. However, it can allow to increase trust significantly: trust in the data as it is willingly given by the employees without forms of misuse[4]; trust in the employees can be perceived as higher as they are given control by the employer; trust in the employer can rise as employees feel heard and respected. Going one step further, the trustworthiness of the employer could increase as the organisation made itself vulnerable to the employees' willingness to collect useful data. These hypotheses are to be validated. Finally, the data collected might be more relevant: instead of relying on data science to identify patterns and potential managerial solutions, such an approach relies on the instincts and needs of workers themselves, who know their own stress and the elements in their work conditions which can be changed (and the ones which cannot). The aggregation of data might help to identify how to compose balanced-teams, or when authorized by the employee, to check whether hypotheses valid for others are also true in their cases. There is a shift between inductive and deductive logic that occurs in how the data is analysed, due to the constraints around consent, which may appeal more to the users as it allows them to use the technology, rather than having the technology use their data to do its work.

4.2 Limits

While such an approach opens new research opportunities as mentioned above, it also raises questions.

Firstly, there are costs due to the organisation of such a process, which are incumbent to the organisation. Indeed, such a scenario comes with more control for users, but is also more demanding of them. It requires that individuals take responsibility about their usage and engage time and cognitive resources in order to understand the information system, its risks and potential advantages, so that they can truly take control over it. This can be a source of stress due to the mismatch between the competencies required from the job and the competencies required to understand the information system. It can be a cost for the organisation as it implies providing training to employees.

Secondly, it needs to be considered whether such an approach is viable through time or if it is meant as a transitory process before gaining a better understanding of how the technology can benefit both employees and the organisation. If it is a transitory process, then the autonomy of employees joining after this initial test phase might be negatively impacted because they had other plans for themselves than the individuals who did participate. As mentioned above, autonomy is socially-embedded and relational, and so mechanisms need to be designed to consider both this social and the

[4] Examples of misuses could be measuring indicators of stress when going up and down the stairs to have a higher heart beat rate and thus give data which might indicate stress which is not related to work.

personal aspects of autonomy. If it is a transitory process, then it becomes difficult to adapt the practice of the technology to the changing environment.

Thirdly, there might be individuals who consent but do not engage fully with the process, so that their point of view is not represented even though from a procedural perspective, these individuals participate.

Fourthly, what about the individuals who decide to not participate and how is their opinion considered?

In such a scenario where the technology might affect work conditions, relationships, is it realistic to proceed without everyone participating? Indeed, we saw earlier that autonomy is socially-embedded, but in many ways also relational. If one person decides to participate, this participation can have an impact on the job conditions of the other person who does not participate, so that non-participation is not a guarantee of status quo. Similarly, when individuals do not participate, the value of the others' participation can decrease as the data is not big enough in order to draw conclusions.

5 Conclusions

We have shown that the simple act of asking for employees' consent to the use of a stress-monitoring wearable is not enough to protect the autonomy of the employees. It is therefore failing at meeting its goal. We introduced the idea of moving from asking consent to giving control to employees themselves by engaging continuously them with the ways in which the data collection occurs and the purposes in which it occurs. This is different from a repeated consent as introducing control gives agency to the individuals. Such approaches can be particularly helpful in order to assess how an existing technology can be used by surveying how it is experienced and best implemented. It is an approach that can be helpful in organisations in order to ensure that technologies are used in a way that solves employee's issues, instead of creating new ones. It might also allow technologies perceived as controlling to help employees to regain more autonomy. The impact of introducing such a process onto trust and trustworthiness need to be assessed, as well as the impact on technology adoption.

Acknowledgements. This project has received funding from the European Union's Horizon 2020 research and innovation programme under the Marie Sklodowska-Curie grant agreement No. 795536. The author also thanks colleagues from 4TU.Ethics Life Science and Healthcare Technology Taskforce and their input during the workshop on the limits of consent.

References

1. Amyx, S.: Privacy dangers of wearables and the internet of things. In: Identity Theft: Breakthroughs in Research and Practice, pp. 379–402. IGI Global (2017)
2. Asveld, L.: Informed consent in the fields of medical technological practice. Techne Res. Philos. Technol. **10**(1), 16–29 (2006)
3. Beauchamp, T.L., Childress, J.F.: Principles of Biomedical Ethics, 6th edn. Oxford University Press, Oxford (2008)

4. D'Arcy, J., Devaraj, S.: Employee misuse of information technology resources: Testing a contemporary deterrence model. Decis. Sci. **43**(6), 1091–1124 (2012)
5. DeSanctis, G., Poole, M.S.: Capturing the complexity in advanced technology use: adaptive structuration theory. Organ. Sci. **5**(2), 121–147 (1994)
6. Engelbart, D.C.: A research center for augmenting human intellect. In: Proceedings of the 9–11 December 1968, Fall Joint Computer Conference, Part I, AFIPS 1968 (Fall, Part I), pp. 395–410. ACM, New York (1962)
7. European Commission. Smart Wearables: Reflection and Orientation Paper (2016)
8. Forsberg, E.-M., Shelley-Egan, C., Thorstensen, E., Landeweerd, L., Hofmann, B.: Evaluating Ethical Frameworks for the Assessment of Human Cognitive Enhancement Applications. SE. Springer, Cham (2017). https://doi.org/10.1007/978-3-319-53823-5
9. Gauttier, S.: I've got you under my skin'–The role of ethical consideration in the (non-) acceptance of insideables in the workplace. Technol. Soc. **56**, 93–108 (2019)
10. Jauréguiberry, F.: Retour sur les théories du non-usage des technologies de communication. In: Proulx, S., Klein, A. (eds.) Connexions: communication numérique et lieu social, pp. 335–350. Presses universitaires de Namur, Namur (2012)
11. Kiran, A.H., Oudshoorn, N., Verbeek, P.P.: Beyond checklists: toward an ethical-constructive technology assessment. J. Responsible Innov. **2**(1), 5–19 (2015)
12. Locke, J.: Two Treatises of Government (1689)
13. Mackay, H., Gillespie, G.: Extending the social shaping of technology approach: ideology and appropriation. Soc. Stud. Sci. **22**(4), 685–716 (1992)
14. More, M.: The philosophy of Transhumanism. In: More, M., Vita- More, N. (eds.) The Transhumanist Reader: Classical and Contemporary Essays on the Science, Technology, and Philosophy of the Human Future, 1 edn. Wiley-Blackwell, Chichester (2013)
15. Muaremi, A., Arnrich, B., Tröster, G.: Towards measuring stress with smartphones and wearable devices during workday and sleep. BioNanoScience **3**(2), 172–183 (2013)
16. O'neill, O.: Autonomy and Trust in Bioethics. Cambridge University Press, Cambridge (2002)
17. O'Connor, S.: Wearables at work: the new frontier of employee surveillance. Financial Times (2015)
18. Parasuraman, R., Riley, V.: Humans and automation: use, misuse, disuse, abuse. Hum. Factors **39**(2), 230–253 (1997)
19. Riemer, K., Johnston, R.B.: Place-making: a phenomenological theory of technology appropriation. In: ICIS Orlando (2012)
20. Sandberg, A.: An overview of models of technological singularity. In: More, M., Vita-More, N. (eds.) The Transhumanist Reader: Classical and Contemporary Essays on the Science, Technology, and Philosophy of the Human Future, 1 edn. Wiley-Blackwell, Chichester (2013)
21. Sarasvathy, S.D.: Effectual reasoning in entrepreneurial decision making: existence and bounds. In: Academy of Management Proceedings, vol. 2001, no. 1, pp. D1–D6. Academy of Management, Briarcliff Manor, August 2001
22. Solberg, B.: Biobank consent models—are we moving toward increased participant engagement in biobanking. J. Biorepository Sci. Appl. Med. **3**, 23–33 (2015)
23. Vijayasarathy, L.R.: Predicting consumer intentions to use on-line shopping: the case for an augmented technology acceptance model. Inf. Manage. **41**(6), 747–762 (2004)
24. Weston, M.: Wearable surveillance–a step too far? Strateg. HR Rev. **14**(6), 214–219 (2015)
25. Yang, H., Yu, J., Zo, H., Choi, M.: User acceptance of wearable devices: an extended perspective of perceived value. Telematics Inf. **33**(2), 256–269 (2016)

Cyber Treat Intelligence Modeling

Adiel Aviad[1,2(✉)] and Krzysztof Węcel[1]

[1] Poznan University of Economics, Poznan, Poland
aaviad@iai.co.il
[2] Begin 9, 54421 Givat-Shmuel, Israel

Abstract. This paper proposes semantic approach to manage cyber threat intelligence (CTI). The economic rational is presented as well as functional needs. Several cases of domain standards, tools and practices are modeled as a representation of the CTI sub-domain. This work focuses on the technical and operational CTI that is common to most organizations.

Keywords: Cyber threat intelligence · Threat intelligence · Threat modeling · Cybersecurity

1 Introduction

1.1 Cyber Threat Intelligence

Intelligence in general is about gaining knowledge. CTI is a sub-domain of cybersecurity and focuses on gaining knowledge about threats. The fast pace of developing threats together with the amount of technologies involved put a heavy burden on organizations trying to establish cybersecurity. CTI deals with "the set of data collected, assessed and applied regarding security threats, threat actors, exploits, malware, vulnerabilities and compromise indicators" [1]. CTI refers to "information such as the different malware families used over time with an attack or the network of threat actors involved in an attack" [2]. OSINT is defined in [3] as open source intelligence, i.e. available in public domain. The importance of CTI is on the rise with many initiatives and commercial activities indicating it. The aim of this article is to propose semantic approach and a model for CTI, focusing on three prominent aspects: information representation, CTI analysis and threat hunting. The term of CTI can be referred to as having several sub-types [4]: strategic vs. tactical and operational vs. technical. In this work we emphasize more the technical and operational types.

1.2 Semantics

Semantic web technology provides means to handle information: organizing knowledge chunks, relating them, sharing and accessing the knowledge. It also provides means to gain further knowledge through machine reasoning. The semantic web aims to provide means to structure a web of interlinked data, while adding meaning to the data. This enables reasoning by drawing conclusions based on rules – processing of data at a higher level of "understanding" by machines. The semantic web technologies include means to classify concepts and their inter-relations – be them super-class and

W. Abramowicz and R. Corchuelo (Eds.): BIS 2019, LNBIP 353, pp. 361–370, 2019.
https://doi.org/10.1007/978-3-030-20485-3_28

sub-class or merely kind of connections between concepts. The relations are used to draw conclusions, producing new knowledge.

Semantic web technology provides flexibility through several qualities: it bridges different terminology by referring to the concepts rather than the terms and the vocabulary building features, as described in [5]: "the relations allow communication and collaboration even when the commonality of concept has not (yet) led to a commonality of terms". This semantic bridging, like the entire semantic web technology, applies to machines as well as humans. The technology enables separate contents to be connected, even if they are not known in advance to each party. Reasoning may be applied to newly connected contents, not defined by a common predetermined schema (as with relational databases). Relationships may also be added on the fly, not necessarily defined up-front. The capability of automatic reasoning is a keystone of semantic technology.

2 Motivation

Cybersecurity in general and CTI in particular are based on knowledge. The CTI body of knowledge is complex, ever evolving and very dynamic. It is more about artefacts, rather than about structured theoretical knowledge. Generating and mastering this knowledge is a heavy burden for defenders and there are economic benefits from sharing of cybersecurity knowledge. Risks can also be reduced by better handling of knowledge. Significant benefits can be achieved by sharing knowledge, thus spreading the burden among the community. Knowledge would also be more complete.

CTI, like cybersecurity, is influenced by economic externalities which cause players to choose lower security. The "tragedy of the commons" effect refers to a case of self-optimization by an individual or an organization through resource consumption, through negligence or by other way. This may lead to self-optimization that may cause great damage to the public, much more than the damage to the single entity [Hardin 1968]. In CTI, this may happen as insufficient handling of CTI acquisition or dissemination by firms, leading to malware infections. The "plums and lemons" effect refers to asymmetric information about situation when buyers are less knowledgeable than sellers when it comes to the quality of the products [Akerlof 1970]. According to this article, plums are assumed to be sweat and desirable while lemons are assumed to be less worthy than plums. If the buyers do not know what fruits are in closed bags while sellers do know, then this will cause a severe downward pressure on both price and quality [Akerlof 1970]. In CTI this may lead to negligence of threat information acquisition due to insufficient knowledge about the quality of CTI information regarding the actual threats.

The need for sharing CTI (which, in essence, is sharing the knowledge about threats) is also driven by functional needs of agility, having more complete information and automation. Sharing is critical, enabling wider scope cooperation and automation [6]. Cyber threat information sharing is of critical importance [2] since without sharing, attackers only need to work once on an exploit and can then reuse it on multiple targets while each defender is forced to work individually in detecting and analyzing all attacks. No organization has resources and knowledge to perform this work independently.

Sharing is important for agility [7] since without sharing, attackers benefit from slow identification and mitigation of threats. Sharing information quickly and often, would give a better chance to anticipate and prepare for an eventual attack [8].

In addition to the need for sharing and agility CTI requires building a "big picture" [3], contextualization [9, 10] and adaptation to the case of each firm by extracting and prioritizing the most relevant threats. Moreover, CTI is challenged by the need for trust and privacy, as firms are reluctant to reveal facts about their weaknesses or that they were attacked. This paper addresses the economical and functional challenges of CTI, focusing mainly on the threat knowledge dealing more with "outside" knowledge rather than internal aspects.

3 Background

3.1 STIX

STIX (Structured Threat Information eXpression) is the state-of-the-art project for CTI representation. It is a language describing and communicating cyber threat information in a standard, automatic manner [11]. It aims for exchange of threat information and "provides a common mechanism for addressing structured cyber threat information" [11]. STIX implies a graph-like format and defines eight core constructs. The constructs are the information entities that are related to other constructs. We argue that it can be represented as an ontological model that is capable of holding the same information, with the constructs represented as concepts. Once represented in ontology, the insights sought after by human analysts may be reasoned automatically.

The eight core constructs are as follows: Observable, Indicator, Incident, TTP, Exploit Target, Course of Action, Campaign and Threat Actor. The core constructs may be represented as ontology concepts, while inter-relations can be defined as ontology properties. Their description appears in [11] and the semantic model in the next chapter considers it for the design of how the STIX constructs and inter-relations can be represented by a semantic model.

3.2 Maltego

The processing of CTI is about identifying connections between entities and using them to build the big picture, then extracting insights. Automated link-analysis tools that identify links in disparate data sets are required to respond to threats [3]. A common tool used by analysts in processing of intelligence is Maltego [12]. Maltego exemplifies the tools used in the gathering and processing of open source intelligence (OSINT) [13, 14]. Data items are gathered and transformations are made to normalize their typing and format. Then various interconnections are tried and established based on common attributes like computer addresses, people identities in social networks. The results are displayed as graphs for processing by human analysts, as in Fig. 1.

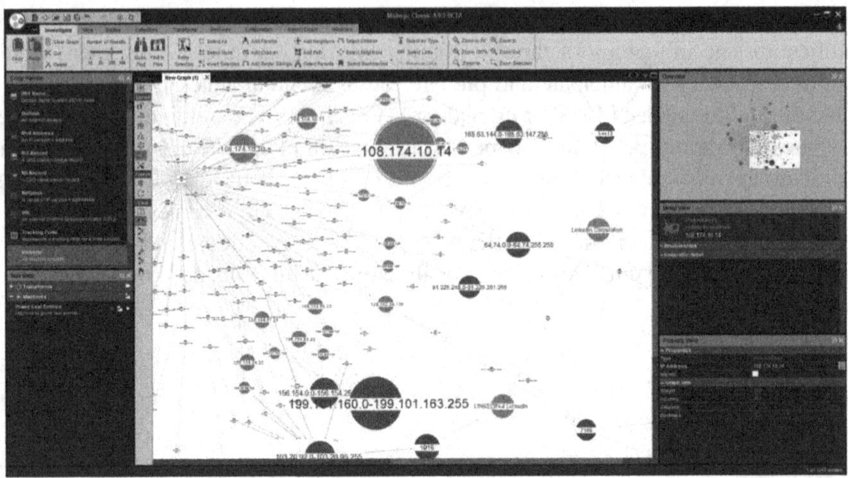

Fig. 1. Maltego image Source: [12]

3.3 Threat Hunting

Threat hunting is a new practice of proactively and iteratively searching for threats assuming that attackers have already entered the defenders networks. The hunters first make hypotheses – a specific assumption of the particular threat that is present and then look for something that proves this to be right or wrong.

CTI is a key for effectiveness – it is an important source for such hypotheses, from which the hunter chooses threats [9]. CTI designates indicators of compromise (IOCs) that are used in formulating the hypotheses and in proving or denying them. The IOCs are usually searched in log files (or messages). The hunter also tries to find the adversaries tactics, techniques and procedures (TTP) gained either directly from CTI or by studying adversary's steps, especially using IOCs to determine these steps. The process is iterative, with the TTPs, IOCs and also the CTI being improved.

4 Semantic Model for CTI

Semantic approach may cope with the above challenges by enabling sharing knowledge and distributing the work among many players, thus reducing the burden of cost as well as speeding the pace of preparing and responding to threats. It is flexible to accommodate new types of threats and relationships, more than the tabular/relational database way. Semantic approach may leverage the family nature of the threats and the relations among the entities. The semantic web technology is aimed for sharing of knowledge (among other goals) [15]. It also enables organizing the knowledge in classes and subclasses, and inferring new knowledge based on the relationships between concepts. Ontology is the mean to represent knowledge by capturing the concepts of the domain and their inter-relationships. In [5] it is referred to as "a taxonomy and a set of inference rules".

Semantic web technology lends itself to capture and handle "families" of threats, which in many cases have lot of derivatives based on technology (e.g. based on versions of operating systems). The family attributes of threats is valuable information [2]. As threats evolve and get more sophisticated, the aspect of families and variants is even more important. Simple threats may be characterized by a signature that can be easily identified even by "old" defensive tools like anti-virus, but more sophisticated threats do not have a signature and require addressing further properties in order to identify them. In [16] this problem is presented as a limit: "automatic analysis technology has limits which are based on the detection of a particular attack pattern as the technique of cyber-attacks becomes more sophisticated". The semantic approach that we propose addresses this limit. For instance, polymorphic threats might use the same encryption/decryption mechanism (while the keys vary) to gain polymorphism or use the same mechanism to communicate with the command and control (C&C) host [17].

Intelligence also requires drawing conclusions based on linking data items together, to reveal connections and build the big picture. Reasoning enables automatic processing of data that is usually done by human consideration, possibly assisted by mere visualization. Cooperation by sharing CTI intensifies the benefits by enabling the automatic processing of larger amounts of data and reasoning over longer ranges of data chunks.

Following Hevner's [18] methodology for design science in information systems, the CTI and its needs is presented from both functional as well as economic aspects and the relevant challenges. The proposed technology is introduced and the contribution is designated. A model is portrayed, referring to CTI representation, analysis and usage for threat hunting (as an example of using CTI).

We demonstrate the semantic approach through the following perspectives: the STIX [11] standard describing cyber threat information, the Maltego tool [12] commonly used for processing CTI by building and analyzing the big picture, and the young practice of threat hunting used for elimination of threats.

4.1 STIX

Semantic representation of STIX introduces the reasoning capability. For example when given a threat, an indicator of compromise may be reasoned, what tools, techniques and procedures are related, through them what exploit target are relevant, through them what course of action may be taken to prevent such threat from being successful. Also, using relationships with other (non-STIX) cybersecurity concepts, proper counter-measures may be reasoned with further information like their incurred costs. Such reasoning may provide the insights that human analysts work to provide, going automatically through more steps than a human analyst is capable of, taking into account more families and variants (of threats or systems) than a human analyst can, using shared pre-prepared examinations (in the form of reasoning) that cover more possibilities than a human analyst is aware of. Such shared knowledge has also the benefit of sharing the cost of research needed to gain the knowledge.

Figure 2 from [11] shows the STIX constructs with their inter relations. The entities are described with structures comprised of several types of fields. Conversion to semantic representation is done field by field, as follows: The STIX constructs themselves are entities that are represented as concepts, while constructs that are included in other constructs or referred by other constructs are represented as relationships with a concept that comes instead of the contained one. Rank or level fields are represented as ordered collections, while fields that hold content with the character of a closed list or enumeration are represented as collections. All other fields with plain text content with no special constraints are represented as properties of the concept.

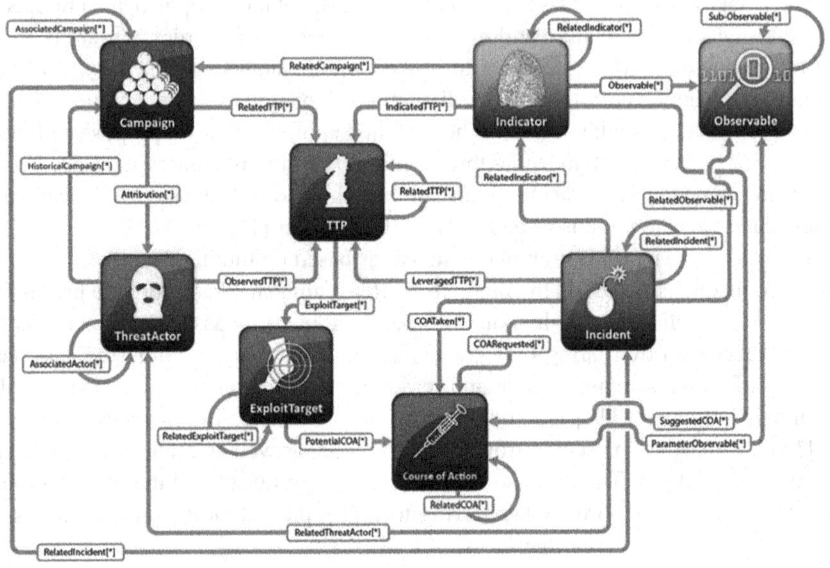

Fig. 2. STIX constructs Source: [11]

The following code in Fig. 3 exemplifies reasoning of likely attack points of an organization, based on CTI that is represented semantically. Implementation was done using Cognitum's Fluent Editor and its controlled natural language [19].

```
every likely-attack-point is a thing .
every stix-construct is a thing .
every exploit-target is a stix-construct .
every ttp is a stix-construct .
every threat-actor is a stix-construct .
if a threat-actor designates a ttp then the ttp is designated-ttp .
if a designated-ttp designates an exploit-target then the exploit-target is designat-
ed-exploit-target .
if a vulnerability is-applicable-to a designated-exploit-target then the vulnerability is
a likely-attack-point .
if a weakness is-applicable-to a designated-exploit-target then the weakness is a
likely-attack-point .
if a configuration is-applicable-to a designated-exploit-target then the configuration
is a likely-attack-point .
```

Fig. 3. Reasoning segment

4.2 Maltego

Semantic web technology may be used to improve the processing of Maltego's analysis graphs. Processing can be automated, gaining agility by utilizing sets of shared, pre-prepared rules to examine such connections between graph nodes. The rules can be written by an organization itself, but can also be contributed by regulators, standardization bodies or any other contributor and be incorporated by the organization in its CTI processing.

Such shared rules may not only improve speed and reduce labor intensive tasks but also provide knowledge going beyond manual work (which in turn may be shared, too) – since reasoning may provide insights that an analyst may not seek unless proper training is comprehensive enough and well up-to-date (e.g. regarding new threats). Machine reasoning may also identify links that human processing is not likely to identify, like links across a chain of many connections. Using shared, pre-prepared rules may also reduce cases of analysts being "locked into a mindset" – a noted problem for the intelligence discipline in general [20].

Such processing could be done better in terms of agility, automation, flexibility for new types of threats and sharing. It can be achieved by semantic queries that are capable of reasoning like SPARQL query language over RDF repositories [21] for representing the CTI information. This will bear the advantages of automated processing and the capability of identifying links that span beyond the capability of a human analyst (e.g. when the link is comprised of many nodes).

Charts like those appearing in [12] may be represented as semantic web models with the graph nodes as concepts and the inter-connections as relationships. Insights are specifically relevant to the organization if they include certain connections between the threats and organizational assets. It is already noted in [2] that "relevancy determination is a manual and complex process that should be facilitated by technology". Such connections may be determined automatically by the same way of reasoning, based on data about organizational assets. Such data the organization may prefer to keep unshared. The reasoning may provide the insights that human analysts spend time and cost to provide. Reasoning also goes through more steps than a human analyst is capable of, using shared pre-prepared examinations (in the form of reasoning) that cover more possibilities than a human analyst is aware of. Such shared knowledge has also the benefit of sharing the cost of research needed to gain the knowledge.

Processing of CTI requires adaptations to the case of each firm, based on the threats that are more relevant (e.g. denial of service or leakage of information). Such adaptation may be done by firms utilizing specific reasoning rules that process the same, shared, big picture. This adaptation is a significant task but the challenge may be smaller with at least the part of processing the common part is improved.

4.3 Threat Hunting

Semantic approach to threat hunting can involve representing the CTI (including the TTP) as well as the IOCs. It can also be used to represent logs. Most attacks involve user credentials. It was found that 63% of confirmed data breaches involved weak, default or stolen passwords [22], so in most log entries that are interesting for security purposes, user identities are present. Log entries may be represented as "object X performed action Y on object Z", with X being a user in the most interesting cases. This can be represented in RDF with Y being the relationship, with the timestamp and other details being properties. This format can be written in real time and processed offline. Network traffic entries have an origin and a destination (broadcast and multicast may also be considered as kinds of destination). The traffic may be considered as messages between origins and destinations, so they may also be represented in RDF.

Semantics may also be used to counter obfuscation of IOCs done by the adversary. Since semantics can provide abstraction, it can be used to overcome differences in IOCs conventional representation and refer to the semantic meaning of the concept. For example, instead of referring to an IP address, it can refer to "known C&C host" including C&C hosts that vary or just change their address.

5 Discussion, Limitations and Conclusions

By referring to the STIX standard and the Maltego tool we demonstrated feasibility of semantic approach for CTI use cases. The CTI information can be represented in RDF repositories and processed by semantic web technologies, being able to cope with new types of threats, even without having to change database schemes like the relational database technology would require. Legacy repositories in the form of tables or relational databases can be converted to RDF [23] or mapped to it [24]. Having this migration path, current resources can be utilized together with future, semantic ones. By having addressed the STIX standard, the Maltego tool and the threat hunting and logs we have a significant representation of the CTI field, so we conclude that the virtues of semantic technology provide further capabilities to CTI over the legacy relational databases.

Sharing CTI may provide the benefits of economy-of-scale, since many organizations are exposed to the same kind of threats and generating and processing the relevant knowledge is beyond the capabilities of a single organization [7]. This is especially right for the external part of CTI and less for the part that deals with organization's assets and vulnerabilities.

Sharing the costs can be done by voluntary contributions of vendors about their products, researchers about their discoveries and regulators or industry bodies about threats. Due to the externalities mentioned, external intervention might be required. In fact, there are already CTI initiatives by governments like USA [25] and by bodies like MITRE's initiatives [6, 26, 27], OWASP [28] and WASC [29].

This work does not refer directly to the trust and privacy issues that are important for CTI, since it emphasizes the common, external CTI rather than the firm internal one. Yet shared resources can be imported into a private environment to be processed with the private internal information, and contributions of information may be exported to a shared environment. This aspect is subject to other works dealing with means to guarantee privacy and trust within a single environment like [6].

Implementation is intended as a future step depending on availability of data source, with data validation depending on the data source. Particular benefits of automated processing and interpretation can be observed in environments where a lot of indicators are produced as it is in the case of Internet of Things (IoT). Big data technology is not addressed here and is a future research direction, raising the issues of data that is not known in advance but rather discovered in real time, with the performance aspect in mind.

References

1. Shackleford, D.: Who's Using Cyberthreat Intelligence and How? (2015)
2. Brown, S., Gommers, J., Serrano, O.: From cyber security information sharing to threat management. In: Proceedings of the 2nd ACM Workshop on Information Sharing and Collaborative Security, pp. 43–49 (2015)
3. Goel, S.: Cyberwarfare: connecting the dots in cyber intelligence. Commun. ACM **54**, 132 (2011)
4. Chimson, D., Ruks, M.: Threat Intelligence: Collecting, Analysing, Evaluating (2015)
5. Berners-Lee, T., Hendler, J., Lassila, O.: The semantic web. Sci. Am. **284**, 34–43 (2001)
6. Connolly, J., Davidson, M., Matt, R., Clem, S.: The Trusted Automated eXchange of Indicator Information (TAXII) (2012)
7. Porche, I.: Emerging cyber threats and implications. Rand Corp. **8**, 14 (2016)
8. Johnson, C., Badger, L., Waltermire, D., Snyder, J., Skorupka, C.: Guide to Cyber Threat Information Sharing NIST Special Publication 800-150 Guide to Cyber Threat Information Sharing (2016)
9. Lee, R.M., Bianco, D.: Generating Hypotheses for Successful Threat Hunting (2016)
10. CERT-UK, CISCP: An Introduction to threat intelligence. Searchsecurity Buyers Guide 7 (2016)
11. Barnum, S.: STIX Whitepaper
12. Paterva: Maltego. https://www.paterva.com/web7/
13. Hayes, D.R., Cappa, F.: Open-source intelligence for risk assessment. Bus. Horiz. **61**, 689–697 (2018)
14. Quick, D., Choo, K.-K.R.: Digital forensic intelligence: data subsets and open source intelligence (DFINT + OSINT): a timely and cohesive mix. Futur. Gener. Comput. Syst. **78**, 558–567 (2018)
15. Shadbolt, N., Berners-Lee, T., Hall, W.: The semantic web revisited. IEEE Intell. Syst. **21**, 96–101 (2006)
16. Kim, N., Kim, B., Lee, S., Cho, H., Park, J.: Design of a cyber threat intelligence. Int. J. Innov. Res. Technol. Sci. **5** (2017)
17. Aviad, A., Węcel, K., Abramowicz, W.: A semantic approach to modelling of cybersecurity domain. J. Inf. Warf. **15**, 91–102 (2016)
18. Hevner, A.R., March, S.T., Park, J., Ram, S.: Design science in information systems research. MIS Q. **28**, 75–105 (2004)
19. Kaplanski, P., Weichbroth, P.: Cognitum ontorion: knowledge representation and reasoning system. Stud. Comput. Intell. **658**, 27–43 (2017)
20. Clark, R.M.: Intelligence Analysis: A Target-centric Approach. CQ Press, Washington (2013)
21. Antoniou, G., Van Harmelen, F.: A Semantic Web Primer (2008)
22. Verizon: 2016 Data Breach Investigations Report (2016)
23. Michel, F., Montagnat, J., Faron-Zucker, C.: A survey of RDB to RDF translation approaches and tools. Informatique, Signaux Et Systèmes, p. 23 (2014)

24. Hert, M., Reif, G., Gall, H.: A comparison of RDB-to-RDF mapping languages. In: Proceedings of the 7th International Conference on Semantic Systems- I-Semantics, pp. 25–32 (2011)
25. The White House, Office of the Press Secretary: Cyber Threat Intelligence Integration Center
26. Kirillov, I.A., Chase, P., Beck, D., Martin, R.: Malware Attribute Enumeration and Characterization (2016)
27. MITRE: CAPEC - About CAPEC
28. OWASP: OWASP Top 10 – 2013 (2003)
29. WASC: The WASC Threat Classification v2.0

The Long Way from Science to Innovation – A Research Approach for Creating an Innovation Project Methodology

Zornitsa Yordanova[1(✉)], Nikolay Stoimenov[2], Olga Boyanova[3], and Ivan Ivanchev[4]

[1] University of National and World Economy, Sofia, Bulgaria
zornitsayordanova@unwe.bg

[2] Institute of Information and Communication Technologies, Bulgarian Academy of Sciences, Sofia, Bulgaria
nikistoimenow@gmail.com

[3] Medical University of Sofia, Sofia, Bulgaria
olga_boyanova@yahoo.com

[4] University of Architecture, Civil Engineering and Geodesy (UACEG), Sofia, Bulgaria
ivanchev_fce@uacg.bg

Abstract. The presented paper aims at proposing a research methodology for creation and approbation of a flexible methodology for development and management of innovative projects in scientific organizations (FMIPSO). For basement, the following flexible methodologies have been used: Lean startup, Agile, Scrum, Design thinking, User centricity and User innovation which all are extremely applied in ICT development. The creation of FMIPSO addresses the weak success of developed and realized innovations by scientific organizations and universities, especially relevant for multidisciplinary innovation projects that include ICT as well as other sciences. This lack of good innovation performance by scientific organizations further increases the distance and integrity of the science-business-related innovation industry. The research approach includes approbation of the FMIPSO by three interdisciplinary innovation projects from science institutions for providing proofs of its relevance and applicability.

Keywords: Project management · Innovation management · Lean startup · Agile · Management · Innovation · Innovation project

1 Introduction

The theoretical basis of project management and innovation management as part of the economics and management science has not yet found an integrative approach and there is no developed and applied methodology to help and support the management of innovative projects in scientific organizations [1]. Often, project management or innovation process management is not applied in scientific projects at any point [2]. These conclusions are very relevant especially for multidisciplinary innovation projects

W. Abramowicz and R. Corchuelo (Eds.): BIS 2019, LNBIP 353, pp. 371–380, 2019.
https://doi.org/10.1007/978-3-030-20485-3_29

which require collaborative work packages between ICT and other sciences. Management techniques, which are only implemented are pre-project planning, implementation, possible achievement of results, and reporting of results and costs. These processes often miss essential elements of project management theory such as management of: quality, scope, time, team, communication, risk, etc., and also do not include some critical phases of the innovation process so to ensure the subsequent realization of the developed innovations such as: market potential analysis, competition, selection of targeted customer segments, validation studies, etc. In addition, some innovative project management approaches support the idea of balancing traditional methods with agile, flexible approaches to managing contemporary projects, especially in science [3].

At the same time, very often the peculiarities of innovative projects and the specificities of scientific projects in scientific organizations make the use of well-known traditional methodologies for project management and innovation development inappropriate. Scientific projects usually have a duration of between two and five years, during which time almost no changes are made to the original innovation development plan. No mechanisms and tools for constant control and validation of the problem and the current level of satisfaction are used. Innovation team do not use any validation techniques so to ensure the innovation hypothesis, which the innovative project is developing goes in the right direction. As a result, scientific projects often end up with doubtful results that do not lead to real innovation.

This problem, identified in many scientific organizations and countries partly influences the overall innovation performance in innovation-related indices and ratings, assessing innovation performance, innovation progress and performance (European Innovation Scoreboard, National Statistical Institute, Innovation.bg, Eurostat, World Bank, Global Innovation Scoreboard, The World Economic Forum, the Global Competitiveness Report, Innovation and the Organization for Economic Mutual Assistance and Development - OECD).

One of the reasons for this disputable performance is not the lack of innovation capacity among scientists, insufficient funding or organizational deficiencies, but it is exactly the lack of effective methodology for development and management of innovative projects by scientific organizations.

The aim of the paper is to be proposed a research approach for creation of a flexible methodology for development and management of innovative projects in scientific organizations. The cross points between the proposed flexible methodology and ICT sector are: first, the special focus of the developing methodology for multidisciplinary projects which require collaboration between ICT and other sciences and second, the utilization of Lean startup, Agile, Scrum, Design thinking, User centricity and User innovation which all usually are extremely used in ICT. The targeted contribution of the proposed methodology in this study is boosting more multidisciplinary projects, including ICT and other sciences.

2 State of the Art

Over the last 40 years it has been established that project organization is an efficient tool for managing complex, complicated and novel activities within organizations. Project organization and project management handle many activities better than any other organizational structure [4]. Projects are the preferred management instrument especially for the implementation of new activities [5, 6]. During the last decades, projects have become a parallel organization structure within almost every organization that has to deal with new activities [7]. Innovations are some of the possible new activities, which might take place in organizations. Innovations are exactly such activities which project organization is extremely appropriate [8]. These kinds of projects are very often innovation projects [9]. All these do support the statement that it is essential first innovation projects to be defined and categorized as innovation projects. After that the second stage is to select a method based on which they are going to be managed since they are not ordinary projects, but innovative ones [2].

An assumption in the study is that innovation projects could be carried out in any industry, for different purposes, targeting diverse customers or needs, but, still, project management of innovation projects faces common issues. Identifying such projects in the first stages of the project work and applying some tools for reducing these projects' unpredictability and uncertainty might improve their performance. The study aims at analysing the specifics of innovation projects from a project management point of view and they should be specifically adapted to the implementation and management of innovative projects in scientific organizations.

The significance of the research comes from the large number of all sorts of project failures, which are especially frequent within innovation projects, and comes also from the growing significance of innovations for the world economy in general [10].

The main goal is to increase the quality and sustainability of the developed innovations. The research has been motivated by the exclusive use of project organization and project management for implementing innovations and the lack of previous studies analysing the specifics of the innovation projects and their impact on the project management. The statistical data also shows extreme increase of project failures – 75%, a tendency which is even more distinct in innovation projects [11, 12]. All these circumstances and their increasing temps, motivate the research in order to figure out what are the specificities of innovative projects in terms of project management and, in particular, innovative projects implemented in scientific organizations.

3 Objectives and Hypotheses

A major hypothesis in the study is that innovative projects implemented in scientific organizations require a specific methodology for their successful management and the creation of sustainable innovations. Flexible methodologies such as Lean startup, Agile, Scrum, Design thinking, User centricity and User innovation are tools that have been successfully deployed in some industries, specifically to develop and manage innovation, and this leads to the assumption that they would be a successful tool for creating and managing innovation in research organizations.

The goal of FMIPSO is to create a unified and common approach to the management of innovative projects in scientific organizations. For the purpose of the testing and creating the final version of the flexible methodology, divergent scientific innovative areas have been selected as a principle of the choice were: high multidisciplinary; scientists interested in developing innovation. These areas are as follows: (1) Innovative Method for 3D Presentation of Plane Culturally-Historical Sites by Tactile Plates for the Disadvantaged (Low Visibility or Visually Impaired People); (2) Detection of Mutations in the Epidermal Growth Factor Receptor (EGFR) Gene in Invasive Urinary Bladder Tumours and (3) Diagnostics, condition assessment and analysis of reinforced concrete elements.

3.1 Innovative Method for 3D Presentation of Plane Culturally-Historical Sites by Tactile Plates for the Disadvantaged (Low Visibility or Visually Impaired People)

The main objective of this innovation is to increase the access to paintings and other plane objects of cultural and historical heritage for disadvantaged people (low-sighted or visually impaired) with the usage of digitization, 3D scanning, 3D modelling and 3D printing. An innovative approach will be used as well as more accessible materials for building paintings and other plane objects which aims a cheaper and more affordable product. Reducing the value of the product will allow galleries and museums to engage in a lifelong mission for increasing the quality of life for disadvantaged people, and that will lead to the development of the culture of society. Also, the availability of the product will make it suitable for application in the learning process for blind or low-sighted students, which is specifically set out in the National Science Program of Bulgaria, recently adopted by the Council of Ministers (ICT) - quote: "9.2.3. Modern tools for digitization in education and working with young talents".

3.2 Detection of Mutations in the Epidermal Growth Factor Receptor (EGFR) Gene in Invasive Urinary Bladder Tumours

Malignant neoplasms are one of the leading causes of morbidity and mortality worldwide. Among them, bladder cancer occupies ninth place, affecting predominantly men over 60 years of the Caucasian race and industrially developed countries. In Bulgaria, bladder cancer is ranked 18th and it is in the middle, both in terms of prevalence and frequency among the European countries. Bladder cancer is a multi-factorial disease. It represents malignant degeneration of tissues that make up the bladder, wherein the cells begin to divide uncontrollably and lead to the occurrence of cell mass forming a tumour.

One of the oncogenes associated with tumour progression from non-invasive to muscle-invasive carcinoma is the gene coding for epidermal growth factor receptor (EGFR). It has been found that thirty to fifty percent of invasive uroepithelial tumors have elevated activity (overexpression) of the EGFR, which is associated with a poor prognosis for the patient [13].

Mutations in the EGFR genes lead to overexpression of the epidermal growth factor receptors and are responsible for the development of a numerous neoplasms, such as breasts, colon, pancreas, lungs, bladder, kidneys etc. Against these receptors are developed special drugs – small molecules tyrosine kinase inhibitors gefitinib and erlotinib [14–16].

According to our results on the topic of bladder tumor, the expression of genes encoding growth factors and hormone receptors in the muscle-invasive tumors is between 7 and 60 times comparing to the healthy control. The use of EGFR-tirosin kinase inhibitors prior to radical cystectomy showed promising results. Thus, EGFR-TK inhibitors may be useful in patients with-out preexisting chemotherapy, with increased expression of EGFR or ERBB2 [17].

The main objective is to identify a spectrum of mutations in the invasive bladder tumour samples and to select those potentially suitable for treatment with novel targeting drugs - Epidermal growth factor receptor (EGFR) - inhibitors. The implementation of such approach in bladder tumors is an extremely cumbersome procedure in terms of existing academic and scientific standards. There is a need for innovative approaches that allow for the implementation of scientific developments in practice.

3.3 Diagnostics, Condition Assessment and Analysis of Reinforced Concrete Elements

The main objective is focused on the study of the current state of reinforced concrete elements. Existing buildings and facilities with reinforced concrete structure are designed for exactly defined period of exploitation. Sometimes after their design and construction, they are subjected to heavy conditions and aggressive impacts. This requires conditional assessment of their actual technical state and, if necessary, prescriptions for their repair or strengthening, which will lead to their greater security and sustainability.

The strength and strain characteristics of the concrete, corrosion of reinforcing steel in reinforced concrete elements will be determined. Homogeneity, internal defects and cracks of concrete in reinforced concrete elements will be assessed. Compared to previous studies, new methods for determining the characteristics will be used by taking test samples – cored specimens of concrete and test pieces of reinforcing steel from the reinforced concrete beams. The measurements will be carried out using different methodologies. The methods used will be compared with an emphasis on their advantages in tests on existing reinforced concrete structures.

4 Research Methodology

Creating a research methodology is the first step of building such a flexible methodology for development and management of innovative projects in scientific organizations. The methods used to achieve the final result of creating FMIPSO are described above, presented in separate sub-sections, given the multidisciplinary of the presented work and the differences in the methods and equipment to be used.

For analysis of flexible methodologies and their transformation into FMIPSO we use implemented:

- System analysis;
- Business Process Reengineering;
- Brainstorming with ICT professionals who apply flexible methodologies;
- Diagnostic analysis;
- Expert evaluation;
- Assessing identified innovations by assessing their core characteristics;
- Profile analysis of potential users of identified innovations;
- Analysis of techniques and tools;
- Retrospective analysis.

For achieving the goal "*Detection of mutations in the EGFR gene in invasive urinary bladder tumors*" we use:

- Analysis of the clinical and pathological information of the DNA samples from bladder tumors, biobanked (stored) in the Department of medical genetics in the Medical university of Sofia and selection of the samples for study in the project: Reviewing existing clinical patients' information, analyzing the results of the completed questionnaire and informed consent.
- Measurement of the concentration and purity of selected candidate samples from uroepithelial tumors and final selection of DNA samples for genetic analysis: - Spectrophotometric determination of DNA concentration with NanoDrop® ND-2000c (ThermoScientific) and creation of stocks with appropriate concentration.
- Perform quantitative real-time polymerase chain reaction (RT-PCR) to detect mutations in the EGFR gene-20 tumors: The RT-PCR method is the most powerful means of quantitative nucleic acid analysis. In this reaction, the amount of product obtained is recorded indirectly during the course of the reaction itself, following the amount of fluorescence-labeled probe. This counts the number of cycles (Ct) needed to obtain a certain number of DNA molecules. Considering that during the PCR reaction the amount of DNA molecules doubled in each amplification cycle, it is possible to calculate the original number of DNA molecules that contain the target sequence. This technology is characterized by high sensitivity, high informative value of the result, reproducibility, speed and reliability.
- Establishment of an exemplary patient database based on new European standards, including the creation of pseudonymization of the patient through a unified code according to the GDPR.

For achieving the goal "*Innovative method for 3D presentation of plane culturally-historical objects by tactile plates for the disadvantaged (low-sighted or visually impaired)*" the following methodology will be applied in order to be followed in 3D modeling and 3D printing of paintings and 2D objects of cultural and historical heritage (see Fig. 1).

- Digitalization of culturally-historical objects by using a 3D scanner;
- Digitalization of culturally-historical objects by using software;
- Producing the digitalized objects by using 3D printers;

- Producing tactile tiles with Braille annotation and symbols for better understanding for visually impaired people;
- Using the 3D printed digitalized objects to represent culturally-historical objects to visually impaired people

Fig. 1. Using a 3D printer for producing cultural-historical objects (figure and tactile tile) for blind or visually impaired people.

For achieving the goal *"Diagnostics, condition assessment and analysis of reinforced concrete elements"* the chosen methodology for the experimental study of characteristics of reinforced concrete structures will provide appropriate conducting of research according to the set purpose, tasks and specific features. Experimental investigations will be the result of an analysis, summary and systematizing of theories, regulations and experiments from previous research. Complex test methods will be used: visual, destructive and non-destructive.

Criteria for evaluating research are scientific significance, relevance of the results, novelty in the used methods, practical application, methodology of research, modern technique, and comparability of results.

The research approach for creating a flexible methodology for development and management of innovative projects in scientific organizations is presented in Fig. 2.

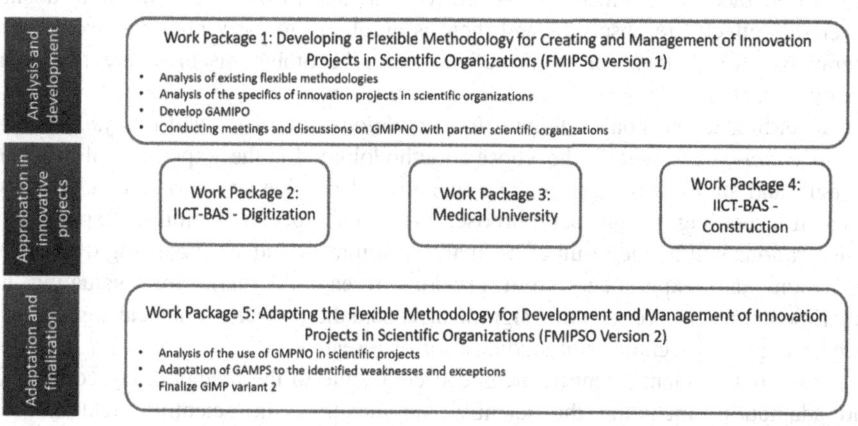

Fig. 2. Research approach for creating a flexible methodology for development and management of innovative projects in scientific organizations

5 Conclusions and Future Developments

The public benefit of creation of such a flexible methodology for the development and management of innovative projects in scientific organizations is essential and of interest to the overall innovation activity of the scientific organizations and the scientific community. It will also enable an instrument to use already successful and proven practices (mostly in the ICT sector at the moment) and adapted to their use in scientific projects, in different scientific fields, for the development and management of innovation. The targeted contribution of the proposed methodology (not yet achieved) is boosting more multidisciplinary projects, including ICT and other sciences. The authors keep the methodology open for adjustments after applying it in different and other projects, specially included ICT project packages.

Approbation of the methodology will take place during the development of the flexible methodology, with the mentioned above independent research areas, which aims at developing innovations in different areas of science. They will use strictly the created flexible methodology, and will appraise it during the innovation and research process.

According to the goal "*Innovative method for 3D presentation of plane culturally-historical objects by tactile plates for the disadvantaged (low-sighted or visually impaired)*", a 3D material which is more wear-resistant, according to the previous used type and works will be used for low-sighted users [18]. This type of material will ensure a longer life of the product. The material will be tested for wear resistance with tribotester [19].

According to the goal "*Detection of mutations in the EGFR gene in invasive urinary bladder tumors*", the main activities for the future are to identify a spectrum of mutations in tumour samples from Bulgarian patients with invasive bladder cancer and to select those potentially suitable for treatment with modern target drugs - EGFR - inhibitors actively used in the therapy of other oncological diseases. This will optimize therapy, reduce hospital stay, and increase the standard of living and quality of life of patients with invasive bladder tumours and the social status of patients. When selecting gene-based therapy, the medications are reduced due to their more accurate dosing, fewer side effects are observed, and there is an ability to carry out organ conserving operations. All of these prevents patients from becoming disabled due to urinary bladder excision.

According to the goal *"Diagnostics, condition assessment and analysis of reinforced concrete elements"*, the chosen methodology for the experimental study of strength and strain characteristics of concrete will provide appropriate conducting of research according to the set purpose, tasks and specific features. Experimental investigations will be the result of an analysis, summary and systematizing of theories, regulations and experiments from previous research. Criteria for evaluating the experimental results are scientific significance, actuality, novelty, practical significance, methodology and technique of study, proof of research.

These independent scientific areas can contribute to the methodology completion and adaptation (including the scientific organization, the scientific field and the specificity of the specific innovation being developed). This is why, in the current paper, several institutions and scientists from completely different scientific areas have joined efforts so to develop and implement a flexible methodology for the development and management of innovative projects in scientific organizations.

Acknowledgments. The paper is supported by the BG NSF Grant No KP-06-OPR01/3-2018, DM 15/4 -2017 and DM 13/4 - 2017.

References

1. Riol, H., Thuillier, D.: Project management for academic research projects: balancing structure and flexibility. Int. J. Project Organ. Manage. **7**(3), 251–265 (2015)
2. Yordanova, Z.: Innovation project tool for outlining innovation projects. Int. J. Bus. Innov. Res. **16**(1), 63–78 (2018)
3. Boehm, B., Turner, R.: Using risk to balance agile and plan-driven methods. IEEE Comput. **36**(6), 57–66 (2003)
4. Avots, I.: Why does project management fail? Calif. Manage. Rev. **12**, 77–82 (1969)
5. Filippov, S., Mooi, H.: Innovation project management: a research agenda. J. Innov. Sustain. **1**, 1–15 (2010). ISSN 2179-3565
6. Ghaben, R., Jaaron, A.: Assessing innovation practices in project management: the case of palestinian construction projects. Int. J. Innov. Sci. Res. **17**(2), 451–465 (2015). ISSN 2351-8014
7. Antoniolli, P.D., Lima, C., Argoud, T., Batista de Camargo Jr, J.: Lean office applied to ict project management: autoparts company case study. IPASJ Int. J. Manage. (IIJM) **3**(6), 5–20 (2015). ISSN 2321-645X
8. Van Lancker, J., et al.: The organizational innovation system: a systemic framework for radical innovation at the organizational level. Technovation **52**, 40–50 (2015). https://doi.org/10.1016/j.technovation.2015.11.008i
9. Maranhão, R., Marinho, M., de Moura, H.: Narrowing impact factors for innovative software project management. Procedia Comput. Sci. **64**, 957–963 (2015). https://doi.org/10.1016/j.procs.2015.08.613
10. Andreassen, H.K., Kjekshus, L., Tjora, A.: Survival of the project: a case study of ICT innovation in health care. Soc. Sci. Med. **132**, 62–69 (2015). https://doi.org/10.1016/j.socscimed.2015.03.016
11. Bonnie, E.: Complete Collection of Project Management Statistics (2015). https://www.wrike.com/blog/complete-collection-project-management-statistics-2015/. Accessed 01 Mar 2019
12. Threlfall, D.: Seven Shocking Project Management Statistics and Lessons We Should Learn (2014). https://www.teamgantt.com/blog/seven-shocking-project-management-statistics-and-lessons-we-should-learn/. Accessed 01 Mar 2019
13. Neal, D.E., et al.: The epidermal growth factor receptor and the prognosis of bladder cancer. Cancer **65**, 1619–1625 (1990)
14. Pirker, R.: What is the best strategy for targeting EGF receptors in non-small-cell lung cancer? Future Oncol. **11**, 153–167 (2015)
15. Dhillon, S.: Gefitinib: a review of its use in adults with advanced non-small cell lung cancer. Target. Oncol. **10**(1), 153–170 (2015)
16. Rolfo, C., et al.: Improvement in lung cancer outcomes with targeted therapies: an update for family physicians. J. Am. Board Fam. Med. JABFM **28**, 124–133 (2015)

17. Moo-so, B.A., Vinall, R.L., Mudryj, M., Yap, S.A., deVere White, R.W., Ghosh, P.M.: The role of EGFR family inhibitors in muscle invasive bladder cancer: a review of clinical data and molecular evidence. J. Urology **193**(1), 19–29 (2015)
18. Cantoni, V., Lombardi, L., Setti, A., Gyoshev, S., Karastoyanov, D., Stoimenov, N.: Art masterpieces accessibility for blind and visually impaired people. In: Miesenberger, K., Kouroupetroglou, G. (eds.) ICCHP 2018. LNCS, vol. 10897, pp. 267–274. Springer, Cham (2018). https://doi.org/10.1007/978-3-319-94274-2_37
19. Kandeva, M., Grozdanova, T., Karastoyanov, D., Assenova, E.: Wear resistance of WC/Co HVOF-coatings and galvanic Cr coatings modified by diamond nanoparticles. In: 13th International Conference on Tribology, ROTRIB 2016 IOP Conference Series: Materials Science and Engineering, vol. 174, p. 012060 (2017). https://doi.org/10.1088/1757-899x/174/1/012060

A COSMIC-Based Approach for Verifying the Conformity of BPMN, BPEL and Component Models

Wiem Khlif[1(✉)], Hanêne Ben-Abdallah[1,2], Asma Sellami[1], and Mariem Haoues[1]

[1] Mir@cl Laboratory, University of Sfax, Sfax, Tunisia
wiem.khlif@gmail.com,
{asma.sellami,mariem.haoues}@isims.usf.tn
[2] Higher Colleges of Technology, Dubai, United Arab Emirates
hbenabdallah@hct.ac.ae

Abstract. Besides its application in the software development lifecycle, COSMIC Functional Size Measurement (FSM) is investigated as a means to measure the size of business processes (BP). This paper proposes a *comprehensive* COSMIC FSM-based framework to verify the conformity of the business process design and run-time models with their aligned information system (IS). It relies on the standard notations BPMN and BPEL to describe the business process and run-time models, respectively, and the component diagram to describe the IS. The paper defines formulas to apply COSMIC on these models and heuristics to verify their conformity. It illustrates the approach through a case study.

Keywords: Functional Size Measurement (FSM) · COSMIC · Component diagram · BPMN · BPEL model (BPEL)

1 Introduction

As a means to efficient governance, enterprises often resort to establishing an Enterprise Architecture framework. This represents the enterprise's blueprint that anchors its strategic goals onto its organizational and operational resources. Among the pillars of any enterprise architecture are the models of the business process and its supporting information system. The Business Process Model (BPM) describes the core business logic in terms of strategies, tasks/activities and policies; it is often described through the ISO standard BPMN [1] at the conceptual/design level, and through BPEL [2] at the runtime level. The Information System (IS) model describes the data manipulated by the BP activities, and it is often represented through the de facto standard UML. Evidently, the alignment between these models (conceptual and runtime BPM and IS model) is key to a coherent governance of the enterprise [3].

In this paper, we use the UML component diagram to align the functionality of the BPM model to its underlying IS model. In addition, we use the ISO standard COSMIC Functional Size Measurement (FSM) method [4] to define a framework where the conformity among these models can be verified. The proposed framework incorporates

W. Abramowicz and R. Corchuelo (Eds.): BIS 2019, LNBIP 353, pp. 381–396, 2019.
https://doi.org/10.1007/978-3-030-20485-3_30

our previously defined measurement methods of the functional sizes of BPMN [5] and the UML activity and component diagrams [6]. In this paper, we extend this framework with a set of heuristics that provide for the verification of the conformity among the BPMN, BPEL and UML component models in terms of the functional size. By model conformity, we mean matching the models' concepts in terms of their functional aspect. In other words, a model *M1* is said to be conforming with a model *M2*, if all the concepts in *M1* can be mapped to concepts in *M2* so that the behavior of *M1* is equivalent to the behavior of *M2*. As a second contribution of this paper, we discuss how the heuristics can be used to estimate a bound on the functional size of a model to be developed from an existing model. Such an estimate can be used for instance in an effort/cost evaluation process [6].

Note that different proposals used FSM methods for BPMN while focusing on either the design level (cf., [6, 7], etc.) and/or requirements specification level (cf., [5], etc.). Except for [5], these proposals treated the BPMN model in an isolated way; that is, they analyzed/measured the BP model design/specification without dealing with neither the runtime business process model nor the underlying information system. In particular, to the best of our knowledge, the functional size of the BPEL model (as a runtime model of the business process) was not examined in spite of its advantage in component reuse especially for the development of complex business processes [2]. The FS measurement of BPEL models is one of the main contributions of this paper; the second contribution is its application to verify the conformity of the three modelling levels: BPMN design, BPEL runtime, and IS component diagram. (We refer the reader to [8, 9] where the conformity of the IS diagrams is analyzed using the COSMIC FSM method.)

The remainder of this paper is organized as follows: Sect. 2 overviews the COSMIC method and existing proposals for COSMIC FSM of business process models. Sections 3 and 4 present, respectively, the proposed measurement procedures and heuristics for measuring the functional size of a BPMN model, a UML component diagram, and a BPEL model. Section 5 illustrates the application of these measurement procedures through the "Patient Admission and Registration for Major Ambulatory Surgery (MAS)" business process example [10]. In addition, this section highlights several validity threats. Finally, Sect. 6 summarizes the presented work and outlines some further works.

2 Related Work

2.1 Overview of COSMIC FSM Method

Functional size measurement (FSM) using COSMIC has the merit of being able to quantify software from a user's point of view (based on the Functional User Requirements). In addition, COSMIC is designed to be applicable to any type of software. These advantages motivated several researchers to investigate the use of COSMIC to determine the functional size of business process models [6, 7, 11].

COSMIC covers four types of data movements (Entry, eXit, Read, and Write). The exchange of data across the boundary between users and software components causes an Entry data movement type (E: from a functional user to the functional process), or an eXit data movement type (X: from a functional process to the functional user). On the other hand, the exchange of data between storage hardware and software component causes a Read data movement type (R: from a persistent storage to the functional process), or a Write data movement type (W: from a functional process to the persistent storage). In the COSMIC measurement phase, every data movement is assigned to 1 CFP (Cosmic Function Point). The software functional size is computed by adding all data movements identified for every functional process [4].

2.2 COSMIC FSM for Sizing BPMN

COSMIC method allows the business designer to measure the functional size of a BPMN model at a high level of granularity (i.e., at a level of functional process where data movements are known) [4]. The functionality of any data manipulation is assumed to be accounted for by the data movement with which it is associated [4]. As mentioned in the introduction, this paper complements our COSMIC FSM of BPMN models [5] to provide for conformity verification among the execution model of the BPMN design and the UML component diagram as a link to the IS model. Most of the researches applied the COSMIC-FSM method to the BPMN model in the design phase. For instance, Monsalve explored the use of COSMIC to measure the functional size of BPMN model [7]. To identify the data movements in a BPMN model, the author mapped the COSMIC 2.0 concepts to BPMN elements as illustrated in Table 1.

Table 1. Mapping of COSMIC concepts with those of BPMN.

COSMIC	BPMN
Functional user	Pool (participant): only those that interact directly with the main Pool
Boundary	Frontier between the Pool/lane representing the software to be measured and the participant's pool
Functional process	Process: A process node presented in the first level
Triggering event	Start or intermediate event
Data group	Name of a message or sequence flow
	Name of a resource or data object
Entry	An incoming message or sequence flow
Exit	An outcoming message or sequence flow
Read	An upstream association with data object, or a data store
Write	A downstream association with data object

Based on the established mapping in [7], we proposed in our previous work [5], COSMIC FSM measurement of a BPMN model that operates iteratively by decomposing the BPMN model into functional level fragments and dynamic level activities.

At the functional level (the first level), the functional size of a BP model, FS (M), is equal to the sum of the sizes of its fragments:

$$FS(M) = \sum_{i=1}^{n} FS(F_i) \tag{1}$$

where:

- n is the total number of fragments in the BP model M (first level); and
- *FS(Fi):* the functional size of a fragment F_i (second, dynamic level).

At the dynamic level, a fragment Fi consists of a set of business activities BAij. Thus, the functional size of a fragment Fi is given by:

$$FS(F_i) = FScond(\Pr econd\ F_i) + \sum_{j=1}^{m} FS(BA_{ij}) \tag{2}$$

where:

- *FS(Fi):* the functional size of the fragment F_i ($1 \leq i \leq n$);
- m is the total number of BA_{ij} detailing the fragment F_i (2nd level: dynamic level);
- *FS(BAij):* the FS of the business activity BA_{ij} (2nd level: dynamic level); and
- *FScond(Precond Fi):* functional size of the precondition F_i(1 CFP if it exists).

To measure the FS(BAij), we use formula (3):

$$FS(BA_{ij}) = FScond(\Pr econd\ BA_{ij}) + \sum_{t=1}^{k} FS(SBA_{ijk}) \tag{3}$$

where:

- *FScond(PrecondBAij):* the FS of the pre-condition of BA_{ij}.
- *FS(SBAij):* the functional size of the sub business activity SBA_{ij} (dynamic level).
- *k:* is the total number of SBA_{ij} detailing the business activity BA_{ij}.

$$FScond(\Pr econd\ BA_{ij}) = \begin{cases} 1\ CFP\ \ if\ BA_{ij}\ \ has\ a\ pre-condition \\ 0\ otherwise \end{cases} \tag{4}$$

To measure the *FS(SBA_{ijk})*, we use formula (5):

$$FS(SBA_{ij}) = FScond(\Pr econd\ SBA_{ij}) + \sum_{l=1}^{p} FS(T_{ijk}) \tag{5}$$

where:

- *FS(SBA_{ijk}):* the FS of the sub business activity ($1 \leq ij \leq p$).
- *p:* the number of tasks detailing the sub business activities SBA_{ij} (dynamic level).
- *FS(T_{ijk}):* the FS of a Task T_{ijk} (dynamic level).
- *FScond(PcondSBA$_{ij}$):* the FS of the pre-condition of SBA_{ij} (1CFP if it exists).

To measure the $FS(T_{ijk})$, we use formula (6):

$$FS(T_{ijk}) = FScond(\Pr econd\ T_{ijk}) + FSDatagp(DatagpT_{ijk}) \tag{6}$$

where:

- $FScond(PcondT_{ijk})$: the FS of the pre-condition of T_{ijk} (1 CFP if it exists).
- $FSdatagp(datagpT_{ijk})$ = 1 CFP if T_{ijk} includes input or output data group.

To measure the functional size of a guard condition, we use the following formula:

$$FScond(\Pr econd\ T_{ijk}) = \begin{cases} 1\ CFP\ \ if\ T_{ijk}\ \ has\ a\ condition \\ 0\ otherwise \end{cases} \tag{7}$$

The functional size of an error (exception) is always equal to 1 CFP (COSMIC, 2017). It is measured according to the following formula:

$$FS(E) = \begin{cases} 1\ CFP\ if\ there\ is\ an\ error \\ 0\ otherwise \end{cases} \tag{8}$$

2.3 COSMIC FSM for Sizing the UML Component Diagram

To measure the UML component diagrams, our framework relies on the COSMIC FSM proposed in [8, 9]. This method overcomes the limits of two previous methods: the one proposed in [12] which focused only the syntactic concepts of the UML component diagram; and the one in [13] which defines data movements independently of the software boundary, which may lead to incorrect results.

Table 2 summarizes the mapping between the COSMIC concepts and those of the component model, as established in [9] and which we use in our framework.

The measurement method of the component diagram (CD) in [8, 9] supposes that data movements are represented by interface's operations across the boundary, and operations in a system component. As such, they define the FS of the CD as follows:

$$FS(CD) = \sum\nolimits_{i=1}^{n} FS(S_i) + \sum\nolimits_{i=1}^{m} FS(I_J) \tag{9}$$

where:

- *FS(CD):* functional size of the component diagram.
- $FS(S_i)$: functional size of operations in the system components.
- *n:* number of the system components.
- *FS(Ij):* functional size of required and provided interfaces.
- *m:* number of the interfaces required and provided in *CD*.

Table 2. Mapping of COSMIC on UML-CD.

COSMIC	Component diagram
Functional user	<Component> External entity directly connected with the system components
Boundary	Frontier between two components (external and system components)
Functional process	Set of <Operation> in one or more interfaces carrying out a process
Triggering event	<Operation> in a system interface invoked directly by an external entity
Persistent storage	<Component> Classes: physical components
Transient data group	<Parameter_int> <Parameter_out> Data across the system boundary, interface's operations or parameter's operations
Entry	<Operation> in a <required interface> directly connected to the system
Exit	<Operation> in a <provided interface> directly connected to the system
Read	<Operation> Get type operation in a system component
Write	<Operation> Set type operation in a system component

The functional size of operations in a system component is given by:

$$FS(S_i) = \sum\nolimits_{J=1}^{y} FS(Op_{ij}) \tag{10}$$

where:

- *FSop(*Op_{ij}*):* functional size of the operation Op_{ij}. (1CFP)
- *y:* number of operations in a component system (i = 1, … n)

The functional size of operations in a system component is given by:

$$FS(I_i) = \sum\nolimits_{k=1}^{z} FSMop(Op_{jk}) \tag{11}$$

where:

- *FSM(Ij):* functional size of required and provided interfaces.
- *z:* number of operations in the interface *Ij*. (j = 1, … m)
- *FSop(*Op_{jk}*):* functional size of the operation Op_{jk}.

Finally, we note that sizing the BPEL model (as a runtime model of the business process) in terms of CFP units was not investigated. However, the application of COSMIC to BPEL has an advantage in component reuse especially for the development of complex business processes. In fact, it is crucial to verify the conformity of the three modelling levels: BPMN design, BPEL runtime, and component diagram. Towards this end, we next present a set of heuristics to map the FSM of a BPMN model to the FSM of a component diagram.

3 BPMN to Component Diagram FSM Mapping

To map the Functional Size (FS) of a BPMN model to the FS of its corresponding component diagram CD, we propose the following six heuristics, which also ensure the conformity between these two models in terms of CFP units:

R1_comp: Each fragment in the BPMN model is transformed into one or more components in the component diagram. Each component expresses a single entity.

Recall that, at a high level of abstraction, a BPMN model is composed of at least one external participant and a functional process, an initial event, an end event, and a set of activities. In the second level of abstraction, a BPMN model represents a functional process. Based on COSMIC concepts, the minimal size of a functional process is equal to two data movements (Entry and Exit or Write) [4]. Therefore, the FS of a BPMN model BPM is at least equal to 2 CFP, *i.e.* $FS(BPM) \geq 2\ CFP$.

On the other hand, the FSM of a component diagram is always less than the FSM of a BPMN model. Hence, the FSM of a component diagram is at least equal to 2 CFP, *i.e.* $FS(CD) \geq 2\ CFP$.

The maximum size of CD depends on the BPMN model size. We explain this by the fact that the CD represents a static view. It does not represent all the details as well as BPM model. Thus, the conformity check between BPM and CD is expressed by Eq. (12):

$$2 \leq FSM(CD) \leq FSM(BPM) \tag{12}$$

Formula 12 compares the FS of the whole BPMN model to the FS of the component diagram. Because the total FS of a BPMN model is given by the total size of its functional processes (fragments), we can therefore apply this formula to the functional processes, one at a time. Consequently, it is possible to compare the FS of a functional process FP in the BPMN model with the functional size of the functional process in the CD. This way, we can identify and localize the error source and determine the cause of error. More specifically, if the FS of a functional process FP_i in the BPMN model is greater than the corresponding one in the CD; we can infer that FP_i does not contain errors since it is possible to have more detail in the BPMN model. In contrast, if the FS of FP_i is less than the FS of its corresponding process in the CD, then there is an error in FP_i.

Furthermore, it is important to note that even if the total size of CD is between 2 and FS(BPM), it is possible to detect and localize errors by examining the FS of the functional processes one by one. When one functional process FS violates the heuristic, we deduce that the BPMN model and its CD are not conforming. However, in the reverse case (when all functional processes' FS satisfy the heuristic), we have no guarantee of the absence of errors in the functional processes of the whole business process.

Besides this high-level FSM boundary confrontation, we propose the following heuristics to ensure the COSMIC FSM conformity between a **BPMN model** and a **CD**:

R2_comp: A public task in the BPMN model is transformed into an operation in interface in the component. We define a public task as a task that sends or receives a message.

R3_comp: A data group from a message between two tasks t_1 and t_2 (sender and receiver) in different pools in the BPMN model is respectively transformed into the output/input parameter's operations in the component.
R4_comp: A data object in the BPMN model is transformed to class's components in the component diagram.
R5_comp: The pre and post-conditions of a fragment/activities/task are transformed to pre and post conditions of a component/operation in a component.

The above heuristics are based on establishing the correspondence between the BPMN elements and those of the component diagram in order to determine the mapping of functional concepts in COSMIC with BPMN and CD models. This mapping can be applied to each functional process in the BPMN and CD models. Therefore, our proposed heuristics allow to calculate the Functional Size (FS) of each FP in a BPMN model and the FS of their corresponding FP in the component diagram CD; and consequently permit to detect and localize errors among these models.

4 Sizing the BPEL Model

4.1 COSMIC to BPEL Concept Mapping

To facilitate the functional size measurement of a BPEL model, we use the COSMIC concepts mapping to those of the BPEL notation shown in Table 3.

Table 3. Mapping of COSMIC on BPEL.

COSMIC	BPEL
Functional user	<partnerLinks>: Participant that interacts with the process
Boundary	Frontier between the <process> and the participants <partnerLinks>
Triggering event	The first <invoke> message without <partnerLinks>
Functional process	<process>: a process node presented in the first level
Data group	<variable>: name of a resource or data object
	operation = "[e-operation]" in <invoke>: name of a message sequence flow: Information provided as part of a flow
Entry	<invoke>: receive message from <partnerLinks>
eXit	<receive>: send message to <partnerLinks>
Read	<toPart part = "[dataInput-name]" fromVariable = "[DataObject name]"/>: read from a data object
Write	<formPart part = "[dataOutput-name]" fromVariable = "[DataObjectname]"/>: write to a data object

4.2 Generating Measurement Formulas

Similar to the top-down decomposition of a BPMN model, we propose to decompose the BPEL model into a set of blocks. Each fragment in the BPMN corresponds to a block in the BPEL. A block hierarchy for a process model is a set of blocks of the process model in which each pair of blocks is either nested or disjoint and which contains the maximal block (i.e., the whole process model). A block A that is nested in another block B is also called a sub-block of B. The sub-block corresponds to a business activity in the BPMN. Each block has an interface representing the public tasks. A public task is defined as a task sending or receiving a message flow.

The functional size of a BPEL model composed of n blocks is given by:

$$FS(BPEL) = \sum\nolimits_{i=1}^{n} FS(B_i) \tag{13}$$

where:

- *FS(BPEL):* functional size of the BPEL model.
- $FS(B_i)$: functional size of block B_i.
- *n:* the total number of blocs in a *BPEL* model

The functional size of a functional process block B_i in *BPEL* is:

$$FS(Bi) = FS(event) + \sum_{j=1}^{m} FS(PTj) + \sum_{y=1}^{z} FS(Oy) \tag{14}$$

where:

- $FS(B_i)$: functional size of block.
- *m:* number of public tasks in a block B_i. (j = 1, … m)
- *FS(event)* is the FS of the event triggering the functional process in block B_i. It is represented by <invoke>, <receive> or <wai>. Its FS is equal to 1 CFP.
- $FS(PT_j)$ is the functional size of the public task PT_j. A public task can be <invoke>, <receive>. The functional size of <invoke> and <receive> task is equal to 1 CFP.
- *FS(Oy):* is the functional size of associations from/to a data object. The functional size of <toPart … from> or <fromPart … to> is equal to 1 CFP.
- *z:* number of associations in a block B_i. (y = 1, … z).

4.3 BPMN and BPEL FS Conformity

We assume that the BPMN model is used at the design phase while the BPEL is used at the implementation phase. The software functional size will appear to grow as we move to a more detailed level of granularity [4]. Consequently, the bounding interval for the functional size of BPMN and BPEL must respect Eq. (15):

$$2 \leq \text{FS (BPMN)} \leq \text{FS (BPEL)} \tag{15}$$

Recall that a BPMN model is composed of at least one actor and a functional system, an initial event and a set of actions (R1, R2, R3, … R7). Equation (15) gives a verification means of the conformity between BPMN and its corresponding BPEL in terms of CFP units. To further refine the verification, we propose the following six heuristics:

ConsR1: Each fragment in BPMN is associated to a block in BPEL.
ConsR2: Each business activity in BPMN is represented by a sub-block in BPEL.
ConsR3: Each incoming message or sequence flow in BPMN is represented by <invoke> public task in BPEL.
ConsR4: Each out coming message or sequence flow in BPMN is represented by <receive> public task in BPEL.
ConsR5: Each upstream association with data object or a data store is represented by <toPart, … , from>.
ConsR6: Each downstream association with data object or a data store is represented by <formPart, … , to>.
ConsR7: Each Timer Intermediate Event is represented by <wait>.

As already presented in Sect. 3, these heuristics are based on a mapping between BPMN elements and those of BPEL model to match functional concepts in COSMIC with BPMN and BPEL models. The established mapping can be applied to each functional process in BPMN and BPEL models and therefore, permit to calculate the Functional Size (FS) of each FP in a BPMN model and their corresponding in the BPEL model. As a result, these heuristics allow to detect and localize errors among these models.

5 Illustrative Example

In this section, we first illustrate the application of the proposed conformity verification. Second, we compare our measurement results with the conformity results obtained with other methods. Finally, we highlight several threats to the validity of our results.

5.1 Measurement Application

To illustrate the application of the proposed measurement formula, we select the "Patient Admission and Registration for Major Ambulatory Surgery (MAS)" BP from the Ciudad Real General Hospital project [10]. Figure 1. describes the selected process. Table 4 presents in detail the measurement results of the BPMN model for the functional process. Based on the presented heuristics in Sect. 3, we elaborate the corresponding CD as illustrated in Fig. 2. This CD includes seven components and five interfaces. The measurement results of its functional size are given in Table 5.

According to Eq. (12), it is ensured that the BPMN model design is conformed to the CD design. In addition, assuming that the consistency heuristics are satisfied, the FSM difference between the CD (10 CFP) and the BPMN (27 CFP) is justified by the difference in the levels of abstraction. Since BPMN model represents process at a more detailed level and CD represents process at a high-level of abstraction, CD does not represent all the process details as BPMN model. In addition, the CD represents the static view of the software. While, the BPMN represents the dynamic behaviour.

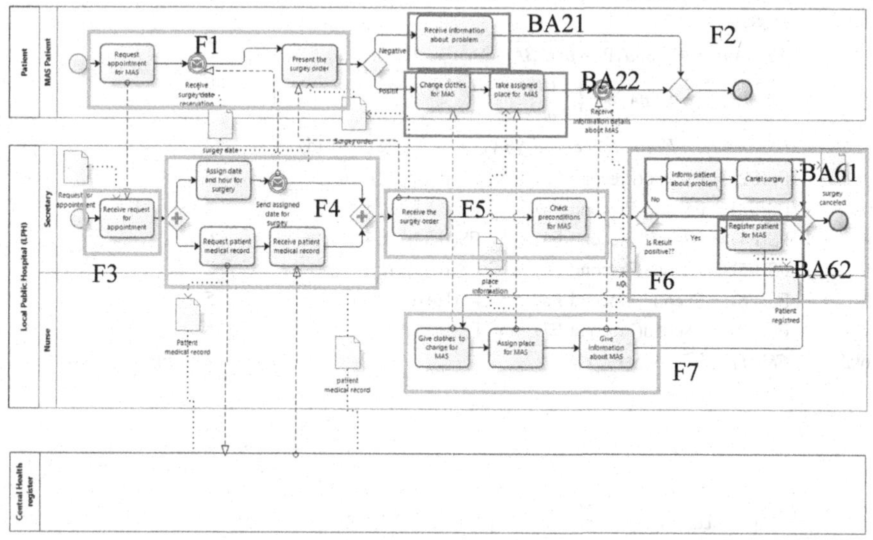

Fig. 1. "Patient Admission and Registration for MAS" Business Process in BPMN.

The functional size obtained from the BPMN (Table 4) and the functional size obtained from the BPEL (Table 6) is both equal to 27 CFP. Actually, the transformation of the BPMN to the BPEL is done based on the standard given by the OMG [1]. Hence, every data movement in the BPMN has its equivalent in the BPEL. Our decomposition makes the model more structured. Hence, it is easier to write the BPEL code for each fragment from its associated BPMN. In comparison to the FS(BPMN) and the FS(BPEL), the FS(CD) is the lowest. This is due to the fact that the CD represents less details in comparison with the BPMN and BPEL models.

Note that the above COSMIC FSM-based conformity result agrees with the result obtained through the BPMN-to-BPEL translation method presented in [14] and applied on the same BPMN example. More specifically, the BPEL generated from the BPMN example (Table 6) through the translation method in [14] is shown to be conforming to the source BPMN model [14]. As discussed above, this conformity result was also derived through our heuristic method which is based on measuring the functional size. While [14] can be used to generate a BPEL model from a BPMN model, our method can be used in a broader context, for instance, to verify the effects of changes in the BPMN model.

Table 4. Measurement results for "Patient Admission and Registration for MAS".

Fragment	Measurement formulas	Measurement results in CFP
F1	FS(F1) = FScond(Precond F1) + $\sum$ FS(BA1j) (2) FS(F1) = FS(Request appointment for MAS) + FS(Present the surgery order)	2
	$FS(BA_{ij}) = FScond(\Pr econd\ BA_{ij}) + \sum_{t=1}^{k} FS(SBA_{ijk})$ (3)	2
	$FScond(\Pr econd\ BA_{ij}) = \begin{cases} 1\ CFP\ \ if\ BA_{ij}\ \ has\ a\ pre-condition \\ 0\ otherwise \end{cases}$ (4)	0
	$FS(SBA_{ij}) = FScond(\Pr econd\ SBA_{ij}) + \sum_{l=1}^{p} FS(T_{ijk})$ (5)	2
F2	FS(F2) = FScond(Precond F2) + $\sum$ FS(BA2j) (2)	5
F3	FS(F3) = FScond(Precond F3) + $\sum$ FS(BA3j) (2)	2
F4	FS(F4) = FScond(Precond F4) + $\sum$ FS(BA4j) (2)	6
F5	FS(F5) = FScond(Precond F5) + $\sum$ FS(BA5j) (2)	2
F6	FS(F6) = FScond(Precond F6) + $\sum$ FS(BA6j) (2)	5
F7	FS(F7) = FScond(Precond F7) + $\sum$ FS(BA7j) (2)	5
Total	$FS(M) = \sum_{i=1}^{n} FS(F_i)$ (1)	27 CFP

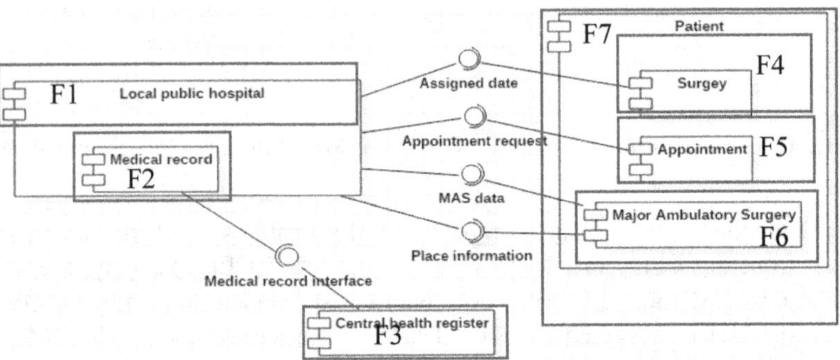

Fig. 2. Component diagram corresponding to "Patient Admission and Registration for MAS".

Table 5. Measurements results (CD of the "Patient Admission and Registration for MAS").

Component	Interfaces in CD diagram	Data movement type	Measurement results in CFP
F1	Assigned date Appointment request MAS data Place information	E E E E	4 CFP
F2	Medical record interface	S	1 CFP
F3	Medical record interface	E	1 CFP
F4	Assigned date	S	1 CFP
F5	Appointment request	S	1 CFP
F6	MAS data Place information	S S	2 CFP
F7	–	–	0 CFP
Total			**10 CFP**

Table 6. Measurements results BPEL of the "Patient Admission and Registration for MAS".

Frag-ment	BPMN		BPEL	CFP	Types
F1	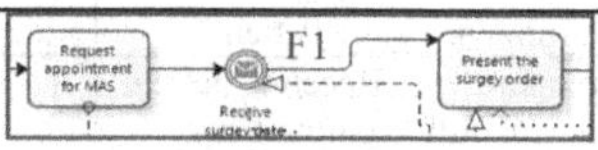		<sequence> <invoke name="Request appointment for MAS"/>	0	--
			<invoke name="Receive surgery date reservation"partnerLink="Patient"> </invoke>	1	E
			<invoke name="Present the surgery order"/>	0	--
			<invoke name="Receive the surgery order" partnerLink="Patient"><toPart part="Surgery order"fromVariable="Surgery"/> </sequence>	1	R
F2	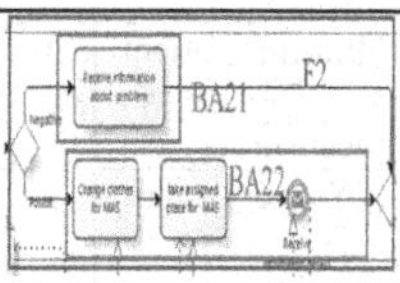	BA21	<if><condition>Negative </condition>	1	E
			<invoke name="Receive information about problem"/>	0	--
		BA22	<elseif><condition>Positive</condition> <sequence>	1	E
			<invoke name="Change clothes for MAS"/>	0	--
			<receive name="Take assigned place for MAS"/> <toPart part="Place information"fromVariable="Place"/>	1	R
			<invoke name="Assign place for MAS" partnerLink="Patient"><invoke>	1	E
			<invoke name="Receive information details about MAS"partnerLink="Patient"> </sequence></elseif>	1	E
F3	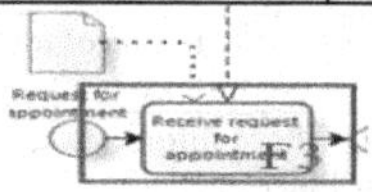		<receive name = "Receive request for appointment" partnerLink = "Patient"> </receive>	1	S
			<toPart part = "Request for appointment" fromVariable = "Appointment"/>	1	R
F4	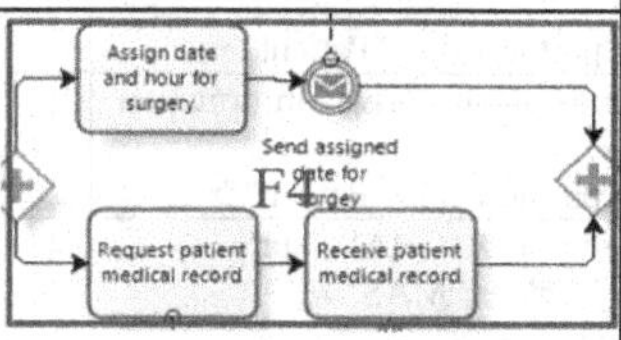		<flow> <sequence> <invoke name = "Assign date and hour for surgery"/>	0	--
			<invoke name = "Send assigned date for surgery" partnerLink = "Patient"> </sequence>	1	E
			<sequence> <invoke name = "Request patient medical record"/>	0	--
			<fromPart part = "Patient medical record" fromVariable = "Patient"/>	1	W
			<invoke name = "Request patient medical record" partnerLink = "Central health register"> </invoke>	1	E

			<invoke name = "Receive patient medical record" partnerLink = "Central health register">	1	E
			<toPartpart = "Patient medical record" fromVariable="Patient"/>	1	R
			<invoke name = "Receive patient medical record" partnerLink = "Central health register"> </invoke> </sequence> </flow>	1	E
F5	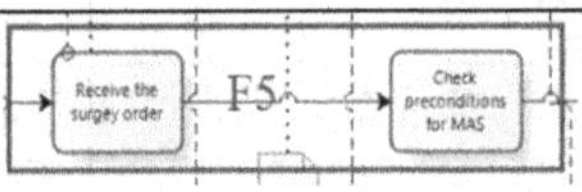		<sequence><invoke name="Receive the surgery order"partnerLink="Patient"/>	1	E
			<fromPart part="Surgery order" fromVariable="Surgery"/>	1	W
			<invoke name="Check preconditions for MAS"/> /sequence>	0	--
F6	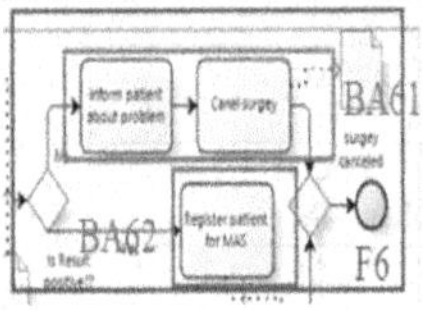	BA61	<if> <condition> No </condition>	1	E
			<invoke name = "Inform patient about problem"/> <invoke name = "Cancel surgery"/>	0	--
			<fromPart part = "Surgery canceled" fromVariable = "Surgery"/>	1	W
			<elseif> <condition> Yes </condition>	1	E
		BA62	<invoke name = "Register patient for MAS"/>	0	--
			<fromPart part = "Patient registred" fromVariable = "Patient"/>	1	W
			<invoke name = "Give clothes to change for MAS" partnerLink = "Nurse"> </elseif>	1	E
F7	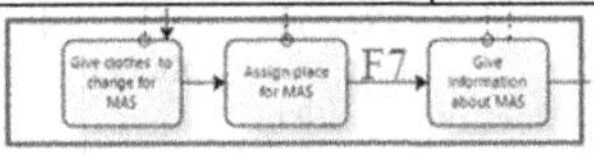		<sequence> <invoke name = "Give clothes to change for MAS"/>	0	--
			<invoke name = "Change clothes for MAS" partnerLink = "Patient">	1	E
			<invoke name ="Assign place for MAS"/>	0	--
			<fromPart part="Place information" fromVariable="Place"/>	1	W
			<invoke name="Take assigned place for MAS"partnerLink="Patient">	1	E
			<invoke name="Give information about MAS"partnerLink="Patient"/>	1	E
			<fromPart part="MAS data" fromVariable="MAS"/> </sequence>	1	W
					Total = 27 CFP

5.2 Threats to Validity

The validity of the above presented results are subject to two types of threats: internal, and construct validity threats [15]. The internal validity threats are related to four issues. The first issue is its dependence on the proposed heuristics, which also ensures the conformity between BPMN model, its corresponding BEPL and CD models in terms of CFP units. However, such heuristics may not always be applicable especially when the BPMN model is not well structured. The second issue is related to the generation of the BPEL model that can be wrong. This case may lead to the non conformity between business process and information system model. In addition, it hinders the BP-IS models alignment, which is a necessity for a better governance. The third issue is the use of COSMIC at a high granularity level and which does not address the detailed measurement of control structures in the business process. In fact, it is possible to rapidly find errors during the conformity verification between BPMN and BPEL describing the BP and run-time models, respectively, and the component diagram to describe the IS. Finally, the fourth issue is related to the quality of the BPMN model (the starting point) and its consistency when we compare it to the functional requirements (being written in natural language). This model should be well elaborated to verify its conformity with BPEL model at the run time and respectively the component diagram describing the IS. More specifically, clear functional requirements allow to generate a good BPMN, BPEL and UML component models.

The threats of construct validity are related to the relation between theory and observation. In our case, this study must be applied on several case studies and in industrial practice. In fact, the COSMIC measurement method should be used to check the conformity of BPMN model, BPEL and the component diagram. However, we believe that the use of our approach with real data is important.

6 Conclusion

Checking the conformity among the BPMN, component, and BPEL models is the main purpose of this paper. To meet this purpose, a COSMIC-based approach is proposed for sizing these models. In addition, bounds on the functional sizes of these models are proposed as a coarse means to verify the conformity between these models in terms of CFP. Furthermore, to provide for a refined verification of their conformity, a set of modeling heuristics and measuring procedures have also been defined based on the mapping between COSMIC concepts and those of BPEL. Finally, the proposed measurement and conformity verification approach was illustrated through the "Patient Admission and Registration for MAS" business process.

References

1. ISO/IEC 19510, Information technology – Object Management Group Business Process Model and Notation (2013)
2. Jordan, D., Evdemon, J.: Web Services Business Process Execution Language Version 2.0. Committee Specification, January 2007

3. Aversano, L., Grasso., C., Tortorella, M.: Managing the alignment between business processes and software systems. J. Inf. Softw. Technol. **72**, 171–188 (2016)
4. COSMIC, Common Software Measurement International Consortium. COSMIC Functional Size Measurement Method, Version 4.0.2 (2017)
5. Khlif, W., Haoues, M., Sellami, A., Ben-Abdallah, H.: Analyzing functional changes in BPMN models using COSMIC. In: International Conference on Software Engineering and Applications (ICSOFT 2017), pp. 265–274. SciTePress, Portugal, July 2017
6. Kaya, M.: E-cosmic: a business process model based functional size estimation approach. MSc Thesis. Ankara, Middle East Technical University, 15 August 2010
7. Monsalve, C., April, A., Abran, A.: On the expressiveness of business process modeling notations for software requirements elicitation. In: Conference of the IEEE Industrial Electronics Society (IECON 2012), Montreal. IEEE, 25–28 October 2012
8. Sellami, A., Haoues, M., Ben-Abdallah, H.: Analyzing UML activity and component diagrams - an approach based on COSMIC functional size measurement. In: Conference on Evaluation of Novel Approaches to Software Engineering, France, pp. 36–44 (2013)
9. Sellami, A., Haoues, M., Ben-Abdallah,H.: Automated COSMIC-based analysis and consistency verification of UML activity and component diagrams. In: Conference on Evaluation of Novel Approaches to Software Engineering, France, pp. 48–63 (2013)
10. Delgado, A., Ruiz, F., de Guzmán, I.G.D., Piattini, M.: Business process service oriented methodology with service generation in SoaML. In: International Conference on Advanced Information Systems Engineering (CAiSE 2011), London, pp. 672–680, 20–24 June 2011
11. Aysolmaz, B., İren, D., Demirörs, O.: An effort prediction model based on BPM measures for process automation. In: International Workshop on Business Process Modeling, Development and Support (BMMDS/EMMSAD 2013), pp. 154–167 (2013)
12. Lavazza, L., Bianco, V.: A case study in COSMIC functional size measurement: the rice cooker evisited. In: International Workshop on Software Measurement (IWSM 2009), Amsterdam, pp. 101–121 (2009)
13. Lind, K., Heldal, R., Harutyunyan, T., Heimdahl, T.: CompSize: automated size estimation of embedded software components. In: International Workshop on Software Measurement (IWSM 2011), Japan, pp. 2183–2192 (2011)
14. Ouyang, C., Dumas, M., Van der Aalst, W.M.: Pattern-based translation of BPMN process models to BPEL web services. Int. J. Web Serv. Res. **5**(1), 42–62 (2009)
15. Wohlin, C., Runeson, P., Höst, M., Ohlsson, M.C., Regnell, B., Wesslén, A.: Experimentation in Software Engineering: An Introduction. Academic Publishers (2000)

Information Systems Planning and Success in SMEs: Strategizing for IS

Maria Kamariotou([✉]) and Fotis Kitsios

Department of Applied Informatics, University of Macedonia, Thessaloniki, Greece
mkamariotou@uom.edu.gr, kitsios@uom.gr

Abstract. Although the Strategic Information Systems Planning (SISP) has been emerging for large firms, family businesses do not develop strategic planning and they do not support business goals using Information Systems (IS). Thus, the implemented plans are not effective, and they do not meet the objectives. The purpose of this paper is to indicate the phases which contribute to a greater extent of success in order to provide conclusions regarding the implementation of this process in SMEs. Data were collected from IS executives in Greek SMEs. Factor Analysis is performed on the detailed items of the SISP process and success constructs.

Keywords: Strategic Information Systems Planning · Phases · Success · SMEs · IT strategy

1 Introduction

As businesses are obliged to deal with the environmental uncertainty and complexity, managers have to develop IS that support business strategy, and accommodate decision making in order to increase competitive advantage [23]. IS could be a source of sustainable competitive advantage only if the IS strategy will be aligned with business strategy [18–20, 22]. This is a crucial challenge for businesses and especially for SMEs that are significant components of the national economy, because they constitute of a large number of businesses in a country. IT executives are not aware about the business objectives and often cannot realize the needs of business decisions [22].

Researchers have suggested that more extensive planning would be more effective because it would support planners understand the impact of the environment and better respond to it. Previous researchers have examined the relationship between the strategic planning process of IS and the success as well as the obstacles that managers face [12–15]. But these studies focused only on large companies. Therefore, the purpose of this paper is to indicate the phases which contribute to a greater extent of success and to provide recommendations regarding the implementation of the SISP process to IT managers. Data were collected using questionnaires to IS executives in Greek SMEs in order to examine which phases are more significant for them as well as which of them have to be improved in order to produce effective IS plans.

The structure of this paper is as following: after a brief introduction to this field, the next section includes the theoretical background regarding the SISP process. Section 3

W. Abramowicz and R. Corchuelo (Eds.): BIS 2019, LNBIP 353, pp. 397–406, 2019.
https://doi.org/10.1007/978-3-030-20485-3_31

describes the methodology, while Sect. 4 shows the results of the survey. Finally, Sect. 5 discusses the results and concludes the paper.

2 Theoretical Background

In complex environments, SMEs tend to formalize processes using certain rules and procedures which support the limitation of environmental uncertainty. Formalization supports the development of aspects which encourage communication among the individuals and sharing of new information. Also, they transform the generation of new ideas through the inflicted structures into real plans, enhancing the growth of innovation. As the environment is getting more and more complex, the need for innovation is increasing if businesses are to be helped to be competitive so as to survive [3, 9, 21].

The findings of studies which examine the relationship of SISP phases and success conclude that IS executives focused their efforts on the Strategic Conception phase. Although planners focus on this phase, they cannot identify the suitable alternative strategies. As a consequence, their efforts do not positively influence SISP success. So, they cannot achieve their objectives. The most common problems which have been raised during the SISP process are the lack of participation and the failure to apply strategic IS plans. Executives cannot be committed to the plan, consequently the members of the team have difficulties to implement the IS strategy. Moreover, results show that executives only focus on the implementation of IS strategy because they consider this process difficult and they ignore its formulation [4–8, 10, 11, 14, 15, 17, 23].

Results from existing studies indicate that many managers put too many efforts to SISP process while others too little. If managers invest too many efforts, many conflicts among team members can be raised as well as the process could be delayed. On the other hand, if managers avoid investing too much time into the process, IS plans could be inefficient so IS goals could not be achieved. Consequently, the assessment of the process is significant because managers can reduce these unsatisfactory results [5, 8, 14].

Many researchers mentioned that managers concentrate more on Strategy Conception and Strategy Implementation and they do not invest time on Strategic Awareness and Situation Analysis. These findings confirm that the implemented plans are ineffective and unsuccessful and they do not meet the objectives [1, 5, 8, 14, 15]. Furthermore, when managers concentrate their efforts on the implementation of the process, they may reduce SISP horizons, but the strategic goals cannot be achieved. IT managers do not understand the strategic objectives and how they can increase business value because they invest time on the horizon of the project and on minimizing its cost due to limited IT budget [1].

The results indicate that executives should pay attention to implementing Situational Analysis with greater meticulousness, so that they can apply Strategy Conception and Strategy Implementation Planning with greater agility rather than now. Planners should focus on the analysis of current business systems, organizational systems, IS, as well as the business environment and external IT environment. If planners understand those elements, they can improve the result of the planning process excluding the increased time and cost which the process is needed. When executives analyze the environment, they can determine important IT objectives and opportunities for improvement, they can evaluate them in order to define high-level IT strategies in their business' strategy conception [13, 16].

3 Methodology

A field survey was developed for IS executives. The instrument used five-point Likert-scales to operationalize two constructs: SISP phases and success. The SISP process constructs measured the extent to which the organization conducted the five planning phases and their tasks. The success constructs measured using four dimensions named alignment, analysis, cooperation and capabilities. The questionnaire was based on previous surveys regarding SISP phases [2, 13–15].

Four IS executives were asked to participate in a pilot test. Each one completed the survey and commented on the contents, length, and overall appearance of the instrument. A sample of IS executives in Greece was selected from the icap list [14, 15]. SMEs which provided contact details were selected as the appropriate sample of the survey. The survey was sent to 1246 IS executives and a total of 294 returned the survey. Data analysis was implemented using Factor Analysis.

4 Results

Tables 1 and 2 provide all measures used for all constructs. Table 3 presents the sample distribution in different industry types. The internal consistency, calculated via Cronbach's alpha, ranged from 0.738 to 0.932, exceeding the minimally required 0.70 level [14]. Factor analysis was implemented on the detailed items of the SISP process and SISP success constructs. Tables 4 and 5 describe the reliability of SISP phases and success constructs. These Tables also present the principal component analysis using the Maximum Likelihood Estimate and the extraction of factors with Promax with Kaiser Normalization method. The factor loadings and cross loadings provide support for convergent and discriminant validity.

Findings indicate that IS executives are not aware of analyzing the external IT environment and evaluating opportunities for IS development. This result stems from the fact that the following variables have low factor loadings in Table 4; Analyzing the current external IT environment and Evaluating opportunities for improvement. This finding is important because it confirms that executives in SMEs do not invest in new technologies and cannot be competitive. Furthermore, a significant obstacle is that managers do not focus on organizing the planning team. The factor loading of this variable is .635. The selection of employees who will participate in the development of

Table 1. SISP phases and activities.

Phases	Activities	References
Strategic awareness	Determining key planning issues (SAw1) Determining planning objectives (SAw2) Organizing the planning team (SAw3) Obtaining top management commitment (SAw4)	[1, 2, 5–8, 11–15]
Situation analysis	Analyzing current business systems (SA1) Analyzing current organizational systems (SA2) Analyzing current information systems (SA3) Analyzing the current external business environment (SA4) Analyzing the current external IT environment (SA5)	
Strategy conception	Identifying major IT objectives (SC1) Identifying opportunities for improvement (SC2) Evaluating opportunities for improvement (SC3) Identifying high-level IT strategies (SC4)	
Strategy formulation	Identifying new business processes (SF1) Identifying new IT architectures (SF2) Identifying specific new projects (SF3) Identifying priorities for new projects (SF4)	
Strategy implementation planning	Defining change management approaches (SIP1) Defining action plans (SIP2) Evaluating action plans (SIP3) Defining follow-up and control procedures (SIP4)	

IS is an important process because managers have to select employees with IT skills, motivation to develop effective IS and cooperation skills. This finding is associated with the lack of management support and the lack of clear guidelines about the IS development. Thus, a new factor named Managers' understanding of IS was developed regarding to the understanding of the importance of SISP by managers. IS executives can define priorities, increase the cooperation among the IS team and provide guidelines regarding in order to support the effectiveness of IS plans.

Another significant finding is that IT managers do not formulate IT strategies and priorities, so they cannot anticipate risks and crises. Executives cannot identify problem areas because they do not redesign business processes. The factor loading of "Identifying new business processes" variable is low. New IS are based on the existing business processes and they cannot meet IS objectives. This finding confirms the negative consequences that SMEs face due to the lack of strategic planning.

Table 2. Success dimensions.

Dimensions	Items	References
Alignment	Maintaining a mutual understanding with top management on the role of IS in supporting strategy (AL1) Understanding the strategic priorities of top management (AL2) Identifying IT-related opportunities to support the strategic direction of the firm (AL3) Aligning IS strategies with the strategic plan of the organization (AL4) Adapting the goals/objectives of IS to changing goals/objectives of the organization (AL5) Educating top management on the importance of IT (AL6) Adapting technology to strategic change (AL7) Assessing the strategic importance of emerging technologies (AL8)	[5–8, 14, 15]
Analysis	Identifying opportunities for internal improvement in business processes through IT (AN1) Maintaining an understanding of changing organizational processes and procedures (AN2) Generating new ideas to reengineer business processes through IT (AN3) Understanding the information needs through subunits (AN4) Understanding the dispersion of data, applications, and other technologies throughout the firm (AN5) Development of a "blueprint" which structures organizational processes (AN6) Improved understanding of how the organization actually operates (AN7) Monitoring of internal business needs and the capability of IS to meet those needs (AN8)	
Cooperation	Developing clear guidelines of managerial responsibility for plan implementation (CO1) Identifying and resolving potential sources of resistance to IS plans (CO2) Maintaining open lines of communication with other departments (CO3) Coordinating the development efforts of various organizational subunits (CO4) Establishing a uniform basis for prioritizing projects (CO5) Achieving a general level of agreement regarding the risks/tradeoffs among system projects (CO6) Avoiding the overlapping development of major systems (CO7)	
Capabilities	Ability to identify key problem areas (CA1) Ability to anticipate surprises and crises (CA2) Flexibility to adapt to unanticipated changes (CA3) Ability to gain cooperation among user groups for IS plans (CA4)	

Table 3. Sample distribution in different industry types.

Industry type	Respondents	Percentage
Agriculture & Food	47	16%
Business services	33	11.3%
Chemicals, Pharmaceuticals & Plastics	18	6.1%
Construction	22	7.5%
Education	4	1.4%
Electrical	11	3.8%
Energy	8	2.7%
IT, Internet, R&D	24	8.2%
Leisure and Tourism	16	5.5%
Metals, Machinery & Engineering	28	9.6%
Minerals	3	1%
Paper, Printing, Publishing	14	4.8%
Retail & Traders	31	10.6%
Textiles, Clothing, Leather, Watchmaking, Jewellery	14	4.8%
Transport & Logistics	20	6.8%
Total	294	100

Previous findings conclude that managers concentrate more on Strategy Conception and Strategy Implementation and they do not invest time on Strategic Awareness and Situation Analysis, as a result the implemented plans are not effective, successful and they do not meet the objectives [2, 5, 8, 14, 15]. Moreover, when managers concentrate on the implementation of the process, shorter SISP horizons are achieved but the strategic goals cannot be met. Executives do not focus on strategic objectives that really concern them and on how they can increase value to the business because they invest time on the horizon of the project and on minimizing its cost due to limited IT budget [2]. The results indicate that executives should pay attention to implementing Situational Analysis with greater meticulousness, so that they can apply Strategy Conception and Strategy Implementation Planning with greater agility rather than now. Planners should analyze their current business systems, organizational systems, IS, as well as the business environment and external IT environment. If planners understand those elements, they can improve the result of the planning process excluding the increased time and cost needed for the process. When executives understand the environment, they can determine important IT objectives and opportunities for improvement and they can evaluate them in order to define high-level IT strategies in their business' strategy conception [5, 8, 13, 23].

Table 4. Factor loadings for SISP phases.

Constructs	Items	Cronbach's alpha of each construct	Factor Loadings
Strategy implementation planning	Evaluating action plans	.910	.929
	Defining action plans		.890
	Defining change management approaches		.661
	Defining follow-up and control procedures		.595
Analysis of IT environment	Analyzing current information systems	.841	.878
	Analyzing the current external business environment		.728
	Analyzing the current external IT environment		.639
	Organizing the planning team		.635
Strategy conception	Identifying opportunities for improvement	.901	.925
	Identifying major IT objectives		.846
	Identifying high-level IT strategies		.556
	Evaluating opportunities for improvement		.504
Strategy formulation	Identifying new IT architectures	.738	.906
	Identifying new business processes		.500
Analysis of internal environment	Analyzing current business systems	.774	.895
	Analyzing current organizational systems		.711

Table 5. Factor loadings for success.

Constructs	Items	Cronbach's alpha of each construct	Factor Loadings
Cooperation	Identifying and resolving potential sources of resistance to IS plans	.922	.835
	Coordinating the development efforts of various organizational subunits		.814
	Achieving a general level of agreement regarding the risks/tradeoffs among system projects		.800
	Establishing a uniform basis for prioritizing projects		.772

(*continued*)

Table 5. (*continued*)

Constructs	Items	Cronbach's alpha of each construct	Factor Loadings
	Maintaining open lines of communication with other departments		.748
	Developing clear guidelines of managerial responsibility for plan implementation		.688
	Ability to anticipate surprises and crises		.523
Analysis	Maintaining an understanding of changing organizational processes and procedures	.925	.795
	Generating new ideas to reengineer business processes through IT		.774
	Improved understanding of how the organization actually operates		.773
	Understanding the information needs through subunits		.746
	Identifying opportunities for internal improvement in business processes through IT		.738
	Monitoring of internal business needs and the capability of IS to meet those needs		.599
	Understanding the dispersion of data, applications, and other technologies throughout the firm		.597
	Ability to identify key problem areas		.562
	Development of a "blueprint" which structures organizational processes		.522
Strategic alignment	Understanding the dispersion of data, applications, and other technologies throughout the firm	.932	.910
	Aligning IS strategies with the strategic plan of the organization		.707
Managers' understanding of IS	Maintaining a mutual understanding with top management on the role of IS in supporting strategy	.788	.979
	Understanding the strategic priorities of top management		.677

5 Conclusion

So far, researchers have paid attention to the effect of SISP phases on success but they concentrated on large firms. The purpose of this paper was to indicate the phases which contribute to a greater extent of success and to provide recommendations regarding the implementation of the SISP process to IT managers. Findings indicated that IS executives are not aware of analyzing the external IT environment and evaluating opportunities for IS development. This finding is important because it confirms that executives in SMEs do not invest in emergent technologies and cannot be competitive. Furthermore, a significant obstacle is that managers do not focus on organizing the planning team. The selection of employees who will participate in the development of IS is an important process because managers have to select employees with IT skills, motivation to develop effective IS and cooperation skills.

Another significant finding was that IT managers do not formulate IT strategies and priorities, so they cannot anticipate risks and crises. Executives cannot identify problem areas because they do not redesign business processes. New IS are based on the existing business processes and they cannot meet IS objectives. This finding confirms the negative consequences that SMEs face due to the lack of strategic planning. The results of this study contribute to IS executives' awareness of the strategic use of IS planning in order to increase competitive advantage.

Understanding those phases may help IS executives concentrate their efforts on organizations' objectives and recognize the greatest value of the planning process in their business. Second, the results of this survey can increase their awareness of the phases of SISP. IS executives should be knowledgeable about the five phases and they should not ignore the tasks of each one because this might be an obstacle which presents the organization from achieving its planning goals and thus from realizing greater value. Finally, the findings contribute to IS executives in Greek SMEs who do not concentrate on strategic planning during the development of IS and they focus only on the technical issues. As a result, they should understand the significance of the SISP process in order to formulate and implement IS strategy which will be aligned with business objectives and increase the success of SMEs.

A limitation of this study stems from the fact that the survey was conducted only in Greece. Future researchers could examine and compare these results with relative ones from large companies and other countries. Apparently, future researchers may use different methodologies for data analysis, such as cluster analysis in order to compare the differences among organizations in different sectors during the implementation of the SISP process.

References

1. Brown, I.: Strategic information systems planning: comparing espoused beliefs with practice. In: ECIS 2010: Proceedings of 18th European Conference on Information Systems, South Africa, pp. 1–12 (2010)
2. Brown, I.T.J.: Testing and extending theory in strategic information systems planning through literature analysis. Inf. Resour. Manage. J. **17**(4), 20–48 (2004)
3. Giannacourou, M., Kantaraki, M., Christopoulou, V.: The perception of crisis by greek SMEs and its impact on managerial practices. Procedia Soc. Behav. Sci. **175**(2015), 546–551 (2015)
4. Kamariotou, M., Kitsios, F.: Strategic information systems planning. In: Khosrow-Pour, M. (ed.) Encyclopedia of Information Science and Technology, 4th Edn., chapter 78, pp. 912–922. IGI Global Publishing, USA (2018)

5. Kamariotou, M., Kitsios, F.: An empirical evaluation of strategic information systems planning phases in SMEs: determinants of effectiveness. In: Proceedings of the 6th International Symposium and 28th National Conference on Operational Research, Greece, pp. 67–72 (2017)
6. Kamariotou, M., Kitsios, F.: Information systems phases and firm performance: a conceptual framework. In: Kavoura, A., Sakas, D.P., Tomaras, P. (eds.) Strategic Innovative Marketing. SPBE, pp. 553–560. Springer, Cham (2017). https://doi.org/10.1007/978-3-319-33865-1_67
7. Kamariotou, M., Kitsios, F.: Strategic information systems planning: SMEs performance outcomes. In: Proceedings of the 5th International Symposium and 27th National Conference on Operation Research, Greece, pp. 153–157 (2016)
8. Kitsios, F., Kamariotou, M.: Decision support systems and strategic planning: information technology and SMEs performance. Int. J. Decis. Support Syst. **3**(1/2), 53–70 (2018)
9. Kitsios, F., Kamariotou, M.: Strategic IT alignment: business performance during financial crisis. In: Tsounis, N., Vlachvei, A. (eds.) Advances in Applied Economic Research. SPBE, pp. 503–525. Springer, Cham (2017). https://doi.org/10.1007/978-3-319-48454-9_33
10. Lederer, A.L., Sethi, V.: Key prescriptions for strategic information systems planning. J. Manage. Inf. Syst. **13**(1), 35–62 (1996)
11. Maharaj, S., Brown, I.: The impact of shared domain knowledge on strategic information systems planning and alignment: original research. S. Afr. J. Inf. Manage. **17**(1), 1–12 (2015)
12. Mentzas, G.: Implementing an IS strategy- a team approach. Long Range Plann. **30**(1), 84–95 (1997)
13. Mirchandani, D.A., Lederer, A.L.: "Less is More:" information systems planning in an uncertain environment. Inf. Syst. Manage. **29**(10), 13–25 (2014)
14. Newkirk, H.E., Lederer, A.L., Srinivasan, C.: Strategic information systems planning: too little or too much? J. Strateg. Inf. Syst. **12**(3), 201–228 (2003)
15. Newkirk, H.E., Lederer, A.L.: The effectiveness of strategic information systems planning under environmental uncertainty. Inf. Manage. **43**(4), 481–501 (2006)
16. Newkirk, H.E., Lederer, A.L., Johnson, A.M.: Rapid business and IT change: drivers for strategic information systems planning? Eur. J. Inf. Syst. **17**(3), 198–218 (2008)
17. Pai, J.C.: An empirical study of the relationship between knowledge sharing and IS/IT strategic planning (ISSP). Manage. Decis. **44**(1), 105–122 (2006)
18. Peppard, J., Ward, J.: Beyond strategic information systems: towards an IS capability. J. Strateg. Inf. Syst. **13**(2), 167–194 (2004)
19. Premkumar, G., King, W.R.: The evaluation of strategic information system planning. Inf. Manage. **26**(6), 327–340 (1994)
20. Rathnam, R.G., Johnsen, J., Wen, H.J.: Alignment of business strategy and IT strategy: a case study of a fortune 50 financial services company. J. Comput. Inf. Syst. **45**(2), 1–8 (2004)
21. Siakas, K., Naaranoja, M., Vlachakis, S., Siakas, E.: Family businesses in the new economy: How to survive and develop in times of financial crisis. Procedia Econ. Financ. **9**(2014), 331–341 (2014)
22. Ullah, A., Lai, R.: A systematic review of business and information technology alignment. ACM Trans. Manage. Inf. Syst. **4**(1), 1–30 (2013)
23. Zubovic, A., Pita, Z., Khan, S.: A framework for investigating the impact of information systems capability on strategic information systems planning outcomes. In: Proceedings of 18th Pacific Asia Conference on Information Systems, China, pp. 1–12 (2014)

An ICT Project Case Study from Education: A Technology Review for a Data Engineering Pipeline

Ioana Ciuciu[1(✉)], Augusta Bianca Ene[2], and Cosmin Lazar[2]

[1] Babes-Bolyai University, Cluj-Napoca, Romania
ioana.ciuciu@cs.ubbcluj.ro

[2] Robert Bosch SRL., Cluj-Napoca, Romania

Abstract. The paper presents a brief technology survey of existing tools to implement data ingestion pipelines in a classical Data Science project. Given the emergent nature of technologies and the challenges associated with any Big Data project, we propose to identify and discuss the main components of a data pipeline, from a data engineering perspective. The data pipeline is showcased with a case study from an ICT university project, where several teams of master students competed towards designing and implementing the best solution for a manufacturing data pipeline. The project proposes a research-based multidisciplinary approach to education, aiming at empowering students with a novel role in the process of learning, that of knowledge creators. Therefore, on the one hand, the paper discusses the main components of a Big Data pipeline and on the other hand it shows how these components are addressed and implemented within a concrete ICT project from education, realized in tight relation with the IT industry.

Keywords: Big Data · Data pipeline · Research-informed education · ICT project management · Virtual teams · Collaborative work

1 Introduction

We are currently facing the Big Data era; being able to exploit data has become vital for every company and business. There are economic sectors where data has always played an important role, however nowadays data seem to play a central role in knowledge discovery and decision making in almost every field. Data Science emerged from the availability of data and the increased computing power. It is an interdisciplinary field combining skills spanning various domains, such as Computer Science, software development, Mathematics and Statistics, Machine learning, domain specific knowledge and traditional research skills. In a nutshell, Data Science is the process of extracting meaning from data using statistical and programming tools, modeling, exploratory data analysis, reporting and teamwork. There are two specific roles defined in Data Science: the data specialist (or data engineer) role and the data analyst role. The *data specialist role* is responsible with all the operations related to data engineering, from data collection and ingestion, to data storage and management, to data processing.

W. Abramowicz and R. Corchuelo (Eds.): BIS 2019, LNBIP 353, pp. 407–420, 2019.
https://doi.org/10.1007/978-3-030-20485-3_32

It is widely accepted that it is the responsibility of the data specialist to prepare the data and make it available for data analysis. The *data analyst role* is responsible with data enriching, filtering, cleansing, quality checking and applying various data analysis algorithms on it, in order to gain valuable insights into the data and to use them for decision making, based on a rigorous understanding of the business question at hand.

This paper addresses the data specialist role, with a focus on the data engineering pipeline. There are currently many data engineering tools and frameworks, so that one could easily get confused in this ever-changing technology landscape. The motivation of this paper is to provide the reader with a concise but useful technology overview on the main components of the data pipeline. We are aiming to help the reader to rapidly get familiar with the technologies that form the building blocks of a data engineering pipeline. Moreover, the paper presents the data engineering pipeline from the perspective of an ICT project originating from Education, and demonstrates the building blocks of the data pipeline with concrete technologies.

The paper is organized as follows: Sect. 2 introduces the Data Engineering pipeline, identifying the main technologies and discussing underlying challenges; Sect. 3 presents a case study from an educational ICT project realized in collaboration with local IT industry around the implementation of a data engineering pipeline, and Sect. 4 concludes the paper.

2 The Data Engineering Pipeline

2.1 The Data Engineering Pipeline

A classical data analytics pipeline is depicted in Fig. 1. The pipeline comprises processes from both data engineering and data analytics. The focus of this paper is the data engineering part of the pipeline. What resides in the data engineering zone of the pipeline is the responsibility of the data engineer. More precisely, a data engineering pipeline addresses several elements and technological areas. According to the data value chain [1], each of these technical areas adds value to the data and ultimately to the society and businesses. The data engineering components are described below:

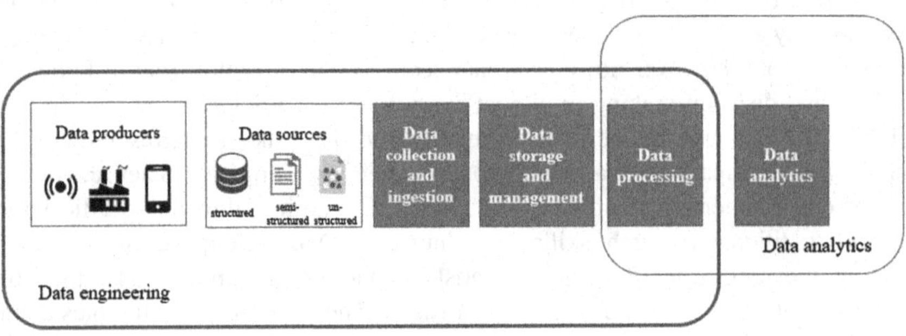

Fig. 1. Big Data analytics pipeline.

- Data producers – range from sensors, machines, processes, to humans (e.g., data from social networks).
- Data sources – are the primary storage of the data, once it has been produced (e.g., a relational database, a log file, etc.); such data sources are heterogeneous, non-integrated and spread across multiple locations/repositories. Data can be either structured, semi-structured or completely unstructured. Typically, a Big Data project would need to integrate, manage and process data from multiple sources.
- Data collection and ingestion – at this stage, data is collected from the data sources, it is filtered and cleansed before being ingested into a data storage solution for further processing and management. Data ingestion can be batch or streaming, depending on the type of data and the business need. Batch data ingestion will periodically upload multiple files into the data store (e.g., manufacturing supplier data delivered once per day), while streaming ingestion will continuously upload the data events (e.g., sensor data).
- Data storage and management – is concerned with storing and managing data in a scalable way. There is no universal solution for (big) data storage. Actually, Big Data storage solutions may be very different depending on the problem that needs to be solved. The choices depend on many factors and the corresponding solutions are designed in an ad-hoc manner. The solutions range from relational databases to non-traditional storage such as NoSQL databases, data warehouses, or even file systems. Data storage might be needed at multiple stages in the data engineering pipeline but data storage generally refers to the process of storing raw data in a versatile format which would allow to use the data in a flexible manner. For example, a common approach in Big Data infrastructures is to store acquired data in a *master data set* that represents the most detailed and raw form of data an organization may have [2]. Further processes (e.g., data curation, data analysis) may then be performed on data for which storage may be used for intermediate results.
- Data processing – at this stage, data processing tools are used in order to model, cleanse, enrich or aggregate data and ultimately get value out of it. The processing tools will, for instance, either query or analyze the data in order to get valuable information from the data, or export the data to another data store. According to the lambda architecture paradigm [2], the data processing can be batch or streaming (speed). Batch processing processes all the collected original data either cyclically or on demand. Speed processing, on the other hand, processes incoming data immediately and as quickly as possible, reducing the delay between the arrival of data and the time when results become available (from the batch processing).
- Data analytics is the second main part of the pipeline, strongly interlinked with the first one, the data engineering pipeline. The main goal of analytics is to extract knowledge from the data which could serve to some defined goals (either business or research). Typically, a data analyst will be involved both in the data preparation (this task can be viewed as an interface between the data engineer and the data analyst) and mainly in the data exploration and modeling. The outcome can be either a report describing relevant findings from the data or a prototype solution that can be further be transformed into a data product.

2.2 Technology Landscape

Data engineering tools are constantly evolving, illustrating a highly dynamic technology landscape. We include in this section a non-exhaustive list of common actual programming languages and technologies under the building blocks of the data pipeline.

Common programming skills for a data engineering pipeline include Java, Scala and Python. Python is presently a popular choice for both data pipelines and data analysis due to the availability of many Data Science libraries. Java and Scala are particularly useful within the Hadoop ecosystem and with Spark (Hadoop is written in Java, while Spark is written in Scala).

As presented in Fig. 2, popular tools for data collection and ingestion include (i) Apache Sqoop, used for ingesting structured (relational) data into Hadoop; (ii) Apache Flume used for collecting, aggregating and ingesting streaming data; (iii) Apache Kakfa, used as a messaging system for reading/writing streaming data; and (iv) Embulk, used for data transfer between various storages, databases, file formats and cloud services.

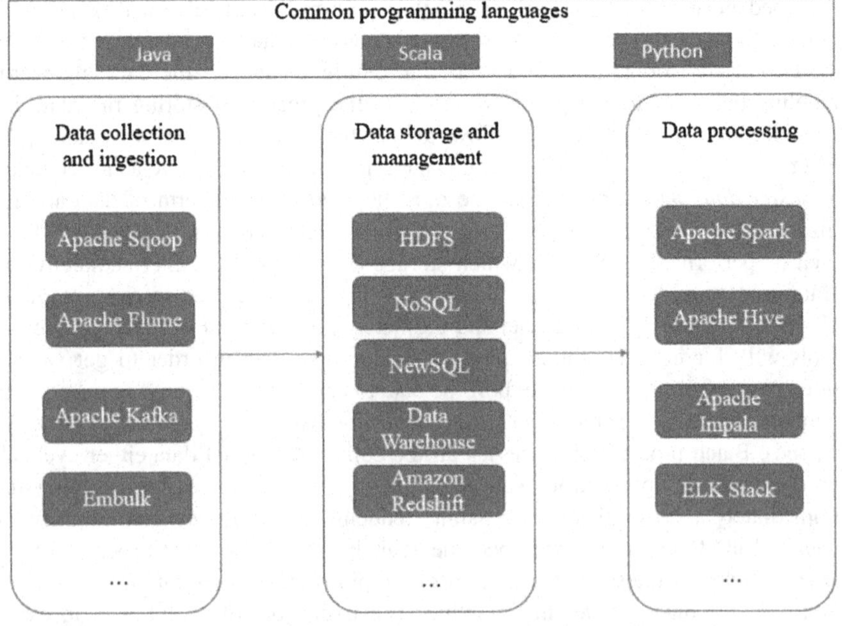

Fig. 2. Data engineering pipeline – technology landscape.

Regarding data storage and data management, common technologies include (i) the Hadoop Distributed File System (HDFS), which provides the capability to store large amounts of unstructured data in a reliable way on commodity hardware. Even though there exist distributed file systems with possibly higher performance, HDFS has reached the level of a *de facto* standard [1], as an integral part of the Hadoop framework [3]. It is worth noting that Hadoop has reached enterprise-grade level, all pure Hadoop vendors directly addressing enterprise needs; (ii) NoSQL databases, probably the most important family of Big Data storage technologies [4], provide a non-traditional storage model as opposed to traditional relational databases, being designed for scalability, often at the cost of sacrificing consistency; (iii) NewSQL databases, a modern form of relational databases that aim at comparable scalability as NoSQL databases, while maintaining the transactional guarantees and ACID compliance of traditional database systems (which are critical for e.g., financial operations); (iv) Data warehouse, an integral part of decision-support systems, constitutes a common repository of integrated data from several sources made available for consumption; and (v) Amazon Redshift, an example of technology that provides fast, scalable data warehousing.

Data processing relies on tools and technologies such as (i) Apache Spark, a large scale data processing engine, suitable for both batch and streaming data. It runs on Hadoop, standalone, in the cloud and accesses various data sources (from Hadoop, Apache Hive, Apache HBase, etc.); (ii) Apache Hive, a data warehouse software that enables reading, writing, and managing large datasets residing in distributed storage using SQL. Due to the batch-driven execution engine, Hive queries have a high latency even for small data sets. Hive provides the benefits of including an SQL-like query interface and the flexibility to evolve schemas easily; (iii) Apache Impala is the native analytic database for Apache Hadoop. It provides an SQL-like query interface for accessing the data. It supports HDFS and HBase (the Hadoop columnar store) as underlying data stores. It uses the same metadata and SQL-like user interface as Hive, but uses its own distributed query engine that can achieve lower latencies than Hive; and (iv) ELK Stack which unifies three open source projects: Elasticsearch (search and analytics engine), Logstash (data processing pipeline that ingests data from multiple sources simultaneously, transforms it and sends it to Elasticsearch) and Kibana (data visualization in Elasticsearch). Technologies under ELK Stack are easy to use, scalable and flexible.

2.3 Challenges and Future Recommendations

An important issue in the process of data collection and ingestion is the integration of data from multiple heterogeneous sources. The challenges in data integration arise from the Vs characterizing Big Data. If we are to refer to the most common Vs characterizing Big Data, the main challenges, as observed in [5], would be: (1) due to the large *volume* of data and the increased number of sources, it is difficult to do schema alignment and the warehousing of all the integrated data is expensive; (2) due to the *velocity* of the data, it is difficult to understand the evolution of semantics and it becomes infeasible to capture rapid data changes in a timely fashion; and (3) data *variety* implies high heterogeneity both at schema level and at instance level. Another practical challenge for companies is how to transform data in assets from which business value can be derived. Data catalogs implementing traditional data management functionalities such as data discovery, data lineage, data curation, as well as new AI and machine-learning based features such as self-generation of metadata, or taxonomy generation are tools that allow data owners to extract the maximum value out of the data they own [6].

A key challenge regarding data storage is related to security and privacy [7]. It is often unclear for companies how NoSQL databases can be securely implemented [1, 8], since many security measures implemented by default in traditional relational database systems are missing in NoSQL databases. Particular challenges for data storage arise due to data distribution. Auto-tiering strategies are needed in order to avoid data being moved to less secure tiers. Data storage and movement need to be governed by monitoring and logging mechanisms. Regarding the access control, data should only be accessible by authorized entities. Cryptographic mechanisms are required in this sense, with key cryptographic material generated and stored at the client side only. Also, fine-grained access control mechanisms are necessary in order to address the problem of data provenance. The work in [1] identifies three key requirements that are expected to govern future Big Data storage technologies: (1) Standardization of query interfaces; (2) Increased support for data security and user's privacy. Users should be aware of how Big Data processes their data. Big Data storage should comply with EU privacy regulations when personal information is being stored. Also, Big Data storage should provide user-friendly provenance mechanisms along the data processing pipeline; and (3) Support for semantic data models, in order to solve the problem of heterogeneous data sources and the development costs they involve.

3 Case Study: An ICT Project from Education

3.1 Concept

The project, titled Research-driven Learning of Big Data and Applications (ReLearn-BigDatA) proposes a multidisciplinary approach rooted in research which aims to empower students with a novel role in the process of learning. The approach enables students to become an active part of the research and learning community, by shifting from just knowledge "consumers" to knowledge "producers", thus contributing to the common wealth of the society. More precisely, the students are allowed to directly practice research and enquiry as part of their learning process.

The project builds around a Big Data course, with a research-based curricula and in tight collaboration with the local IT industry and similar EU initiatives (such as, the European Data Science Academy [9]). The students were involved in a research and innovation project where they faced real world challenges to which they proposed (designed and implemented) solutions that integrate the strategic infrastructure of the Babes-Bolyai University. More, precisely the project's aim is to design and implement a data engineering pipeline for production data, also involving a few simple data visualization tasks. Students undertake tasks involving data generation, data collection and integration, data storage, data processing and data visualization and propose a solution to integrate these tasks into a data engineering pipeline.

The aim of the project is dual: (1) from a didactical perspective, the students autonomously practice research tasks, critical thinking, and learn the latest technologies from the emergent field of Big Data in an applied context; (2) from a technical perspective, this was realized by encouraging students to propose technical solutions, which exist but are not fully covered during the course, for implementing solutions in the domain of Big Data integration, processing and analysis. The goal was to come up with and implement a valid solution for a data engineering pipeline for a manufacturing production line. A future goal would be to enable future analysis of the production data in view of error discovery and prediction. The students' prototypes were presented and demonstrated within a student workshop organized at the end of the semester.

The proposed action is inline with the principles of the Leagues of European Research Universities (LERU, [10]) with respect to developing strategies that enhance synergies between research and education, mainly with respect to: (i) Implementing a research-rich curricula, in collaboration with students and the IT industry (see section Implementation for details); (ii) Acknowledging the importance of research in education and vice-versa; (iii) Offering an active research experience to students, via real-world case studies and challenges; (iv) Empowering students to become leaders and agents of change, via practicing autonomous team-work with precise roles assigned and mini research projects/tasks.

3.2 Implementation

The ReLearnBigDatA experiment was realized as a semester project. Students undertook a team-based project (with maximum 5 team members with precise roles) to accomplish the following goals, based on a research-informed approach, as illustrated in the previous section:

1. Identify a real-world problem/challenge related to Big Data,
2. Design a Big Data solution to the identified problem/challenge,
3. Implement a prototype of the proposed design,
4. Prepare a demo of the prototype.

Project Topic and Requirements. This year, the problem definition was guided by researchers of Robert Bosch SRL Romania. Our partners provided real-world problems and challenges to be solved by the students during the project. Each team of students designed and implemented its own solution (prototype). At the end of the semester, the prototypes were demonstrated and evaluated during a workshop organized in the context of this course.

Regarding the project topic, students were asked to design and implement a data analytics pipeline for data from a manufacturing production line. They were provided with clear specifications of the input data. These data are closely designed to fit a manufacturing plant scenario. Students were asked to implement a production machine simulator that generates toy production data to be stored in a relational database. These data are then combined with sensor data measuring environment and inertial parameters. The sensor data source were the sensor log files containing the sensor's parameters. The students used Bosch sensors of type XDK Node, a connected sensor technology for IoT use cases [11] provided by our collaborators. The sensor generated various data related to light, temperature, pressure, humidity, accelerometer, gyroscope, acoustic and magnetometer data. All these data were combined, according to the scenario proposed by each team of students. The students were given the challenge of data integration between the simulator and the XDK node sensor. More precisely, they had to find an integration solution taking into account relational and non-relational data, batch and streaming data. The two data sources were joined using a timestamp parameter. The data were passed through all the phases of a data analytics pipeline and in the end were correlated and visualized in real-time according to specified requirements. Students were asked to store all data coming from the data sources on the Hadoop Distributed File System in a format of their choice. Finally, each team had to implement: (i) a visualization dashboard for real time monitoring of sensor parameters; and (ii) a visualization dashboard that allows the users to create reports.

It is important to note that the focus was on the learning experience of the students and therefore their autonomy and creativity played a central role in the experiment (project). They were encouraged to select and experiment as many technologies as needed from the emergent field of Big Data.

Project Phases. The project consisted of four main phases, which were inline with the above-mentioned project goals. The project phases are depicted in Table 1.

Table 1. Semester project phases.

Project phase	Planning and delivery	Details
Phase 1	week 1 – week 3	**Problem definition** Deliverables: - Teams announce on the course page or by email their composition (names of the team members) - Team members roles are also announced together with the team composition
Phase 2	week 3 – week 7	**Big Data solution design** Deliverables: - Each team sends by email to the instructor a one page document with the description of the proposed solution and the technologies to be used; it should be clear from the document the contribution of every team member, according to their roles - Demonstration of the intermediate (mid-term) solution
Phase 3	week 7 – week 11	**Prototype implementation** Tutorials provided in order to help students get familiar with important aspects of Hadoop Deliverables: - The instructor checks the prototype progress of every team; - Each team prepares a short demo of the application in its current status
Phase 4	week 11 – week 13 examination session	**Prototype demonstration and project evaluation (Workshop)** Deliverables: - Final version of the application is demonstrated - Evaluation of the prototype during the workshop

Project Implementation. The status of the project is currently finalized. In *Phase 1* of the project, a visit to a Robert Bosch manufacturing plant in Jucu was organized together with our industrial partners (the logistics of the visit were arranged by Robert Bosch SRL.), in order to familiarize the students with the application domain of their project (i.e., the production line).

As a result of this phase, project teams with precise roles were formed. There was a constraint to have the teams formed around the six intelligent sensors provided by Robert Bosch SRL. for the implementation of the students' projects. Six teams have been formed with a maximum of five people. In order to ensure the smooth management of the project, each team member had a specific role assigned (one team leader, several team members), and acted according to this role during the semester project. Tasks were precisely assigned among the team members. The project manager, together with team members, were responsible of respecting the project planning and delivery. The project manager was responsible of the communication and delivery of the project results to the course coordinator.

In *Phase 2* of the project, the student teams have designed and presented their proposed (conceptual) solutions in front of their colleagues, at the practical work class. A researcher from Robert Bosch SRL. was present as well in order to provide feedback and advice (if needed). We did not intend to influence the students in their choices, but only wanted to have a verification in order to make sure that all the proposed solutions were on track towards the desired objectives.

In *Phase 3* of the project, the students had one presentation per team with the instructor to illustrate the progress of their implemented solution. Most of the teams demonstrated good progress of their implementation. Five out of six teams managed to visualize the sensor data in real time, and practically all the teams implemented the manufacturing machine simulator. The teams showed good collaboration among team members, but also inter-team collaboration. Still within *Phase 3* of the project, the students presented an intermediate demo of their implemented solution, in front of the class, the instructor and a representative from the industrial collaborators.

In *Phase 4* of the project, the instructor, together with the students organized a workshop where all the implemented solutions were presented and demonstrated in their final version. The best solution was awarded by the industrial partners after the workshop. Also, the semester projects were evaluated by the instructor during the workshop, a percentage of the final mark being from the students' inter-evaluation. There is an open opportunity for the best solutions to be further improved in collaboration with the industry in order to be presented within a Big Data Meetup in Cluj-Napoca.

3.3 Results, Evaluation and Lessons Learnt

An important aspect in the context of this project was the fact that the students have used the university's competitive infrastructure High Performance Computing System and Private Cloud IBM IntelligentCluster [12, 13] for the implementation of their solutions. This enabled them to practice real-world Big Data processing and storage tasks in a collaborative manner on a Hadoop (Big Data Infrastructure) cluster. This, together with the industrial collaboration from which the project benefited, provided the students with a real setting for their work. As expected, the technical results (the prototypes) obtained by the student teams were based on a multitude of emerging technologies. The technological landscape from the student projects is illustrated in Table 2.

Table 2. Technology landscape as resulted from the students' project.

	Simulator (data producer)	Data collection and ingestion	Data storage and management	Data processing	Data visualisation
Team 1	Bogus C# NuGet.Net SQL DB	Apache Sqoop Apache Spark	HDFS JSON	Apache Spark Apache Hive	ASP.NET
Team 2	Bogus C# VB.NET Faker.js MySQL	Apache Kafka	HDFS	Apache Spark Apache Storm Apache Hive	Power BI
Team 3	Python	Java app	IBM Cloudant IBM Cloud Object Storage HDFS	Apache Hive Apache Impala IBM Watson Studio	IBM Watson IoT Node.js
Team 4	JavaScript MySQL	Apache Sqoop	HDFS	Apache Spark Apache Hive	Angular HTTP Node.js
Team 5	MySQL	Apache Sqoop Apache Kafka MQTT	HDFS MongoDB	Apache Spark Apache Hive	Tableau IoT MQTT Panel
Team 6	C# MySQL	Apache Sqoop	MongoDB HDFS	Node.js Apache Impala Apache Hive	Angular NVD3

An important aspect in this context was the way students have learned and have generated new knowledge in return, via the practice of research-guided tasks in an autonomous manner. The collaboration we managed to have with the industry along the learning path was of high impact for the students' learning process. They had the chance to learn from researchers and professionals that employ these emergent technologies in industry on a daily basis. The project topic was related to real-world case studies, which has motivated the students in their project. The industrial collaboration was materialized with two invited lectures (on the topics of "Industrial Standards for Data Mining Projects" and "Big Data Visualization"), a visit to the production line and the involvement in the students' projects (from problem definition to prototype implementation and demonstration) with a proposal for a project topic, with advice and feedback when needed and six intelligent sensors provided to the students.

Academic Evaluation. The evaluation criteria considered from the perspective of the teaching responsible were the following: criterion 1 - data producers; criterion 2 - data sources; criterion 3 - data storage; criterion 4 - data visualization; criterion 5 - the technologies used (the solution design); criterion 6 - the maturity of the prototype; criterion 7 - the presentation of the projects; and criterion 8 - the team (collaborative) work. The individual contribution of the students was also considered, both from the teacher perspective and from the perspective of the team leaders of the respective teams.

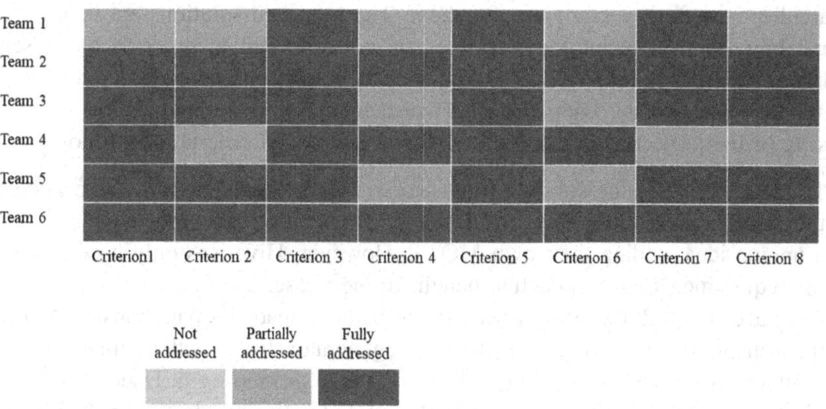

Fig. 3. Results of the evaluation of student teams along the eight evaluation criteria.

In order to conduct the evaluation, every criterion received a score corresponding to the degree to which it was addressed in the project: 0 – meaning not addressed (missing), 0.5 – meaning partially addressed, and 1 – meaning fully addressed (complete). The eight criteria evaluated for each of the six teams of students are illustrated in Fig. 3. The conclusions that can be drawn from these figures with respect to the strengths and the weaknesses of the criteria observed are that the real strengths of the implemented solutions were with respect to the data storage, and the technologies used. The students took responsibility in making decisions on the choice of data storage solutions and their choices materialized in a wide range of the available technologies. The presentations were of high quality, even though the degree of involvement of all the team members needs to be improved, as also observed by the industrial partners. There were minor issues with respecting the data producers specifications and the data sources integration, observed within two student teams. The visualization and the prototype maturity criteria are the ones that need improvement in half of the cases. However, these constitute two major challenges of the project. Improvement is also needed regarding the team work and the student motivation. Even though the teams showed good collaboration along the semester and during the progress reports, their involvement as a team during the workshop presentation leaves place for improvement. Regarding student motivation, it was observed that the students addressed all the required criteria. However, only one team went beyond what was required and addressed speed processing.

Evaluation from the Industry. The industrial partners evaluated and ranked the student projects with respect to the following criteria: (1) the implementation of the requirements; (2) the quality of the presentation during the student workshop; (3) the automation of the proposed solution; and (4) the originality. Overall, the student teams managed to implement the project requirements, and the solutions they had chosen were based on a variety of technologies, which, in the vast majority of cases, led to a functional solution. The industrial partners have appreciated the team spirit demonstrated by the students during their work. The fact that the final solution was the result of a collective effort was obvious for them. The proposed solutions are largely functional, however, the technology readiness is not sufficient in order to be presented within an external event (e.g., a Data Science meetup). As a general recommendation for the students was the improvement of presentation and communication skills. The ranking of the projects according to the above-mentioned criteria is as follows:

- 1st place, Team 6: the solution proposed was automatized; it was appreciated that the students made efforts towards optimizing the data aggregation and came up with a unique idea, coding their own MQTT. However, Hive was only used to respect the requirements, with no actual benefit in their case.
- 2nd place, Team 2: they developed three separate threads for data transmission from the sensor, the simulator and the error generator for a better efficiency; good dashboards for sensor data but with no real-time processing in Hadoop.
- 3rd place, Team 4: a rather complete solution, but with large data simplification with a 1:1 relation implemented for the time stamp matching of the sensor and machine data.
- 4th place, Team 3: interesting approach based on IBM tools; however, the jobs were not automatized; the update (refresh) of the data visualization was not sufficiently implemented. MapReduce was used instead of Hive.
- 5th place, Team 5: job automation and data update for the visualisation were missing.
- 6th place, Team 1: the project did not succeed with data ingestion from the sensor; the automation was not implemented and the required histograms were missing.

As a final result, it is worth mentioning that the course "Big Data Processing and Applications", which constitutes the context of the present study, was awarded with the ANIS scholarship [15] at first edition, on the topic of Big Data. This demonstrates the appreciation that the industrial actors in the Romanian IT landscape have showed for the initiative reported here.

3.4 Impact

The ReLearnBigDatA initiative had a major impact on the learning process of the students, and on the corresponding teaching methods. The students have learned to be active part of their learning process, to employ their critical thinking and problem-solving capabilities in a research-informed and industry-oriented project. The method employed aimed towards sustainable learning, with the focus on knowledge generation, knowledge sharing and reuse. The knowledge gained along this course will serve as a basis for the students' future projects, internships and their dissertation thesis. Students have become aware of the importance of their work at a social level as well. They have demonstrated responsible attitudes in their work and ideas towards the research and education processes, being aware of the impact their work can have on their society. This has actually motivated students in their creative ideas.

The work obtained under this project has potential for scientific impact. It has already been presented within a scientific workshop, and will be further disseminated in scientific publications.

During this project, we have been constantly in contact with similar initiatives (EU European Data Science Academy [9], BYTE Big Data Community [14], etc.), have used learning materials from them and intend to continue this direction in the near future, for the benefit of the students.

4 Conclusion

The main goal of the paper is to illustrate how the collaboration between academia and industry enables students to get into design of solutions for data engineering and empowers them to take their own decisions. It is an interesting way of teaching a topic which is extremely wide and constantly changing. For this, the paper presents in the second part a real use case from Education, illustrating how an ICT project enabled students to learn how to implement a data pipeline. In conclusion, we would like to stress that the ReLearnBigDataA initiative is a sustainable one, both for students and for the teaching process. As such, students will be able to use their experience with this initiative in future projects/work. While at a teaching level, the benefit is mainly from the bidirectional flow education-research [16].

The future iterations of the master course "Big Data Processing and Applications" will be continually improved and updated. The next academic year, a more tutorial-oriented teaching method will be targeted, together with various application domains, such as Green University/Green Campus. Research-wise, we are aiming for a comprehensive technology survey as future work emerging from this study.

Acknowledgement. The present study was realized within the Advanced Fellowships 2018 offered by the STAR-UBB Institute from the Babes-Bolyai University [17]. The sensors and manufacturing use case were provided by Robert Bosch SRL. Romania.

References

1. Cavanillas, J.M., Curry, E., Wahlster, W. (eds.): New Horizons for a Data-Driven Economy. A Roadmap for Usage and Exploitation of Big Data in Europe. Springer, Cham (2016). https://doi.org/10.1007/978-3-319-21569-3
2. Marz, N., Warren, J.: Big Data. Principles and Best Practices of Scalable Realtime Data Systems. Manning, New York (2015)
3. White, T.: Hadoop: The Definitive Guide, 4th edn. O'Reilly Media Inc., USA (2015)
4. Sadalage, P.J., Fowler, M.: NoSQL Distilled: A Brief Guide to the Emerging World of Polyglot Persistence. Addison-Wesley, Pearson Education, Inc., Boston (2013)
5. Dong, X.L., Srivastava, D.: Big Data Integration, Synthesis Lectures on Data Management. Morgan & Claypool, San Rafael (2015)
6. Gartner: Data catalogs are the new black in data management and analytics. https://www.gartner.com. Accessed 22 Mar 2019
7. Cloud Security Alliance Homepage. https://cloudsecurityalliance.org/articles/csa-big-data-releases-top-10-security-privacy-challenges/. Accessed 22 Mar 2019
8. Curry, E., Ngonga, A., Domingue, J., Freitas, A., Strohbach, M., Becker, T., et al.: D2.2.2. Final version of the technical white paper. Public deliverable of the EU-Project BIG (318062; ICT-2011.4.4) (2014)
9. EDSA Homepage. http://edsa-project.eu/. Accessed 22 Mar 2019
10. LERU (League of European Universities): Excellent education in research rich universities (2017)
11. XDK Cross Domain Development Kit. https://xdk.bosch-connectivity.com/. Accessed 22 Mar 2019
12. High Performance Computing System and Private Cloud IBM IntelligentCluster, UBB. http://hpc.cs.ubbcluj.ro/. Accessed 22 Mar 2019
13. Bufnea D., et al.: Babes-Bolyai University's High Performance Computing Center. Studia Universitatis Babes-Bolyai, Informatica Series, vol. LXI, no. 2, pp. 54–69 (2016)
14. BYTE Project Homepage. http://new.byte-project.eu/. Accessed 7 Jan 2019
15. Employers' Association of the Software and Services Industry, ANIS Scholarships. https://www.anis.ro/burse-anis. Accessed 22 Mar 2019
16. Harland, T.: Teaching to enhance research. High. Educ. Res. Dev. **35**(3), 461–472 (2016)
17. STAR-UBB Institute (Institute of Advanced Studies in Science and Technology). http://starubb.institute.ubbcluj.ro/. Accessed 22 Mar 2019

Smart Infrastructure

Business Challenges for Service Provisioning in 5G Networks

Alexandros Kostopoulos[1](✉), Ioannis P. Chochliouros[1], and Maria Rita Spada[2]

[1] Hellenic Telecommunications Organization S.A. (OTE), Marousi, Greece
{alexkosto, ichochliouros}@oteresearch.gr

[2] WIND Tre, Rome, Italy
MariaRita.Spada@windtre.it

Abstract. 5G networking is expected to induce more changes in the era of mobile communications, since new players will appear, including the new vertical industries, while the old ones will undertake new roles. 5G ESSENCE [1] addresses the paradigms of Edge Cloud computing and Small Cell as-a-Service (SCaaS) by fuelling the drivers and removing the barriers in the Small Cell (SC) market. In this paper, we investigate current market status in 5G networks and small cells, as well as we apply the business model canvas methodology for the 5G ESSENCE approach in order to explore the 5G business environment.

Keywords: 5G networks · Market analysis · Business model canvas · Cloud-enabled small cell (CESC) · Mobile edge computing (MEC) · Network functions virtualization (NFV) · Software defined networking (SDN)

1 Introduction

Up to now several visions of 5G have been proposed and their basic features, apparently, converge to the idea that any person or item can connect at arbitrarily high data rates, from any place, and with extremely low latency [2]. How these traits can be realised depends on several factors, including combinations of existing types of communication networks, as well as new and ground-breaking implementations. 5G solutions envisage consolidation of cellular, Internet of Things (IoT), and Wi-Fi networks; this list may be enriched with broadcast networks and automotive systems. Furthermore, in order to achieve higher transmission rates, new wireless solutions, such as exploitation of millimetre-Wave (mmWave) bands, are expected to be utilised. The reasons behind these developments are well grounded in practical requirements. Separate radio interfaces are required for the different solutions, such as cellular over IoT and additionally, the exigency for extremely low latency drives inexorably to ultra-dense deployments and usage of higher frequencies [3].

The main problem of the proposed 5G solutions is that they neither have been adequately tied to a solid business case, nor well integrated to the legacy infrastructure of network operators and the rest of actors within the communications ecosystem. This is not surprising, considering the lack of new monetisable revenue streams [4].

W. Abramowicz and R. Corchuelo (Eds.): BIS 2019, LNBIP 353, pp. 423–434, 2019.
https://doi.org/10.1007/978-3-030-20485-3_33

For years, operators have been making efforts to launch new services, such as image messaging, video calls, location based services and so on (and even some examples of Over-The-Top (OTT) players like Skype have achieved to deliver added value solutions); however, the average revenue per user remains low.

Forecasts of new services that will be rolling resulting in large scale consumer uptake and a corresponding increase in revenues remain vague, and as history has shown, this remains a difficult task. For example, users are able to, and they do, revert to Wi-Fi usage when the prices from the cellular network increase. Many do not value data rates above 4G levels – for example only around 22% of UK households have chosen to upgrade their home broadband to BT's "Infinity" package despite it being available [5]. Moreover, the benefit of low-latency appears to a small number of professional applications and may only be available to relatively few locations.

Alternatively, if there are no new revenues, there arises the ultimate necessity that 5G reduces operators' costs. To address this requirement, 5G ESSENCE introduces innovations in the fields of network softwarisation, virtualisation, and cognitive network management. It also provides a highly flexible and scalable platform, capable of supporting new business models and revenue streams by creating a neutral host market and ultimately reducing operational costs by providing new opportunities for ownership, deployment, operation and amortisation.

5G ESSENCE [1] addresses the paradigms of Edge Cloud computing and Small Cell as-a-Service (SCaaS) by fuelling the drivers and removing the barriers in the Small Cell (SC) market. 5G ESSENCE provides a highly flexible and scalable platform, able to support new business models and revenue streams by creating a neutral host market and reducing operational costs, by providing new opportunities for ownership, deployment, operation and amortisation.

This paper extends our previous work in [6], and focuses on the market and business aspects of the 5G ESSENCE. The paper is organized as follows; Sect. 2 gives a brief overview of the 5G ESSENCE architecture and use case scenarios. Section 3 investigates the market status of 5G networks and small cells. In Sect. 4, we apply the business model canvas methodology for 5G service provisioning. We conclude our remarks in Sect. 5.

2 Architecture and Use Case Scenarios

The Small Cell concept is evolved as not only to provide multi-operator radio access, but also, to achieve an increase in the capacity and the performance of current Radio Access Network (RAN) infrastructures, and to extend the range of the provided services while maintaining its agility. To achieve these ambitious goals, 5G ESSENCE leverages the paradigms of RAN scheduling and additionally provides an enhanced, edge-based, virtualised execution environment attached to the small cell, taking advantage and reinforcing the concepts of Mobile Edge Computing (MEC) and network slicing.

The architecture combines the current 3rd Generation Partnership Project (3GPP) framework for network management in RAN sharing scenarios and the ETSI Network Function Virtualization (NFV) framework for managing virtualised network functions.

The Cloud Enabled Small Cells (CESC) offer virtualised computing, storage and radio resources and the CESC cluster is considered as a cloud from the upper layers. This cloud can also be "sliced" to enable multi-tenancy. The execution platform is used to support VNFs that implement the different features of the Small Cells as well as to support for the mobile edge applications of the end-users.

As shown in Fig. 1, the 5G ESSENCE architecture allows multiple network operators (tenants) to provide services to their users through a set of CESCs deployed, owned and managed by a third party (i.e., the CESC provider). In this way, operators can extend the capacity of their own 5G RAN in areas where the deployment of their own infrastructure could be expensive and/or inefficient, as it would be the case of e.g., highly dense areas where massive numbers of Small Cells would be needed to provide the expected services.

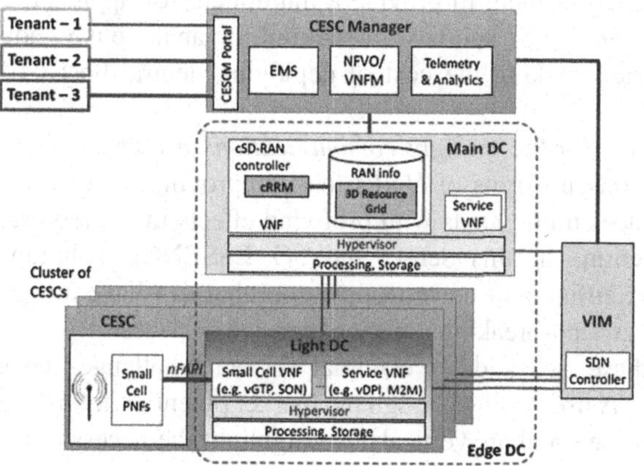

Fig. 1. 5G ESSENCE architecture [6]

In addition to capacity extension, the 5G ESSENCE platform is equipped with a two-tier virtualised execution environment, materialised in the form of the Edge Data Centre (DC) that allows also the provision of MEC capabilities to the mobile operators for enhancing the user experience and the agility in the service delivery. The first tier, i.e., the Light DC hosted inside the CESCs, is used to support the execution of Virtual Network Functions (VNFs) for carrying out the virtualisation of the Small Cell access. In this regard, network functions supporting for example traffic interception, or tunneling encapsulation/decapsulation, are expected to be executed therein. VNFs that require low processing power could also be hosted here. The connection between the Small Cell Physical Network Functions (PNFs) and the Small Cell VNFs can be realised through, e.g., the network Functional Application Platform Interface (nFAPI). Finally, backhaul and fronthaul transmission resources will be part of the CESC, allowing for the required connectivity.

The second cloud tier, i.e., the Main DC, will be hosting more computation intensive tasks and processes that need to be centralised in order to have a global view of the underlying infrastructure. This encompasses the centralised Software-Defined

Radio Access Network (cSD-RAN) controller which will be delivered as a VNF running in the Main DC and makes control plane decisions for all the radio elements in the geographical area of the CESC cluster, including the centralised Radio Resource Management (cRRM) over the entire CESC cluster. Other potential VNFs that could be hosted by the Main DC include security applications, traffic engineering, mobility management, and in general, any additional network end-to-end services that can be deployed and managed on the 5G ESSENCE virtual networks, effectively and on demand.

The considered use cases target important vertical industries. For the use case *5G edge network acceleration for stadium*, it is evident that such venues could offer new and outstanding services to their customers, opening up new business models, significantly improving network quality and user experience, and sustaining continuous profitability for both operators and enterprises. Moreover, deploying intelligence at the network's edge enables them to provide a multiplicity of applications relevant to the venue's thematic area (i.e., sports in a stadium, or analogously, advertisements in a shopping mall, etc.) making the cost of deploying neutral host CESCs increasingly viable.

The same holds for the *in-flight communications* use case, where deploying edge network applications in forms of VNFs (e.g., improving compression of video data) substantially reduces traffic loads with profound effects in the required data rate of the backhaul connection. In this sense, the 5G ESSENCE solution remains totally affordable for the airline and network operators that can have access to new revenue streams through ground-breaking, still inexpensive services.

On top of these, we address essential societal challenges through the *mission critical applications* use case. Although the market potential in this case could not be as tremendous as in the previous verticals, we highlight the necessity of 5G technologies in the community not only business related but also pertaining to the good of the wider society.

3 5G and Small Cells Market Status

Soundness of 5G ESSENCE solutions and market potential 5G is a fundamental shift and a key enabler in the future of the digital world, changing the way people innovate, collaborate, and socialise, and this shift to 5G introduces novel challenges and approaches. The introduction of 5G technologies in economic and societal procedures of everyday life is a key asset that supports transformation of the vertical industries and focuses on the development of cloud infrastructure and the concomitant development of mobile broadband digital access networks.

This rising demand for high data volume and bandwidth-greedy applications will certainly introduce new requirements in the existing infrastructure. The services market offered by the network edge is estimated to grow up to €7.1 Billion by 2021, at a compound annual growth rate of 32.6% [7]. The advent of IoT and the predictive and real-time intelligence on network devices act as a catalyst to the growing adoption of edge cloud solutions and services. To respond, the next generation of communication and services includes virtualised applications that run much closer to mobile users ensuring network flexibility, economy and scalability. The delivery of services from an edge cloud represents a new synergistic business model between network service providers or edge cloud providers and enterprises. Reference [8] presents an estimate of the potential revenue associated with providing these edge cloud hosted services. For the edge cloud provider it represents an opportunity to nearly double the revenues associated with the new services, while for an Information Communication Technology (ICT) services enterprise it shows a potential opportunity to add a new revenue stream to its portfolio by using and delivering over the top services.

The small cell market, which is at the core of 5G ESSENCE, is exhibiting a very substantial growth in the last years, and will continue to grow, to the point that, according to [9], the value of Long Term Evolution (LTE) small cells will represent more than 85% of the small cell equipment market. The Small Cell Forum (SCF) market status report [10] indicated that up to 13.3 million small cells were shipped up to the end of 2015, and that during 2015, the urban small cell shipments grew 280%. The analysis in [11] forecasts that the global small cell networks market will grow up to €3.6 billion by 2020 at a Compound Annual Growth Rate (CAGR) of 29.80%. Similar forecasts presented in [10] predict even higher revenues, reaching €6.2 billion by 2020. It is also worth emphasising that more than 90% of this total revenue will be associated to small cells in non-residential areas, like it is the case of the small cells considered in the 5G ESSENCE scenarios.

5G ESSENCE realises a two-tier cloud architecture enabling the provision of dynamically repurposed virtual network infrastructures with tailored computing and flexible networking capabilities. This will greatly benefit the Mobile Network Operators (MNOs) and Over-the-top (OTT) players to deploy and offer cutting-edge services to specific customers, with increased cost savings (e.g., energy efficiency thanks to the Edge DC design and the portability of functionalities closer to the mobile network edges) and allowing optimal reuse of the deployed virtual infrastructures.

At the same time, in order to promote greater network capacity, scalability, and reduced latency, ultra-dense networks with numerous small cells have been proposed. They require advanced interference mitigation and advanced automation of management and operations, all at the edge of the network. In this context, the Small Cell as-a-Service (SCaaS) approach (enabled also through multi-tenancy and the network slicing model in 5G ESSENCE) will be substantially more efficient in highly dense scenarios than solutions based on independent deployments on a per-operator basis, thus bringing significant cost reductions. A study by Nokia [12] investigated how many small cells would be needed to deliver the same Quality of Experience (QoE) using the neutral host model compared to each of three operators deploying its own small cell network. The results showed that deploying a small cell network provided a substantial boost of up to 35% in the "baseline" QoE measured in terms of the number of subscribers receiving 1.8 Mbps downlink throughput. Even more, only 25 shared sites were needed in the neutral host deployment compared to 57 sites required if the operators used the "go-it-alone" method. That represents a 56% saving in the number of sites, leading to a significant reduction of costs. In similar terms, the study by Ericsson in [13], analysed the benefits of a wholesale model in which operations and assets are shared among multiple players through a third party. The financial benefits of this model were quantified to be around 40% in terms of asset savings and up to 31% in terms of cash-flow improvements.

Based on the aforementioned references, we clearly observe that the 5G ESSENCE solutions based on the neutral host model for facilitating the sharing of infrastructure among many tenants can lead to an important impact in terms of cost savings for network operators.

In [14], it is identified that resource management is one of the fundamental approaches to kick-start the multi-operator model. In particular, the resource management model of the small cells should enable greater operator's control over his resources, evolving this approach towards full network slicing in virtualised environments, so that the different tenants can access their own resources. In this respect, it is expected that the new enhanced operation achieved thanks to the cSD-RAN controller of 5G ESSENCE will become fundamental, impacting towards a wider adoption of the multi-operator model.

Under all the perspectives mentioned above, the core challenge in 5G ESSENCE develops an ecosystem to sustain network infrastructure openness, built on the pillars of network virtualisation, mobile-edge computing capabilities and cognitive network management. A further challenge addressed in 5G ESSENCE is that it dramatically decreases network management operating expense (OPEX) through automation, whilst increasing user perceived QoE and security.

4 A Business Model for 5G Services

In order to take clear decisions regarding the commercialisation future of 5G ESSENCE, the viability and sustainability of different exploitation schemes need to be explored. Business Model Canvas (BMC) [15] is a widely used methodology for business and market analysis. This framework includes a set of elements in order to help a company in aligning its activities. As a first step, we apply the BMC methodology in order to analyse the current gaps and potential opportunities for 5G services:

Value Proposition: The Value Proposition block provides the necessary information about what the service is actually offering. It solves a customer problem or satisfies a need in a way that gives customers a reason to choose a company's service over another.

Key Partners: The key partners block refers to other technical partners that ensure the smooth operation of the business unit. In particular, hardware providers and manufacturers are partners who provide, support or/and maintain the existing infrastructure. Software providers could also provide the new releases, as well as the necessary software updates.

Key Activities: The key activities are the crucial operations that the company offering the new service needs to do in order to deliver its services and make the rest of the business work. The main activity is the new service provision. Apart from such activity, additional activities may include support and maintenance, as well as promotion actions to the stakeholders using the available customer channels, etc.

Key Resources: Key resources are the strategic assets that are needed for the company to keep its competitive advantage. The key resources mainly include the personnel and know-how needed for the aforementioned key activities.

Furthermore, they may include any type of hardware infrastructure, as well as software technology.
Customer Segments: The customer segments block describes the potential customers of the provided service.
Customer Relationships: The customer relationships block describes the relationship of the customers with the company providing the new service.
Customer Channels: The customer channels block indicates how the provided service can be reached by the customers.
Revenue Streams: The revenue streams may vary depending on how a specific service is offered within the market by employing specific charging schemes (e.g., usage fees, subscription fees, leasing, licensing, brokerage fees, etc.).
Cost Structure: The cost structure is aligned with the key activities. It includes any type of fixed and variable costs.

The business model canvas can be split into four major groups of building blocks: A central aspect is the *value proposition*, which defines what a business model is built around. The *customer group* includes customer relationships, segments, and channels. The third group focuses on the *involved stakeholders* of a business model, namely the key partners and their activities, as well as the key resources. Underlying to the business model is the *financial structure* with revenue streams and costs. The financial structure analysis is not included in this paper. Our future work will be focused on conducting a techno-economic analysis for each use case scenario.

4.1 Value Proposition

In the first use case "*5G edge network acceleration for a stadium*", 5G ESSENCE delivers benefits to media producers and mobile operators, enabling them to offer a highly interactive fan experience and to optimise operations by deploying key functionalities at the edge (i.e., evolved Multimedia Broadcast Multicast Services (eMBMS) together with multitenancy support from small cells). By leveraging the benefits of small cell virtualisation and radio resource abstraction, as well as by optimizing network embedded cloud, it becomes possible to ease the coverage and capacity pressure on the multimedia infrastructure, and also to increase security since content will remain locally. Furthermore, additional benefits for the operators and the venue owners arise: (a) *lower latency*, due to shortening the data transmission path; (b) *increased backhaul capacity*, due to playing out the live feeds and replays locally that puts no additional strain on the backhaul network and upstream core network components.

In the second use case "*mission critical applications for public safety*", 5G ESSENCE common orchestration of radio, network and cloud resources significantly contributes to the fulfilment of the requirements of the Public Safety (PS) sector, bringing new tools to share both radio and edge computing capabilities in localised/temporary network deployments between PS and commercial users. The challenge consists on allocating radio, network and cloud resources to the critical actors (e.g., the first responders), who by nature require prioritised and high quality services. For an allocation like this to be efficient and thus to enable E2E network slicing, the 5G ESSENCE solution is applied.

The third scenario is "*next-generation integrated in-flight connectivity and entertainment systems*". In order to offer cost-effective mobile broadband Internet connectivity, it is imperative to integrate on-board a neutral host solution that will allow multi-operator connectivity to the passengers, also accounting for variable service offerings. Such host-neutral services are important for European airliners since they traverse several regional boundaries served by a large variety of mobile operators. Regarding backhauling, terrestrial direct-air-to-ground (continental airspace) and satellite (oceanic airspace) solutions can used. The unique architecture proposed in 5G ESSENCE, which combines efficiently the virtualised and multi-tenant small cell networks with a multi-tier cloud edge infrastructure, is an essential step for integrating a pioneering integrated In-Flight Entertainment and Connectivity (IFEC) system that will jointly deliver the required communication and network infrastructure for the wireless IFEC (to both the embedded IFE devices and the wireless "Bring Your Own Devices" (BYODs)).

4.2 Key Stakeholders, Activities and Resources

Communication Service Providers (CSP) and Application Providers (AP): 5G ESSENCE enhances the rapid deployment of new services for consumer and enterprise business segments which can help them differentiate their service portfolio, benefitting from the CESC framework. Moreover, it adds new revenue streams from innovative services delivered closer to the user, together with offering the user a better service-oriented QoE, leveraging the Edge DC and the CESCM entities, and furthermore, improving revenue opportunities by sharing the infrastructure for specific service providers. Furthermore, it introduces new applications which are aware of the local context in which they operate (RAN conditions, localised information, density information, etc.) through the integration of the CESC virtual small cells functionalities, which open up new service categories and enrich end-user offerings. Additionally, it will result in drastic reduction of OPEX costs by offloading management related functionalities closer to the edge and by developing smarter management techniques, and further limiting the Total Cost of Ownership (TCO)/capital expenditure (CAPEX) costs by promoting shared infrastructures enabled by the multi-tenancy and the virtualised multi-service management framework (CESCM). Furthermore, it boosts the flexible development of market innovative and ground breaking services and applications that take advantage of the contextual information provided by the CESC on the radio network conditions and other information at the edges (e.g., edge caching, critical services). Last but not least, it creates new market entrants by opening up the shared infrastructures to new software and application providers, infrastructure vendors and other CSPs, thereby increasing revenues and also promoting regulatory support.

Equipment Manufacturers and Vendors: Using the 5G ESSENCE approach, the equipment manufacturers will be able to greatly enhance their product portfolio and to develop novel offerings in virtualised and cloud-enabled small cells platforms. The 5G ESSENCE solutions and especially the CESC offerings will allow the manufacturers to place themselves as key proponents in the mobile edge-computing arena, which is critical towards 5G network deployments. It will enable equipment manufacturers to offer competitive solutions that can co-locate computing and small cells through an Edge DC, towards delivering computational and networking resources via VNFs empowered by hardware accelerators.

IT Service Providers and Solutions Suppliers: IT service providers and solutions suppliers get the maximum benefit and advantage out of the 5G ESSENCE project by being able to closely work with leading mobile market vendors and ICT companies, and so they have the opportunity to position themselves strategically and gain an early-entrant advantage within the industry, also keeping in consideration the rapid evolution of the mobile edge computing market, expected to be a €14bn market by 2020 [10]. In particular, by contributing to the design and prototyping of the CESC, Edge DC, and CESCM functionalities, the IT industry will get a clear roadmap and exploitation opportunity to be placed as key proponents in the open source development communities, promoting the enhancement of solutions which can be developed further through collaboration with application and open programmable platforms developers. This will enable the IT suppliers to even directly approach equipment vendors, to offer competitive solutions at lower costs, enabling new business avenues and increased revenues. The virtualisation and "cloudification" of the network architectures often result in transformation of the networks and increase the demand of a broad range of expertise of ICT.

Enterprise Customers, Home and Mobile Users: European vertical industries are seeking enhanced technical capacity in order to differentiate themselves at international level and to strengthen their brands. As regarding the selected use cases for demonstration 5G ESSENCE is in position to address both the increasing operational traffic at events and to fulfil the high expectations from spectators. Telecom operators will be able to differentiate their products and increase brand awareness with large volume of spectators. In this context, the event organisers can also have access to new, and potentially big, revenue streams. The public safety professionals can benefit from 5G ESSENCE solutions since lower cost and more interoperable, with improved functionalities, equipment will be available to them. This will lead to safer working and to better operational efficiency. Finally, the airline companies through 5G ESSENCE IFEC system can also expect new revenue streams, better service differentiation and improved passenger experience.

Small Medium Enterprises (SMEs): The decoupling of software and hardware via SDN and NFV technologies, and the introduction of successful, open source software stacks for networks leads to open network ecosystems that are no longer limited to the large manufactures and their telecom customers. With the use of commercial off-the-shelf (COTS) instead of current proprietary technologies, as proposed by 5G ESSENCE, network functionalities will not remain restricted in monolithic "boxes", but will become totally virtualisable and thus easily reconfigurable. This will greatly influence SMEs that develop network services, since they will be able to innovate and launch new applications leveraging the new capabilities of 5G. In the new, unlocked ecosystem of 5G ESSENCE, SMEs can take the role of the network application developers and maintainers. Lowering the barrier for new market entrants is a recognised benefit for software network technologies, and is of particular importance for the telecom market, which is traditionally dominated by a few “big” players.

4.3 Customer Segments, Relationships and Channels

The *Customer Segments* building block identifies the different groups or entities that an enterprise may intend to reach. It is vital for every business model that a conscious decision is made about which customer groups to attend and which to ignore. The *mass market* focus, basically makes no distinction between different customer segments. Value proposition, distribution channels and customer relationships are aimed at all types of consumers with similar needs and problems. Business models that focus on a *niche market* have specific and specialized customer segments. The other building blocks of the business model necessarily follow the particular needs of the niche market. In a *segmented* approach, a business model distinguishes between customer groups with different preferences and problems. Again, this differentiation has consequences for all building blocks of the business model. The *diversified* customer segment approaches serve two customer segments at once that are not directly related to each other. These customer groups have different preferences and demands. Finally, businesses can serve two or more customer segments, called *multi-sided platforms*, which depend on each other. All segments are necessary to make the business model work. Providers of credit cards or free newspaper are good examples for multi-sided platforms. In the stadium use case, we may identify either a *segmented market* where a set of differentiated services is provided to the end users, or even a *niche market* for providing innovative 5G services (e.g., virtual/augmented reality). For the public safety use case, the mission critical services are focused on a *mass market*. Similarly to the stadium use case, the in-flight connectivity and entertainment services could be offered to a *segmented market*.

The *Customer Relationships* building block describes the different styles of relationships to be established with the various customer segments considered, in order to engage them to the product. Those relationships can range from very personal to automated and follow different motivations, such as customer acquisition, retention or increasing sales via upselling. The *self-service* has no direct relationship to its customers while they use the service. The *communities* generated by the firm enable communication between members and learn about their needs. *Personal customer care* enables personal interaction when the customer needs help during or after the purchase. In *individual customer care*, each customer has his own representative for individual needs and problems. We believe that *personal*, *individual customer care*, as well as *community-based* relationships could be present for all the 5G ESSENCE services. It should be noted here that the aircraft crew could also provide support to the end-users in case the latter face connectivity issues (hence, they will not be able to access any other customer care service, e.g., call centers). *Self-service* does not seem to be appropriate for such services.

The *Channels* in a business model can be understood as the company’s “interface” with their customers. They comprise the means of communication, distribution and sales with and to their customers. Channels are the points of contact with the customers and, therefore, play a very important role for the customer experience. For these purposes, there are different direct and indirect channel types that can be provided by the business promoter or a partner. *Direct channels* may include *sales force* and *web sales*, while indirect channels may include *wholesaler, partner stores* and *own stores*. The web sales seem to be the core channel for boosting the adoption of the 5G ESSENCE services. End users could access a webstore in order to perform the initial installation of the corresponding service application. *Partners and own stores* (e.g., a telecom operator's store offering connectivity services, end devices, etc.) could also act as an alternative channel for marketing and advertising purposes (e.g., offering a 5G service for free for a limited period of time when a customer purchases a new sim card/connection).

5 Discussion

5G networking is expected to induce more changes in the era of mobile communications. In detail, new players will appear, including the new vertical industries, while old ones will undertake new roles. This could be exemplified by end-users who tend to be transformed to content producers and providers. As a first step, in this paper we investigated current market status in 5G networks and small cells, as well as we applied the business model canvas methodology for the 5G ESSENCE approach in order to explore the 5G business environment. Our future work will explore novel business models starting from the business scenarios analysed from the verticals' use cases addressed, and further, identifying both transparent and collaborative business models to exploit 5G ESSENCE deployments. Such market models, identified gaps, and business scenarios will contribute to identify commercially exploitable elements.

Acknowledgements. The paper has been based on the context of the 5G-PPP phase 2 "5G ESSENCE" ("Embedded Network Services for 5G Experiences") Project (GA No. 761592), funded by the European Commission.

References

1. 5G ESSENCE H2020 5G-PPP Project. http://www.5g-essenceh2020.eu
2. European Parliament Think Tank: 5G network technology: Putting Europe at the leading edge, January 2016
3. Ericsson White Paper: 5G radio Access, Uen 284 23-3204 Rev C, April 2016
4. Bock, W., et al.: Five Priorities for Achieving Europe's Digital Single Market. The Boston Consulting Group, Boston (2015)
5. Ofcom: The Communications Market 2016: Internet and online content (2016)
6. Chochliouros, I., et al.: Using small cells for enhancing 5G network facilities. In: IEEE NFV-SDN NFVPN, Berlin, Germany, November 2017
7. Edge Analytics Market by Component (Solutions and Services), Analytics Type, Business Application (Marketing, Sales, Operations, Finance, and Human Resources), Deployment, Vertical, and Region – Global Forecast to 2021. MarketsandMarkets (2016)
8. Weldon, M.: The Future X Network: A Bell Labs Perspective. CRC Press, Boca Raton (2016)

9. Hill, K.: 6 predictions for the small cell market, February 2016. http://www.rcrwireless.com/20160227/featured/6-predictions-for-the-small-cell-market-tag6-tag99
10. SCF: Small cell deployments. Market status report, February 2016
11. Small Cell Network Market worth 3.92 Billion USD by 2020. http://www.marketsandmarkets.com/PressReleases/small-cell.asp
12. Small cell deployments: you don't have to learn the hard way Study for Small Cells deployment. http://resources.alcatel-lucent.com/asset/200248
13. Ericsson: Wholesale Network Sharing, white paper, May 2012
14. SCF 017.06.01: Market drivers for multi-operator small cells, January, 2016
15. Muhtaroglu, F., Demir, S., Obali, M., Girgin, C.: Business model canvas perspective on big data applications. In: IEEE International Conference on Big Data, CA, USA (2013)

Towards an Optimized, Cloud-Agnostic Deployment of Hybrid Applications

Kyriakos Kritikos[1(✉)] and Paweł Skrzypek[2]

[1] ICS-FORTH, Crete, Greece
kritikos@ics.forth.gr
[2] AI Investments, Skierniewice, Poland
pskrzypek@aiinvestments.pl

Abstract. Serverless computing is currently taking a momentum due to the main benefits it introduces which include zero administration and reduced operation cost for applications. However, not all application components can be made serverless in sight also of certain limitations with respect to the deployment of such components in corresponding serverless platforms. In this respect, there is currently a great need for managing hybrid applications, i.e., applications comprising both normal and serverless components. Such a need is covered in this paper through extending the Melodic platform in order to support the deployment and adaptive provisioning of hybrid, cross-cloud applications. Apart from analysing the architecture of the extended platform, we also explain what are the relevant challenges for supporting the management of serverless components and how we intend to confront them. One use case is also utilised in order to showcase the main benefits of the proposed platform.

1 Introduction

Serverless computing [1] is currently uptaken and leads to a kind of a revolution with respect to how cloud applications are developed and managed. This is due to the core benefits it delivers which include zero administration and reduced operational costs as well as the capability to scale indefinitely the serverless components. In this respect, various frameworks, like Serverless.com[1] and Fission[2], have been proposed to support devops in the deployment and provisioning of serverless components, while serverless platforms are offered by major cloud providers, like Amazon, Google and IBM, to support the deployment and scalable provisioning of such components under existing cloud infrastructures.

Unfortunately, there are various limitations in the context of making applications and their components serverless. Such limitations are grounded in the very nature of these components as well as the current serverless platform capabilities. In particular, not all application components can be made serverless as serverless computing is applicable only in certain situations (e.g., related to functional

[1] https://serverless.com/framework/.
[2] https://fission.io/.

W. Abramowicz and R. Corchuelo (Eds.): BIS 2019, LNBIP 353, pp. 435–449, 2019.
https://doi.org/10.1007/978-3-030-20485-3_34

reactive programming and stateless computations) and is recommended mainly for certain application domains [2] (e.g., image processing, social media analysis, IT automation and financial monitoring). Further, there are currently major obstacles in transforming applications into a serverless form although FaaSification tools do exist for particular programming languages [3,4]. This ends up in the ability to easily faasify only certain application components while for the rest a great re-engineering effort might be required. This is worsen by the current serverless platform limitations, including restrictions [5] particularly impacting the size of serverless components as well as their execution duration.

Based on the above analysis, it is apparent that only a certain application portion can be made serverless. This means that the rest of the components can still follow the micro-service paradigm but need to be deployed by the well-known VM/container abstractions. Thus, we move towards the advent of hybrid applications which conjunctively adopt different component deployment models. Such applications also need to become cross-cloud so as to exploit the following benefits: (a) avoid the lock-in effect by changing on-demand a cloud provider; (b) exploit the best of all available cloud service offers in accordance to the application requirements; (c) enhance their security level by incorporating various kinds of security services that could be offered by different cloud providers.

To support cross-cloud and hybrid applications, there is a great need to abstract away from current peculiarities of existing platforms/services and focus mainly on the application requirements and deployment structure. This is evident due to the diversity in both the VM/container and serverless worlds. Further, there is the need to properly integrate different kinds of application components as well as monitor and reconfigure the serverless ones.

Unfortunately, we currently evidence either management frameworks that focus on micro-service [6,7] or serverless [8] (e.g., serverless.com) but not both types of components. There also exist platforms like AWS which, via the use of respective languages, like CloudFormation, enable the deployment of hybrid applications. However, in the latter case, there is vendor lock-in plus the application's reconfiguration is not supported, while required in the current highly dynamic cloud environments. In this respect, to the best of own knowledge, there is currently no framework or platform able to support the complete management of cross-cloud, hybrid applications in a provider-agnostic manner. As such, this paper moves towards covering this gap by extending an existing platform, the Melodic one [6], dedicated to the management of big-data as well as normal cross-cloud applications, to enable it to also support the management of serverless components. Such an extension has the following benefits: (a) it saves effort by not re-inventing the wheel but re-using existing work; (b) it enables to satisfy various requirements related to the proper management of cross-cloud applications, like transparent deployment and execution as well as big data management; (c) it allows to deal with certain challenges pertaining to the management of serverless components which have been unveiled in our previous work [9].

The extended platform has been carefully designed by considering both the above generic requirements and serverless challenges, as well as the main issues introduced by a new but emerging application area, generated by merging the Artificial Intelligence (AI) and Big Data domains. As such, after presenting the architecture of the proposed platform as well as its current status of realisation, we validate its design through two different ways: (a) *requirement-oriented validation*: we validate that the platform both satisfies the requirements posed to it as well as the AI & Big Data application challenges; (b) *use-case validation*: the platform is applied on a AI & Big Data use-case which exhibits certain characteristics and structure as well as poses particular non-functional requirements. Such an application showcases the main benefits of our platform and well justifies the need for the management of hybrid, cross-cloud applications.

The remainder of this paper is structured as follows. Section 2 introduces the main challenges that AI and Big Data applications pose. The main requirements that drove the development of our extended platform are analysed in Sect. 3 along with their correlation with the identified challenges. The platform's high-level architecture as well as an analysis of its supported application deployment process is supplied in Sect. 4. The platform's validation in the context of requirement satisfaction and a certain use case is covered in Sect. 5. Finally, the last section concludes the paper and draws directions for further research.

2 Modern Big Data and AI Applications Challenges

The rising importance of AI- and Big Data-based applications creates completely new challenges for AI-based ICT systems, whose market value is forecast at 169 mld $ until 2025[3]. The computing power need of such applications is much higher, especially for training complex neural networks (NN) models that use an advanced NN architecture or processing huge datasets. For Convolutional NNs [10] there are advanced and improved architectures, such as Residual NNs [11], Inception [12], Visual Geometry Group (VGG) [13], and DenseNet [14]. Each of aforementioned architectures contains many, even over one hundred layers that need to be trained through hundreds of thousands or even millions epochs. Such a training requires huge computing power as well as specialised (e.g., GPU) or dedicated hardware with machine learning capabilities.

Further, modern AI and Big Data applications usually cannot exist without integration with an already existing, typical IT system. The purpose of such integration is to fetch data from these systems and provide real-time or near real-time predictions for various business functions like customer offer recommendation or advertisement selection on web portals [15].

The aforementioned analysis indicates that new challenges have been raised for ICT systems. These challenges are actually validated by the real use case presented in Sect. 5 and can be analysed as follows.

---

[3] https://www.alliedmarketresearch.com/press-release/artificial-intelligence-market.html/.

2.1 Data Pre-processing

The necessary element of each AI and Big Data based system is the ability to rapidly and efficiently pre-process input data, both for training and prediction. Usually such data are gathered from different sources, joined, transposed and finally pre-processed using advanced algorithms. For specific use cases, the data are prepared based on simulations, so it is required to repeatedly run many simulations to get a representative set of input data. Specifically for an AI based application, we could notice two different sub challenges for that issue:

1. Processing of data for training purposes—a very big amount of data needs to be pre-processed to allow model training. This requires high computing power, but usually the execution time is not strictly limited.
2. Data processing for prediction—this usually involves less amount of input data, but should be executed very rapidly, quite often in near real-time.

2.2 Distributed Model Training

Employing an optimal approach for training models is one of the most important challenges for modern AI and Big data applications. The modern AI system architecture usually requires training a significant number of models. In the presented use case, there is a need to train even hundreds of models. It requires a huge computing power and usually is conducted in a distributed manner via cloud computing and using Big Data frameworks like Hadoop or Spark, due to very high cost of High Performance Computing (HPC) solutions. On the other hand, a distributed architecture is usually more complex and difficult to optimise than a monolith architecture. The combination of the aforementioned facts creates a very important challenge which needs to be properly addressed.

2.3 Optimisation of Resource Usage

Due to the high computing power necessary for AI and Big Data based applications to support data pre-processing and model training, it is usually most optimal to use cloud-based solutions for this purpose. The Cloud computing model allows for a very flexible creation of the virtual infrastructure and its scaling based on available budget and cloud providers' offers. The proper use of a single- or even multi-cloud computing model and the optimisation of cloud resource efficiency are some of the most important challenges for distributed, AI and Big data applications. Due to the use of a distributed architecture, the diversity of cloud providers and offers as well as the versatility in the service level and pricing offered, choosing the optimal solution is very difficult. The cost saving and performance gain (in terms of number of trained models, pre-processed data and simulations run) could be really significant and reach as far as hundred percent, as shown in Sect. 5. So, the proper addressing of this challenge is crucial for most of the distributed AI and Big Data applications.

2.4 Real Time Predictions

The main objective of a significant number of AI and Big Data based systems is to generate predictions and analysis in real or near real time. Good examples are financial applications used for investment portfolio optimisation, dynamic pricing models for vehicle-sharing operators, and e-commerce retail offer suggestions. In these use cases, the predictions need to be quite rapidly and reliably calculated as delayed calculations would make them outdated in particular situations (i.e., customer waiting for suggested offers). So, a proper architecture design with sufficient performance and reliability as well as multi-instance deployment (deployment of more than one instance per component) are crucial for that type of applications. However, using bigger and multiplied resources is very costly, so it is very important to find the best balance between performance and reliability, on one side, and cost on the other.

3 Platform Requirements

To support our vision for cross-cloud, hybrid application management, there is a need for addressing particular challenges, either of a generic nature or mapping to the serverless domain. They were derived from our recent review of serverless frameworks [9]. According to the context of our project, Functionizer, those challenges were confined to certain phases of the application lifecycle (see Fig. 1) excluding those related to the application requirement analysis, design and development. The challenges maintained are summarised as follows:

- C_1: Support deployment reasoning for mixed cross-cloud applications. This is broken down into:
  - $C_{1.1}$: Matching any kind of application component with platform/infrastructure capabilities.
  - $C_{1.2}$: Developing and solving an optimisation problem covering the deployment of both normal and serverless components into matched VM/container and serverless platform offers by respecting application requirements.
- C_2: Rich description of both hybrid applications as well as respective platforms and cloud infrastructures by re-using and extending existing cloud modelling languages, like CAMEL [16].
- C_3: Support the dynamic management of serverless provisioning frameworks on top of cloud infrastructures.
- C_4: Extend the support level for unit testing and support integration testing for hybrid applications by possibly extending existing integration methods.
- C_5: Support measurement of custom metrics, the aggregation of measurements in multiple levels plus the detection of complex event patterns.
- C_6: Support the dynamic adaptation of hybrid, cross-cloud applications.

Apart from the above requirement/challenge set, we also advocate that there is a need to support the orchestration of the different cloud services for supporting the adaptive deployment and provisioning of cross-cloud, hybrid applications.

In addition, certain generic requirements relevant for the management of big data cross-cloud applications have been considered in order to support the design and development of our envisioned platform. Such requirements were derived based on our experience and knowledge as well as our involvement in the parent project of Functionizer called Melodic. These requirements are summarised as follows:

- R_1 – *Transparent deployment and execution*: there is a need to support the transparent deployment and execution of cross-cloud applications with the right automation level based on the user's application requirements and deployment structure given in the form of an application model.
- R_2 – *Data management*: especially in the context of big data applications, there is the need to appropriately manage the data manipulated by them. A special focus lies on the following two aspects: (a) appropriate capturing and maintenance of meta-data about the data to be managed; (b) appropriate placement and migration of data based on application requirements, where location and security requirements represent the norm via which compliance to national and international laws and legislation can be achieved.
- R_3 – *Privacy & Confidentiality*: there is a need to properly handle private/sensitive data to disable their disclose to unauthorised parties. This must be achieved by securing the management platform and validating that privacy requirements are respected during application deployment and provisioning.
- R_4 – *Application Availability*: there is a need to attain high levels of application availability by dynamically adapting its provisioning and placement.
- R_5 – Application Scalability: there is a need to scale the cross-cloud application on-demand according to certain events or event patterns by exploiting the capabilities of the respective cloud infrastructures/platforms.

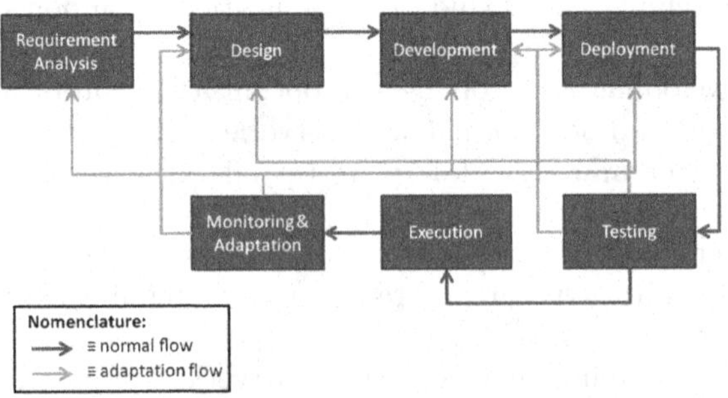

Fig. 1. The application management lifecycle

The agglomeration of the above requirement sets makes up the main requirements for the Functionizer platform. This agglomeration directly relates to the AI and big data application challenges identified in the previous section, as depicted in Table 1. As it can be seen, all requirements seem to be relevant for the challenges identified in Sect. 2. This indicates that the requirements are not only valid but definitely need to be realised in the development of our platform.

Table 1. Mapping of AI & Big Data challenges to the platform challenges

AI & Big Data challenges/requirements	C_1	C_2	C_3	C_4	C_5	C_6	R_1	R_2	R_3	R_4	R_5
Data pre-processing	✓	✓	✓	✓	✓	✓	✓	✓	✓	✓	✓
Distributed model training	✓	✓	✓	✓	✓	✓	✓	✓			✓
Handling data streams		✓		✓			✓	✓		✓	✓
Optimisation of resource usage	✓	✓	✓	✓	✓	✓	✓	✓		✓	✓
Real time predictions	✓	✓	✓	✓	✓	✓	✓	✓		✓	✓

4 Platform Architecture

4.1 High-Level Architecture

To manage cross-cloud hybrid applications, there is a need to minimise the respective effort by extending existing work as much as possible. As such, we have decided to extend the Melodic platform, being developed in the context of the Melodic H2020 project[4]. This enabled us to satisfy all generic requirements (from R_1 to R_5) in the context of normal and Big-Data, cross-cloud applications as well as partially cover the challenges $C_1 - C_6$, as identified in Sect. 3. In this respect, our focus is mainly on extending this platform to support the deployment and adaptive provisioning of serverless components within such cross-cloud applications, thus also enabling to fully address the challenges $C_1 - C_6$.

By extending Melodic, the Functionizer platform also follows a model-driven approach to manage cross-cloud applications in which CAMEL, a state-of-the-art cloud modelling language, plays a central role. CAMEL covers multiple aspects of an application, including deployment, requirement, monitoring and scalability. It has been extended to support the modelling of serverless components as well as their configuration and requirements, including now both (serverless) platform, infrastructure, location and scaling requirements. CAMEL extension was realised in its textual editor while the web-based editor implementation is under way.

The Functionizer platform includes two planes for integration: the control and monitoring plane. The control plane includes an enterprise service bus (ESB) on top of which sits a business process management (BPM) layer. On the other hand, the monitoring plane offers a message-based mechanism via which monitoring feedback can be forwarded to appropriate platform components. The BPM layer flexibly enables supplying and executing all management functions (e.g., the application deployment one) in form of business processes which encapsulate respective calls to the platform components, offered in the form of micro-services under the unified ESB. This critical platform feature enables its easy and flexible extension according to the following ways: (a) existing platform components can be extended to cover the handling of serverless application components. In case of interface update, this can be handled through updating the respective task

[4] melodic.cloud.

description in the business process; (b) new components can be put in place by just integrating them in the respective management process concerned.

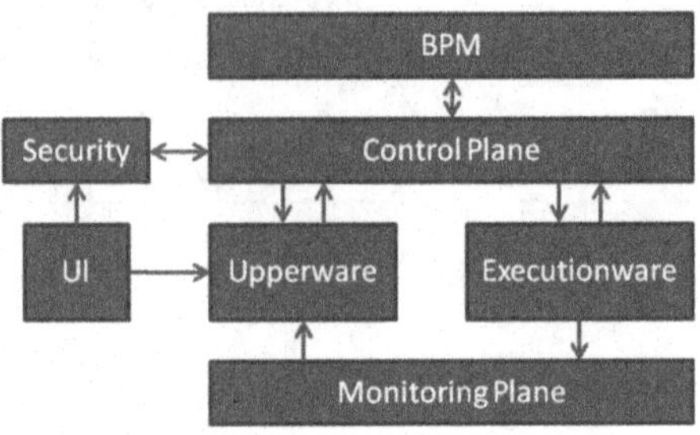

Fig. 2. High-level overview of Melodic/Functionizer platform architecture

By following a model-driven approach, a high-level view of the Functionizer's platform architecture is depicted in Fig. 2 (see also [6] for a more detailed analysis). As it can be seen, there are mainly 4 platform modules and two control planes (as already discussed). These 4 modules are shortly analysed as follows:

- *Upperware*: This module is responsible for the transformation of the users' CAMEL model, characterising their application, into a deployment plan, by employing application deployment reasoning. It is also in charge of orchestrating this plan with the assistance of the *Executionware* module. The *Upperware* is also in charge of detecting the opportunities for reconfiguring the user application and enforcing them. This covers both local and global reconfiguration, where the former takes the form of the execution of certain scalability actions in the context of triggering particular scalability rules. The global reconfiguration of the application is triggered when the application's Service Level Objectives (SLOs) are violated through the production and orchestration of a new deployment plan.
- *Executionware*: It handles the overall resource management and is responsible for executing each step involved in the application deployment plan. In the context of application component deployment, this module is also in charge of the lifecycle management of application components and the installation of sensors useful for their monitoring. Indirectly, it also offers big data processing facilities through the by-default management of manager components and the corresponding deployment of slave-based big data processing components for big data applications.
- *UI*: It includes editors and user interfaces that can assist users in exploiting the main facilities and services offered by the platform. The editors include a web-based editor that can be used for creating CAMEL (application) models.
- *Security*: It includes some security services that focus on enforcing a controlled access to the platform resources as well as the proper management of cloud credentials. These services are well integrated with the rest of the platform modules and planes.

All four modules, originally part of the Melodic platform, have been or are being extended. In a nutshell, the *Upperware* was enhanced to obtain a richer description of an application model including also the specification of serverless components and their requirements while it will be soon capable of producing deployment plans which denote to which (serverless) platforms these components are deployed. On the other hand, the *Executionware* has been updated to support the deployment and lifecycle management of serverless components. Currently, we are investigating how the monitoring of such components can be performed in a platform-independent manner. The UI is under modification to support the editing of CAMEL models based on its serverless extension. Finally, the security module is being extended with the capability to handle the credentials related to serverless platforms as well as the secured access to novel components.

An additional element planned in the Functionizer project at a later stage is the framework for testing serverless components. It would be implemented as an extension to the most appropriate, existing testing frameworks. This will enable our platform to also satisfy requirement/challenge C_4.

4.2 Application Deployment Process

A multi-cloud application's (re-)deployment is realised by executing a management process. This process includes the following sequence of activities, depicted in Fig. 3, where the extensions to the Melodic platform are especially highlighted.

1. The user describes the mixed application via the CAMEL language, including requirements for both normal and serverless components.
2. The initial deployment plan for the application is calculated by executing components of the platform's *Upperware* module. This involves:
   - The *CP Generator* conducts application profiling and finally produces a CP (constraint optimisation problem) model, which imprints the deployment alternatives for all application components. This component will be extended to support serverless platform matching to enrich the CP model with the deployment alternatives for serverless components;
   - The *Metasolver* (especially the constraint solvers that it incorporates and manages) performs deployment reasoning over the CP model. The reasoning part of the platform will also be extended to support the optimised deployment/configuration of serverless components.
3. The deployment is executed across multiple clouds based on the initial deployment plan calculated. The deployment is orchestrated by the *Adapter* and step-wise executed by the *Cloudiator* (*Executionware*) platform components. In particular, the *Adapter* transforms the initial deployment plan in a form of a graph with dependencies between resources and application components. After that, it executes the transformed deployment plan in a step by step fashion by invoking a respective *Cloudiator* method, which results in performing cloud agnostic deployment to the selected cloud provider. Both components have been adjusted to support the transparent and cloud agnostic deployment of serverless components in the context of hybrid, multi-cloud applications.

The Cloudiator was extended to support deployment of serverless components to AWS and Azure, in addition to the already implemented deployment of VMs, Docker containers and Spark Big Data framework.
4. After its deployment, the application is monitored by the *Event Processing* subsystem. This sub system will be also extended to handle and aggregate the measurements of metrics related to the application's serverless components.
5. Based on gathered and aggregated metric measurements, the application adaptation at runtime will be performed. The platform components responsible for this adaptation will be extended to support also the reconfiguration of serverless components. The reconfiguration of serverless components will contain, but not be limited to, the following possibilities:
   - Migration to another cloud provider
   - Migration to another region of already selected cloud provider
   - Setting the max and min limit of serverless component instances.
6. The runtime application reconfiguration will be run, based on the new deployment plan generated, again via the *Adapter* and *Cloudiator* components.

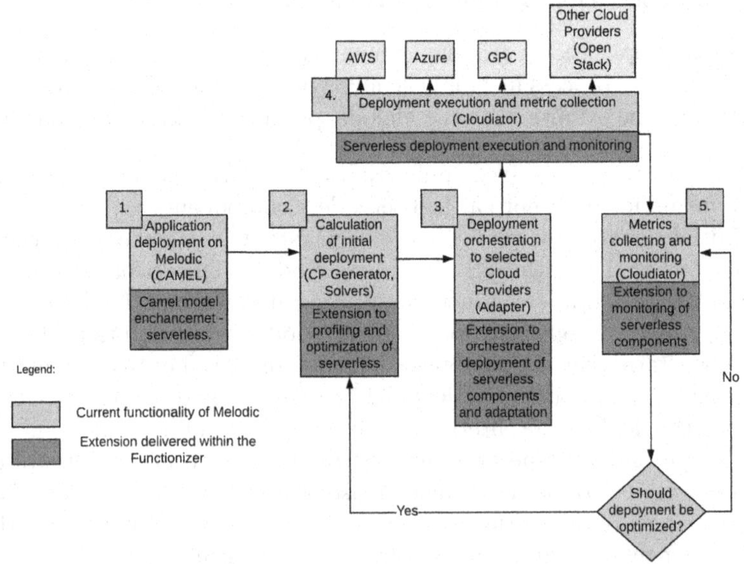

Fig. 3. Functionizer's deployment process

5 Validation

The Functionizer platform validation comes in two ways: (a) through the satisfaction of the requirements and challenges set; (b) its application in a certain use case. These two ways are elaborated in the next two sub-sections, respectively.

5.1 Requirement-Oriented Validation

Table 2 showcases how the planned extensions in the Melodic platform, as clarified in Sect. 4, address the requirements identified in Sect. 3.

Table 2. Mapping of Melodic platform extensions to the serverless-related platform challenges (from Sect. 3)

Melodic platform components/challenges	C_1	C_2	C_3	C_4	C_5	C_6	R_1	R_2	R_3	R_4	R_5
CAMEL		✓	✓		✓	✓	✓	✓	✓	✓	✓
CP Generator (Upperware)	✓		✓				✓		✓		
Meta-Solver & Solvers (Upperware)	✓		✓			✓	✓	✓		✓	✓
Adapter (Upperware)	✓		✓			✓	✓	✓	✓	✓	✓
Cloudiator (Executionware)			✓		✓	✓	✓	✓		✓	✓
Framework for serverless testing				✓							
Melodic/Functionizer platform	✓	✓	✓	✓	✓	✓	✓	✓	✓	✓	✓

As it can be seen from Table 2, the Functionizer platform satisfies all the requirements that have been set. This satisfaction can be related to the utilisation of a single or multiple components. In the latter case, there is a complementarity between the components which enables the satisfaction of the respective requirements. By correlating this table content with that of Table 1, it can be highlighted that Functionizer has the potential to support the cross-cloud deployment of mixed applications which operate on both Big Data and AI domains.

5.2 Use Case Validation

Problem Description. This use case relates to using AI for investment portfolio optimisation. It is required by the AI Investments start-up, aiming at creating a complete trading solution, including a diversified way of signalling transactions, determining market conditions and managing exposition, with the goal to develop a software platform for advanced investments in the global markets. To this end, advanced NNs like Capsule [17], residual [18] and dilated LSTM [19] are used, while a reinforcement learning approach will be applied for exposure, risk and position sizing management to build a self-improving system.

Extracting value from Big Data will become a key differentiator in the financial market. Systems using AI can analyse much more data than human beings; thus, are expected to react quicker and make better investment decisions. So, the solution will offer to investment funds and companies a complete advanced investment platform that will make their investment decisions more effective.

The most critical issue to resolve for that type of AI applications is conducting predictions in near real time. Approximate total time for prediction is maximum ten seconds per lower intervals between transactions (5 min and

15 min). Within this, data must be gathered from the markets, transposed and transformed while the respective prediction must be also performed. Further, the costs of the infrastructure need to be minimised.

The above issue requires a unified, fast and reliable architecture for data fetching/pre-processing and a distributed architecture for predictions. As thousands prediction requests could be executed in parallel, this requires many models to be available in that time. The prediction execution sequence is as follows:

1. Predictions are conducted by given interval of time. Time intervals are 5 min, 15 min, 1 h, 4 h, 1 day, and 1 week.
2. Before each interval, pre-warming requests are issued to load model and start component (for serverless components).
3. At each interval, pricing data are collected from the external source and pre-processed.
4. After data pre-processing, prediction requests are sent to the prediction component in parallel. This could raise up to 1000 of that component instances.

Application Structure and Technology. The application structure (as presented in Fig. 4) contains the following elements:

1. *Prices stream collector*: it includes the complete transformation pipeline, in which one type of component is responsible for one type of transformation. Each type of component could be deployed in multiple instances. The component has been developed in Java on Karaf platform, as an OSGI component.
2. *Cloud training distribution* for supervising training: this component is deployed in active-passive configuration. It orchestrates the model (re-)training. The Flower framework was used as a model training orchestration engine.
3. *Training model*: component used for training a neural network model. It is a Tensorflow backed by Python implementation.
4. *Cloud portfolio optimisation*: Orchestration of prediction components invocation and calculation of investment strategies. It is written in Java using serverless components for parallel processing.
5. *Prediction model*: it makes one-time prediction using a given model. It is a light Tensorflow version customised to be run on serverless components.
6. *Exposure, risk & position size management* - optimisation of the exposure for given markets, total risk and positions size using the Monte Carlo Tree Search method. The component is written in Java, as a Karaf OSGI module.
7. *Trade execution & monitoring* - execution and monitoring of trades on broker platform. It is written in Java, as a separate component using JForex Duckascopy API.

Application is fully distributed, while both pre-processing data, training and prediction could be done in a distributed manner. Application is also ready to deploy in a hybrid mode (on-premises and cloud) and in a multi-cloud way.

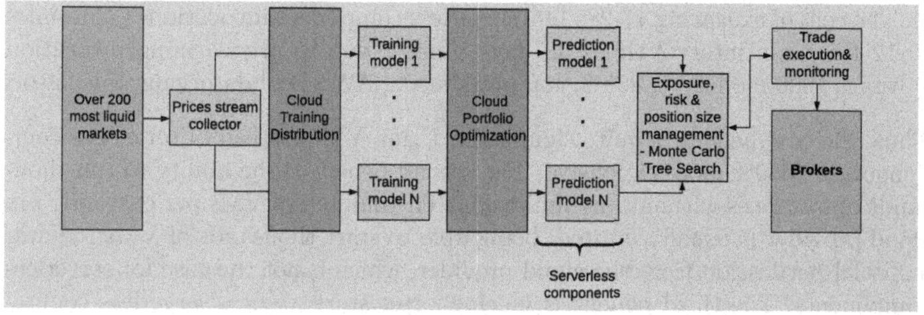

Fig. 4. AII architecture overview

Based on above description, the following challenges described in Sect. 3 are applicable here:

1. C_1 (including $C_{1.1}$ and $C_{1.2}$) - Presented application has hybrid architecture with VM/container components and serverless components.
2. C_2 - due to its complexity the ability to fully describe application in one place is an obvious advantage in this case.
3. C_3 - dynamic management of the serverless components provisioning is a must for this type of application.
4. C_4 - unit and integration testing is especially important for complex and distributed applications, so this challenge also applies here.
5. C_5 - metrics are used to monitor and scale the application as well as its training and predictive components.
6. C_6 - dynamic application adaptation allows for fully optimising the resources without degrading the performance of the application.

Performance with Functionizer. Serverless components were used to optimise the prediction sub-system cost and scalability. We considered (2) scenarios:

1. Deploy up to one thousands of VMs with prediction components. The machines should live all the time, as the distance between the lowest interval is shorter than the time to start up the VM.
2. Using serverless components with machine learning models to make predictions. The serverless components are started one minute before the interval time using pre-warming requests. Typical time to start components is up to 30 s. After the interval time, the right prediction request is issued.

The key benefit of using a hybrid architecture for the described application is cost savings. The difference in cost of resources is presented below:

1. The cost of deploying 1000 VMs of c5.large flavour (the smallest possible flavour to be used for predictions in production grade) is 61.200 $ per month - 0,085 $ cost of 1 h per VM times 24 h times 30 days times 1000 VMs. Even using the cheapest flavour t2.small with 2 GB of RAM, the cost is 14.976 $ (the cost of 1 h per VM is 0,0208 $). These flavours are not recommended for use in production due to working time per hour limitation.

2. The cost of executing 17.280.000 serverless component invocations (24 h times 12 for 5 min interval times 30 days times 2 due to pre-warming invocation times 1000 models) is 868 $, computed using AWS lambda pricing calculator.

Thus, the cost benefit is quite significant. Using VMs instead of serverless components is 7050% more expensive. The second benefit is the ability to run thousands of instances without any limitation. The number of VMs per customer per cloud provider is usually limited: being able to start thousands of VMs requires a special permission from the cloud provider, which is not the case for serverless components. The third benefit is much shorter start up of a serverless component. For the cold start up, it is up to 30 s and for warm start up it is usually 5 s. While a typical VM start usually takes about 5 min. The benefits listed above (espec. cost) are quite important and could be critical in terms of using cloud computing for AI applications, especially for startups with limited funding. The employment of serverless components for prediction could give them significant savings and enable to build AI applications at scale.

6 Conclusions and Future Work

This paper clearly highlighted the need for managing hybrid, cross-cloud applications. It also presented an extension of an existing cross-cloud application management platform with the implanting of serverless component management functionality. This extended platform was validated via a use case, coming from a recent but novel application area related to the merging of the AI and Big Data domains, which clearly shows its added-value and main benefits. Such a use case also well highlights the added-value of serverless computing adoption.

As part of our future work, the following directions are planned. First, the extension to the CAMEL language will be validated in the context of certain use cases. Second, the platform implementation will be finalised. Under this context, there will be a major focus on how the monitoring and adaptation of serverless components can be supported, through the use of the right abstraction mechanisms, in a platform-neutral manner. Third, a thorough validation of the platform will be performed in order to obtain the right feedback for optimising it. Finally, we will investigate the development of new or the extension of existing integration testing methods in the context of hybrid, cross-cloud applications.

Acknowledgements. This work has received funding from the Functionizer Eurostars project and from AI Investments Fast Track to innovation project.

References

1. Baldini, I., et al.: Serverless computing: current trends and open problems. In: Chaudhary, S., Somani, G., Buyya, R. (eds.) Research Advances in Cloud Computing, pp. 1–20. Springer, Singapore (2017). https://doi.org/10.1007/978-981-10-5026-8_1
2. Fox, G.C., Ishakian, V., Muthusamy, V., Slominski, A.: Status of serverless computing and function-as-a-service(FaaS) in industry and research. CoRR abs/1708.08028 (2017)
3. Spillner, J., Dorodko, S.: Java code analysis and transformation into AWS lambda functions. CoRR abs/1702.05510 (2017)
4. Spillner, J.: Transformation of python applications into function-as-a-service deployments. CoRR abs/1705.08169 (2017)
5. Spillner, J., Mateos, C., Monge, D.A.: FaaSter, better, cheaper: the prospect of serverless scientific computing and HPC. In: Mocskos, E., Nesmachnow, S. (eds.) CARLA 2017. CCIS, vol. 796, pp. 154–168. Springer, Cham (2018). https://doi.org/10.1007/978-3-319-73353-1_11
6. Horn, G., Skrzypek, P.: MELODIC: utility based cross cloud deployment optimisation. In: 32nd International Conference on Advanced Information Networking and Applications (AINA) Workshops, Krakow, Poland, pp. 360–367. IEEE Computer Society (2018)
7. Ardagna, D., et al.: MODAClouds: a model-driven approach for the design and execution of applications on multiple clouds. In: MiSE, Zurich, Switzerland, pp. 50–56. IEEE Press (2012)
8. Aske, A., Zhao, X.: Supporting multi-provider serverless computing on the edge. In: ICPP, Eugene, OR, USA, pp. 20:1–20:6. ACM (2018)
9. Kritikos, K., Skrzypek, P.: Review of serverless frameworks. In: Fourth International Workshop on Serverless Computing (WoSC), Zurich, Switzerland. IEEE Computer Society (2018)
10. LeCun, Y., Bengio, Y.: Convolutional networks for images, speech, and time series. In: Arbib, M.A. (ed.) The Handbook of Brain Theory and Neural Networks, pp. 255–258. MIT Press, Cambridge (1998)
11. He, K., Zhang, X., Ren, S., Sun, J.: Deep Residual Learning for Image Recognition. CoRR abs/1512.03385 (2015)
12. Szegedy, C., et al.: Going Deeper with Convolutions. CoRR abs/1409.4842 (2014)
13. Simonyan, K., Zisserman, A.: Very Deep Convolutional Networks for Large-Scale Image Recognition. CoRR abs/1409.1556 (2014)
14. Zhu, Y., Newsam, S.D.: DenseNet for Dense Flow. CoRR abs/1707.06316 (2017)
15. Kumar, A., Boehm, M., Yang, J.: Data management in machine learning: challenges, techniques, and systems. In: SIGMOD, Chicago, Illinois, USA, pp. 1717–1722. ACM (2017)
16. Rossini, A., et al.: D2.1.3—CAMEL documentation. Deliverable, PaaSage European Project, October 2015
17. Sabour, S., Frosst, N., Hinton, G.E.: Dynamic routing between capsules. In: Advances in Neural Information Processing Systems, Montreal, Canada. IEEE Computer Society (2017)
18. Kim, J., El-Khamy, M., Lee, J.: Residual LSTM: design of a deep recurrent architecture for distant speech recognition. CoRR abs/1701.03360 (2017)
19. Chang, S., et al: Dilated Recurrent Neural Networks. CoRR abs/1710.02224 (2017)

Deep Influence Diagrams: An Interpretable and Robust Decision Support System

Hal James Cooper[1(✉)], Garud Iyengar[1], and Ching-Yung Lin[1,2]

[1] Columbia University, New York, NY 10027, USA
{hal.cooper,gi10,c.lin}@columbia.edu
[2] Graphen Inc., New York, NY 10110, USA
cylin@graphen.ai
http://www.graphen.ai

Abstract. Interpretable decision making frameworks allow us to easily endow agents with specific goals, risk tolerances, and understanding. Existing decision making systems either forgo interpretability, or pay for it with severely reduced efficiency and large memory requirements. In this paper, we outline DeepID, a neural network approximation of Influence Diagrams, that avoids *both* pitfalls. We demonstrate how the framework allows for the introduction of robustness in a very transparent and interpretable manner, without increasing the complexity class of the decision problem.

Keywords: Agent systems and collective intelligence · Robust systems · Interpretable systems

1 Introduction

The goal of decision making frameworks is to provide a tool for representing key elements affecting agents decisions' and their relationships, and to provide assistance in selecting good decisions. In most decision making frameworks, there is a trade-off between interpretability, i.e. the ability to clearly identify the relationships between the key elements, and computational efficiency. There are many decision making settings, in particular, those involving financial or regulatory decisions, where interpretability often takes precedence over efficiency.

Influence Diagrams (IDs) [12] were one of the earliest quantitative approaches for decision making with a single agent. IDs represent the elements relevant to a decision making problem using a directed graph consisting of three kinds of nodes (see the graph on the left side of Fig. 1): "oval" chance nodes that represent exogenous random variables, "rectangular" decision nodes where the decision maker chooses a strategy (i.e. a distribution over available actions), and one or more "rhombus" utility node that output a utility for the chosen strategy[1].

[1] Initially, IDs had one utility node, but this was later relaxed.

W. Abramowicz and R. Corchuelo (Eds.): BIS 2019, LNBIP 353, pp. 450–462, 2019.
https://doi.org/10.1007/978-3-030-20485-3_35

The dependence between the exogenous random variables, the strategy, and the utility is encoded by directed edges. The objective in a (single-agent) ID is to compute the strategy that maximizes expected utility. An ID reduces to a Bayesian network once a strategy is chosen at each of the decision nodes, thus inheriting the conditional dependence structure of the Bayesian network. This allows for the relationships between key elements impacting the decision problem to be defined in a clearly interpretable manner. The ID in Fig. 1 represents a decision problem with three decisions and two chance nodes. The conditional independence $C_2 \perp D_1 | D_2$ is clearly apparent from the ID representation.

Although IDs facilitate interpretable decision analysis, IDs are not able to efficiently represent and integrate over distributions associated with the chance and decision nodes. IDs typically require a "no-forgetting" condition where all future decisions must retain knowledge of, and thus, in general, be dependent on all past decisions, with complexity therefore growing exponentially with network size. Consequently, IDs have been limited to relatively small decision problems with small discrete chance and decision distributions, and only limited support for continuous distributions [3,14]. As a result, they have lost prominence as a method of choice for decision analysis, which has led to a lack of new academic literature in the field (though still find some use in the medical domain [6] due to the particular importance of interpretability in the field).

In direct contrast to IDs, Markov Decision Processes (MDPs) and their extensions [4] must satisfy the Markov property wherein the state encodes all relevant past information, and the optimal action is only a function of the current state. Consequently, the optimal policies can be computed efficiently. However, in many decision problems, the Markov property is achieved only by defining a very complex state such that the resulting MDP lacks any interpretability [18]. The graph on the right hand side of Fig. 1 is an MDP representation for the ID on the left hand side. In order to create the MDP we have assumed that each chance node C_i, $i = 1, 2$, and decision node D_j, $j = 1, 2, 3$, has a binary outcome $\{0, 1\}$. The gray nodes are the states, and the white nodes are the allowed values of actions in a particular state. The MDP has a larger graphical representation because one has to track all past decisions in the state definitions to satisfy the Markov property; often making it impossible to characterize the optimal policy because of the curse of dimensionality. But perhaps more importantly, the conditional independence structure is completely obscured.

An increasingly important requirement for decision problem is robustness to uncertainty in the elements of the associated decision problems. The tension between computational efficiency and an interpretable representation becomes even more stark when one is uncertain about some of the parameters and wants to be robust to these perturbations. The decision analysis framework must allow agents to model uncertainty in a more granular manner, e.g. uncertainty in distribution of exogenous noise, or execution uncertainty, or uncertainty in the risk tolerance of the agent. For example, it is possible that the distribution of the chance node C_1 (see the ID on the left side of Fig. 1) is uncertain; however, the distribution of the chance node C_2 has no error. We propose a framework that allows for the modeling of such targeted uncertainty; moreover, the resulting framework retains the macro-level structure of an ID, albeit with a few more nodes. Targeted uncertainty is very hard to introduce in MDP models for decision problems because the requirement to maintain the Markov property completely obscures the conditional independence structure of the decision process. The chance nodes C_1 and C_2 are conflated into the state S_1, since the outcomes of both these nodes define the decision D_3 (see the MDP on the right hand side of Fig. 1). Thus, the targeted uncertainty in C_1 has to be represented as a structured uncertainty in the transition matrices. It is well known that introducing structured uncertainty in the transition matrices is computationally intractable [13,20].

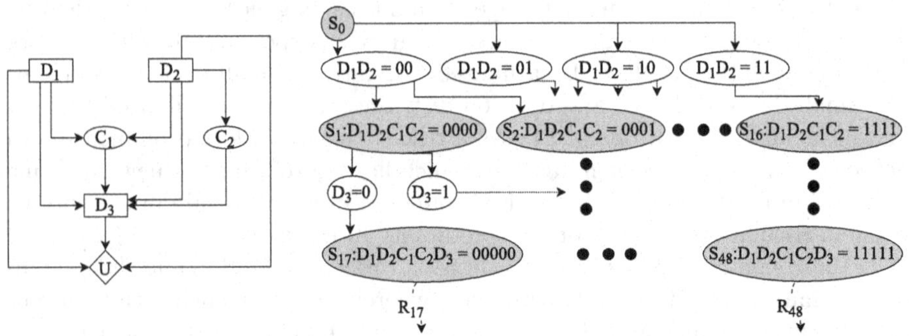

Fig. 1. Example ID (LHS) and a corresponding MDP formulation (RHS).

To summarize, IDs are interpretable and allow for the targeted modeling of uncertainty; however, the difficulties associated with representing distributions and integrating over these distributions are prohibitive. Recent developments in the deep learning community precisely address this computational difficulty. Differentiable generator nets (DGNs) essentially convert the task of generating samples from (or equivalently, integrating over) complicated distributions into a function approximation task [9]. The function is represented by a deep neural network that can be efficiently trained using gradient descent. Generative adversarial networks (GANs) [10], have made the task of such a function approximation more efficient by removing the need to maintain a balance between the generator and the discriminator, and also reduced mode dropping. Variational auto-encoders [15] extensively employ DGNs to efficiently learn high dimensional posterior distributions. Conditional GANs [19], Wasserstein GANs [11] and autoregressive networks e.g. [16] extend function approximation to learning conditional distributions.

2 The DeepID

We propose a new representation for robust decision making called DeepID (note that this work is a significant expansion of a concept previously published by the authors as an extended abstract [5], with new experimentation detailing the functioning concept) that retains the Bayesian-like conditional independence interpretability of IDs, but does not suffer from the associated computational difficulties. In our representation, each chance, decision, and utility node is represented by a DGN. Thus, selecting strategies in decision nodes reduces to optimizing over the parameters of the neural net. GANs can efficiently encode both a very large class of continuous distributions, as well as discrete distributions via concrete distributions [17]. We show that the DeepID framework allows us to introduce targeted uncertainty in a very flexible manner by appropriately introducing additional chance, decision, and utility nodes. In the DeepID framework, we model each node of an ID as a separate deep network, and connect the corresponding networks according to the macro-level structure of the original ID; the larger deep network we construct thus retains the interpretability associated with IDs. From DGNs and GANs, we inherit computational efficiency, as we are able to use all of the associated tools.

In the resulting network, the chance and utility nodes can be trained apriori as GANs since they approximate conditional distributions; whereas the decision nodes are collectively trained as feedforward neural networks to optimize the sample mean. This is in contrast to traditional end-to-end feed-forward deep networks constructed by connecting a series of layers in an ad-hoc fashion, which lack interpretability.

We focus primarily on the case of a single agent in a one-shot game [7], (which covers important contexts like medical decisions [8]). Robust decision making is particularly important in one-shot settings since agents do not have the ability to repeat the game or update models of uncertainty. Our main contribution here is to show that robustness can be introduced in a very flexible and interpretable manner in IDs, that can then be trained by constructing and optimizing the corresponding interpretable DGN. We propose DeepID as a tool for interpretable robust decision making. In the DeepID approach, the nodes of an ID are replaced by DGNs with free parameters for decision nodes, and fixed parameters for chance or utility nodes.

A DGN transforms the task of generating samples from (or, integrating over) a complex distribution into a function approximation task. Let $f(\mathbf{x}|\mathbf{y})$ be a given conditional distribution. A DGN generates samples from this distribution by first generating a sample ϵ from a known distribution and transforming the sample using a differentiable nonlinear function $g_\theta(\mathbf{y}, \epsilon)$. Thus, the task of representing the distribution $f(\mathbf{x}|\mathbf{y})$ reduces to learning the parameter θ. This task is non-trivial; however, there is now a mature literature on how to efficiently and robustly learn the parameter θ [10,19]. In this work, we leverage this technology to argue the computational efficiency of DeepIDs.

Recall that an ID consists of set chance nodes C, decision nodes D, and utility nodes U, connected by directed arcs representing the conditional independence relations. In the DeepID framework, each chance node and utility node is replaced by a DGN, i.e. the function $g_{\bar{\theta}_c}(\pi(c), \epsilon)$, where $\pi(c)$ denotes the outputs at the direct parents of the node c in the ID representation[2], the fixed parameter $\bar{\theta}_c$ is learned by matching conditional distributions using a GAN, and $\boldsymbol{\epsilon}$ are samples from a given fixed distribution. In contrast, at a decision d, the strategy is represented by the DGN $g_{\theta_d}(\pi(d), \boldsymbol{\epsilon})$ where the parameter θ is chosen to maximize the expected utility. Thus, DeepID decision nodes do not directly represent strategies over actions; rather, parameters of DGNs control the distribution of the generated samples. To the best of the authors' knowledge, this is the first ID approximation framework where decision nodes are replaced by a sample generating procedure. Some distributions (such as those in the location-scale family) can be easily implemented using DGNs. For example, suppose $f(\mathbf{x}|\mathbf{y}) = \mathcal{N}(\boldsymbol{\mu}+\mathbf{C}\mathbf{y}, \mathbf{R}\mathbf{R}^\top)$ is a multivariate Gaussian distribution with mean $\boldsymbol{\mu} + \mathbf{C}\mathbf{y}$ and covariance $\mathbf{R}\mathbf{R}^\top$. This distribution can be generated by the DGN $g_\theta(\mathbf{y}, \boldsymbol{\epsilon}) = \boldsymbol{\mu} + \mathbf{C}\mathbf{y} + \mathbf{R}\boldsymbol{\epsilon}$, where $\epsilon \sim \mathcal{N}(\mathbf{0}, \mathbf{I})$, which is a simple linear layer with an offset. We can use this "reparameterization trick" [15] to create simple but exact chance nodes (where parameters are trained or set apriori) or decision nodes, where parameters are free to be trained during joint learning of the full DeepID DGN. Whenever a known differentiable reparameterization exists, we can significantly reduce the complexity of learning appropriate DGNs (and even eliminate it entirely, e.g. for a chance variable in the location-scale family with *known* mean and variance).

The DeepID reduces the ID into a interpretable DGN – differentiability is very important to ensure that the strategy space can be searched efficiently. In many applications of IDs, discrete distributions play a very important role. In order to be able to model these in the DeepID, we need a differentiable approximation of discrete distributions. The Concrete distribution [17], allows us to approximate discrete distribution using DGNs. In the Concrete relaxation, samples from a discrete set of size n are approximated by samples from the continuous simplex $\Delta^{n-1} = \{x \in \mathbb{R}^n | x_k \in [0,1], \sum_{k=1}^n x_k = 1\}$, with vertices of the simplex being the one-hot vectors that can be mapped to the elements of the discrete set. A sample $X \in \Delta^{n-1}$ is generated by sampling $G_k \sim \mathit{Gumbel}$ IID $\forall k \in \{1, \ldots, n\}$ and setting $X_k = e^{\lambda^{-1}(\log \alpha_k + G_k)} / \sum_{i=1}^n e^{\lambda^{-1}(\log \alpha_i + G_i)}$ with temperature λ controlling the degree of the approximation and with samples drawn according to the unnormalized probabilities α. The key feature here is that the parameter α is allowed to be a differentiable function of the parameters or outputs of the parent nodes in the DeepID. Thus, complex probability tables can be approximated as conditionally dependent functions of parent nodes that output the appropriate values of α leading to a significant reduction in storage.

Clearly, the DeepID framework is of interest only if we one can guarantee that the optimal strategy can be represented by the associated DGN. We show that a large class of IDs can be arbitrarily closely approximated by DeepIDs, with optimal solutions that correspond to one another. Consider an ID with chance nodes C, decision nodes D and utility nodes U. Suppose the outputs of all nodes $i \in C \cup D$ take values in a compact set, the inverse conditional

[2] For example, in Fig. 1 $\pi(C_2) = D_2$, and $\pi(U) = \{D_1, D_2, D_3\}$.

CDF $F_i^{-1}(x_i|\pi(i)\}$ is continuous for all $i \in C \cup D$, and the utility functions are differentiable and bounded.

(a) Let $\bar{\sigma} = \{\bar{\sigma}_d\}_{d \in D}$ denote any strategy profile across all decision nodes D in the ID, and let $\mathbb{P}_{\bar{\sigma}}$ denote the corresponding joint distribution over actions, chance and utility node outcomes. Then there exists a sequence of DGNs $g^{(n)}$ and parameter vectors $\theta^{(n)}$ such that the corresponding joint distribution over actions, chance, and utility outcomes $\mathbb{P}_{g^n_{\theta^{(n)}}} \xrightarrow{D} \mathbb{P}_{\bar{\sigma}}$.

(b) Let σ^* denote the optimal strategy for the ID, with expected utility $\mathbb{E}[u(\sigma^*)]$. Then there exists a sequence of DGNs g^n with input size m such that $\mathbb{E}[u(g^n_{\theta^{(n)}_{\max}}(X))] \rightarrow \mathbb{E}[u(\sigma^*)]$, for $X \sim \text{Uniform}[0,1]^m$, where the components of the parameter vector $\theta^{\max}$ corresponding to the chance and utility nodes are defined by matching conditional distributions apriori, and the parameters corresponding the decision nodes are computed by the maximization of the expected utility.

3 Interpretable Robustness

Parameter and distributional uncertainty has been a focus of research in a number of different fields, with numerous techniques proposed for ensuring that the solutions are robust to the underlying uncertainties. In neural network training, one employs drop-out techniques, or adds noise to network parameters or the input data to ensure that the parameters converge to values that ensure robustness with respect to perturbations [9]. This goal can also be achieved by suitably regularizing the network parameters. Robustness in many problems is achieved by explicitly modeling the uncertainty as part of the problem. In the financial risk management context, coherent risk measures [2] ensure robustness to uncertainty in the distributions of the underlying risk factors. Conditional value at risk (CVaR) is a very popular coherent risk measure that ensures robustness by reweighting the worst quantile of losses. Clearly, robustness to uncertainty is equally important in decision problems. Decision problems, typically, involve a complex set of interacting elements, some more uncertain than others. Thus, it is beneficial for decision making frameworks to be flexible enough to allow for the targeted introduction of robustness.

In this section, we show that DeepIDs allow for targeted robustness to be introduced in an interpretable manner. In a DeepID, the purpose of each component deep network is retained; therefore, one can separately control the robustness of each component network. Introducing such targeted robustness is nearly impossible in a standard deep network where we have relatively limited understanding of the function of particular nodes or sub-networks. MDPs also do not easily accommodate targeted robustness since the definition of state, and corresponding transitions, often obscure the underlying independence relations. State aggregation further makes such a task difficult. DeepIDs allow us to model the following different classes of uncertainties.

Distributional Uncertainty for Specific Chance Nodes: For a chance node $c \in C$ with the DGN $g_{\bar{\theta}_c}(\pi(c), \epsilon)$ we can modify the parameters $\bar{\theta}_c$ or the exogenous samples ϵ to model distributional uncertainty. For example, consider the case where the network $g_{\bar{\theta}_c}(\pi(c), \epsilon) = \mu_c + \sigma_c \epsilon_1$, where $\bar{\theta}_c = (\mu_c, \sigma_c)$ and ϵ_1 is distributed according to a location-scale family. Then we can encode an agents uncertainty in the mean of the output of node c by setting $\mu_c \leftarrow \mu_c + \gamma_c \epsilon_2$, where ϵ_2 is sampled from another zero mean density. This is represented graphically by adding a new chance node c' (representing ϵ_2), adding a directed arc (c', c), and updating $g_{\bar{\theta}_c}(\pi(c), \epsilon)$ accordingly.

Regularization for Specific Decisions: Consider an agents decision node d with the DGN $g_{\theta_d}(\pi(d), \epsilon)$, we can regularize the parameters θ_d to encourage certain properties at node d. A particularly noteworthy application of decision level regularizers is to encourage particular discrete decisions to have pure or mixed strategies. This can be achieved by adding a utility node u with $g_{\bar{\theta}_u}(g_{\theta_d}(\pi(d), \epsilon), \epsilon')$ a p-norm regularization penalty on θ_d (see Sect. 4 for an example employing the Concrete distribution).

Execution Uncertainty for Specific Decisions: For an agents decision node d with DGN $g_{\theta_d}(\pi(d), \epsilon)$, we can add noise at any level – to θ_d, to ϵ, or the output of d – to encourage gradient descent to compute a stable decision strategy. We can also interpret this as encoding that decision execution isn't exact, with the agent sometimes making mistakes (e.g. a financial trading strategy where executing a trade at a desired price is not possible, or takes time). Introducing such an uncertainty will ensure that the chosen strategy is less sensitive to execution errors.

Custom Risk-Tolerances: We can modify the network $g_{\theta_u}(\pi(u), \epsilon)$ at a utility node u by adding a new layer that represents an agents' risk-reward tolerance, i.e. setting the output to $g_{\theta_{u'}}(g_{\theta_u}(\pi(u), \epsilon), \epsilon')$. This is represented by introducing a new node u' and adding an arc (u, u').

Note that all these modifications to a DeepID are equivalent to adding new chance or utility nodes, changing the objective of the training algorithm. Consequently, such modifications do not increase the complexity class of the problem.

4 Experiments

Here we demonstrate how interpretably robust modifications of an ID can be reformulated as DeepIDs. For purposes of demonstrating this translation, we use a relatively simple ID, and straight-forward DGN formulations for all distributions. We focus on the Reactor Problem [21] that involves an agent choosing between constructing a conventional or an advanced reactor. The performance of both the conventional and advanced reactors are uncertain; however, the performance of the advanced reactor is less certain. The state of the reactor in the future takes one of the following three values: **C**atastrophic failure, **S**mall failure, and **N**o failure. The probability distributions for the future states, and the corresponding utilities are in Table 2. The agent conducts a test to predict the

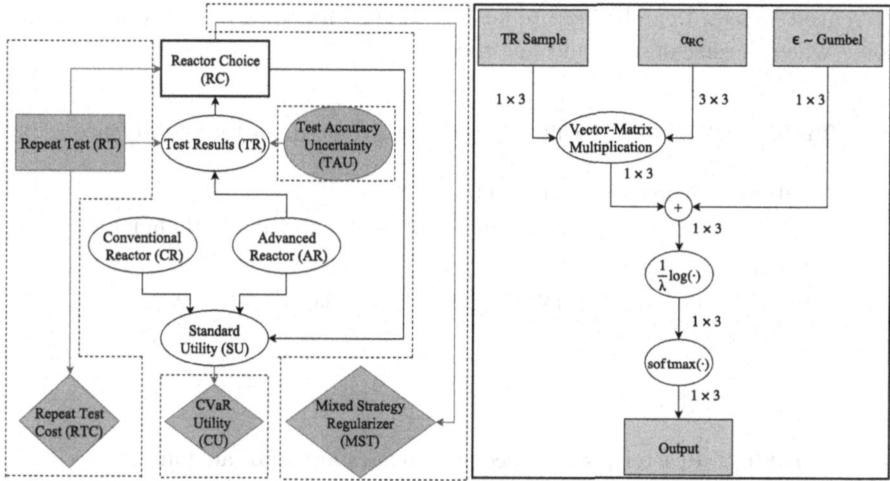

Fig. 2. Reactor problem DeepID and modifications (LHS) and internal DGN of RC (RHS).

performance of the advanced reactor. The conditional probability table for the test outcome is given in Table 1.

The DeepID formulation for this problem is displayed in Fig. 2, with the macro-level ID structure shown on the LHS, and example DGN internal structure shown on the RHS (where shaded nodes are inputs or outputs, and non-shaded nodes are operations). It consists of 3 chance nodes, Conventional Reactor (CR), Advanced Reactor (AR) and Test Results (TR), 1 decision node, Reactor Choice (RC), and 1 utility node, Standard Utility (SU). Note that the SU node is not a rhombus. This is because it can be fed into a CVaR utility node (CU) to add robustness to the utility outcome. The RC decision node chooses between constructing a conventional or advanced reactor as a function of the output from the TR chance node that take three possible values: **C**atastrophic failure, **S**mall failure, and **N**o failure. The utility of the decision at RC node is a function of the true future state of the conventional and advanced reactors, and hence, the SU node has inputs from the CR, AR and RC nodes.

Since the outcomes of all nodes of the ID are discrete, we model the associated discrete distributions using the Concrete DGN as outlined in Sect. 3, with distinct $\alpha_{i|\pi(i)} \forall i \in C \cup D$ parameters for each possible combination of input and output. We let α_i denote the matrix of parameters that define the distribution at node i. The α parameters for chance nodes are fixed to the corresponding values, e.g. $\alpha_{TR=\mathbf{N}|AR=\mathbf{S}} = 0.288$, whereas the α corresponding to the decision node is a free parameter. For a chance node $c \in C$ (resp. decision node $d \in D$) with parent outcome $\pi(c)$ (resp. $\pi(d)$), a sample of the outcome from c is drawn according to the Concrete DGN using $\alpha = \pi(c)\alpha_c$ (see RHS of Fig. 2). Though not necessary for a problem of this size, we again note that the function approximation literature referenced in Sect. 1 can be used to let $\alpha = f(\pi(c))$ for some approximating function f when the size of α_c would be prohibitive.

Table 1. CPT of test results given future advanced reactor performance

Future performance	Test results		
	Catastrophic failure	Small failure	No failure
Catastrophic failure	0.9	0.0	0.1
Small failure	0.147	0.565	0.288
No failure	0.0	0.182	0.818

Table 2. Reactor performance probabilities and associated utilities

Reactor	Catastrophic failure		Small failure		No failure	
	Prob.	Utility	Prob.	Utility	Prob.	Utility
Advanced	0.14	−50	0.2	−6	0.66	12
Conventional	N/A	N/A	0.02	−4	0.98	8

Thus far we have described the standard DeepID for the Reactor Problem. Next, we demonstrate how to introduce robustness to the standard DeepID in an easily interpretable fashion. Each (optional) change is represented by additional shaded nodes and lines in the LHS of Fig. 2. In our experiments, we explore the effects of making each robust addition separately. We can, of course, also mix and match combinations of these robustness concepts. We can replicate a CVaR like custom risk-tolerance by taking advantage of the sample-based nature of the DeepID. To set $CVaR_p$ as the objective function of the DeepID, we simply sort the output samples and take the mean of the pth quantile. This corresponds to a deterministic utility node with the function $g_{\theta_{u'}} = \frac{1}{k}\sum_{i=1}^{k} g_{\theta_u}(\pi(u), \epsilon_{[i]})$ where $k = \lfloor pN \rfloor$ and $g_{\theta_u}(\pi(u), \epsilon_{[i]})$ is the ith smallest standard utility sample calculated through a sort function. Note that although $g_{\theta_{u'}}$ is in general not differentiable, there exist standard techniques to smooth this function to any degree of accuracy [1]. We demonstrate how this affects the DeepID layout in the LHS of Fig. 2, where the regular utility node now feeds into the CU node at the output. This change was incorporated at the end of the network where we have a clear concept of good and bad utilities, though such a sample-based transformation could be applied anywhere in the network.

In the standard formulation, we are quite sure about the ability of the test to correctly predict a catastrophic failure, and so we choose to build the advanced reactor if the test result is good. However, we may be interested in observing how our suggested behavior changes if we introduce distributional uncertainty to our test accuracy. In this experiment, we add $Q \sim \text{Uniform}(0, a)$ noise as

Table 3. Test results CPT given future catastrophic performance and repeat testing modification

Num. tests	Test results — C		
	Catastrophic failure	**Small failure**	**No failure**
1	0.7	0	0.3
2	0.91	0	0.09

Table 4. Experiment results summary

Test	Parameter	Strategy before	Strategy after	Switchpoint
CVaR	dec. p	$BAoN$	ABC	0.7
Test accuracy	inc. a	$BAoN$	ABC	0.3
Rep. Tests	inc. tc	Two tests, $BAoN$	Single test, ABC	0.8
MS Reg.	dec. δ	$\delta = -0.01$, $\alpha_{RC\|TR} = \begin{bmatrix} 0.93 & 0.07 \\ 0.55 & 0.45 \\ 0.41 & 0.59 \end{bmatrix}$	$\delta = -0.06$, $\alpha_{RC\|TR} = \begin{bmatrix} 0.65 & 0.35 \\ 0.51 & 0.49 \\ 0.49 & 0.51 \end{bmatrix}$	N/A

$\alpha_{TR=\mathbf{C}|AR=\mathbf{C}} \leftarrow \alpha_{TR=\mathbf{C}|AR=\mathbf{C}} - Q$ and $\alpha_{TR=\mathbf{N}|AR=\mathbf{C}} \leftarrow \alpha_{TR=\mathbf{N}|AR=\mathbf{C}} + Q$ for increasing values of a. In the LHS of Fig. 2, this corresponds to the additional chance node TAU representing the noise Q that is fed into $\alpha_{TR|AR}$.

In some decision problems, we may have the ability to repeat actions to improve certainty. For instance, we may choose to repeat the advanced reactor test for some additional cost, resulting in an overall testing procedure that is more accurate. We show in the LHS of Fig. 2 how a new node can be added to represent this ability. The new decision node is connected to both TR (as it changes the accuracy of the test to be more accurate) and RC (because we must know when making the decision if our test results correspond to the more accurate situation where we have run an additional test), changing the α_i matrices accordingly. We experiment with this modification for increasing test costs (tc), with the additional test cost corresponding to a new utility node. Note that in this experiment, TR uses the conditional probability table of Table 3, where the Catastrophic failure prediction corresponds to the probability of *any* individual test in the overall testing procedure suggesting Catastrophic failure. In the standard ID formulation of the reactor problem, learning encourages the use of pure strategies. However, we may have a preference for mixed strategies in certain situations. Here, we demonstrate how regularizers can be added to individual decisions to encourage or discourage mixed strategies for particular decisions. We add a p-norm regularizer on α_{RC} through a new utility node MST with $g_{\theta_{MST}}(\pi(MST)) = \delta \left\|\alpha_{RC|TR=\mathbf{C}}\right\|_p + \delta \left\|\alpha_{RC|TR=\mathbf{S}}\right\|_p + \delta \left\|\alpha_{RC|TR=\mathbf{N}}\right\|_p$ for $p = 0.1$, where we can force RC to have a mixed strategy or pure strategy by varying the parameter δ.

5 Discussion

The effects of our ID modifications in Sect. 4 are summarized in Table 4, where $BAoN$ is shorthand for the strategy "Build Advanced on a No failure test result, otherwise build Conventional", and ABC is shorthand for the strategy "Always Build Conventional". In Fig. 3 we display convergence behavior for training the DGN associated with the DeepID using gradient descent for a selected set of instances demonstrating convergence to the before (blue) and after (black) strategies. We observed that the convergence was not sensitive to the particular choice of algorithm or the learning rate. We note with the CVaR modification that lowering p causes the learned strategy to change from $BAoN$ to ABC. With the advanced reactor having significantly worse consequences for failure, and a higher chance of failure, such samples dominate when the strategy allows for the advanced reactor to be built (due to imperfections in the testing procedure). As p is decreased, such samples increasingly dominate the utility. In contrast, ABC means we are indifferent to test inaccuracy, and $CVaR_p$ is relatively constant in p due to the small failure rate of 0.02. A similar trade-off is observed when increasing a for $Q \sim \text{Uniform}(0, a)$. As a increases, the test is considered increasingly unreliable until the recommended strategy likewise becomes ABC. Though the result is ultimately similar, the mechanisms of robustness that these tests represent differ, with CVaR representing risk-tolerance, and Q representing distributional uncertainty in a specific component of the model. As we increase tc in the RT modification, we switch from being willing to pay for increased accuracy to allow the $BAoN$ strategy to be effective, until the increase in cost is greater than the difference between the $BAoN$ and ABC strategies. With the mixed strategy regularizer, we observe the same standard recommendation for $BAoN$, with the pure strategy suggestion becoming increasingly mixed as δ becomes more negative. Though it is not necessary in this example, we could also add a regularizer with $\delta > 0$ to encourage a pure strategy.

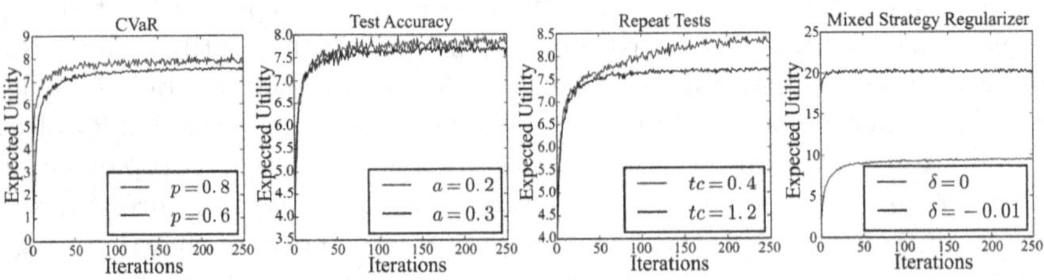

Fig. 3. Gradient training on the DeepIDs. (Color figure online)

6 Future Work

This paper presented an existence proof for the arbitrary accurateness of the DeepID and its ability to be trained with gradient based methods. We did not present formal guarantees of convergence or convergence rates. Developing such guarantees, as well as identifying classes of DeepIDs where such guarantees can be made, is an important area of future research. In this paper, our examples emphasized interpretability and robustness benefits. We are thus motivated to demonstrate scalability and correctness through the application of our method to more complex real-world scenarios with large numbers of easily modified agents.

References

1. Abad, C., Iyengar, G.: Portfolio selection with multiple spectral risk constraints. SIAM J. Financ. Math. **6**(1), 467–486 (2015)
2. Artzner, P., Delbaen, F., Eber, J.M., Heath, D.: Coherent measures of risk. Math. Financ. **9**(3), 203–228 (1999)
3. Bielza, C., Gomez, M., Shenoy, P.P.: A review of representation issues and modeling challenges with influence diagrams. Omega **39**(3), 227–241 (2011)
4. Boucherie, R.J., Van Dijk, N.M.: Markov Decision Processes in Practice, vol. 248. Springer, Cham (2017). https://doi.org/10.1007/978-3-319-47766-4
5. Cooper, H., Iyengar, G., Lin, C.Y.: Interpretable robust decision making. In: International Conference on Autonomous Agents and Multiagent Systems, Stockholm (2018)
6. Diez, F.J., et al.: Markov influence diagrams: a graphical tool for cost-effectiveness analysis. Med. Decis. Making **37**(2), 183–195 (2017)
7. Everitt, T., Ortega, P.A., Barnes, E., Legg, S.: Understanding agent incentives using causal influence diagrams, Part I: single action settings. arXiv preprint arXiv:1902.09980 (2019)
8. Gomez, M., Bielza, C., del Pozo, J.A.F., Rios-Insua, S.: A graphical decision-theoretic model for neonatal jaundice. Med. Decis. Making **27**(3), 250–265 (2007)
9. Goodfellow, I., Bengio, Y., Courville, A.: Deep Learning. MIT Press (2016). http://www.deeplearningbook.org
10. Goodfellow, I., et al.: Generative adversarial nets. In: Advances in Neural Information Processing Systems, pp. 2672–2680 (2014)
11. Gulrajani, I., Ahmed, F., Arjovsky, M., Dumoulin, V., Courville, A.: Improved training of Wasserstein GANs. In: Proceedings of the 31st International Conference on Neural Information Processing Systems, pp. 5769–5779. Curran Associates Inc. (2017)
12. Howard, R.A.: Readings on the Principles and Applications of Decision Analysis, vol. 1. Strategic Decisions Group (1983)
13. Iyengar, G.N.: Robust dynamic programming. Math. Oper. Res. **30**(2), 257–280 (2005)
14. Jensen, F.V., Nielsen, T.D.: Probabilistic decision graphs for optimization under uncertainty. Ann. Oper. Res. **204**(1), 223–248 (2013)
15. Kingma, D.P., Welling, M.: Auto-encoding variational Bayes. arXiv preprint arXiv:1312.6114 (2013)
16. Larochelle, H., Murray, I.: The neural autoregressive distribution estimator. In: AISTATS, vol. 1, p. 2 (2011)
17. Maddison, C.J., Mnih, A., Teh, Y.W.: The concrete distribution: a continuous relaxation of discrete random variables. arXiv preprint arXiv:1611.00712 (2016)
18. Magni, P., Quaglini, S., Marchetti, M., Barosi, G.: Deciding when to intervene: a Markov decision process approach. Int. J. Med. Informatics **60**(3), 237–253 (2000)

19. Mirza, M., Osindero, S.: Conditional generative adversarial nets. arXiv preprint arXiv:1411.1784 (2014)
20. Nilim, A., El Ghaoui, L.: Robust control of Markov decision processes with uncertain transition matrices. Oper. Res. **53**(5), 780–798 (2005)
21. Virto, M.A., Martin, J., Insua, D.R., Moreno-Diaz, A.: Approximate solutions of complex influence diagrams through MCMC methods. In: Probabilistic Graphical Models (2002)

A Reference Architecture Supporting Smart City Applications

Stanimir Stoyanov(✉), Todorka Glushkova, Asya Stoyanova-Doycheva, and Vanya Ivanova

Plovdiv University "Paisii Hilendarski", 24 Tzar Asen Street, 4000 Plovdiv, Bulgaria
stani@uni-plovdiv.net

Abstract. This paper briefly presents a reference architecture called Virtual Physical Space (ViPS). The purpose of the architecture is to be able to adapt to the development of various Cyber-Physical-Social applications. In the paper, a possible adaptation for a smart seaside city is discussed. The need for virtualization of things from the physical world in a formal way is also considered. Furthermore, the virtualization and modeling of spatial aspects through the AmbiNet formalism is demonstrated by an example.

Keywords: Internet of Things · Cyber-Physical-Social System · Virtual Physical Space · Calculus of Context-aware Ambients

1 Introduction

The Internet of Things (IoT) is an extension of the Internet into the real physical world, in which physical entities (things) are interconnected. An IoT application is composed of things that can sense the environment changes, analyse these changes based on shared gained knowledge, and act or make plans accordingly for achieving a personal or a shared goal. IoT is closely related to the Cyber-Physical System (CPS) and the Cyber-Physical-Social System (CPSS). A basic feature of these technologies is the integration of the virtual and the physical worlds. Due to revealing new horizons and opportunities, CPS, CPSS and IoT are areas of growing scientific and practical interest.

The reference architecture, called ViPS, did not emerge suddenly. It is the result of an improvement of two of its predecessors – the Distributed eLearning Center (DeLC) and the Virtual Education Space (VES). Fifteen years ago, to support e-learning at the Faculty of Mathematics and Informatics at the University of Plovdiv, the environment DeLC was implemented and has been used for years [1]. Although it was a successful project for applying information and communication technologies in education, one of its major drawbacks is the lack of close and natural integration of its virtual environment with the physical world where the real learning process takes place. Considering the material world is especially important for disabled learners. At the same time, the CPSS and IoT paradigms reveal entirely new opportunities for taking into account the needs of people with disabilities, in our case disabled learners. For these reasons, the environment has recently been transformed into VES that operates as an IoT ecosystem [2].

W. Abramowicz and R. Corchuelo (Eds.): BIS 2019, LNBIP 353, pp. 463–474, 2019.
https://doi.org/10.1007/978-3-030-20485-3_36

Summarizing the experience of constructing VES, we have begun developing a reference architecture known as the Virtual Physical Space (ViPS), which can be adapted to various CPSS-like applications. The idea of this architecture and an initial prototype are presented in [3]. The original idea for adapting this architecture to a smart seaside city is given in [4]. In [5] is demonstrated the adaptation of ViPS to develop a smart tourist guide.

When implementing a CPSS application, the focus is placed on assisting the user, who has to interact with a complex system that integrates a virtual and a physical world. In ViPS, the user is supported by personal assistants. An essential task of the space is the virtualization of "things" from the physical world, which are of interest to the user. Each of these "things" has spatial characteristics and can be related to a certain point in time and a specific event. The article discusses in more detail the spatial aspects of the things virtualization by an ambient-oriented modeling approach. Our idea is to model possible scenarios in a seaside city in which we can explore the possibilities of delivering different services to users.

The rest of the paper is organized as follows: a short review of IoT, CPS and CPSS is considered in Sect. 2, which is followed by an overall description of the ViPS architecture in Sect. 3. Section 4 addresses an adaptation to a smart city and a demonstration to model a simple scenario in a smart city. Finally, Sect. 5 concludes the paper.

2 Related Works

Typical CPSS-like applications are smart cities. A smart city integrates a distributed sensor network, government initiatives on openly available city-centric data and citizens sharing and exchanging city-related messages on social networks [6]. In [7], an extremely useful review of the literature on the subject of smart cities has been made. A comprehensive analysis of the concept of a smart city and the existing technologies and platforms has been carried out. A model for designing an intelligent urban architecture has been proposed. The reader receives a clear understanding of the services that a Smart City has to provide. The article identifies the flaws of smart cities. In addition, three typical case studies have been proposed – a simulator for studying the application of various services and technologies, incident management in a Smart City and monitoring the implementation of sensors in a Smart City. The vast amounts of data obtained by sensing of the physical world and those contributed by city inhabitants through their sensor-enabled smartphones and online social networks can offer near real-time settings [8]. Extracting knowledge out of the data, typically through big data analytics techniques, can help to build a picture of urban dynamics, which can enable intelligent applications and services and guide decision-making, both for city authorities and city inhabitants [9]. CPSS applications are emerging in everyday life in smart urban systems, in domains as varied as command and control, smart environments, smart transportation, smart social production systems, and so on [10]. Such applications rely on efficient monitoring of the urban physical infrastructure and ambient environment and they combine the collected data through intelligent cyber processes to deliver improved services to citizens. The resulting urban big data system offers the potential of creating more sustainable and environment-friendly future cities [11]. In this sense, an essential aspect of a smart city is the efficient use of energy resources. In [12], a multi-agent system for energy optimization is presented. The multi-agent system is viewed as a modeling tool that is easily extrapolated in the field of power engineering and can be further developed by using the IoT paradigm.

3 ViPS in a Nutshell

ViPS is being built as a reference infrastructure, which can be used to develop CPSS applications for different domains. In particular, we are pursuing three main objectives:

- Building a formal environment for “virtualization” of real-world “things”. In line with the IoT concept, the architecture provides virtualization of real “things”. In the context of software technologies, this means creating digitized versions of real physical objects that can be specified and digitally interpreted.
- Creating an interface between the virtual and the real worlds.
- A genetic personal assistant – we assume that from an architectural point of view a CPSS-like application will be based upon a personal assistant that will help users work with this application.

The logic scheme of ViPS is presented in Fig. 1. Practically, the virtualization of “things” is supported by the ViPS middleware. In addition to the relevant to us inherent attributes of the real “things”, the virtualization has to take into account complementary factors such as events, time, space, and location. The components implementing the virtualization are located in two subspaces called “Virtualized Things Subspace” (VTS) and “Digital Objects Subspace” (DOS). DOS is implemented as open digital repositories, which store objects of interest to the particular field of application with their characteristics. VTS consists of three modeling components – AmbiNet, ENet and TNet. In AmbiNet the spatial characteristics of the “things” can be modeled as ambients. The core of AmbiNet is the Calculus of Context-aware Ambients (CCA [13]); it is a formalism for the presentation of ambients along with their boundaries, contexts and channels of communication. ENet models various types of events and their arguments as identification, type, conditions for occurrence, and completion. In our event model there are three types of events – base, system and domain-specific ones. Base events are a date, time of day and location. System events are an operation, related to the system, such as sending and receiving messages or generating and removing objects. Domain-specific events represent user-relevant events representative of the field of interest. TNet (Temporal) provides an opportunity to present and work with the temporal aspects of things. TNet is based on the formal specifications Interval Temporal Logics (ITL [14]). In order to demonstrate the terms used (concepts) and especially the links between them, the two subspaces use the services of OntoNet. The OntoNet layer is built as a hierarchy of ontologies. The operative assistants, implemented as rational intelligent agents, provide access to the resources of both subspaces and accomplish interactions with the personal assistants and web applications. They are architectural components suitable for providing the necessary dynamism, flexibility and intelligence. However, they are unsuitable to deliver the necessary business functionality in the space. For this reason, the assistants work closely with the services or micro-services. Furthermore, they interact with the guards to supply ViPS with data from the physical world.

The guards operate as a smart interface between the virtual and the physical worlds. They provide data about the state of the physical world transferred to the virtual environment of the space (the two subspaces). There are multiple IoT Nodes integrated in the architecture of the guards that implement access to sensors and actuators of the "things" located in the physical world. The sensors-actuators sets are configured in accordance with the application. The communication in the guard system operates as a combination of a personal network (e.g. LoRa) and the Internet.

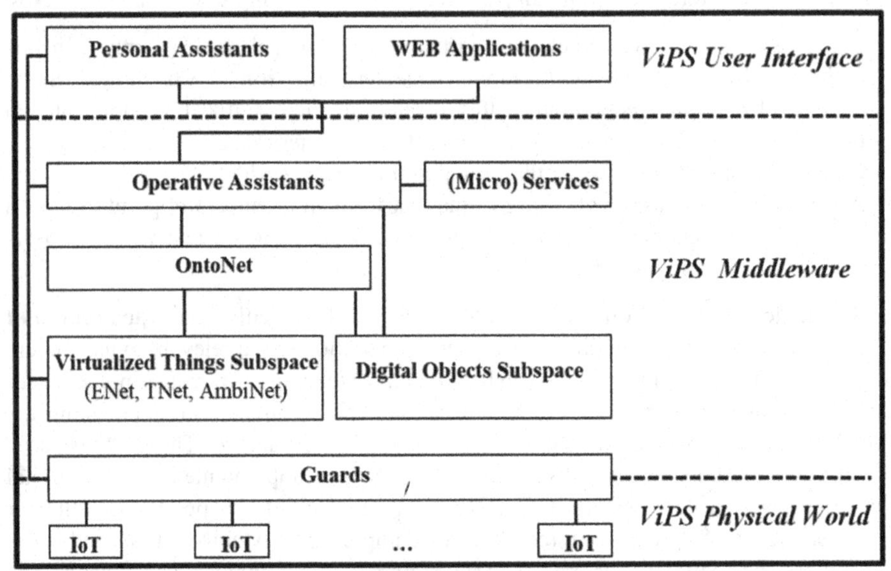

Fig. 1. ViPS architecture

The Personal Assistants (PA), operating on behalf of users and aware of their needs, will conduct scenarios for the execution of the users' requests and will manage and control their execution by interacting with the ViPS middleware. A PA is implemented as a rational agent with a BDI-like architecture consisting of two basic components – Deliberation and Means-Ends Reasoning [15]. During the deliberation the PA forms its current goal while in the next component a plan for achieving the goal is prepared. The central structure for determining the current intention of the PA is a "personal time-table" named PS (Personal Schedule). The records of the PS are domain-specific events that present potential activities, in which the user expects aid from a personal assistant. Depending on the user's wish and taking into account the situation in the physical world (presented as a context), the personal assistant determines the actual intention and initiates the implementation of a suitable scenario (plan) in ViPS. In doing so, the personal assistant is supported by the VTS components. Usually, a plan is launched when a corresponding event occurs. However, the PA also has to be able to operate in an "overtaking action" mode, i.e. it has an early warning system. In this way, certain operations must be performed before the event occurs. Therefore, the PA cannot rely on ENet in the timeframe before the event occurs. Then, the PA is supported by the CCA and ITL components. The interaction between the PA and the three supporting components is implemented by a specialized protocol under development.

The public information resources of the space are openly accessible through appropriate web applications that are usually implemented especially for the particular domain.

The main technology for development of the individual components is agent-orientated with opportunities for interaction with services, including intelligent ones as well. The personal assistants are usually located on mobile devices and they operate on them. The two subspaces and the other middleware components form the server part of the application, which is built with ViPS (the server part can be distributed). Specific guards are positioned on the server, while others, depending on the infrastructure of the particular application, can be placed in the frog and edge layers, respectively, in the reference architecture proposed in [16]. The architecture has been fully implemented in Java. In ViPS, core components are assistants implemented as BDI rational agents by the help of JADE [17] and Jadex [18] environments.

4 An Adaptation for a Smart City

A possible application of ViPS is to model the infrastructure (or a part of it) of a smart city. With that end in view, we have chosen a smart seaside city because we have some experience in IoT monitoring of open water basins. Moreover, in our opinion, the sea coast, territorial waters and possibly the adjacent islands offer interesting scenarios to apply ViPS. In [3], the virtualization of a major "thing" for this city is discussed, namely the open water spaces. In order to continue working on the model, it is necessary to virtualize other "things" too such as those related to the possibilities of people to travel by different types of transport, special facilities for people with disabilities, and specific public places.

In this model we can study the behavior of implementable services that the smart city may offer. Each user will have his/her own personal assistant that will help them individually to use the services provided by the smart city. This can be done using the following procedure:

- By means of the mobile device, the user will be able to choose the desired service from the number of options provided by the Smart City;
- The Smart City interacts with the personal assistant to prepare a personalized service plan for the user, taking into account his/her location, the desired destination, the time (including the amount of time that the user has available), and unforeseeable events.

Adapting ViPS to a smart seaside city will be demonstrated with a simple but practical scenario.

4.1 A Simple Scenario

Let us consider the following sample scenario: A tourist in a wheelchair is at his/her hotel in a seaside city and wishes to visit a famous museum on a nearby island. The tourist has an intelligent wheelchair and a personal assistant on his/her mobile device. Upon arrival on the island, the wheelchair recognizes some life-threatening indicators and, through communication with the smart city, the personal assistant provides transportation of the tourist to the hospital. We will examine the two parts of this scenario:

- Scenario 1: Searching for a suitable route to visit the island museum. The trip from the hotel to the port will be made by bus, and from the port to the island – by ship;
- Scenario 2: Providing the fastest route to transport the tourist from the island to the local hospital. The travel options are: by medical helicopter; by medical shuttle boat to the port and then by ambulance to the hospital; or by ship to the port and from there by ambulance.

4.2 Ambient-Oriented Modelling

To model these scenarios we will use the capabilities of the ViPS AmbiNet Subspace and the CCA formalism. The syntax and formal semantics of CCA can fully implement the π-calcule modeling by providing much better opportunities for Context-Aware Ambient modeling.

A CCA ambient is an identity that is used to describe an object or a component such as a process, a device, location, etc. Each ambient has a specific location, borders, and a name to be identified by in the environment. It may contain other ambients in itself, thus allowing the creation of ambient hierarchies. There are three possible relationships between two ambients: parent, child, and sibling. Ambient communication is realized by sending and receiving messages. In the CCA notation, when two ambients exchange a message, we use the "::" symbol to describe an interaction between sibling ambients; "↑" and "↓" are symbols for parent and child; "<>" means sending a message, and "()" – receiving a message. Ambients are mobile. With CCA, there are two ways to move: inwards (in) and outwards (out), which allow ambients to move from one location to another. CCA can distinguish between four syntax categories: location α, opportunities M, processes P, and context expressions k.

To model the proposed scenario, we will examine a smart seaside city where every change, essential to the current context, is registered by a network of physical sensing devices and is processed dynamically. The tourist's personal assistant must ensure that the scenario is implemented by means of communication with the operational assistants and the components of the Virtualized Things Subspace. As the tourist with his/her wheelchair changes their location, the management will be realized mainly by AmbiNet via communication with ENet to manage the individual events (such as departure and arrival of the boat, opening and closing of the museum, etc.), and by TNet for the temporal characteristics of the separate stages in the scenario implementation. AmbiNet is supported by the following run-time and development tools:

- AmbiNet ccaPL Interpreter – a run-time interpreter of the formal modeling language ccaPL based on the Calculus of Context-aware Ambients (CCA);
- AmbiNet Route Generator – used for routes generation on the network by user-defined criteria;
- AmbiNet Route Optimizer – applied to optimize routes depending on the instant state of the participating ambients;
- AmbiNet Editor – a visual modeling editor;
- AmbiNet Repository – a repository for storing ambients, ambient structures, and physical space templates.

Depending on their location, ambients can be static and mobile. Static ambients have a constant location in the physical world or in the modeled virtual reality, for example hotels, hospitals, museums, ports, bus stops, etc. Mobile ambients have a variable location, e.g. buses, people, ships, wheelchairs, ambulances, and so on. Each ambient can contain a hierarchical structure of ambient-children, which can be static or mobile itself. Mobile ambients can move by the options "coming in" and "going out" from other static or mobile ambients. Typically, if they change their location, then all the "ambient-children" move as well.

In the AmbiNetEditor are created templates of the physical city with static ambients, which are stored in the AmbiNet Repository. According to the current scenario, a particular template is used (for example, a map of the city with static ambients placed on it), on which mobile ambients are modeled and interactions and messages in the ambient network are established. The AmbiNet Route Generator component is used to generate routes and the AmbiNet Route Optimizer is applied to find the optimal route (whenever that is necessary). After the scenario modeling, the interactions between the participating ambients are presented as a ccaPL program and the AmbiNet cpPL Interpreter is launched to simulate and visualize the model under consideration.

4.3 CCA- Modeling of a Simple Scenario

The following table presents the main ambients used in the CCA model of the described scenario (Table 1).

Table 1. CCA Ambients.

Notation	Description	Notation	Description
PA	Personal Assistant	Hotel	City Hotel
WCh	Intelligent Wheelchair	Museum	Museum
VTS	Virtualized Things Subspace	Port	Ports
ANet	Ambient Net-AmbiNet	GA	Guard Assistants
ENet	Event Net	Hosp	City Hospital
RG	AmbiNet Route Generator	MedCop	Medical Helicopter
RO	AmbiNet Route Optimizer	Ambul	Ambulance
BS	Bus Stops	MedSh	Medical Shuttle

The PA Ambient communicates with the other ambients and delivers the necessary information to the user. This ambient represents an intelligent assistant; it interacts with the Operational Assistants and the Guard Assistants, which constitute the remaining ambients in the multi-agent system of ViPS. In order to model the first part of the scenario, the purpose of the PA is to provide the tourist with information about the opportunity to visit the island museum and to enable him/her to choose a route to travel. The CCA process of the PA is illustrated by (1). In the second part of the scenario, the PA launches bidirectional communication in order to provide the fastest route to move the patient from the island to the hospital. The process is described in (2).

$$P_{PA} \cong \left(\begin{array}{l} WCh :: < getLocation > .WCh :: (location).0| \\ VTS :: < PAi, location, WantToVisitIslandMuseum > .0| \\ VTS :: (hasShip, hasBus, openMuseum). \\ VTS :: < PAi, location, getRouteToIslandMuseum > .0| \\ VTS :: (ListRoutes).WCh :: < choosenRoute, startMoving > .0| \\ WCh :: (location, LifeThreateningIndicators). \\ GA :: < PAi, location, necessaryTransportationToHospital > .0| \\ GA :: (OptimalRoute).WCh :: < OptimalRoute > .0 \end{array} \right) \quad (1)$$

WSh and GA Ambients – the tourist participates in the scenario through his/her intelligent wheelchair, which has a set of sensors that render an account of both the current location and the human condition (blood pressure, pulse, etc.). When life-threatening indicators are reported, it sends a supplementary message to the PA, which launches the second part of the scenario. The GA Ambient communicates with the physical world and dynamically provides information about the state of those areas, which are important for the wheelchair movement, such as elevators, ramps, opening doors, etc. After the launch of the emergency scenario, the GA assists the PA in providing information to generate the optimal route to travel. The processes of the two ambients are presented in (2) and (3).

$$P_{GA} \cong \left(\begin{array}{l} VTS :: (PAi, getActiveZones).VTS :: < PAi, ListAZ > .0| \\ PA :: (PAi, location, NecessaryTransportationToHospital). \\ Hosp :: < PAi, location, needMedCop, needAmbul > . \\ Port :: < PAi, location, needMedSh > .0| \\ Hosp :: (PAi, MedCopStatus, AmbulStatus).0| \\ Port :: (PAi, MedShStatus).0| \\ VTS :: < PAi, location, MedCopStatus, AmbulStatus, MedShStatus > .0| \\ VTS :: (PAi, OptimalRoute).PA :: < OptimalRoute > . \\ Hosp :: < PAi, OptimalRoute > .Port :: < PAi, OptimalRoute > .0 \end{array} \right) \quad (2)$$

$$P_{WCh} \cong \begin{pmatrix} PA :: (getLocation).PA :: \ <location> .0| \\ PA :: (choosenRoute, startMoving).! :: \ <PAi, MovementByRoute> .0| \\ PA :: \ <location, LifeThreateningIndicators> .0|PA :: (OptimalRoute).0 \end{pmatrix} \tag{3}$$

The VTS Ambient contains ENet and AmbiNet. It communicates with the other ambients and supplies the information needed to identify upcoming events and to generate a route to travel. In the second part of the scenario, it provides the search for and selection of an optimal route for moving the tourist-patient to the hospital. The process of this ambient is presented in (4).

$$P_{VTS} \cong \begin{pmatrix} PA :: (PAi, location, WantToVisitMuseum).Port :: \ <PAi, visitIsland).0| \\ Port :: (PAi, ShipTimeTable).Museum :: \ <PAi, visitMuseum> .0| \\ Museum :: (PAi, MuseumWorkingTime).BS :: \ <PAi, TravelByBus> .0| \\ BS :: (PAi, BusTimeTable). \\ ENet \downarrow \ <PAi, ShipTimeTable, MuseumWT, BusTimeTable, time> .0| \\ ENet \downarrow (PAi, hasShip, hasBus, openMuseum). \\ PA :: \ <hasShip, hasBus, openMuseum> .0| \\ PA :: (PAi, location, getRouteToIslandMuseum).GA :: \ <PAi, getActiveZones> .0| \\ GA :: (PAi, ListAZ).ANet \downarrow \ <PAi, ListAZ, getRoutes> .0| \\ ANet \downarrow (PAi, ListRoutes).PA :: \ <ListRoutes> .0| \\ GA :: (PAi, location, MedCopStatus, AmbulStatus, MedShStatus). \\ ANet \downarrow \ <PAi, location, MedCopStatus, AmbulStatus, MedShStatus> .0| \\ ANet \downarrow (PAi, OptimalRoute).GA :: \ <PAi, OptimalRoute> .0 \end{pmatrix} \tag{4}$$

Museum, Hosp, BS and Port Ambients – this group of ambients is responsible for providing information about the timetables of buses and ships as well as the working hours of the island museum. Upon the realization of the second part of the scenario, the Port ambient supplies information about the status of the medical shuttle and sends it to the island if this is the optimal route to travel. The Hosp ambient gives information about the status of the ambulances and the medical helicopter by means of communication with the GA. After finding the optimal route, it provides transportation for the patient. These processes are presented in (5), (6), (7), and (8).

$$P_{Museum} \cong (VTS :: (PAi, visitMuseum).VTS :: \ <PAi, MuseumWorkingTime> .0) \tag{5}$$

$$P_{BS} \cong (VTS :: (PAi, TravelByBus).VTS :: \ <PAi, BusTimeTable> .0) \tag{6}$$

$$P_{Hosp} \cong \begin{pmatrix} GA :: (PAi, location, needMedCop, needAmbul). \\ GA :: \; <PAi, MedCopStatus, AmbulStatus> .0| \\ GA :: \; <PAi, OptimalRoute> .0 \end{pmatrix} \tag{7}$$

$$P_{Port} \cong \begin{pmatrix} VTS :: (PAi, visitIsland).VTS :: \; <PAi, ShipTimeTable> .0| \\ GA :: (PAi, location, needMedSh).GA :: \; <PAi, MedShStatus> .0| \\ GA :: (PAi, OptimalRoute).0 \end{pmatrix} \tag{8}$$

ANet Ambient is a child of VTS and its functionalities are related to the processing of information about the location of the other ambients and the generation and optimization of routes for mobile ambients. Its subambients are RG and RO. In the second part of the scenario, the role of ANet and its subambients becomes the key to identifying the fastest route for transportation from the island to the hospital. Among all the possible routes generated by RG, RO needs to find the optimal one by time. To implement this search, a modified GraphSearch algorithm is used, in which certain nodes, important to the route, are involved, such as a medical shuttle, an ambulance, a medical helicopter, etc. The processes of ANet, RG and RO are demonstrated in (9), (10) and (11).

$$P_{ANet} \cong \begin{pmatrix} VTS \uparrow (PAi, ListAZ, getRoutes).RG \downarrow \; <PAi, ListAZ> .0| \\ RG \downarrow (PAi, ListRoutes).VTS \uparrow \; <PAi, ListRoutes> .0| \\ VTS \uparrow (PAi, location, MedCopStatus, AmbulStatus, MedShStatus). \\ RG \downarrow \; <PAi, location, MedCopStatus, AmbulStatus, MedShStatus> .0| \\ RO \downarrow (PAi, OptimalRoute).VTS \uparrow \; <PAi, OptimalRoute> .0 \end{pmatrix} \tag{9}$$

$$P_{RG} \cong \begin{pmatrix} ANet \uparrow (PAi, ListAZ).ANet \uparrow \; <PAi, ListRoutes> .0| \\ ANet \uparrow (PAi, location, MedCopStatus, AmbulStatus, MedShStatus). \\ RO :: \; <PAi, ListRoutes> .0 \end{pmatrix} \tag{10}$$

$$P_{RO} \cong (RG :: (PAi, ListRoutes).ANet \uparrow \; <PAi, OptimalRoute> .0) \tag{11}$$

The ccaPL language is a computer-readable version of the CCA syntax. The interpreter of ccaPL has been developed as a Java application. Based on the main version [19], we have developed a simulator for verification of the scenario described above. The notation "A === (X) ===> B" means that Ambient "A" sends an "X" message to Ambient "B". "Child to parent", "Parent to child" and "Sibling to sibling" provide information about the relationship between sender A and recipient B according to the hierarchy of ambients. An animator has been created to present graphically the ambients and their processes (Fig. 2).

Fig. 2. ccaPL simulator and animator

5 Conclusion

The article presents the general architecture of ViPS. Currently, our efforts are focused on adapting the ViPS architecture to a smart seaside city. There are several reasons for choosing a seaside city. Firstly, our team has some practical experience in monitoring large open water areas. Secondly, we are in contact with specialists developing a concept for a smart city by the sea. Thirdly, there are interesting scenarios relating to the aquatic environment, the coastline and the facilities of the adjacent smart city. Our idea is to model possible scenarios in such a city and to study the impact and effects of the delivery of e-services to different groups of users. It is a special feature that scenarios, services and users are in an integrated physical and virtual space. Initial experiments show some deficiencies in the architecture. For example, developing a visual editor for the CCA models would greatly facilitate the preparation of the scenarios to be modeled.

Acknowledgement. The authors wish to acknowledge the partial support of the National Program "Young Scientists and Postdoctoral Students" of the Ministry of Education and Science in Bulgaria, 2018-2019 and of the MES by the Grant No. D01-221/03.12.2018 for NCDSC – part of the Bulgarian National Roadmap on RIs.

References

1. Stoyanov, S., et al.: DeLC educational portal, cybernetics and information technologies (CIT). Bul. Acad. Sci. **10**(3), 49–69 (2010)
2. Stoyanov, S.: Context-Aware and Adaptable eLearning Systems, Internal Report, Software Technology Research Laboratory, De Montfort University, Leicester (2012)

3. Stoyanov, S., Stoyanova-Doycheva, A., Glushkova, T., Doychev, E.: Virtual Physical Space – an architecture supporting internet of things applications. In: XX-th International Symposium on Electrical Apparatus and Technologies SIELA 2018, 3–6 June 2018, Bourgas. IEEE (2018)
4. Stoyanov, S., Orozova, D., Popchev, I.: Internet of things water monitoring for a smart seaside city. In: XX-th International Symposium on Electrical Apparatus and Technologies SIELA 2018, 3–6 June 2018, Bourgas, Bulgaria. IEEE (2018)
5. Glushkova, T., Miteva, M., Stoyanova-Doycheva, A., Ivanova, V., Stoyanov, S.: Implementation of a personal internet of thing tourist guide. Am. J. Comput. Commun. Control **5**(2), 39–51 (2018)
6. De, S., Zhou, Y., Abad, I.L., Moessner, K.: Cyber–physical–social frameworks for urban big data systems: a survey. Appl. Sci. (2017)
7. Chamoso, P., González-Briones, A., Rodríguez, S., Corchado, J.M.: Tendencies of technologies and platforms in smart cities: a state-of-the-art review. Wirel. Commun. Mob. Comput. (2018)
8. Guo, B., et al.: Mobile crowd sensing and computing: the review of an emerging human-powered sensing paradigm. ACM Comput. Surv. **48**, 1–31 (2015)
9. Zhou, Y., De, S., Moessner, K.: Real world city event extraction from Twitter data streams. Procedia Comput. Sci. **98**, 443–448 (2016)
10. Guo, W., Zhang, Y., Li, L.: The integration of CPS, CPSS, and ITS: a focus on data. Tsinghua Sci. Technol. **20**, 327–335 (2015)
11. Guo, B., Yu, Z., Zhou, X.: A data-centric framework for cyber-physical-social systems. IT Prof. **17**, 4–7 (2015)
12. González-Briones, A., De La Prieta, F., Mohamad, M., Omatu, S., Corchado, J.: Multi-agent systems applications in energy optimization problems: a state-of-the-art review. Energies **11** (8), 1928 (2018)
13. Siewe, F., Zedan, H., Cau, A.: The calculus of context-aware ambients. J. Comput. Syst. Sci. **77**, 597–620 (2011)
14. Moszkowski, B.C.: Compositional reasoning using interval temporal logic and tempura. In: de Roever, W.-P., Langmaack, H., Pnueli, A. (eds.) COMPOS 1997. LNCS, vol. 1536, pp. 439–464. Springer, Heidelberg (1998). https://doi.org/10.1007/3-540-49213-5_17
15. Wooldridge, M.: An Introduction to MultiAgent Systems. Wiley, Hoboken (2009)
16. Hanes, D., Salgueiro, G., Grossetete, P., Barton, R., Henry, J.: IoT Fundamentals: Networking Technologies, Protocols, and Use Cases for the Internet of Things. Cisco Systems Inc., San Jose (2017)
17. Bellifemine, F.L., Caire, G., Greenwood, D.: Developing Multi-Agent Systems with JADE. Wiley, Hoboken (2007)
18. Pokahr, A., Braubach, L., Lamersdorf, W.: Jadex: implementing a BDI-infrastructure for JADE agents. Search Innov. (Spec. Issue JADE) **3**(3), 76–85 (2003)
19. Al-Sammarraie, M. H.: Policy-based Approach For Context-aware Systems. Software Technology Research Laboratory, De Montfort University, Leicester, UK (2011)

Explaining Performance Expectancy of IoT in Chilean SMEs

Patricio E. Ramírez-Correa[1(✉)], Elizabeth E. Grandón[2], Jorge Arenas-Gaitán[3], F. Javier Rondán-Cataluña[3], and Alejandro Aravena[4]

[1] Escuela de Ingeniería, Universidad Católica del Norte, Larrondo 1281, 1781421 Coquimbo, Chile
patricio.ramirez@ucn.cl

[2] Departamento de Sistemas de Información, Universidad del Bío-Bío, Av. Collao 1202, 4051381 Concepción, Chile
egrandon@ubiobio.cl

[3] Departamento de Administración de Empresas y Marketing, Universidad de Sevilla, Av. Ramón y Cajal 1, 41018 Seville, Spain
{jarenas,rondan}@us.es

[4] Bioforest S.A, Camino a Coronel Km. 15 S/N, Coronel, Chile
alejandro.aravena@arauco.cl

Abstract. The purpose of this paper is to validate a research model that explains performance expectancy of IoT from psychological and cognitive variables: personal innovativeness of information technology (PIIT) and social influence respectively. Data were collected from small and medium-sized enterprises (SMEs) in Chile. A confirmatory approach using PLSc was employed to validate the hypotheses. The conclusions of the study are (a) Chilean SMEs companies do not use IoT massively, (b) goodness of fit indicators allowed to validate the proposed research model successfully, (c) both constructs, social influence and personal innovativeness of information technology, explain 61% of performance expectancy of IoT.

Keywords: IoT · PIIT · Performance expectancy · Social influence · SMEs

1 Introduction

While the number of Internet of Things service users has increased in developed countries lately, there is a massive opportunity for growth in developing countries. The Internet of Things (IoT) is defined as a network that connects everyday objects to the Internet, where objects have sensors and programmable capabilities that allow to collect information and change its status from anywhere and at any time [1]. The IoT is emerging as a significant development in information technology, with the potential to increase convenience and efficiency in daily life. It is considered the fourth industrial revolution and it is believed that in 2020 there will be around 50,000 million connected devices [2]. However, this prediction may not be precise for developing countries. As Miazi et al. [3] pointed out, developing countries face problems to have access to

W. Abramowicz and R. Corchuelo (Eds.): BIS 2019, LNBIP 353, pp. 475–486, 2019.
https://doi.org/10.1007/978-3-030-20485-3_37

communication technologies in terms of poverty, lack of Internet speed, low levels of expertise, and overall lack of infrastructure. Government is these countries face enormous challenges to improve the current systems to make the infrastructure capable of deploying IoT as a whole. Chile, although it ranks first in terms of adoption of digital technologies in Latin America, still presents significant gaps to climb the digital revolution. As stated by Fundación País Digital [4], the main gaps to implement IoT are related to infrastructure, human capital, and cybersecurity.

The lack of IoT adoption in developing countries, in turn, makes managers of SMEs and people in general, to ignore the usefulness of this technology. For instance, and as stated by [5], farmers in developing countries could use remote sensors to monitor moisture levels and soil conditions in the fields to avoid crop failure. These sensors can provide remote control of micro-irrigation and water pumps, improving functionality and reducing repair intervals.

Following Karahamma and Straub line of thought, even though there is no doubt that technological developments occur at a fast rate, it is not evident that individual users of new technology can adopt and use new technology at the same pace [6]. Therefore, it is necessary to explore what drives perceptions of usefulness in order to better understand the technology adoption phenomenon. As proposed and validated by Davis [7] and Venkatesh et al. [8], and supported in numerous studies, perceived usefulness (performance expectancy) is a strong antecedent of behavioral intention which, in turn, predicts actual behavior (use of technology). Psychological characteristics of individuals along with social influence might play an essential role in the explanation of the performance expectancy belief.

Even though the positive predictions regarding the growth of IoT, and the benefits it could bring not only to companies but also to the wellbeing of individuals, few researchers have focused on studying this emerging technology. This study is the first to analyze the impact of innovativeness and social influence on performance expectancy by those responsible for information and communication technologies (ICTs) in small and medium-sized enterprises (SMEs) in Chile. In particular, this paper aims to validate a research model that explains performance expectancy of IoT from cognitive and psychological variables.

The contribution of this study is twofold. First, the study increases knowledge about the adoption of IoT, particularly in Chile. Second, and following the recommendations of [9] who suggest focusing on a specific type of products, the results of the analysis allow to deepen the effect of personal innovativeness on perceptions about IoT technology.

The paper is organized as follows: a definition and background of IoT and the constructs considered in the research model are given in the subsequent section. This section also presents the hypotheses to be tested along with the research model. Then the methodology of the study and the results of the statistical analyses are presented followed by a discussion of the findings and conclusions.

2 Background

2.1 Internet of Things

The international consultancy, IDC FutureScape [2], indicated that IoT along with other technologies (cloud computing, big data, and cognitive systems) is part of the third technological platform that will support a digital transformation. Also, it made predictions for the year 2020, where it mentions that this technology is in full development and that the business world, specifically South America, is in an exploration stage, moving to the implementation stage in different use cases.

Hsu and Lin [10] developed a conceptual framework from the perspective of network externalities and privacy to provide a theoretical understanding of the motivations that drive continued use of IoT services. The authors validated the model by interviewing 508 users regarding their perceptions of IoT services. The results indicated that network externalities play a significant role in influencing consumers' perception of usage benefits and thus adoption, whereas privacy concerns have a relatively weak effect on adoption. Mital et al. [11, 12], on the other hand, based their research on exploring IoT technology based on cloud computing in India. The authors mention that there is a shortage of exploratory studies that may explain the adoption of the IoT-based cloud. Therefore, more studies are required to understand the adoption of IoT phenomena better.

Following the above request, Grandón et al. [13] performed a study to determine the factors that influence the adoption of IoT using the unified model of acceptance and use of technology (UTAUT) as the theoretical framework. An electronic survey was applied to managers of small and medium-sized companies in Chile. The results showed that the intention to adopt IoT is explained in 68% by the constructs of UTAUT, being the social influence and facilitating conditions the key factors to explain this phenomenon.

2.2 Personal Innovativeness of Information Technology

Agarwal and Prasad [14] proposed the measure of a new psychological trait by developing the Personal Innovativeness of Information Technology (PIIT) construct in the domain of information technology. Initially, they stated that PIIT has a moderating effect on the antecedents as well as the consequences of individual perceptions about new information technology. They defined PIIT as "the willingness of an individual to try out any new information technology" (p. 206). The construct was operationalized and validated in a sample of 175 MBA students in the context of the World Wide Web.

In recent years, PIIT has been increasingly used in the IT literature due to its predictive capacity [15]. It has been mainly used in the context of the adoption of technologies such as electronic commerce [15], knowledge management systems [16], ERP systems [17, 18], and social network sites [19]. In some cases, PIIT has been complemented by theories of technology adoption, such as the technology acceptance model (TAM) and UTAUT as in the cases of [20] and [21].

2.3 Social Influence and Performance Expectancy

There are two important cognitive constructs developed in the models of technology adoption, social influence and performance expectancy. Venkatesh et al. [8] define social influence as "the degree to which an individual perceives that important others believe he or she should use the new system" (p. 451). Social influence is a significant antecedent of behavioral intention of technology when dealing with SMEs. For instance, [22] as well as [23] and [24] found that social influence (named as the subjective norm by [25]) positively and significantly predicts the intention to adopt e-commerce among managers/owners of SMEs in Chile. Other researchers have revised the impact of social influence on performance expectancy in the context of user continuance intention toward mobile commerce in the USA and found significant relationships between both [26].

Correspondingly, Venkatesh et al. [8] define performance expectancy as "the degree to which an individual believes that using the system will help him or her to attain gains in job performance" (p. 447). Performance expectancy (named as perceived usefulness by [7]) has also been found to be a significant predictor of behavioral intention to adopt information technology [7]. Koivisto et al. [27] focused on the role of personality traits in technology acceptance and compare weather personal innovativeness in the domain of information technology (PIIT) or technology readiness index (TRI), performs better in terms of promoting the explanatory power of the technology acceptance model (TAM). They found that both, PIIT and TRI, significantly impact perceived usefulness and use of online services offered by electric suppliers.

2.4 Hypotheses and Research Model

Individual beliefs about technology use have been shown to have a profound impact on subsequent behaviors toward information technology. For instance, Lewis et al. [28] conducted a study to examine factors that influence critical individual beliefs, perceived usefulness and perceived ease of use. Their findings suggest that personal innovativeness in the domain of information technology has a significant positive influence on individual beliefs about the usefulness of technology. Opposed to what they hypothesized, social influences from multiple sources exhibited no significant effects.

Built on the Technology Acceptance Model, Liu et al. [21] proposed a model of mobile learning adoption in China. They found that personal innovativeness is a predictor of both the perceived ease of use and perceived long-term usefulness. Of all variables, the perceived long-term usefulness contributes to the most influential predictor of m-learning adoption. The model accounts for approximately 60.8% of the variance of behavioral intention.

In a more recent study and focused on IoT technology, Mącik and Sklodowska [29] incorporated PIIT as an antecedent of the intention to adopt IoT by college students in Poland. They found a significant relationship between PIIT and performance expectancy towards IoT, which in turn, impact behavioral intention.

Consequently, in order to validate the proposed research of this study model and based on previous research, the following hypotheses are suggested:

H1: The degree of personal innovativeness in the domain of information technology is positively associated with performance expectancy of IoT

H2: The degree of social influence is positively associated with performance expectancy of IoT

Figure 1 below depicts the research model.

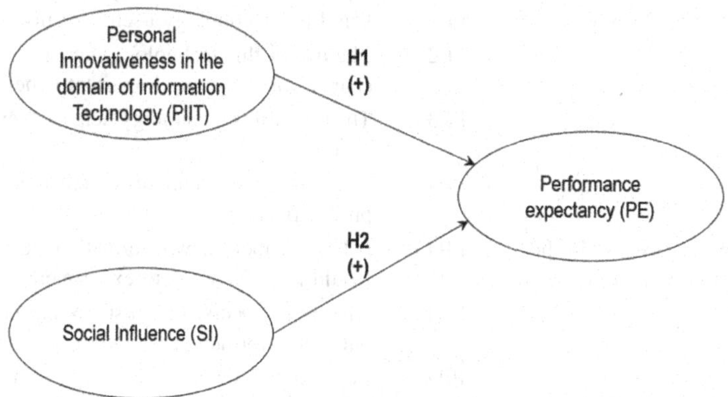

Fig. 1. Research model

3 Methodology

3.1 Subjects and Data Collection Process

A random sample of Chilean SMEs was chosen from different business directories. The Chilean Ministry of Economics defines a small and medium size business as one that employs between 10 and 200 employees [30]. The "Online Survey" tool was used to design, create, and monitor electronic survey responses. The perceptions of social influence, performance expectancy, and innovativeness personality trait were measured using a five-point Likert scale where 1 represents strongly disagreeing with the statement, while five indicates to be in complete agreement. The questionnaire was sent to 2000 contacts with the profile of ICT responsible for SMEs in Chile. After three weeks, follow-up telephone calls were made to the same non-respondents. One hundred and twenty-one contact addresses rebounded as "invalid address," and of the remaining, one hundred and three contacts validly answered the survey in two months.

3.2 Instrument Development

The definition of IoT was given in the first paragraph of the questionnaire so that respondents understood the meaning of the term and answered the questions accordingly. The survey consists of two parts. The first part included questions regarding perceptions of social influence, performance expectancy, and personal innovativeness of information technology. The second part consisted of demographic questions regarding the manager who answered the questionnaire. Social influence and performance expectancy questions were taken directly from the study of [8]. Questions regarding personal innovativeness of information technology were adapted from the study of [14]. Table 1 below shows all the items used to measure the three constructs.

Table 1. Items used to measure the variables

Variable	Item	
Performance Expectancy	PE1	I find this technology useful for my daily life
	PE2	The use of this technology increases the chances of achieving important benefits
	PE3	The use of this technology helps achieve things faster
	PE4	The use of this technology increases my productivity
Personal Innovativeness in the domain of Information Technology	PIIT1	If I heard about new information technology, I would look for ways to experiment with it
	PIIT2	Among my peers, I am usually the first to try out new information technologies
	PIIT3	In general, I am hesitant to try out new information technologies (reverse code)
	PIIT4	I like to experiment with new information technologies
Social Influence	SI1	People who influence the behavior of my company think that we should use this technology
	SI2	People who are important to my company think that we should use this technology
	SI3	In general, management would support the use of this technology
	SI4	The manager of my company has collaborated when using this technology

3.3 PLSc

Latent variables of the research model were measured as reflective constructs based on scales validated previously. Given these measurement characteristics, the consistent PLS algorithm (PLSc) was selected for the statistical analysis of the model. In particular, PLSc available in SmartPLS 3.2.8 was used. The PLSc algorithm performs a correction of reflective constructs' correlations to make results consistent with a factor-model [31].

4 Results

4.1 Descriptive Analysis of the Sample

Table 2 shows a description of the sample. Of the total sample, 89% correspond to men. Most of the respondents are between 30 and 49 years old, and they have an education in engineering or technology. Regarding IoT, the results show that the use of this technology is not massive.

Table 2. Distribution of the variables of interest in the sample.

Variable		N	%
Gender			
	Female	11	10.7
	Male	92	89.3
	Total	103	100
Age range			
	20–29	8	7.8
	30–39	36	35.0
	40–49	34	33.0
	50–59	14	13.6
	60+	11	10.7
	Total	103	100.0
Academic background			
	Engineering	49	47.6
	Technology & Communications	26	25.2
	Management & Business	18	17.5
	Others	10	9.7
	Total	103	100.0
IoT implemented		30	29.1

4.2 Evaluation of Global Model Fit

To assess the overall fit of the model the standardized root mean square residual (SRMR) is used. The literature considers that values below 0.08 are favorable in this case. The estimation of the model with PLSc reveals for the saturated model an SRMR value of 0.0513 (HI95: 0.0658, HI99: 0.840), and for the estimated model an SRMR value of 0.0513 (HI95: 0.0655, HI99: 0.846).

4.3 Evaluation of the Measurement Model

Table 3 indicates outer loadings and properties of measurement scales. For the analysis PIIT3 item was reversed. Cronbach's Alpha, Dijkstra and Henseler's rho (pA), Composite Reliability and the average variance extracted (AVE) are above the common criterions of 0.8, 0.7, 0.7 and 0.5 respectively. Additionally, the VIF value is below the threshold of 3 for the exogenous constructs.

Table 3. Outer loadings and properties of measurement scales.

Construct	Indicators	Outer loadings	Cronbach's alpha	ρA	Composite reliability	AVE	Inner VIF
Performance Expectancy (PE)	PE1	0.9049	0.9400	0.9430	0.9406	0.7989	
	PE2	0.9233					
	PE3	0.9284					
	PE4	0.8137					
Personal Innovativeness in the domain of Information Technology (PIIT)	PIIT1	0.9721	0.8362	0.8804	0.8412	0.5809	1.1659
	PIIT2	0.6500					
	PIIT3R	0.5498					
	PIIT4	0.8085					
Social Influence (SI)	SI1	0.8515	0.9374	0.9400	0.9379	0.7909	1.1659
	SI2	0.9094					
	SI3	0.9472					
	SI4	0.8454					

To analyze the discriminant validity the heterotrait-monotrait ratio of correlations (HTMT) was used. Table 4 shows the result of the examination, all the values are below 0.9, and therefore, the discriminant validity has been established between all the constructs.

Table 4. Heterotrait-monotrait ratio (HTMT).

Construct	PE	PIIT	SI
PE			
PIIT	0.6379		
SI	0.6544	0.3599	

4.4 Evaluation of Structural Model

Figure 2 and Table 5 show the results of the structural model analysis using PLSc. The p-values were obtained using a consistent bootstrapping procedure with a resample of 5,000. The values of R^2 and Q^2 are greater than 0.6, which indicates an appropriate level of prediction of the exogenous variables. On the other hand, path coefficients with values over 0.4 and statistically significant, support the two hypotheses of the study.

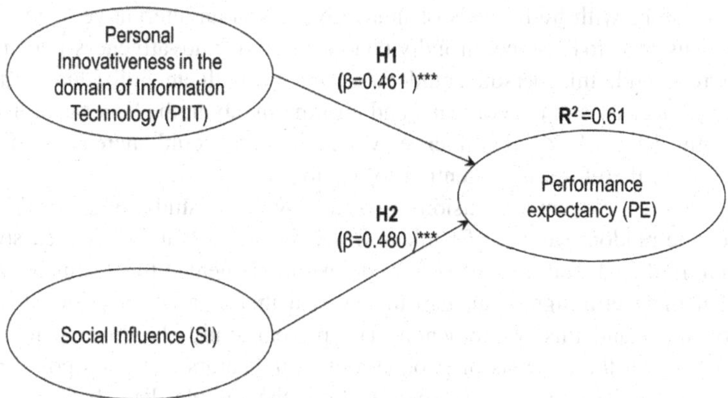

Fig. 2. PLSc model results.

Table 5. Structural model results.

Dependent variable: PE		Path coefficient	f^2	p-Value
R^2	0.6108			
Q^2	0.6020			
PIIT		0.4610	0.4672	0.0000
SI		0.4799	0.5062	0.0000

5 Discussion and Conclusion

This research was oriented to validate a research model that explains performance expectancy of IoT from two variables, a psychological and a cognitive. From data collected randomly from Chilean SMEs, the results of the study indicate that both the cognitive variable, social influence, and the psychological variable, PIIT, are relevant to explain the performance expectancy of IoT. The effect size of each of these variables is greater than 0.35 which allows to denominate it as large.

There are two implications of these results that should be highlighted. First, and from a theoretical point of view, while the relationship between social influence and performance expectancy has been widely studied in the literature, the validation of the relationship between PIIT and performance expectancy is a novel finding that should be studied in contexts where new technologies are starting in the diffusion process, for example, blockchain or biochips. Second, and from a practical point of view, although the distribution of the PIIT variable in the sample has a negative skew, there are more respondents with higher levels of innovativeness, which could be explained by their academic background. Most of them are in engineering or technology and communication areas. It is important to emphasize that the behavior of this variable should have a normal distribution [32]; that is, a large group of individuals would have an average level of innovativeness. Therefore, providing information about IoT to influencers of business decisions with high levels of innovativeness is an alternative to increase the intention to use the IoT. Since an individual with high innovativeness perceives that IoT is more useful, this person could communicate with an individual that makes technological decisions in the company and inform him about the benefits of using IoT, increasing his level of social influence, which in turn, could increase performance expectancy, and therefore, the intention to use IoT.

There are three relevant conclusions associated with this study. First, the descriptive analysis of the random sample of Chilean SMEs indicates that IoT is used sparingly. Less than a third of the companies surveyed have implemented this technology; which implies that there is a significant gap to cover in this area of adoption of IoT, and studies to understand this phenomenon. The present study should help to face this challenge. Second, the analysis of goodness-of-fit indicators of the proposed research model shows that its validation is possible from the sample data. It is important to emphasize that for this analysis, it was used consistent PLS algorithm. The traditional PLS algorithm shows the inconsistency of the path coefficient estimates in the case of reflective measurements, which may have some issues for hypothesis testing. To remedy this potential problem, PLSc offers a correction for estimates when PLS is applied to reflective constructs, as is the case of the research model proposed in this study. Third, the analysis of the structural model indicates that social influence and personal innovativeness in the domain of information technology explain 61% of performance expectancy of IoT. Also, it can be pointed out that the effect sizes of the exogenous variables are large and of very similar values.

Two limitations of this study must be recognized. First, although the sample was random, its small size indicates that the results should be treated with caution if they are extrapolated at the country level. Second, data were collected only at one point in time. Given that the IoT is an emerging technology, a longitudinal study that encompasses both early and late adopters would be very useful to corroborate the results of this study.

Future research may, in addition to performing longitudinal analyzes, incorporate SMEs from other Latin American countries with similar realities to those of Chile, in order to validate the findings of this study.

References

1. Ibarra-Esquer, J.E., González-Navarro, F.F., Flores-Rios, B.L., Burtseva, L., Astorga-Vargas, M.A.: Tracking the evolution of the internet of things concept across different application domains. Sensors (Switzerland) **17**(6), 1379 (2017)
2. Parker, R., et al.: IDC FutureScape : Worldwide IT Industry 2017 Predictions
3. Miazi, M.N.S., Erasmus, Z., Razzaque, M.A., Zennaro, M., Bagula, A.: Enabling the internet of things in developing countries: opportunities and challenges. In: Proceedings of the 2016 5th International Conference on Informatics, Electronics and Vision, ICIEV 2016 (2016)
4. Fundación País Digital Cómo emprender en Internet de las Cosas: Conceptos Prácticos; Gobierno de Chile. Santiago, Chile (2018)
5. Hasenauer, C.: The Internet of Things in Developing Countries. https://borgenproject.org/internet-of-things/. Accessed 5 Jan 2019

6. Karahanna, E., Straub, D.W.: The psychological origins of perceived usefulness and ease-of-use. Inf. Manag. **35**, 237–250 (1999)
7. Davis, F.D.: Perceived usefulness, perceived ease of use, and user acceptance of information technology. MIS Q. **13**, 319–340 (1989)
8. Venkatesh, V., Morris, M.G., Davis, G.B., Davis, F.D.: User acceptance of information technology: toward a unified view. MIS Q. **27**, 425–478 (2003)
9. Ramírez-Correa, P.E., Grandón, E.E., Arenas-Gaitán, J.: Assessing differences in customers' personal disposition to e-commerce. Ind. Manage. Data Syst. (2019, In press). IMDS622482
10. Hsu, C.L., Lin, J.C.C.: An empirical examination of consumer adoption of Internet of Things services: Network externalities and concern for information privacy perspectives. Comput. Human Behav. **62**, 516–527 (2016)
11. Mital, M., Chang, V., Choudhary, P., Pani, A., Sun, Z.: Adoption of cloud based Internet of Things in India: a multiple theory perspective. Int. J. Inf. Manage. (2016)
12. Mital, M., Chang, V., Choudhary, P., Papa, A., Pani, A.K.: Adoption of Internet of Things in India: a test of competing models using a structured equation modeling approach. Technol. Forecast. Soc. Change **136**, 339–346 (2018)
13. Grandon, E.E., Aravena, A.A., Guzman, S.A., Ramirez-Correa, P., Alfaro-Perez, J.: Internet of Things: factors that influence its adoption among Chilean SMEs. In: Proceedings of the Iberian Conference on Information Systems and Technologies, CISTI (2018)
14. Agarwal, R., Prasad, J.: A conceptual and operational definition of personal innovativeness in the domain of information technology. Inf. Syst. Res. **9**, 204–215 (1998)
15. Jackson, J.D., Yi, M.Y., Park, J.S.: An empirical test of three mediation models for the relationship between personal innovativeness and user acceptance of technology. Inf. Manage. **50**(4), 154–161 (2013)
16. Prasetya, W.Y.S., Shihab, M.R., Sandhyaduhita, P.I.: Exploring the roles of personality factors on knowledge management system acceptance. In: Proceedings of the 2015 3rd International Conference on Information and Communication Technology, ICoICT 2015 (2015)
17. Hwang, Y.: User experience and personal innovativeness: an empirical study on the enterprise resource planning systems. Comput. Human Behav. **34**, 227–234 (2014)
18. Wang, W., Li, X., Hsieh, J.P.A.: The contingent effect of personal IT innovativeness and IT self-efficacy on innovative use of complex IT. Behav. Inf. Technol. **32**(11), 1105–1124 (2013)
19. Zhong, B., Hardin, M., Sun, T.: Less effortful thinking leads to more social networking? the associations between the use of social network sites and personality traits. Comput. Human Behav. **27**(3), 1265–1271 (2011)
20. Wong, C.H., Tan, G.W.H., Tan, B.I., Ooi, K.B.: Mobile advertising: the changing landscape of the advertising industry. Telemat. Inform. **32**(4), 720–734 (2015)
21. Liu, Y., Li, H., Carlsson, C.: Factors driving the adoption of m-learning: an empirical study. Comput. Educ. **55**(3), 1211–1219 (2010)
22. Nasco, S.A., Grandón, E., Mykytyn Jr., P.P.: Predicting electronic commerce adoption in Chilean SMEs. J. Bus. Res. **61**, 697–705 (2008)
23. Grandon, E.E., Nasco, S.A., Mykytyn Jr., P.P.: Comparing theories to explain e-commerce adoption. J. Bus. Res. **64**, 292–298 (2011)
24. Grandón, E., Ramirez-Correa, P.E.: Managers/Owners' innovativeness and electronic commerce acceptance in Chilean SMEs: a multi-group analysis based on a structural equation model. J. Theor. Appl. Electron. Commer. Res. **13**, 1–16 (2018)
25. Ajzen, I.: The theory of planned behavior. Organ. Behav. Hum. Decis. Process. **50**, 179–211 (1991)

26. Lu, J.: Are personal innovativeness and social influence critical to continue with mobile commerce? Internet Res. **24**(2), 134–159 (2014)
27. Koivisto, K., Makkonen, M., Frank, L., Riekkinen, J.: Extending the technology acceptance model with personal innovativeness and technology readiness: a comparison of three models. In: BLED 2016 Proceedings 29th Bled eConference Digit. Econ. (2016). ISBN 978-961-232-287-8
28. Lewis, W., Agarwal, R., Sambamurthy, V.: Sources of influence on beliefs about information technology use: an empirical study of knowledge workers. MIS Q. (2003)
29. Mącik, R.: The adoption of the internet of things by young consumers – an empirical investigation. Econ. Environ. Stud. **17**(42), 363–388 (2017)
30. MINECOM Antecedents for the revision of the classification criteria of the SME Statute; Santiago, Chile (2014)
31. Dijkstra, T.K., Henseler, J.: Consistent partial least squares path modeling. MIS Q. **39**, 297–316 (2017)
32. Parasuraman, A., Colby, C.: Techno-ready marketing: how and why your customers adopt technology. J. Consum. Mark. (2002)

Blockchain-Based Platform for Smart Learning Environments

Rawia Bdiwi[1(✉)], Cyril de Runz[2(✉)], Arab Ali Cherif[1(✉)], and Sami Faiz[3(✉)]

[1] University of Paris 8, LIASD Research Laboratory, Saint-Denis, France
bdiwi.rawya@gmail.com, aa@ai.univ-paris8.fr

[2] University of Reims Champagne-Ardenne, CReSTIC Laboratory, Reims, France
cyril.de-runz@univ-reims.fr

[3] LTSIRS Laboratory, National Engineering School of Tunis, Tunis, Tunisia
sami.faiz@insat.rnu.tn

Abstract. The Internet of Things (IoT) is making significant advances, especially in smart environments, but it still suffers from security issues and vulnerabilities. Hence, the security approach seems to be inappropriate for IoT-based workspaces due to the centralised architecture used by most connected objects. However, BlockChain (BC) is a technology that has been recently used to enhance security in mainly peer-to-peer (P2P) networks. The main research goal of this study is to determine whether BC, IoT, and decentralised models can be used to secure Smart Learning Environments (SLEs). In this paper, we propose a new secure and reliable architecture for connected workspaces that eliminates the central intermediary, while preserving most of the security benefits. The system is investigated on an environment containing several connected sensors and uses the BC application, offering a hierarchical network that manages transactions and profits from the distributed trust method to enable a decentralised communication model. Finally, an evaluation is conducted to highlight the effectiveness in providing security for the novel system.

Keywords: Smart Learning Environments · Blockchain technology · Internet of Things · Decentralised architecture

1 Introduction

The Internet of Things (IoT) is a recent communication paradigm that represents the future, where our everyday lives are merged with devices equipped with a wide range of sensors that make objects able to transfer information among the connected users. The generated information provides useful knowledge, taking advantage of different services using IoT that works with different web-based platforms. Some examples of these platforms help teachers automatically monitor the progress of groups of students, so they can provide real-time feedback to improve the learning experience. Learners can be automatically counted as present or absent using wearable devices. That is where IoT comes in. Thus, this concept aims at making our environments even more ubiquitous [1]. IoT vision is focusing on improving SLEs and context-aware

W. Abramowicz and R. Corchuelo (Eds.): BIS 2019, LNBIP 353, pp. 487–499, 2019.
https://doi.org/10.1007/978-3-030-20485-3_38

applications for users anywhere worldwide [2]. SLEs include the interactions between learners, teachers, and the workspace. The smart learning environment (SLE) is considered a technology-supported environment that provides appropriate information in a determined place and time. These needs can be determined by analysing the learning, performance, and behaviours of students. The complexity of recent architectures has led to the development of new technological tools, such as BC [3], which has evolved to face numerous challenges, such as developing secure platforms that automatically check the provenance data from connected objects [4]. These platforms facilitate the management of data collection and verification regarding privacy, interoperability, and security, which are the most important issues [5]. BC offers new methods to establish this process, and its main functionalities are the privacy, accessibility, and security information database where there is no confidence between users [6]. Despite the significant benefits provided by ubiquitous environments based on IoT for education, these technologies can also overlay serious security and privacy issues, if the data collected by objects reveal confidential information. However, data produced by devices are processed and stored via centralised systems controlled by a single authority [7]. In fact, if centralised data are distributed among several servers, there is always only one point of control. In the context of IoT, since multiple objects are continuously sending information, the risk is even more concrete [8]. In this paper, we present our proposal for decentralised architecture for an IoT-based SLE. The innovative system consists of three main elements. The first one is the network of IoT, where objects and sensor data are securely collected and stored. The second component is BC that provides a distributed platform used to manage transactions securely. The third component is the method of access that enables the management of collected data to define which data to share and with whom. This system preserves the benefits of IoT and enables the use of various services offered by the BC to secure the processing data. Before using the BC, some imperative technical problems must still be addressed, such as the scalability of the BC with the huge number of transactions [9]. The proposed type of communication is shared among multiple nodes of the network, and the BC has the highest potential to answer the interoperability problems currently present in smart infrastructures. It facilitates the everyday lives of distant students and teachers and securely shares educational services. In this work, a new secure platform for SLEs based on the BC is proposed. The discussed model allows the migration from centralised architecture to the decentralisation trend that uses a distributed manner to manage data. The elimination of a central authority is essential to solving many issues. The distributed network provides carrier-grade security to manage heterogeneous IoT devices and collects numerous data. The novel platform should support recent technology to be able to support IoT control via the distributed nodes. Therefore, the evaluation of this new decentralised architecture highlights its performance in improving security benefits for SLEs. The rest of this paper is structured as follows. Section 2 presents some previous work of the use of the IoT-BC. Section 3 describes the design of the decentralised architecture based on the BC, which enables secure communication between heterogeneous devices. This part will be followed by the presentation of our experimentation results (Sect. 4). Finally, (Sect. 5) the paper ends with a discussion of the fundamental results and the conclusion (Sect. 6).

2 Related Work

In this section, we present some recent research papers that describe SLEs. Then, we focus our research on IoT communication models dedicated to the design of connected workspaces, where the IoT allows things and people to be linked and connected anytime, anyplace, using appropriate services. Finally, we present how the BC can lead to the improvement of different platforms and efficient ecosystems according to security, reliability, etc.

2.1 Technology for Smart Learning Environments

In [10], the authors proposed a framework for applying intelligent learning environments in the context of a smart city. Despite the significant advance in smart cities, the use of technology for learning and its different fields, such as learning analytics and big data, still lacks solutions to ensure the connectivity to intelligent city governance to meet stakeholder demands. The concept of the SLE has drawn attention from researchers recently, so the authors in [11] proposed a system that consists of enhancing the quality of service in smart classrooms. The whole designed architecture is controlled using a smart phone. Researchers have developed a mobile application dedicated to the administrator and students. It includes modules that enable the upload of documents for learners, and the administrator can send notifications for the meeting schedule, events, alerts, etc. These architectures face issues of security. In this context, the design architecture for the SLE remains challenging, considering the heterogeneous equipment, as noted by [12]. Ubiquitous learning (u-learning) has become very useful in the education area. It is considered a physical world that is related to sensors and actuators in the everyday objects of our lives and is principally connected via a continuous network. The major goal of this work is the development of a ubiquitous learning environment (ULE) that provides educational content to students. The results of this work showed that ULE with a multi-agent architecture can improve student learning significantly for both individual and collaborative learning modes. The majority of these environments are designed to exploit new technology to deliver several personalised services to users once they interact with devices. Otherwise, in smart environments, there is a strong need for solutions offering more security and reliability. In the same context, the possible risks are more important due to the interaction and communication between numerous users and devices. These systems are based on a centralised network that deploys a central authority that controls data. Security is a crucial and critical component of any network system, and a centralised model is easier to attack than a decentralised system.

2.2 Internet of Things for Smart Learning Environments

Integrating the IoT with SLEs can transform the way of learning. In [13], a proposal of a practical implementation of an IoT gateway aiming to monitor data in real time via the central authority enables bidirectional communication among users and the wireless sensor network. This system offers the control of a smart swimming pool remotely. In the area of education, technology can change the learning process and even the students' behavioural analysis. The paper [14] describes student collaboration and interaction using an IoT-based interoperable infrastructure. The main goal of this study is to implement and develop an IoT-based interaction framework, and then the analysis of the student experience of e-learning is evaluated. We notice that the potential and benefits of the IoT on SLEs are substantial. However, the enormous amount of data sent from sensors to the network generates many issues. This literature survey enables us to gain a better understanding of our surroundings. Moreover, among the biggest challenges of security are those of IoT architectures. The above solutions are not required to deal with heterogeneous networks. Likewise, the use of many recent technological tools, such as BC, can scale and overcome several challenges of constrained SLEs.

2.3 Blockchain Technology and Applications

Incorporating the BC into the revolution of the SLEs will make new technology more effective to deliver various personalised services, and even add a better method of transparency and security. In this new trend of evolution, BC can extremely enhance a wide variety of fields, such as education, health, etc. In [15], the authors introduce the main features of the BC by presenting some of the current BC applications for the education field. This paper determines the potential of BC-based educational applications that solve some learning problems. Therefore, the BC is also coupled with the IoT in [16], where the authors introduced a new method using the BC for IoT devices to accomplish industrial grade reliability in the transfer of data from wireless sensor systems to production systems. Their essential findings reveal that BC mechanisms can highly secure the wireless communication of connected devices. Hence, the original innovation of this work is the effective application of BC mechanisms to increase the level of security of wireless systems based on sensors. In [17], the researchers explain the concept of the BC to adapt security issues to cloud computing. Further, the BC can be applied beyond environments based on the IoT and its applications. It must be considered that cloud computing is also adopted in all SLEs for its availability. In the same context, [18] researchers have shown the emergence of the IoT and BC using P2P approaches that play a significant role in the change of decentralised and data-intensive applications used on numerous connected devices, while preserving the privacy of users. The main goal of this work is to determine whether the BC and P2P methods can be used to foster the design of a decentralised and private IoT system. However, the BC is gaining and sustaining the attention of many researchers, especially after the foundation of Bitcoin [19]. For this reason, the increase of BC-based applications is very prominent regarding the numerous fields that use a great number of services and data. Nevertheless, this technology still has numerous challenges and issues. Moreover, [20] presented a comprehensive overview on BC architecture and compared some related typical consensus algorithms. Furthermore, [21] introduced the cryptocurrency bitcoin and the related advantages of the security of a bitcoin transaction. The benefits and limits of the BC are discussed, considering that it is used in different areas, such as finance, IoT, etc. Consequently, the results reveal that the primary advantages of this technology are security on the network and transactions. The decentralised technique is explored in this context due to the effectiveness of bitcoin transactions, which is known as a ledger or a database distributed via the network. In this study, the BC removes the central authority of the centralised architecture to enhance security in making a transaction. Thus, it is verified by special nodes named miners that solve a complex cryptography to add a block to the chain. Each block is hashed and contains the hash of the previous one. Thus, changes in the history of the transactions are impossible. The discussed papers in this section demonstrate that the BC provides secure and transparent communication via the control and manipulation of several transactions on the network. Hence, the foremost issues of IoT-based SLEs are the use of a centralised architecture, interoperability, security, etc.

These systems are not recommended for managing numerous data generated by smart ecosystems. Currently, most architectures suffer from many problems, such as the deployment cost of central servers based on cloud computing and the related networking equipment. Likewise, the distributed techniques presented in some research works based on the BC face many of the cited challenges. Obviously, the distributed model that is dedicated to processing an enormous number of transactions among connected devices is more effective than the supervision of a central authority. The emergence of the BC leads us to assess the benefits that can be brought by this technology, especially for the SLE, where students can access the workspace anywhere and anytime. Hence, the lack of security is one of the main issues that makes the IoT and connected systems less attractive for users and companies.

3 Design Solution

In this research work, SLE based on the IoT and BC is designed and evaluated via the use of the BC application. The proposed solution determines how the combination of the BC via distributed ledger methods can be used to ensure secure communication among connected objects. The paper involves several technological tools that serve for the migration from a centralised system towards a distributed architecture to manage transactions. In addition, the BC is a recent technology that deals with security issues for connected objects. The designed system must be aware of the requirements and needs for security, scalability, etc., regarding the number of heterogeneous devices connected to the system. Thus, the new technological improvement in the IoT and ubiquitous computing allows us to investigate the development of more secured and efficient systems that are easier to install, control, and process using a decentralised communication model. This paper discusses how we can merge the IoT and BC to create and deliver secure shared educational services to enhance the learning process. Our novel system must ensure the control of a huge number of objects and enable a great number of transactions and the coordination of processing among devices and participants (users) or named peers. Certainly, the distributed approach would eliminate the lack of security via the creation of a more flexible ecosystem [22]. The use of cryptographic algorithms adopted by BC applications makes the consumption and control of data more secured and reliable. The evaluation of the newly designed architecture highlights its potential to produce more flexibility and security in the management of transactions on the network. Information is stored into a non-relational database and is accessible by all authorised users. In this case, the database is distributed to exchange and store information securely. In this section, we present the collection of data from sensors via our novel system, and we explain how transactions are processed. The platform is used to gather data from connected objects, and then the BC acts to secure the exchange to deliver services to the smart devices of students.

3.1 Internet of Things-Based Smart Learning Environments

We present different modules of the proposed IoT-based SLE. The main part of our architecture is the management of transactions among objects in each network node. Hence, the system, as shown in Fig. 1, contains connected objects, sensors, and smart devices. Each of them reads a particular aspect of the smart space and sends the data to the cloud platform, which is attached to the system. The collected data are allowed to pass via several stages, depending on several requirements. The network and BC-based cloud platform is deployed to process, collect, and control information. In this work, we eliminate the central authority since we use the distributed model, so the collector is replaced by the gateway in the case of the decentralised architecture. Likewise, the transfer of data from the different entities of the distributed architecture is performed securely to be sent to the platform for analysis and processing. The BC records all transactions of the network, which are determined and verified by all authorised users. Moreover, all transactions are transparent, and any modification can be monitored and traced. The improvement of security and reliability in our system is approved by deploying the BC in the communication among the connected objects and IoT devices. The SLE includes IoT devices and a BC-based cloud platform. This platform monitors all incoming and outgoing transactions. Local IoT devices connected to smart environments generate transactions to share, request, and store information. We implement the BC network using the *IBM Bluemix* service to run an IoT application developed by the *IBM Watson team.*

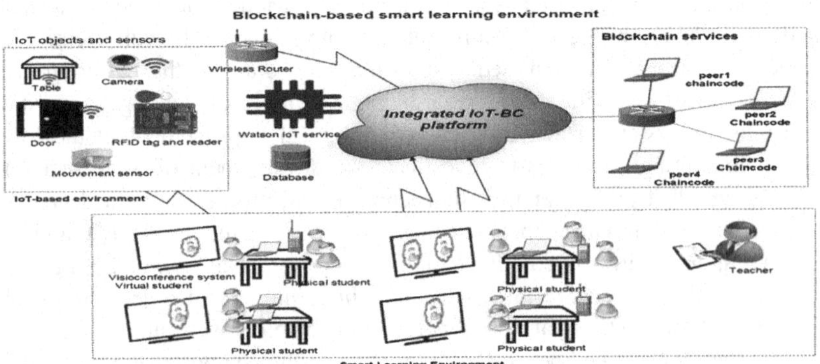

Fig. 1. Blockchain-based architecture for smart learning environment (SLE).

In this study, we use a distributed architecture to ensure the validity and security of the received transactions and even the blocks. It works as follows:

- To validate a transaction, the user requests a transaction to the defined ledger. The comparison of the current and next transaction is verified by the hash.
- Every computer in the BC network validates the transaction using the validation rules defined by the BC application. Once, it is validated, this transaction will be stored in the block and hashed.
- The user signature is proved using the hash of each transaction. Then, the verification enables control so that only one of the outputs of the existing transaction can confirm it. The number of successful or rejected transactions is only augmented one by one. If the steps are validated successfully, the transaction is confirmed. The transaction becomes part of the BC and cannot be modified.

3.2 Blockchain-Based Architecture

In this subsection, we explain the functions of the BC used in this study. The BC is known as a distributed ledger technology (DLT) [23] that provides advanced processes by creating new levels of trust and transparency for innovative generation of transactional applications. We use the *IBM BC* in view of the fundamental exchanges and applications. It reduces the complexity and cost of transactions via an efficient secure network where objects and sensors can be tracked without dependence on a centralised point of control. In a BC network, the records of transactions are detained on a collective ledger that is replicated across peers, so if a peer is not subscribed to a BC channel, the peer will not have access to the channel transactions. All records of collected transactions, whether valid or not, are stored in blocks and then added to the *hashchain* for each appropriate channel. The verified transactions update the state of the database, while an unconfirmed transaction will not. The smart contracts, also known as *chaincodes,* are a set of functions enabling the mode of read/write on the ledger using the *getstate* and *putstate* functions.

4 Experimentation and Results

Based on the design, analysis, and implementation of the proposed architecture, we reach a number of experiment results based on data values from the sensors of the SLE. We test the system with a BC application that enables the creation of new blocks and the control of some transactions to check for any issue. In this experimental section, we explain the aim of this work that defines the creation of new levels of security and reliability based on DLT [24]. In our case, the shared key among the miner and IoT objects is kept in the transaction. Furthermore, miners validate and record new transactions on the global ledger. Thus, it solves a complex mathematical problem based on a cryptographic hash algorithm used to secure transactions. When a block is solved, the transaction is considered confirmed. The miner in our context is the *IBM BC platform* that offers several services, such as collecting data from connected devices, creating new blocks, validating transactions, etc. The BC-based service used in this experimentation contains four peers or users for testing by default.

4.1 Transaction Management

The process of recording and verifying transactions is instantaneous and permanent. The ledger is distributed among different nodes, then data are replicated and stored immediately on each node in our system. Once a transaction is recorded in the BC, the details of this transaction are verified in all nodes. A change is checked on any ledger and is registered on all copies of the ledger [25]. The transaction is transparently confirmed among all ledgers and is open for any peer to see. Consequently, there is no need for third-party verification. Our BC network is based on transactions that are detained on a shared ledger, and if a peer is not subscribed to a network, the peer did not have access the channel transactions. In fact, IoT objects in the designed platform can transfer information directly with each other or with student devices. The achievement of user control via the SLE is verified via a shared key that is allocated by the BC-based system to devices. The system validates the owner and distributes a shared key across IoT objects. If the shared key is validated, different devices can communicate without any trusted intermediary since we eliminate the centralised model. Likewise, to validate access and permissions, the component (miner) indicates that the shared key is invalid. The significance of these methods is checked as follows: the miner can identify the IoT object that has the shared data, and then the connection among IoT devices is secured and protected with a shared key. Indeed, the *transactor* (known as a network connected to the BC via different nodes) is in charge of submitting the transaction from a client using the API of *Bluemix*. It ensures the request to perform some functions on the BC network. All transactions are either deployed or invoked as a query, which are executed via the *chaincode* functions. In this study, we use the *chaincode* application deployed on the network. As user, we invoke the *chaincode* via the offered client-side application that interfaces with the BC network peer or node. The smart contracts or *chaincode* executes the network transactions, which is already validated, and then appends to the common ledger and consequently modifies the state.

In Fig. 2, we show that, after the confirmation of the ledger, the reliable and trusted transaction is invoked and submitted to the BC network successfully. The Invoke() function is usually called to do the verification in the BC network. The update of the ledger is accomplished only by the invocation of the *chaincode*. It is a software that includes a set of functions that enable the reads and writes to be made against the ledger. The client-side application is used to initially interface with peers and finally the call functions on a specific *chaincode*.

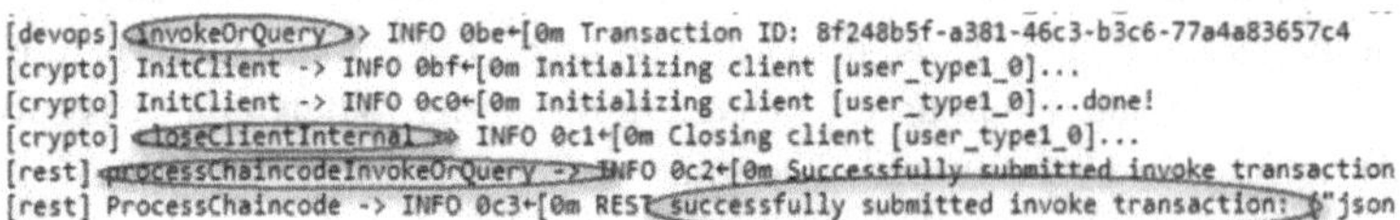

```
[devops]InvokeOrQuery -> INFO 0be←[0m Transaction ID: 8f248b5f-a381-46c3-b3c6-77a4a83657c4
[crypto] InitClient -> INFO 0bf←[0m Initializing client [user_type1_0]...
[crypto] InitClient -> INFO 0c0←[0m Initializing client [user_type1_0]...done!
[crypto] closeClientInternal -> INFO 0c1←[0m Closing client [user_type1_0]...
[rest] processChaincodeInvokeOrQuery -> INFO 0c2←[0m Successfully submitted invoke transaction
[rest] ProcessChaincode -> INFO 0c3←[0m REST successfully submitted invoke transaction: {"json
:1}
```

Fig. 2. Transaction validation

4.2 Block Activity

As shown in Fig. 3, we present the activity of blocks that specifies several pieces of information, such as the number of invoked transactions, number of new blocks appended in the *hashchain*, and the new message. A block in the BC is a page of a ledger that is checked and verified to allow the system to move to the next block in the BC channel.

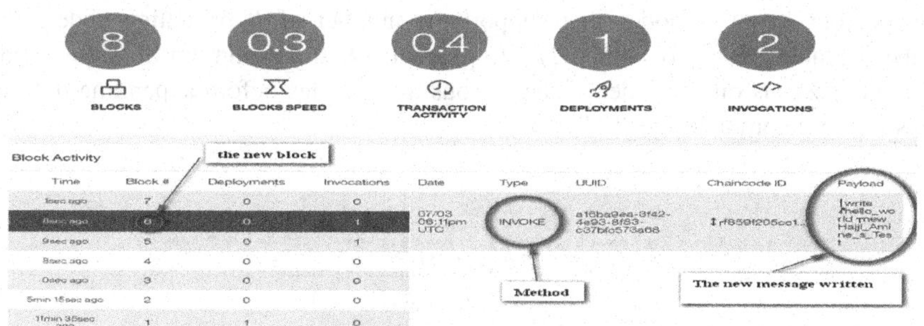

Fig. 3. Block invocation

In Fig. 4, we observe the deployment of the *chaincode* that aims to verify the peer node. The message has a unique ID in the *chaincode* of the BC network.

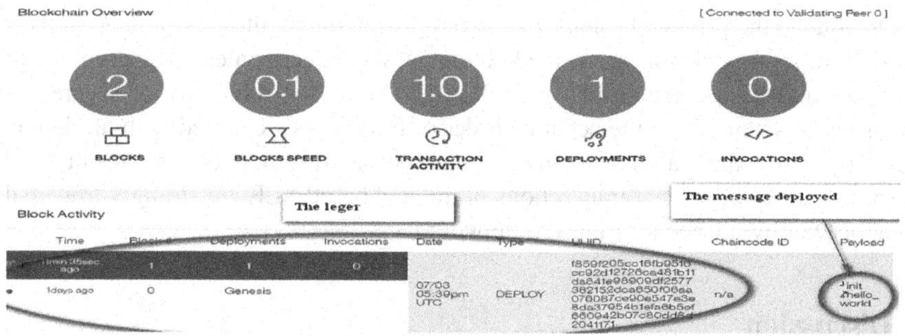

Fig. 4. Block activity

The complete block is composed of a record of all previous transactions and the new invoked transactions that should be stored in the present record. Thereby, blocks have the records and the history of all transactions. We use this program in Fig. 5 to test the addition of new blocks, which has the related payload. The BC is a list of different records that are assembled into blocks and preserved in a distributed way. Each block in our network includes a cryptographic hash of the previous block, which avoids changes to the current blocks. In conjunction with some protocols for distributed consensus, the BC is presently used as a platform that supports various transactions. This experimental study presents how the consensus protocol can verify the sequencing of invoked transactions on the BC network. Therefore, the consensus network contains several nodes known as validating and non-validating nodes. In Fig. 5, the practical Byzantine fault tolerance (*PBFT*) [26] that is one of the main entities of the consensus method until it preserves the sequence of transactions. The IoT objects in the proposed SLE use *PBFT* to confirm the validity of information nodes. All the validated information is appended to the BC, and a consensus protocol guarantees that nodes agree on a unique sequence or order in which entries are added. The consensus protocol is coupled with cryptographic tools to ensure the success of this process. Thus, BC network consists of *N* nodes that support the number of *B* byzantine nodes. This number is calculated as $B = (N - 1)/3$ to guarantee that a minimum of $2*B+1$ nodes achieve consensus on the order of the various transactions before appending different nodes to the common ledger [27].

Fig. 5. Consensus protocol

Obviously, the principal role of *PBFT* is to maintain the integrity of data. In the BC network, nodes use the consensus method to approve ledger content. The cryptographic signatures and hashes are similarly deployed to settle the reliability of secure transactions. The system saves the accurate order of transactions coupled with all deployed nodes to ensure the validity of the current transactions. Indeed, the result of this experimentation shows that transactions in the IoT-based SLE are strongly irreversible and approved by all peers in the BC network.

5 Discussion

In this experimental study, the use of the BC and distributed architecture in the context of an IoT-based SLE reduces the cost and complexity of transaction management via the implementation of a reliable and highly secure network, where IoT objects can be traced without any central intermediary called a gateway. Moreover, the distributed architecture improves the handling of the secure data channel gathered from connected devices and then delivers educational applications. The distributed IoT network faces several problems related to the centralised communication model. In this work, the authors recommend the use of a decentralised platform based on the BC to prevent any central authority control of the entire system. Furthermore, the consensus method reveals numerous advantages, such as the consistency of exchanged data with reduced errors, prevention of attacks, and flexibility for peers to modify the description of their assets. The security and trust across all elements of SLEs are the most important criteria. Recently, technology is made up of many of these tools, like the BC to solve some problems. Further, according to our researchers, the distributed platform based on the BC and IoT devices enhances the reliability regarding the security challenges by tracing sensor information and avoiding duplicated data. The combination of the IoT and BC seems very complex. Nonetheless, the use of the *chaincode* and distributed ledger that are provided by *IBM* applications improves device security. Instead of using a decentralised model, all devices can exchange and transfer data securely via the deployed BC channel. The BC-based SLE allows the autonomy of the objects via smart contracts that ensure self-verifying tracking performance and bring tremendous cost savings. Consequently, the deployment and costs of all elements of the smart workspace are reduced via the BC since there is no central intermediary. All connected sensors are directly interrelated to the BC platform to make the delivery of educational services more secure. Based on the results, the reliability and security investigations are improved using security requirements known as integrity, confidentiality, and availability. First, the test of integrity was proved since it ensures the handling of the whole transaction without any alteration. The criterion of confidentiality is demonstrated regarding the authorised peer who can access the encrypted message. Finally, the availability is confirmed via making services and data available at the request of users. In summary, the BC-based decentralised architecture enforces reliability and security properties via handling several transactions. This communication model is based on critical success factors for facing numerous expectations of this technology to change many aspects of smart education.

6 Conclusion

The BC is a new technology with potential power. It deals with security and reliability issues. Moreover, applying the BC in SLEs is very complex due to the numerous associated challenges, such as the high resource consumption of connected devices, scalability, and data processing time. In this paper, we proposed a BC-based SLE that ensures the trust and security of the data collected by smart devices. The discussed idea of using the BC application as a demonstrative use case enhances the exchange of secure transactions using the consensus method. The results demonstrate that data can be securely managed using transactions and methods to handle IoT devices in the SLE. Future work will determine the evaluation of the effect of the BC on the privacy and security of all components of the proposed architecture.

References

1. Abdennadher, I., Khabou, N., Rodriguez, I.B., Jmaiel, M.: Designing energy efficient smart buildings in ubiquitous environments. In: 15th International Conference on Intelligent Systems Design and Applications (ISDA), Marrakech, pp. 122–127 (2015)
2. Alam, F., Mehmood, R., Katib, I., Albogami, N.N., Albeshri, A.: Data fusion and IoT for smart ubiquitous environments. IEEE Access **5**, 9533–9554 (2017)
3. Huckle, S., Bhattacharya, R., White, M., Beloff, N.: Internet of Things, blockchain and shared economy applications. In: The 7th International Conference on Emerging Ubiquitous Systems and Pervasive Networks, vol. 98, pp. 461–466 (2016)
4. Vishal, V., Kumar, M., Rathore, M., Vijay Kumar, B.P.: An IOT framework supporting ubiquitous environment. Int. J. Sci. Eng. Res. (IJSER) **4**, 2347–3878 (2015)
5. Huichen, L., Neil, W.B.: IoT privacy and security challenges for smart home environments. Information **7**, 1–15 (2016)
6. Banerjee, M., Lee, J., Choo, K.-K.R.: A Blockchain future to Internet of Things security: a position paper. Digit. Commun. Netw. **4**(3), 149–160 (2018)
7. Dorri, A., Kanhere, S.S., Jurdak, R.: Towards an optimized blockchain for IoT. In: The Second IEEE/ACM Conference on Internet of Things Design and Implementation, IoTDI, USA, pp. 173–178 (2017)
8. Yang, Y., Wu, L., Yin, G., Li, L., Zhao, H.: A survey on security and privacy issues in Internet-of-Things. IEEE Internet Things J. **4**, 1250–1258 (2017)
9. Christidis, K., Devetsikiotis, M.: Blockchains and smart contracts for the Internet of Things. IEEE Access **4**, 2292–2303 (2016). Special Section: The Plethora of Research in Internet of Things (IoT)
10. Hammad, R., Ludlow, D.: Towards a smart learning environment for smart city governance. In: 9th International Conference on Utility and Cloud Computing (UCC), Shanghai, pp. 185–190 (2016)
11. Kamble, A., Chavan, P., Salagare, A., Sharma, K.: Cloud based ubiquitous computing for smart classroom using smart phone. Int. J. Eng. Sci. Comput. **7**(3), 5993–5996 (2017)
12. Temdee, P.: Ubiquitous learning environment: smart learning platform with multi-agent architecture. Wireless Pers. Commun. **76**, 627–641 (2014)
13. Glória, A., Cercas, F., Souto, N.: Design and implementation of an IoT gateway to create smart environments. Procedia Comput. Sci. **109**, 568–575 (2017)
14. Farhan, M., et al.: IoT-based students interaction framework using attention-scoring assessment in eLearning. Future Gener. Comput. Syst. **79**, 909–919 (2018)
15. Chen, G., Xu, B., Lu, M., Chen, N.-S.: Exploring blockchain technology and its potential applications for education. Smart Learn. Environ. **5**(1), 1–10 (2018)
16. Skwarek, V.: Blockchains as security-enabler for industrial IoT-applications. Pac. J. Innovation Entrepreneurship **11**, 301–311 (2017)
17. Ho, P.J., Hyuk, P.J.: Blockchain security in cloud computing: use cases, challenges, and solutions. Symmetry **9**, 1–13 (2017)
18. Conoscenti, M., Vetro, A., De Martin, J.C.: Blockchain for the Internet of Things: a systematic literature review. In: 13th International Conference of Computer Systems and Applications (AICCSA), Agadir (2016)
19. Bonneau, J., Miller, A., Clark, J., Narayanan, A., Kroll, J.A., Felten, E.W.: Sok: research perspectives and challenges for bitcoin and cryptocurrencies. In: IEEE Symposium on Security and Privacy (SP), San Jose (2015)

20. Zheng, Z., Xie, S., Dai, H., Chen, X., Wang, H.: An overview of blockchain technology: architecture, consensus, and future trends. In: IEEE International Congress on Big Data (BigData Congress), Honolulu, pp. 557–564 (2017)
21. Songara, A., Chouhan, L.: Blockchain: a decentralized technique for securing Internet of Things. In: Conference on Emerging Trends in Engineering Innovations & Technology Management (ICET: EITM-2017) (2017)
22. Bdiwi, R., De Runz, C., Faiz, S., Cherif, A.A.: Towards a new ubiquitous learning environment based on blockchain technology. In: IEEE 17th International Conference on Advanced Learning Technologies (ICALT), Timisoara (2017)
23. Parkin, A., Prescott, R.: Distributed ledger technology: beyond the hype. J. Digit. Bank. **2**, 102–109 (2017)
24. Khan, C., Lewis, A., Rutland, E., Wan, C., Rutter, K., Thompson, C.: A distributed-ledger consortium model for collaborative innovation. Computer **50**, 29–37 (2017)
25. Kuzuno, H., Karam, C.: Blockchain explorer: an analytical process and investigation environment for bitcoin. In: Symposium on Electronic Crime Research (eCrime), APWG, Scottsdale, pp. 1–10 (2017)
26. Sukhwani, H., Martínez, J.M., Chang, X., Trivedi, K.S., Rindos, A.: Performance modeling of PBFT consensus process for permissioned blockchain network (hyperledger fabric). In: 36th IEEE Symposium on Reliable Distributed Systems, pp. 253–255 (2017)
27. Sankar, L.S., Sindhu, M., Sethumadhavan, M.: Survey of consensus protocols on blockchain applications. In: 4th International Conference on Advanced Computing and Communication Systems (ICACCS), Coimbatore, pp. 1–5 (2017)

Improving Healthcare Processes with Smart Contracts

Aleksandr Kormiltsyn[1(✉)], Chibuzor Udokwu[1], Kalev Karu[2], Kondwani Thangalimodzi[1], and Alex Norta[1]

[1] Department of Software Science, Tallinn University of Technology, Akadeemia tee 15A, 12816 Tallinn, Estonia
alexandrkormiltsyn@gmail.com, {chibuzor.udokwu,kthang}@ttu.ee, alex.norta.phd@ieee.org

[2] Medicum AS, Punane 61, 13619 Tallinn, Estonia
kalev.karu@medicum.ee

Abstract. Currently, we are on the brink of a period of fundamental change for the healthcare expertise of specialists, i.e., existing know-how becomes less available to patients. One of the main reasons is an economic problem: most people and organisations cannot afford the services of first-rate professionals; and most economies are struggling to sustain their professional services, including schools, court systems, and health services. Another reason for the change in healthcare is the rapid growth of evidence-based medical knowledge, where a new research paper is published, on average, every forty-one seconds. Thus, evidence-based medicine is malfunctioning, and delayed, missed, and incorrect diagnoses occur in 10 to 20% of the time. Innovative IT technologies can solve the critical challenges in the healthcare domain. One of such technologies is smart contracts that manage and enforce contracts (business rules) without the interference of a third party. Smart contracts improve interoperability and privacy in cross-organisational processes. In this paper, we define problematic processes in healthcare and then provide a smart contract-based mapping for improving these processes. This paper proposes the way to overcome inequality in services accessibility, inefficient use of services and shortcomings in service quality, using smart contracts and blockchain technology.

Keywords: Know-how distribution · Knowledge overload · Smart contracts · Blockchain · eHealth

1 Introduction

The development of healthcare information systems can improve medical quality and reduce costs [13]. The existing problems result in low accessibility of healthcare due to high costs. At the same time, the amount of medical knowledge has rapidly grown [2,25] and thus, the cognitive overload of healthcare professionals has increased. Delivering healthcare knowledge and know-how to

W. Abramowicz and R. Corchuelo (Eds.): BIS 2019, LNBIP 353, pp. 500–513, 2019.
https://doi.org/10.1007/978-3-030-20485-3_39

a patient with the help of information technologies can increase the accessibility of healthcare services for people with lower income. Patient-centric systems require information technologies that support information security, trust and transparency between different stakeholders [14].

This paper proposes to use information-technological means as a way to overcome the inefficiency and inequalities in health-services utilisation. The core of the proposal is the use of blockchain technology [12] to create a protected way to establish a meaningful connection between a patient's medical data and good medical practice guidelines, i.e., know-how. Briefly, a blockchain [18] is a growing list of blocks that are linked comprising a cryptographic hash to the previous block, a timestamp, and recorded transaction data. The blockchain records are immutable and resistant to modification. A smart contract [5] is a machine-readable programming code to digitally facilitate, verify, or enforce the negotiation, or performance of a contract. Smart contracts extend blockchains to establish well-defined protocols for governing peer-to-peer (P2P) collaboration. Similar methods are also instrumental for releasing the cognitive overload of healthcare professionals. Understanding the current state of healthcare problems and smart-contracting practical usage is a focal point of this paper [22]. Much progress has been made in developing technologies that support smart contracts [16] while little research exists to understand the practical usage possibilities to solve healthcare problems and propose solutions to address the significance of blockchain.

This paper fills the gap by answering the primary research question of how to use smart contract-based systems to reduce the costs of healthcare services and increase their accessibility. The answer to this question helps to build patient-centred processes effectively. We deduce the following sub-questions from the primary research questions to establish a separation of concerns. What are the processes that result for the problems of healthcare? Understanding the reasons helps to focus on the practical technical solutions. What are the solutions for implementation with information technologies? Describing identified processes in details helps to focus on specific tasks that are possible to improve with information technologies. What are the blockchain technologies that support the defined solutions? When blockchain technologies are specified, we map them with tasks identified in the previous step.

The remainder of the paper is structured as follows. Section 2 provides literature review and Sect. 2.1 identifies problematic processes. Section 3 presents identified processes in details and Sect. 4 describes a proposed solution. Section 4.1 provides the definition of identifying top health-issue process. Next, Sect. 4.2 describes smart-contract based solution for booking- and consultation processes. Section 4.3 describes the medical knowledge-update process and Sect. 5 provides a mapping to technologies with healthcare functions. Section 6 compares the results with similar researches, describes the limitations and sets directions for future research. Section 7 concludes by giving research results and a discussion of future work.

2 State of the Art

The authors [25] explore two significant problems faced by healthcare systems. First, most people, organisations, and economies cannot afford the services of first-rate professionals. Only the rich and robustly insured can have access to top-level medical specialists. On the other hand, there exists a lack of experts, not the knowledge. System limitations are due to the way the professional work is currently organised, often requiring face-to-face interaction. Second, there exists a rapidly growing body of new medical literature. Authors estimate that remaining aware of relevant medical literature containing evidence-based knowledge, would take a doctor at least twenty-one hours of reading every day. This task is beyond human capabilities and the aim is to have computer systems scan these volumes, connecting existing knowledge with practical know-how.

In research [9], an equitable system of health care delivery remains a core objective in most of the Organisation for Economic Co-operation and Development (OECD) member states with comprehensive and universal coverage. Research analyses inequity in the utilisation of general practitioner (GP) and healthcare specialist services in 10 European Union (EU) member states. The paper proposes the existence of a significant pro-rich inequity in specialist services for all countries. Concerning GP services, the pro-rich inequity exists in 7 countries out of 10. Another study [8] has demonstrated the same results: in 10 EU countries wealthier people are expected to seek and use more specialist services. Furthermore, in three states – Austria, Portugal, and Finland, stronger socioeconomic gradients towards specialists' services utilisation are found by education rather than by income.

Research [7] analyses the data from national health surveys of 18 OECD countries and demonstrates that among people, having the same needs for specialist, or dentist services, those who have higher income visit specialists, or dentists, significantly more than these having lower income. Additionally, the author discover that inequities and inequalities in service utilisation remain the same when they compare data from two studies, 2006–2009 vs. 2000. In [3], the authors describe the dynamics of the income-distribution degree of income-related health inequality (IRHI) in China, where rapid economic progress has occurred during the last decades. The authors conclude that over 50% of raise in IRHI occurs in a population group (female 50+) having a weak replacement-income system for when reaching retirement age.

Another challenge the healthcare systems are struggling with is the amount, relevance and availability of medical literature containing evidence-based new knowledge that should be taken into account by the everyday practice of physicians. Research [2] estimates that 7287 articles relevant only to primary care practice are published monthly, and for specially trained physicians in epidemiology, it takes by authors' estimation 627,5 hours per month to be aware of newly published findings, i.e., ca 3,7 full-time work hours. In citation [25], the authors estimate that it would take a human 21 hours a day to read newly published medical literature relevant to one's medical specialty. In review article [23], the authors study the use of evidence-based literature in nursing practices, conclude that the main barriers are time restrictions, searching skills, access requirements, computer literacy, interface characteristics, quality of information, a users' personal preferences and organisational settings.

Another side of knowledge distribution is the level of health literacy among consumers. Research [15] concludes that health literacy is one of the first communication issues relevant to the practical use of eHealth. The author finds that many eHealth applications are demanding for audiences to utilise because they provide health information that is not easy for many consumers to understand and apply. Therefore, eHealth-application developers have to design and implement applications to meet health literacy levels of different audiences.

2.1 Problem Identification

In this research, we use a literature review to identify problematic health-provision related processes. According to the performed literature review, we define high costs for healthcare services as the main problem in the healthcare domain. High prices are due to the lack of healthcare specialists and patient demand. Some researches define booking- [4,17,26] and consultation [27,28] processes as key processes in healthcare systems and provide their improvements. We consider these processes as an entry point for every healthcare service. Smart-contract technology is required to increase the healthcare-system efficiency. Currently, healthcare providers are overloaded with consultation and booking requests and the usage of smart contracts helps to ensure an even distribution of healthcare needs among healthcare providers. We use smart contracts in the consultation process to eliminate the need to visit doctors in common cases.

3 Booking and Consultation Processes

We consider booking- and consultation processes in detail with the Business Process Model and Notation (BPMN) [1] and define specific tasks that require improvements with information technologies. The described processes are based on the common practices used in Estonian healthcare organisations, such as Medicum AS[1] and are validated by medical experts in Estonia.

The booking process starts as per Fig. 1 with a patient's intention to visit a doctor, or if one doctor recognises the need to consult the patient with another doctor. When the patient has found an available healthcare provider, he can either contact a provider's booking department, or use a booking service in a web portal, if such an option is available. After contacting the booking department, the patient's personal information is collected by the booking department's employee and the patient is provided with available time slots. If the time slot suits the patient then the booking is created. Otherwise, the patient can either continue searching for available time slots with other healthcare providers, or

[1] https://www.medicum.ee/.

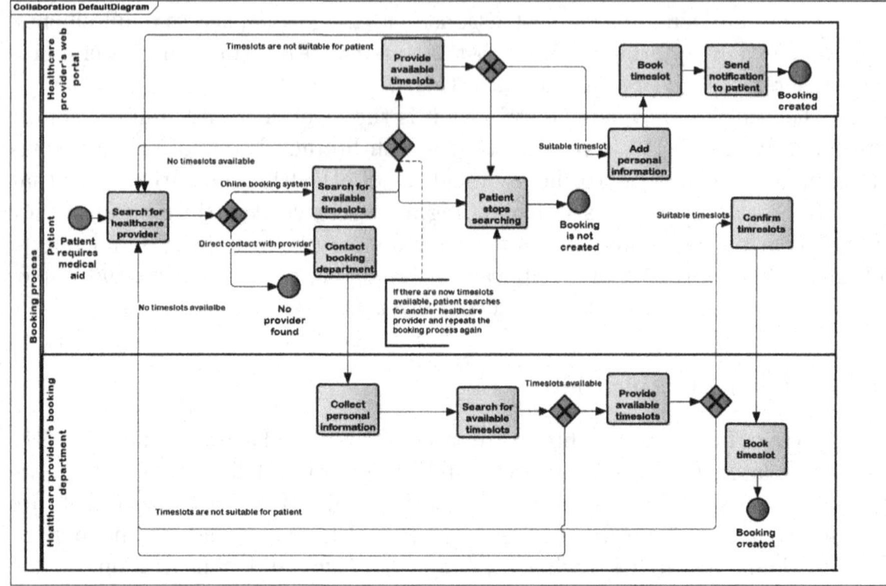

Fig. 1. The BPMN diagram of the booking process.

stop searching. The booking process via the web portal includes searching available time and creating booking tasks. After a booking is created, a notification is sent to the patient.

Still, there are projects such as MedicalBooking[2] that provide booking services for different healthcare providers in one place as mostly the booking process is specific to the healthcare provider. Interoperability problems result from long search times as the patient needs to contact different healthcare providers. Another consequence of the interoperability issue is an uneven distribution of the healthcare providers' resources, e.g., when the patient does not have information about available time slots of all healthcare providers. In this case, he can book a first available time slot even when the waiting time is not the shortest one. Increasing the accessibility of available time slots from different healthcare providers, reduces the time needed for registration, as well as the visit waiting time.

As shown in Fig. 2, the consultation process starts when a patient enters the doctor's room. The patient describes his medical history and if a patient requires immediate treatment, then the doctor asks additional information about a patient's complaints and defines the final diagnosis. Otherwise, the doctor defines an initial diagnosis, required examinations, initial treatment and books the next visit. To define the final diagnosis, the doctor defines medications and a treatment regime, if needed. Finally, the doctor closes the medical case.

[2] https://medicalbooking.nl.

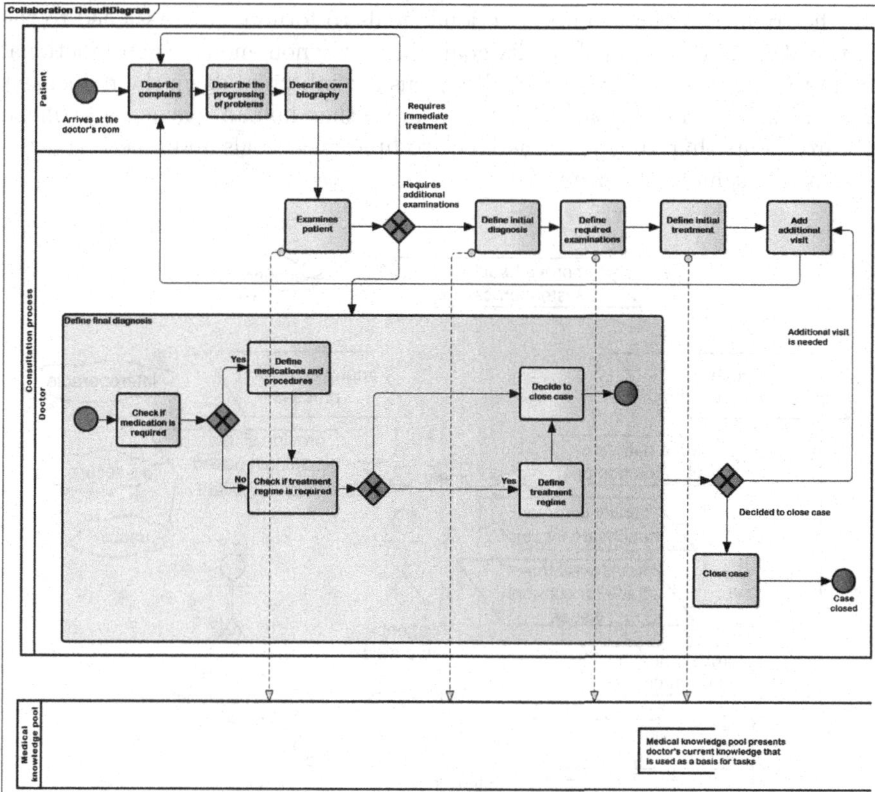

Fig. 2. The BPMN diagram of the consultation process.

The doctor uses his knowledge when examining a patient and defining diagnosis, examinations and treatment.

During the consultation, the doctor uses a limited amount of medical knowledge as new knowledge is not easily accessible and trustable. The tasks in the consultation process connected to the medical knowledge pool are related to the clinical decision making and influence the quality of consultation. Adding new medical knowledge to the pool increases the quality of consultation and therefore, reduces costs for the patient.

4 Improving Healthcare Processes with Blockchain Technologies

In order to design a system that improve the healthcare processes to address the issues identified, the goal model [24] in Fig. 3 shows our proposed solutions in improving the efficiency of processes that support healthcare. Briefly, parallelograms denote functional goals have a value proposition as a root. The latter is further refined by lower-level functional goals to form a tree. Assigned to the functional goals are so-called quality goals that are synonymous to non-functional system requirements. Note that quality goals are inherited down the refinement hierarchy of functional goals. Stick men are either human agents or artificial software agents that are equally assigned to functional goals and inherited down the refinement hierarchy of the latter.

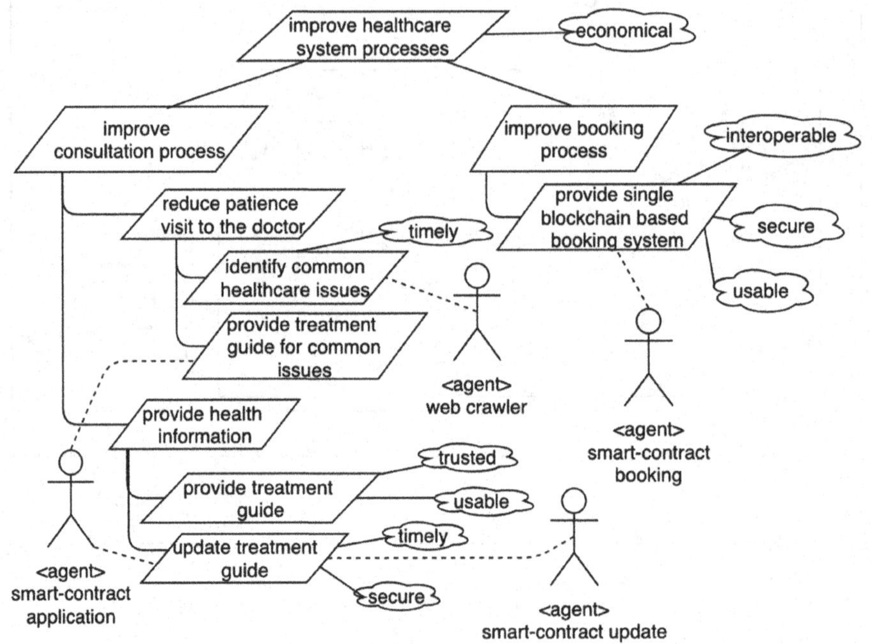

Fig. 3. Healthcare-processes improvement goal model.

The main functional goal and value proposition in Fig. 3 is to reduce the cost of healthcare by making the healthcare booking and consultation processes more efficient. To improve the consultation process, we need to reduce the number of visits to medical doctor. This is achieved by providing patients with treatment information for common health issues as in such cases, they do not need to visit the doctor. To identify common health issues, we use a WebCrawler agent as shown in Sect. 4.1 in detail. We also need to assure that treatment information provided to patients stems from trusted source and is up-to-date as shown in the goal model diagram in Fig. 3. We use blockchain enabled smart-contracts to achieve trust by ensuring that only information provided by trusted sources is used for treatment.

For the booking part, the sub-goal is to make sure that all healthcare booking systems are interoperable. This ensures that no lost time in the booking process and few hospitals are not overloaded with booking requests as shown in Fig. 3.

To achieve interoperability, our proposed solution uses blockchain network to store booking information of healthcare providers.

4.1 Common Health Problem Identification

The first step in our proposed solution of improving a patient's consultation process is to identify top health issues that a blockchain enabled smart-contract application addresses. To identify top health issues that require medical consultation, the system uses health information collected by automated web crawlers. Next, the information is further filtered to obtain top common health issues with standard treatments and information about the body location where the health problem is an issue. The typical health-issue-identification process is shown in Fig. 4 below. A web crawler is the primary function of any search engine in the World Wide Web and a powerful technique that passes through web pages, indexes, saves web content and updates the websites' pages [10].

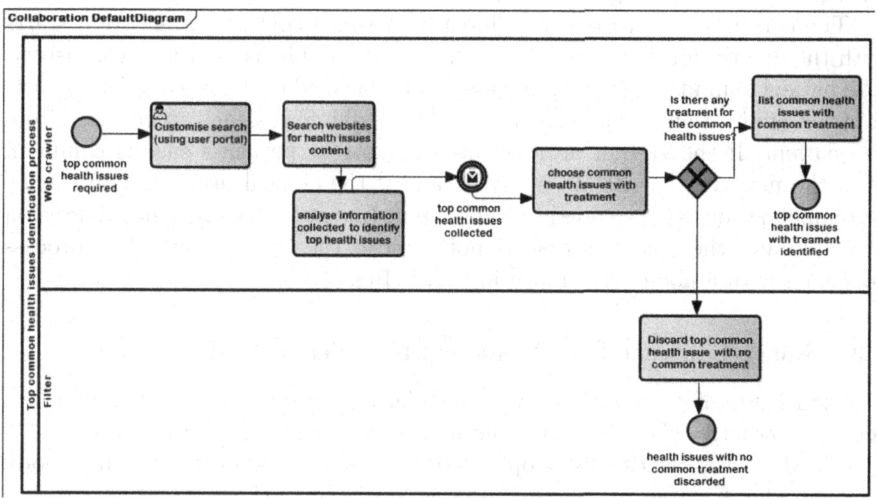

Fig. 4. Identifying top health issues.

As shown in Fig. 4, the identification of the top health-issues process starts with the task of customising the automated crawler using the user portal. The second task is for the crawler to search websites with typical health issues that require medical consultation, analyse the collected content to get the top common health issues, then discard the top common health issues without common treatment to remain with top common health issues with standard treatment, and also save their location information.

The crawlers traverse the web pages belonging to sound health-service providers, organisations that publish health reports and statics, among others [10]. Web pages with health reports provide reliable health data with information about methods and measurement issues [6]. Web pages with peer-reviewed healthcare publications may contain information about top common health issues that require medical consultation.

4.2 Smart-Contract Base Booking and Consultation

In the second step, we provide the integration of the smart-contract application to existing booking- and consultation processes to solve the accessibility and quality issues. New medical knowledge accessible by smart contracts is used to improve the patient logistics in a booking process. First, the patient's data is analysed according to existing knowledge patterns, followed by agreeing on a clinical decision. If a patient requires a doctor visit, then a search is performed for available time slots for doctors of specific expertise. If time slots are available and suitable for the patient, then he can perform a booking. If the patient does not use the smart-contract application, then he uses the booking process as described in Fig. 1. As a result, the patient is directed to the correct specialist with the shortest waiting time (Fig. 5).

The consultation process with integrated smart-contract application starts with the description of the patient's complaints, problems, and medical history. The patient's input data is mapped with available medical knowledge by a smart-contract application and if needed, the additional information can be asked from the patient. If the system decides that the current patient's case is common, then the mapping results are analysed and if the clinical decision is available, recommendations are provided to the patient. In case there is no clinical decision available and the patient's case is not typical, then the consultation process continues with a doctor, as shown in Fig. 2 (Fig. 6).

4.3 Smart-Contract Based Medical Knowledge-Update Process

The third step of our solution involves updating the smart contract with a new set of knowledge when they become available as per Fig. 7. Since we already identified the top issues why people want to visit the doctor, when a new body of knowledge from an approved source is available to this field, the system is updated with this new set of information. Although the information entered in a blockchain is immutable, our system adopts best practices in smart-contract management lifecycles and updates the obtainable information from the patients via the smart contract as described in [11,19–21].

5 Mapping Blockchain Technologies to the Specific Functions

After specifying blockchain technologies that support the defined solutions, we map them as shown in Table 1 to specific process functions identified in the previous steps. We use blockchain technology for storing information about healthcare data for the booking process and medical-knowledge data is used in the consultation process.

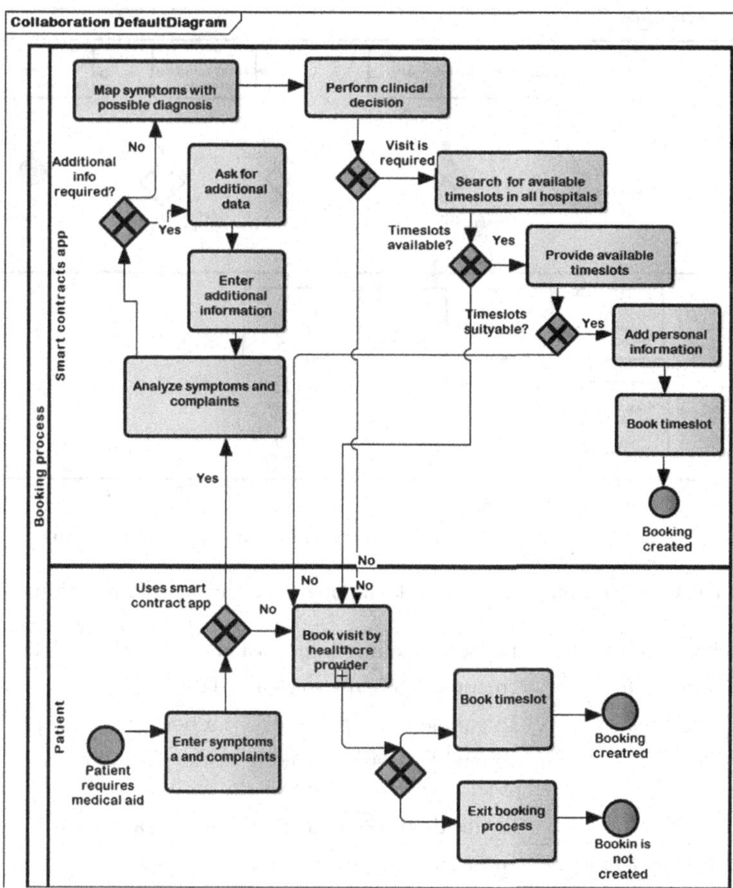

Fig. 5. Booking process with integrated smart-contract application.

Smart contracts are used to enable interoperability of a booking process and provide patients with time slots of different healthcare providers in one system. Furthermore, smart contracts support the decision making processes during the booking process that reduces the workload of healthcare providers. During the consultation process, when a doctor is not involved, smart-contract technology provides the decision making functionality supporting a patient with available medical knowledge. Finally, we use smart-contract lifecycle management to update the medical knowledge base.

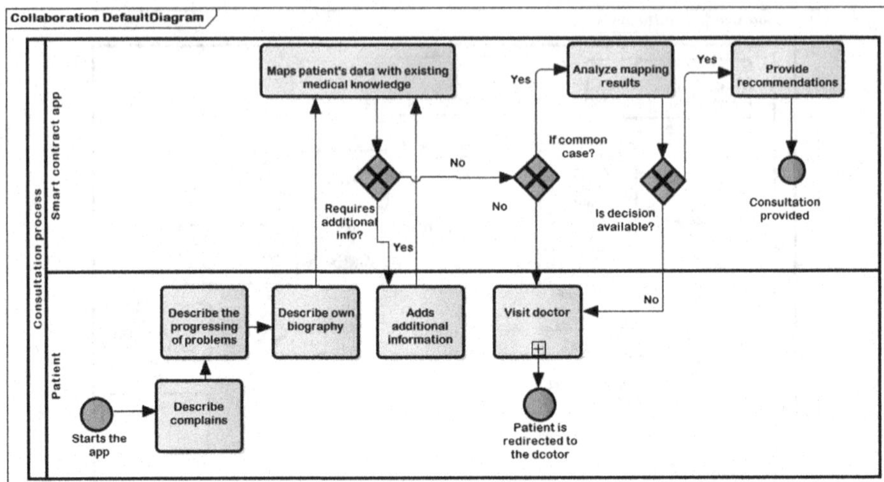

Fig. 6. Consultation process with a smart-contract application.

Table 1. Mapping blockchain technologies to the specific functions

Blockchain node	To store booking information of healthcare data
Blockchain node	To store medical knowledge data
Smart contract	To enable interoperability of a booking process
Smart contract	To enable decision making in the consultation process without doctor
Smart contract	To decide if a patient needs to visit a doctor
Smart contract lifecycle management	To update the medical knowledge base

6 Discussion

Earlier research work proposes blockchain-based solutions [29,30] that are focused on the security issues of patient medical data. Citation [29] provides an architecture for storing and sharing a patient's medical data using blockchain technology. In paper [30], the authors provide software patterns for access control to patient data.

Our solution focuses on improving the efficiency of specific healthcare processes such as booking and consultation by implementing interoperability in healthcare information systems while [30] is focused more on the interoperability of patient healthcare data to avoid duplication. Also, citation [29] focuses on patient healthcare-data privacy and sharing while our study considers blockchain technology for the secure sharing of medical knowledge to eliminate the need for hospital visits for common healthcare issues.

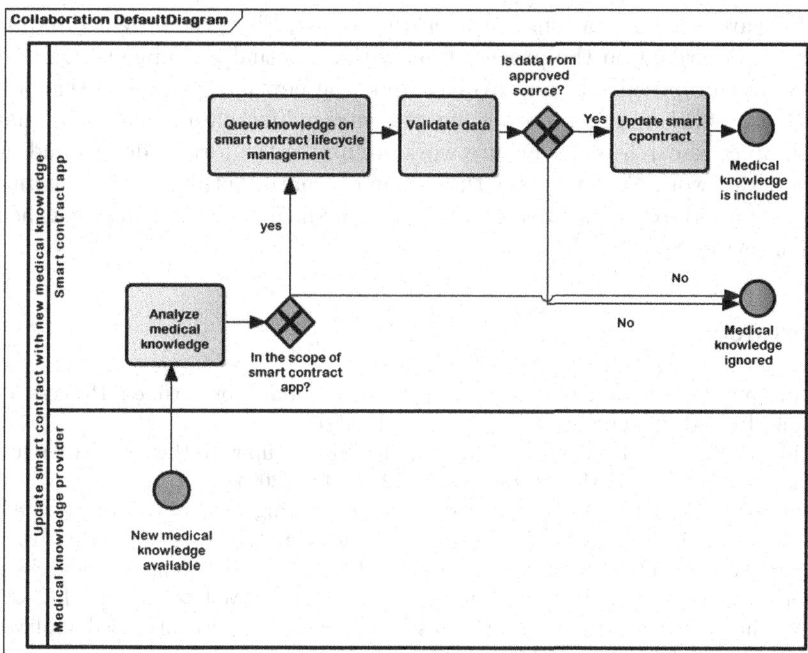

Fig. 7. The BPMN diagram of the smart-contract medical-knowledge update process.

7 Conclusion

In this paper, we define a problematic healthcare processes and propose the use of blockchain technology to improve them. We first conduct a literature review to define the reasons for current healthcare problems. These reasons include high costs and low accessibility of medical services. Additionally, booking- and consultation processes are problematic in healthcare and we model the identified processes in detail.

The paper also proposes a smart-contract based system that is integrated into booking- and consultation processes and specified with business-process diagrams. The updated-booking process provides patients with time slots from different healthcare providers and thus, reduces the waiting time. The improvements for the consultation process yield a medical knowledge sharing system employing blockchain technology that enables a patient to use consultation services without a doctor. These improvements increase the accessibility of healthcare services and reduce their costs.

Mapping blockchain technologies to the specific booking- and consultation-process functions shows that smart contracts are used to solve interoperability issues, to support decision making without a doctor and to optimise the logistics of patients. Blockchain nodes are used to store booking information and medical knowledge. At the same time, smart contract management is required when updating the medical knowledge base.

There are some limitations in the current paper. First, the provided improvements are described on the process level without actual implementation. Additionally, our proposed solution involves updating new medical knowledge on the blockchain, which requires proper smart-contract lifecycle management that is still an open research issue. Future work comprises the formal design and evaluation of our work. Additionally, the design of smart-contract lifecycle management is required to successfully integrate new medical knowledge into healthcare-decision processes.

References

1. Allweyer, T.: BPMN 2.0: Introduction to the Standard for Business Process Modeling. BoD-Books on Demand, Norderstedt (2016)
2. Alper, B.S., et al.: How much effort is needed to keep up with the literature relevant for primary care? J. Med. Libr. Assoc. **92**(4), 429 (2004)
3. Baeten, S., Van Ourti, T., Van Doorslaer, E.: Rising inequalities in income and health in China: who is left behind? J. Health Econ. **32**(6), 1214–1229 (2013)
4. Baldassarre, F.F., Ricciardi, F., Campo, R.: Waiting too long: bottlenecks and improvements-a case study of a surgery department. TQM J. **30**(2), 116–132 (2018)
5. Buterin, V., et al.: A next-generation smart contract and decentralized application platform. White paper (2014)
6. Chew, K.: Publications and Information Products from the National Center for Health Statistics (2018). https://hsl.lib.umn.edu/biomed/help/health-statistics-and-data-sources
7. Devaux, M.: Income-related inequalities and inequities in health care services utilisation in 18 selected OECD countries. Eur. J. Health Econ. **16**(1), 21–33 (2015)
8. d'Uva, T.B., Jones, A.M.: Health care utilisation in Europe: new evidence from the ECHP. J. Health Econ. **28**(2), 265–279 (2009)
9. d'Uva, T.B., Jones, A.M., Van Doorslaer, E.: Measurement of horizontal inequity in health care utilisation using European panel data. J. Health Econ. **28**(2), 280–289 (2009)
10. Elyasir, A.M.H., Anbananthen, K.: Web crawling methodology, vol. 47 (2012)
11. Eshuis, R., Norta, A., Roulaux, R.: Evolving process views. Inf. Softw. Technol. **80**, 20–35 (2016)
12. Hawlitschek, F., Notheisen, B., Teubner, T.: The limits of trust-free systems: a literature review on blockchain technology and trust in the sharing economy. Electron. Commer. Res. Appl. **29**, 50–63 (2018)
13. Hoyt, R.E., Yoshihashi, A.K.: Health Informatics: Practical Guide for Healthcare and Information Technology Professionals. Lulu.com, Morrisville (2014)
14. Kormiltsyn, A., Norta, A.: Dynamically integrating electronic - with personal health records for ad-hoc healthcare quality improvements. In: Alexandrov, D.A., Boukhanovsky, A.V., Chugunov, A.V., Kabanov, Y., Koltsova, O. (eds.) DTGS 2017. CCIS, vol. 745, pp. 385–399. Springer, Cham (2017). https://doi.org/10.1007/978-3-319-69784-0_33
15. Kreps, G.L.: The relevance of health literacy to mHealth. Inf. Serv. Use **37**(2), 123–130 (2017)
16. Lu, Y.: Blockchain: a survey on functions, applications and open issues. J. Ind. Integr. Manage. **3**(04), 1850015 (2018)

17. Mans, R.S., van der Aalst, W.M.P., Vanwersch, R.J.B.: Process Mining in Healthcare: Evaluating and Exploiting Operational Healthcare Processes. SBPM. Springer, Cham (2015). https://doi.org/10.1007/978-3-319-16071-9
18. Nakamoto, S., et al.: Bitcoin: a peer-to-peer electronic cash system (2008)
19. Narendra, N.C., Norta, A., Mahunnah, M., Ma, L., Maggi, F.M.: Sound conflict management and resolution for virtual-enterprise collaborations. SOCA **10**(3), 233–251 (2016)
20. Norta, A.: Establishing distributed governance infrastructures for enacting cross-organization collaborations. In: Norta, A., Gaaloul, W., Gangadharan, G.R., Dam, H.K. (eds.) ICSOC 2015. LNCS, vol. 9586, pp. 24–35. Springer, Heidelberg (2016). https://doi.org/10.1007/978-3-662-50539-7_3
21. Norta, A., Othman, A.B., Taveter, K.: Conflict-resolution lifecycles for governed decentralized autonomous organization collaboration. In: Proceedings of the 2015 2nd International Conference on Electronic Governance and Open Society: Challenges in Eurasia, pp. 244–257. ACM (2015)
22. Pham, H., Tran, T., Nakashima, Y.: A secure remote healthcare system for hospital using blockchain smart contract. In: 2018 IEEE Globecom Workshops (GC Wkshps), pp. 1–6. IEEE (2018)
23. Sadoughi, F., Azadi, T., Azadi, T.: Barriers to using electronic evidence based literature in nursing practice: a systematised review. Health Inf. Libr. J. **34**(3), 187–199 (2017)
24. Sterling, L., Taveter, K.: The Art of Agent-Oriented Modeling. MIT Press, Cambridge (2009)
25. Susskind, R.E., Susskind, D.: The Future of the Professions: How Technology will Transform the Work of Human Experts. Oxford University Press, Cary (2015)
26. Tavakol, O.D.K., Massoumi, C.E.: Data synchronization for booking of healthcare appointments across practice groups. US Patent 8,688,466, 1 April 2014
27. Tozzo, P., Mazzi, A., Aprile, A., Rodriguez, D., Caenazzo, L.: Certification ISO 9001 in clinical ethics consultation for improving quality and safety in healthcare. Int. J. Qual. Health Care **30**(6), 486–491 (2018)
28. Wilson, A., Childs, S.: The relationship between consultation length, process and outcomes in general practice: a systematic review. Br. J. Gen. Pract. **52**(485), 1012–1020 (2002)
29. Yue, X., Wang, H., Jin, D., Li, M., Jiang, W.: Healthcare data gateways: found healthcare intelligence on blockchain with novel privacy risk control. J. Med. Syst. **40**(10), 218 (2016)
30. Zhang, P., White, J., Schmidt, D.C., Lenz, G.: Applying software patterns to address interoperability in blockchain-based healthcare apps. arXiv preprint arXiv:1706.03700 (2017)

Energy Efficient Cloud Data Center Using Dynamic Virtual Machine Consolidation Algorithm

Cheikhou Thiam[1] and Fatoumata Thiam[2](✉)

[1] Université de Thiès, Thies, Senegal
cthiam@univ-thies.sn
[2] Université Gaston Berger, Saint-Louis, Senegal
fatoumata.thiam@univ-thies.sn

Abstract. In Cloud Data centers, virtual machine consolidation on minimizing energy consumed aims at reducing the number of active physical servers. Dynamic consolidation of virtual machines (VMs) and switching idle nodes off allow Cloud providers to optimize resource usage and reduce energy consumption. One aspect of dynamic VM consolidation that directly influences Quality of Service (QoS) delivered by the system is to determine the best moment to reallocate VMs from an overloaded or undeloaded host. In this article we focus on energy-efficiency of Cloud datacenter using Dynamic Virtual Machine Consolidation Algorithms by planetLab workload traces, which consists of a thousand PlanetLab VMs with large-scale simulation environments. Experiments are done in a simulated cloud environment by the CloudSim simulation tool. The obtained results show that consolidation reduces the number of migrations and the power consumption of the servers. Also application performances are improved.

Keywords: Cloud · Consolidation · Energy · Scheduling · Virtualization

1 Introduction

Until recently, high performance has been the sole concern in data center deployments, and this demand has been fulfilled without paying much attention to energy consumption. To address this problem of minimizing energy and drive Green Cloud computing, data center resources need to be managed in an energy-efficient manner.

In recent years, there is increasing concern that Cloud's high energy consumption issues are a disadvantage for various institutions. However, until now, little attention has been given to the various methods to reduce energy consumption in cloud environments while ensuring performance. The obligation of providing good quality of service (QoS) to customers leads to the necessity in dealing with the energy-efficiency trade-off, as aggressive consolidation may lead

W. Abramowicz and R. Corchuelo (Eds.): BIS 2019, LNBIP 353, pp. 514–525, 2019.
https://doi.org/10.1007/978-3-030-20485-3_40

to performance degradation. There are many algorithms on how to reduce the total power consumption, one of the most efficient being to turn off a maximum number of nodes without impacting jobs running on the system.

Furthermore the increasing demand of computation resources has led to new types of cooperative distributed systems for cloud computing. Cloud management tools have standalone controllers that handle the VM-to-PM mapping. This standalone management of the virtual machine involves both the initial placement of the virtual machine and adaptation to changes in the cloud environment through dynamic reallocation (consolidation) of the virtual machines. The initial placement of the virtual machine does not react to changes in the cloud environment through the live migration of virtual machines. Dynamic consolidation of virtual machines is essential to efficiently utilize the resources of a data center by reacting or proactively acting on changes to the cloud data center. Dynamic consolidation of virtual machines can be based on reservation or on demand. In reservation-based placement, the placement decision of the VM depends on the amount of reserved resources defined by the size of the VM in terms of processor capacity, memory, and bandwidth rather than actual demand. Dynamic VMs consolidation generally requires the detection of overloaded and underloaded hosts in the Cloud datacenter. Beloglazov and Buyya [3] presented an algorithm which sets an under load threshold and an overload threshold to control all hosts' load between the two thresholds.

One of main contributions in this article is to propose an efficient algorithm that migrate VMs to reduce energy-consumption while preserving performance and preventing hot spots. Khan in [6] proposed a review of Dynamic Virtual Machine Consolidation Algorithms for Energy-Efficient Cloud Resource Management.

Our contribution is to design, implement, and evaluate a novel cloud management system which provides a holistic energy-efficient VM management solution by integrating advanced VM management mechanisms such as underload mitigation, VM consolidation, and power management. This article is structured as follows. Section 1 introduces the design principles. Section 2 presents some research in the domain. Section 3 describes the model and objectives. Section 4 introduces a new approach of Dynamic Consolidation of Virtual Machines. Section 5 presents results. Finally, Sect. 6 summarizes the contributions.

2 Related Work

Some works addressed optimization of energy consumption in data through VM consolidation [6,8]. In such cases, for a node load ($C_{i,j}$) two thresholds were used: under-loaded threshold (γ) and overloaded threshold (ϵ). These thresholds determines whether a host is overloaded or underutilized, respectively.

Recently, many energy-aware scheduling algorithms have been developed primarily using the dynamic voltage-frequency scaling (DVFS) capability which has been incorporated into recent commodity processors. Studies by [5] provide a better understanding of the approaches and algorithms that have been produced to ensure better allocation policies for virtual machines in the context of cloud data center and to identify future directions. A better understanding of existing approaches and algorithms that ensure better VM placement in the context of cloud computing and to identify future directions is provided by [13]. In [11], Shirvani et al. give a review which is conducted to serve as a roadmap for further research toward improvement in VM migration, server consolidation and DVFS approaches which are utilized in modern Data Centers. In [2], authors present a Dvfs-aware dynamic consolidation of virtual machines for energy efficient cloud data centers. Their approach produces relevant contributions on the optimization of Cloud data centers from a proactive perspective. They present a Frequency-Aware Optimization that combines a novel reactive DVFS policy with a proactive Frequency-aware Consolidation technique. They have achieved competitive energy savings of up to 41.62% at the data center level maintaining QoS, even improving slightly the SLA violations of around 0.01%.

However, these techniques are unsatisfied with optimizing both schedule length and energy consumption. In [7], using dynamic voltage frequency scaling with VM migration helps in reducing the energy consumption. However, they note that the model can lead to more SLA violations. A critical problem for these ideas is that in order to turn lightly loaded machines off or to assign workload to newly turned-on machines, the task needs to be transferred from one machine to another. But almost all the operating systems used in the real clusters, e.g. Windows, Unix and Linux, cannot support such kind of operations. So in their research specific OS features have to be developed and applied, which in turn limits the practicability of their approaches. A number of good methods and ideas in these studies could be introduced to the energy saving schemes in cluster systems. [9] proposes an algorithm which is executed during the virtual machine consolidation process to estimate the long-term utilization of multiple resource types based on the local history of the considered servers. The mixed use of current and predicted resource utilization allows for a good characterization of overloaded and underloaded servers, thereby reducing both the load and the power consumption after consolidation.

To guarantee the demand for computer resources, VMs need to be migrated from over utilized host. However performance requirements must be accomplished. Besides, VMs need to be migrated from underutilized host for saving power consumption. In order to solve the problem of energy and performance, efficient dynamic VM consolidation approach is introduced in literature. [12] proposed multiple redesigned VM allocation algorithms and introduced a technique by clustering VMs to migrate by taking into account both CPU utilization and allocated RAM. Alsadie et al. in [1] proposed an approach, named the 'dynamic threshold-based fuzzy approach' (DTFA). It is a fuzzy threshold-based approach used for adjusting the threshold values of PMs in a cloud environment. Authors present an approach for the dynamic adjustment of threshold values that aims to minimize the number of migrations invarying workload environments.

In all these studies the objective is to minimize the energy consumption of the servers, while satisfying given performance-related bounds on the period. Proposed work considers the Dynamic Virtual Machine Consolidation Algorithms to improve tasks management, taking into account the cost of migrations.

3 Model and Objectives

In this section, we discuss the model, hypotheses, constraints and objectives. This section presents the system model of energy used for an energy-efficient IaaS cloud management for private clouds. First, the distributed system model is introduced. Then, objectives are presented.

3.1 Model

In the following, the terms work, task and virtual machine are the same. Nodes are physical machines that host virtual machines.

We consider a cloud data center environment consisting of $H = \sum_{i=1}^{N} H_i$ heterogeneous physical nodes. Each site i has H_i nodes. There are N sites. Each node is characterized by the CPU performance defined in Millions Instructions Per Second (MIPS). We consider T_i tasks associated to VM_i VMs, that run on the site i.

- R_i: load of the site i. This load depends on the number of VM (VM_i) executed by the site and their load ($l_{i,j,k}$ is the requested load of VM k in site i on node j). $r_{i,j}$ is the aggregated requested load of all VM on node j in site i. Note that if VM k is not running on node j in site i, then $l_{i,j,k} = 0$.
  $R_i = \sum_{j=1}^{H_i}(r_{i,j})$, $r_{i,j} = \sum_{k=1}^{T_i}(l_{i,j,k})$
- Load C_i and speed V_i of site i
  $C_i = \sum_{j=1}^{H_i}(c_{i,j})$, $V_i = \sum_{j=1}^{H_i}(v_{i,j})$
  $c_{i,j}$ Actual load of node j in site i, $v_{i,j}$ Maximum speed of node j in site i in Mips
- Job satisfaction S_i of site i (same for $s_{i,j}$, job satisfaction of node j in site i)
  $S_i = \frac{C_i}{R_i}$ $s_{i,j} = \frac{c_{i,j}}{r_{i,j}}$
- $VM^{pe}_{i,j,k}$: number of processing elements (PEs) requested by a VM k in node j (CPUs) on site i. "Note: It will be assumed in experiment part that this value is 1".
- $VM^{RAM}_{i,j,k}$ is constant across the node and only depends on k. j and i are kept for coherence of notations.
- $D_{i,j,k}$: the estimated execution duration of VM $VM_{i,j,k}$
- $w_{i,j,k} = VM^{pe}_{i,j,k}.D_{i,j,k}$: represent the weight of VM $VM_{i,j,k}$
- The load of $VM_{i,j,k}$ depends on the speed estimation $VM^{mip}_{i,j,k}$ of the $VM_{i,j,k}$ in node j of site i, its estimated execution duration: $l_{i,j,k} = w_{i,j,k}.VM^{mips}_{i,j,k}$.

3.2 Hypothesis

- Communications are modeled by a linear model of latency and bandwidth.
- For each scenario there is at least one VM.
- Migration cost. To migrate a VM, only RAM has to be copied to another node. The migration time depends on the size of RAM of $VM_{i,j,k}$ in node j of site i and the available network bandwidth.
  VM migration delay = $VM_{i,j,k}^{RAM}$/bandwidth + C (C is a constant).
  Bandwidth is considered as constant.
- Nodes have two different power states: Switched on and switched off. While switched on, power consumption is linear in function of load between $P_{i,j}^{min}$ and $P_{i,j}^{max}$.

We will use the classical linear model of power consumption in function of load: $\forall i, j \quad P_{i,j} = P_{i,j}^{min} + c_{i,j}(P_{i,j}^{max} - P_{i,j}^{min})$ if node j is switched on, $P_{i,j} = 0$ otherwise. Therefore the total power consumption of the system is:

$$P = \sum_{i=1}^{N} \sum_{j=1}^{H_i} P_{i,j}$$

To obtain energy consumed during a time slice, instantaneous power is integrated over time $\int_{t1}^{t2} \mathrm{P(t)}\, dt$. Total energy is then obtained by adding all the energy of those time slices.

3.3 Objectives

The main objective of our approach is to improve cloud's total energy efficiency by controlling cloud applications' overall energy consumption while ensuring cloud applications' service level agreement. Therefore our work aims to satisfy several objectives:

- Ease of task management: we design a system which is flexible enough to allow for dynamic addition and removal of nodes.
- Energy Efficiency: One of our goals is to propose task placements management algorithms which are capable of creating idle times on nodes, transitioning idle nodes in a power saving state and waking them up once required (e.g. when load increases).

4 Dynamic Consolidation of Virtual Machines in Cloud Data Centers

The proposed algorithm works by associating a credit value with each node. For a node, the credit depends on the affinity it has for its VMs, its current workload, and VM communications. Energy savings are achieved by continuous consolidation of VMs according to current utilization of resources, virtual network topologies established between VMs and thermal state of computing nodes. The rest of the section is organized as follows. Subsection 4.1 discusses related Credit based Anti load-balancing model, followed by the algorithm description presented in Subsect. 4.2.

4.1 Credit of Node

The algorithm proposed aims at maximizing *Credit* which is a value used when calculating the energy-efficiency of the system behavior.

The proposed algorithm in this section works by associating a credit value with each node. The credit of a node depends on the node, its current workload, its communications behavior and history of task execution. When a node is under-loaded ($c_{i,j} < \gamma$), all its VMs are migrated to a comparatively more loaded node. In dynamic load unbalancing schemes, the two most important policies are *selection policy and location policy.* Selection policy concerns the choice of the node to unload. Location policy chooses the destination node of the moved VMs. An important characteristic of selection policy is to prevent the destination node to become overloaded. Also, migration costs must be compensated by the performance improvement. Each node has its own *Credit*, which is a float value. The higher a node Credits, the higher the chance its VMs to stay at the same node. It is equivalent to say that chances of its VMs to be migrated is lower. The credit of a node increases if:

- Its workload or the number of VMs in the node increases
- Communication between its VMs and other nodes increases
- Its load increases while staying between the under-loaded threshold γ and the over-loaded threshold ε

On the contrary, the credit of a node decreases in the cases below:

- Its workload or its number of VMs decrease
- It has just sent or received a message from the scheduler which indicates that the node will probably become empty in a short while.

The Credit of a node will be used in the selection policy: the node which credit is the lower is selected for VMs migration.

The location policy identifies the remote node with the highest credit which is able to receive the VMs selected by the selection policy without being over-loaded.

However, as shown in Fig. 1, such consolidation is not trivial, as it can result in overloading a node. Also keeping one particular node heavily loaded for a large period of time can lead to a heating point. It has a negative effect on the cooling system compared to several colder points.

In Fig. 1 we have two nodes N_1 and N_2. Node N_2 is under-loaded because its load is below the underload threshold ($\gamma = 50\%$). In this situation the load N_2 migrates to the node N_1. Then N_2 is switched to sleep mode.

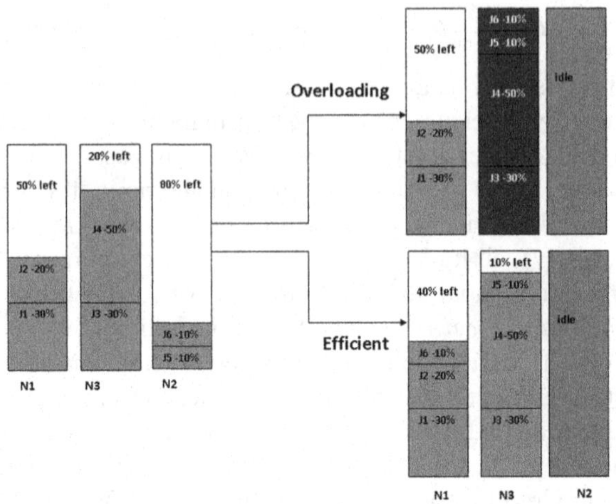

Fig. 1. Different contexts for a migration. Both cases are possible but the first one creates a hot spot on N_3

4.2 Algorithm

We work with VMs which migrate depending on the load of the node. We apply the Credit concept to the migration of VMs. ($\sigma_{i,j}$) of each node j (in site i). With our algorithm, the following formula is used:

$$\sigma_{i,j} = c_{i,j} - t_{i,j}s_{i,j} + \varepsilon - \gamma$$

$$t_{i,j} = (100 * (\varepsilon - c_{i,j}) * (c_{i,j} - \gamma) * (\lambda * c_{i,j} + \delta))/s_{i,j}$$

Where $c_{i,j}$ is the actual load of node j in site i, $r_{i,j}$ is its requested computation load, $s_{i,j}$ is its VM satisfaction, and γ and ε are respectively under-load and over-load threshold. $t_{i,j}$ is the attractivity of the node, λ and δ are two constants respectively 10 and 20.

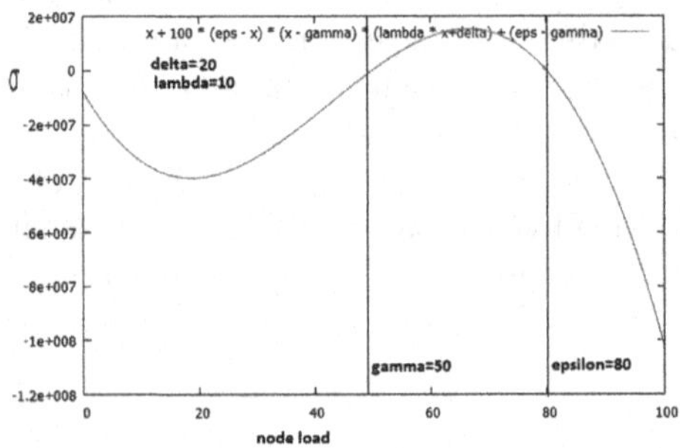

Fig. 2. Credit of node. $\gamma = 50\%$ and $\varepsilon = 80\%$

The values 10 and 20 were chosen in order to favor the avoidance of hotspots (overload) compared to avoiding underload nodes. Indeed, the $\sigma_{i,j}$ value is used for sorting the nodes to consider in the following Algorithm 1. A smaller value of $\sigma_{i,j}$ leads a node j in site i to be considered first, and actions are firstly held on this one. On Fig. 2, one can see that overloads leads to smaller values than underloads. Other values than 10 and 20 would lead to different behaviour of the algorithm: It could favor underload versus overload nodes first, for instance. In the experiment part, we only considered the previous scenario. The algorithm provides task scheduling strategy, which dynamically migrate tasks among computing nodes, transferring tasks from underloaded nodes to loaded but not overloaded nodes. It balances load of computing nodes as far as possible in order to reduce program running time.

Algorithm 1. Dynamic Consolidation of Virtual Machines (DCVM) in Cloud Data Centers for each node $H_{i,j}$

```
Calculate $c_{i,j}$ , $\sigma_{i,j}$    // Load, credit of node j in site i
Sort in ascending order other nodes ($\forall(i,j) \neq (i',j')$) according to the value of their
  credit (all, but $H_{i,j}$)
if ($c_{i,j} < \gamma$) then
   for (all nodes $j'$ in all sites $i'$ sorted by their credits) do
      if ($H_{i,j} \neq H_{i',j'}$) and ($c_{i',j'} \geq \gamma$) and ($c_{i,j} + c_{i',j'} < \varepsilon$) then
         Add $H_{i',j'}$ to potential destination set Potential
      end if
      Migrate all $VM_{i,j}$ from $H_{i,j}$ to the element with lower credits weighted by
        migration cost in Potential )
   end for
else
   while ($c_{i,j} > \varepsilon$) do
                                  // $c_{i,j}$ and $\sigma_{i,j}$ have changed after migration of $VM_{i,j,k}$
      Calculate $c_{i,j}$
      Calculate $\sigma_{i,j}$
      Calculate $l_{i,j}^{min}$    // load of the lightest task
      Let $VM_{i,j,k}$ the task with $l_{i,j,k}^{min} = l_{i,j,k}$
      for ( all nodes $j'$ in all sites $i'$ sorted by their credits) do
         if (( $H_{i,j} \neq H_{i',j'}$) and ($c_{i',j'} + l_{i,j}^{min} < \varepsilon$)) then
            Migrate $VM_{i,j,k}$ from $H_{i,j}$ to $H_{i',j'}$.
         end if
      end for
   end while
end if
```

The decision making algorithm behaves globally as follows: if $\sigma_{i,j} < 0$, the node j of site i is over-loaded or under-loaded; if $\sigma_i < r_{k,z}$? the node i has affinity of node k; if $c_{i,j} > \varepsilon$, the node j of site i is over-loaded; if $c_{i,j} < \gamma$, the node j of site i is under-loaded. This algorithm is described in Algorithm 1. For the sake of simplicity, corner cases such as all nodes over-loaded are not included. Selection policies take into account credits and migration cost. The selected node (node j' in site i') is the one with the minimum $\sigma_{i',j'}$ weighed by the migration cost between the current position of the VM and the potential node. If $\tau_{i,j,i',j'}$ is the migration cost between the node j in site i and the node j' in site i'. the selected node is the one that minimize:

$$\sigma_{i',j'}.\frac{\tau_{i,j,i',j'}}{Max_{i'',j''}(\tau_{i,j,i'',j''})}. \tag{1}$$

To reduce the load of an overloaded node, it begins to migrate the lightest task. Selection policy will choose the task that will stay the longest on the node. Policy of localization will then identify the node that will receive the task without exceeding its capacities (i.e. its load after migration will still be under ε). So this node will be the new destination of the task.

5 Experiments and Results

In order to evaluate the gains of the algorithm we implemented this algorithm using CloudSim [3].

5.1 Simulation Environment

Experiments are done in a simulated cloud environment by the CloudSim simulation tool. We conduct experiments using PlanetLab workload traces [10]. In the following we compare DCVM with a dynamic *Anton Beloglazov et al.'s algorithm* [4] *(belgo)* where host are sorted according to their maximum power consumption.

5.2 Experimental Results

In this subsection, we describe the simulation study performed to evaluate the performance of our algorithms in terms of energy minimization as well as the execution time and the number of migrations.

The first observation is that DCVMB consumes less energy than the algorithm described in [4] (see Fig. 3). The second observation is that DCVM algorithm is able to reduce the energy consumption by 5% to 20% when jobs increase. f switched on host as a function of task number. When jobs increase then the number of nodes switched on also increase, leading to a higher power consumption. This is particularly true if there is no migration after the initial placement of tasks. Hence, the gain of our algorithm increases power-wise with the number of tasks because migration is activated. Computing resources are fully used in both cases at the start of the experiment. In the case of the consolidation, less hosts are switched-on because we can adapt to the workload dynamism (Fig. 4).

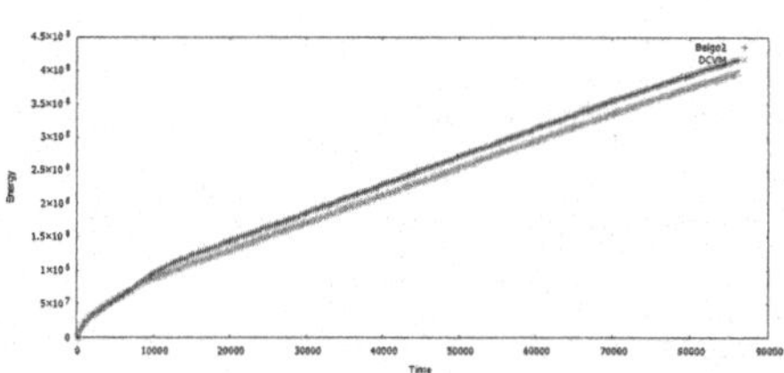

Fig. 3. Energy of algorithms compared to Anton Beloglazov et al.'s algorithm [4] with unsorted hosts. Lower is better

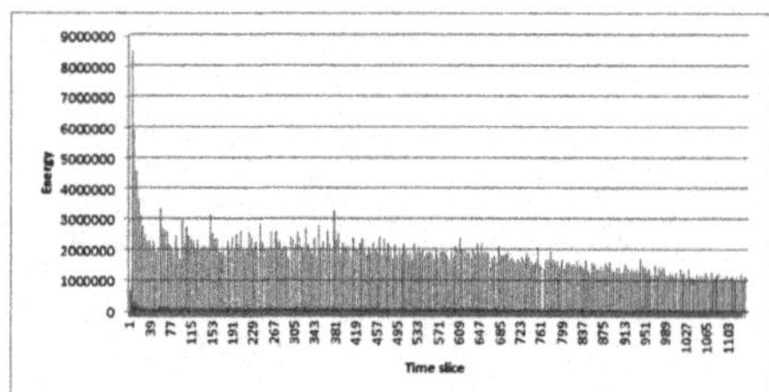

Fig. 4. Energy of algorithms by time slice.

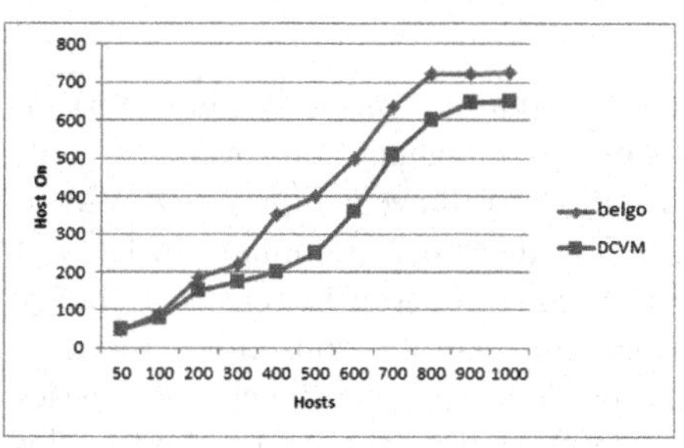

Fig. 5. Host switched on DCVM vs belgo

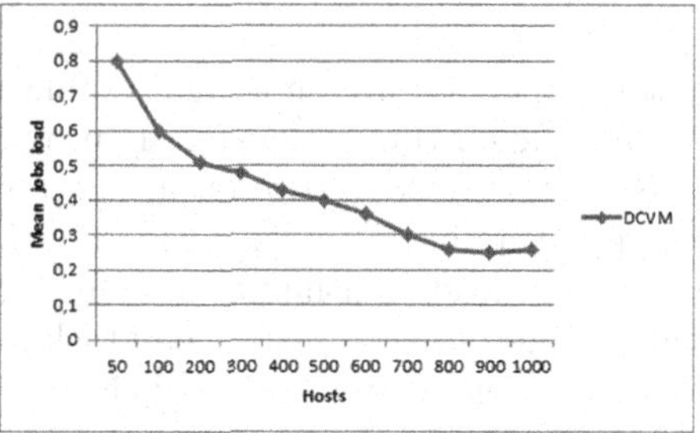

Fig. 6. Job mean load with DCVM

Figure 5 shows the number o The observed gain increases with the number of VMs and becomes constant when hosts are saturated. The best results of DCVM comes from the fact that with the increase of the number of VMs, it has more possibilities to migrate tasks. It can then better allocate VMs on computing resources, reducing the number of switched-on hosts. Due to the thresholds of DCVM, it would be possible to reduce further the number of switched on hosts but it would overload remaining hosts. Those hosts would become hot points and would have a negative impact on cooling. Figure 7 shows that our algorithm is better than classical *Dynamic First Fit* regardless of the makespan when the number of tasks is important. For small number of tasks, energy consumption increases because of the many migrations. To choose this value it is important to consider the dynamism of the tasks. As said previously, increasing γ, reduce energy consumption at the cost of consolidating more and more the tasks. DCVM has the shortest execution time when the number of jobs increases. The result implicates that the scheduling algorithm such as DCVM can leverage interconnects with migrations to achieve high performance and energy efficiency (Fig. 6).

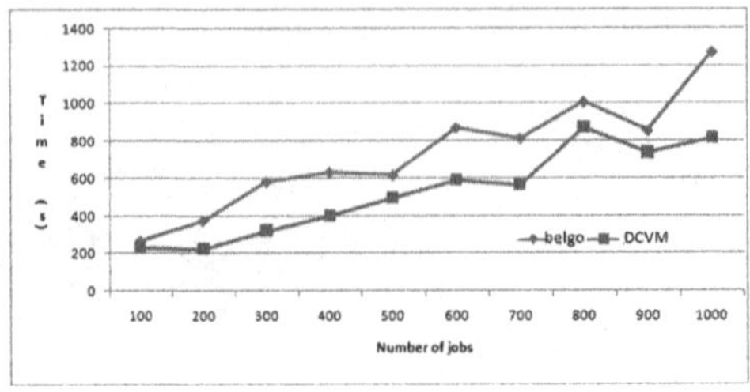

Fig. 7. Number of migrations

6 Conclusion

In this paper, we presented and evaluated our dynamic consolidation of virtual machines algorithm for clouds. It provides energy-efficiency improvement compared to Anton Beloglazov et al.'s algorithm in [4] We have compared DCVM to the algorithm in [4] over a range of problem instances using simulation. DCVM parameters lead to a family of heuristics that perform well in terms of energy savings while still leading to good task performance. DCVM consolidate tasks on a subset of the cluster hosts judiciously chosen depending on the characteristics and state of resources. Overall, the proposed DCVM algorithm can compute allocations effectively with an important energy gain. Experiments showed that with our algorithm we obtained a 20% gain over standard algorithms. However, it is important to investigate further how to improve the quality of service, but also the optimization algorithm. Future version of DCVM will also take int account other measures to compute Credit such as network communication patterns.

References

1. Alsadie, D., Alzahrani, E.J., Sohrabi, N., Tari, Z., Zomaya, A.Y.: DTFA: a dynamic threshold-based fuzzy approach for power-efficient VM consolidation. In: 2018 IEEE 17th International Symposium on Network Computing and Applications (NCA), pp. 1–9. IEEE (2018)
2. Arroba, P., Moya, J.M., Ayala, J.L., Buyya, R.: Dynamic voltage and frequency scaling-aware dynamic consolidation of virtual machines for energy efficient cloud data centers. Concurrency Comput. Pract. Experience **29**(10), e4067 (2017)
3. Beloglazov, A., Buyya, R.: Energy efficient allocation of virtual machines in cloud data centers. In: 2010 10th IEEE/ACM International Conference on Cluster, Cloud and Grid Computing, pp. 577–578. IEEE (2010)
4. Beloglazov, A., Buyya, R.: Optimal online deterministic algorithms and adaptive heuristics for energy and performance efficient dynamic consolidation of virtual machines in cloud data centers. Concurrency Comput. Pract. Experience **24**(13), 1397–1420 (2012)
5. Challita, S., Paraiso, F., Merle, P.: A study of virtual machine placement optimization in data centers. In: 7th International Conference on Cloud Computing and Services Science, CLOSER 2017, pp. 343–350 (2017)

6. Khan, M.A., Paplinski, A., Khan, A.M., Murshed, M., Buyya, R.: Dynamic virtual machine consolidation algorithms for energy-efficient cloud resource management: a review. In: Rivera, W. (ed.) Sustainable Cloud and Energy Services, pp. 135–165. Springer, Cham (2018). https://doi.org/10.1007/978-3-319-62238-5_6
7. Kumar, N., Kumar, R., Aggrawal, M.: Energy efficient DVFS with VM migration. Eur. J. Adv. Eng. Technol. **5**(1), 61–68 (2018)
8. Laili, Y., Tao, F., Wang, F., Zhang, L., Lin, T.: An iterative budget algorithm for dynamic virtual machine consolidation under cloud computing environment (revised December 2017). IEEE Trans. Serv. Comput. (2018)
9. Nguyen, T.H., Di Francesco, M., Yla-Jaaski, A.: Virtual machine consolidation with multiple usage prediction for energy-efficient cloud data centers. IEEE Trans. Serv. Comput. (2017)
10. Park, K., Pai, V.S.: CoMon: a mostly-scalable monitoring system for PlanetLab. ACM SIGOPS Operating Syst. Rev. **40**(1), 65–74 (2006)
11. Shirvani, M.H., Rahmani, A.M., Sahafi, A.: A survey study on virtual machine migration and server consolidation techniques in DVFS-enabled cloud datacenter: taxonomy and challenges. J. King Saud Univ. Comput. Inf. Sci. (2018)
12. Shrivastava, A., Patel, V., Rajak, S.: An energy efficient VM allocation using best fit decreasing minimum migration in cloud environment. Int. J. Eng. Sci. **4076** (2017)
13. Silva Filho, M.C., Monteiro, C.C., Inácio, P.R., Freire, M.M.: Approaches for optimizing virtual machine placement and migration in cloud environments: a survey. J. Parallel Distrib. Comput. **111**, 222–250 (2018)

Safety Is the New Black: The Increasing Role of Wearables in Occupational Health and Safety in Construction

João Barata(✉) and Paulo Rupino da Cunha

Department of Informatics Engineering, University of Coimbra,
Pólo II, Pinhal de Marrocos, 3030-290 Coimbra, Portugal
{barata, rupino}@dei.uc.pt

Abstract. As wearable technologies are gaining increased attention in construction, we present an integrated solution for their adoption in occupational health and safety (OHS). Research methods include a structured literature review of 37 articles and a year-long design science research project in a construction group. The main results are (1) the identification of new wearable solutions made available by industry 4.0 to prevent hazards, and (2) a wearable model for voluntary regulations compliance. For theory, our research identifies key application areas for integrated smart OHS in construction and highlights the importance of continuous monitoring and alerts to complement the traditional sampling techniques. For practice, we offer recommendations for managers wishing to implement continuous compliance checking and risk prevention using wearable technology. Our findings help improve health and safety audits supported by digital evidence in the sector with most risks of accidents in the European Union.

Keywords: Occupational health and safety · Industry 4.0 · Construction · Internet-of-Things · Wearables · Regulatory compliance

1 Introduction

Occupational health and safety (OHS) is a priority for the construction sector and one of the areas with more potential for improvements with the technological transformation of industry 4.0 (I4.0) [1]. This claim is confirmed by the positive impact of sensors in construction equipment and wearable devices available to construction workers, allowing real-time alerts to prevent accidents [1–5]. Nevertheless, there are also barriers to the adoption of smart devices in construction, including privacy issues and the perceived usefulness and ease of use of each device [6]. Despite the extensive research in augmented reality, wearables, and sensors for construction safety management [2, 7–9], most studies focus on specific applications for preventing accidents (e.g. collision alert), safety training [10], or monitoring specific parameters of work conditions using biosensors [7]. Other authors address the reduction of consequences in accidents via internet-of-things (IoT), detecting falls and ensuring the earliest possible assistance [11]. More recently, [12] explores continuous monitoring of environmental

W. Abramowicz and R. Corchuelo (Eds.): BIS 2019, LNBIP 353, pp. 526–537, 2019.
https://doi.org/10.1007/978-3-030-20485-3_41

factors such as carbon monoxide and noise in factories. Interestingly, we could not find any studies that integrate these important sources of data for compliance, as it happens in audits required by OHSAS 18001 safety standards and legal regulations.

Our project started with a European co-funded research project involving a Portuguese construction group. Two companies in the consortium (C1: consulting, training, safety inspections; C2: construction equipment supplier) had industry 4.0 in their agendas. They were aware of experiments in academic projects, such as the use of sensors to avoid collisions between vehicles and workers in construction sites [1]. Yet, according to them, such an application only scratches the surface of what's possible for OHS. For example, the same system used to identify the worker in collision detection could be important to avoid falls – using the same wearable to detect proximity to danger areas or to prevent unauthorized access to the construction site. Accordingly, we formulated two main research questions:

- RQ1. What are the opportunities for using wearable technologies in industry 4.0 for occupational health and safety in construction?
- RQ2. How do we design an information system for OHS in construction, for real-time integrated prevention (e.g. user alerts), correction (e.g. minimize accident consequences), and compliance (e.g. ensuring the correct adoption of regulations)?

The remainder of our paper is presented as follows. Section 2 describes the research approach. Subsequently, we present the literature survey, and, in Sect. 4, we detail the different phases of our project. Section 5 discusses the findings and the implications for theory and practice. The paper closes with opportunities for future research.

2 Research Approach

To address the two research questions we have selected a design-science research (DSR) approach [13], having its foundations in the work of [14]. DSR enables the creation and evaluation of artifacts to solve specific organizational problems, which can "be in the form of a construct, a model, a method, or an instantiation" [13], integrating informational, technological, and social aspects [15]. Figure 1 outlines our DSR.

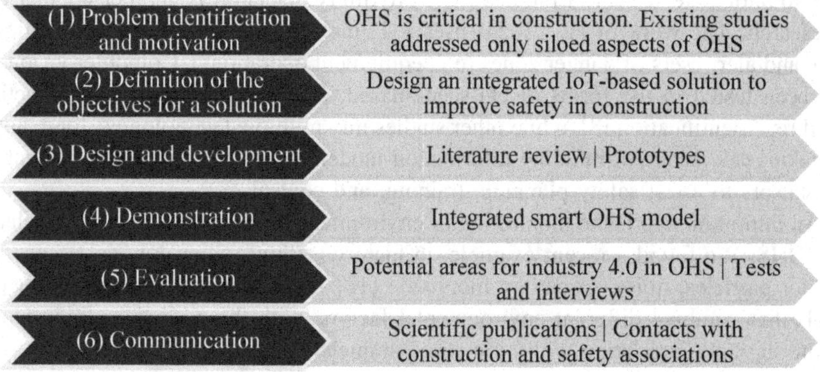

Fig. 1. Design science research approach (adapted from [16]).

The sequence of DSR activities is particularly suitable for our research and the reasons are fourfold. First, the necessity to conduct a comprehensive literature review to identify wearable industry 4.0 solutions for OHS, which contributes to RQ1 and guides the field activities in RQ2. Second, the development of artifacts justified by business needs and tests in a real environments [13]. Third, new wearable solutions for OHS should be supported by theory and practice (RQ2). Fourth, the examples obtained through the field work, contributing to demonstrate and communicate industry 4.0 opportunities to construction practitioners. Having identified the problem and the objectives for a solution, in the next section, we review key contributions from the literature.

3 Literature Review: What Are the Opportunities?

We have followed the recommendations made by [17] to guide the steps of the literature review and used a concept-centric approach to summarize the results [18]. First, we made searches in Google Scholar using different combinations of keywords, for example, “safety management” + “industry 4.0” + construction, returning 278 results (excluding patents and citations) and wearable + “occupational health and safety” + construction yielding 611 results. Then, we screened the title and abstract to identify studies about wearable industry 4.0 technologies applied to construction. Afterwards, we made searches in other databases such as IEEEXplore (e.g. 16 results using “construction safety iot”) and EBSCOhost (e.g. 2 results with construction + “internet of things” + safety). A total of 37 papers were classified in three main concepts presented below.

3.1 Site Oriented OHS

Significant research has been conducted to improve safety in the relation between humans and the environment in construction. For example, the work of [11] to detect falls using wearable technologies and reduce response time. Other authors focus on prevention. Examples include the creation of safety barriers [19], the detection and alert of unsafe conditions of moving objects [20], and monitoring the inclination of retaining walls to anticipate structural failures [21]. Proximity detection is another popular area of research. The work of [1, 22–24] uses IoT, mobile, and wearable technologies to sense and alert users of danger zones (e.g. equipment operations). Computer vision has also been tested in the forms of (1) scene-based; (2) location-based; and (3) action-based risk identification [8], while other studies mix multiple technologies, for example [2], taking advantage of building information modeling, augmented reality, wearables, and sensors to assist safety planning, training, and control.

An important gap is the monitoring of environmental parameters and its correlation with OHS. We found a recent example, in factory settings, using low cost sensors to monitor particles, noise, or carbon monoxide [12]. This research is inspiring for proposals that combine different environmental factors, with the potential of continuous monitoring and correlation with biometric parameters and logging of risk alerts.

3.2 User Oriented OHS

The interest of monitoring health parameters in construction is increasing, as revealed by [25] for stress recognition, [26] for psychological status based on heart rate, energy expenditure, metabolic equivalents, and sleep efficiency, or [27] using biosensors. It is now clear that continuously collected data can be used, for example, using a photoplethysmography (PPG) sensor embedded in a wristband-type activity tracker [7].

Personal protective equipment (PPE) is key in OHS activities and the risks increase when the staff does not comply with its use [28]. Taking advantage of cloud, wireless, and wearable technologies, [29] presents a proof of concept to ensure the use of PPE. The results were positive, although privacy (e.g. using technology to monitor performance, number of steps taken, and heart rate) was pointed as a major concern of the workers. Other barriers include "employee compliance, sensor durability, the cost/benefit ratio of using wearables, and good manufacturing practice requirements" [30]. Perceived usefulness, social influence, and perceived privacy risk are key aspects that influence the intention to use equipment, such as smart vests and wristbands [6]. Therefore, additional sociotechnical research is necessary in business information systems.

A group of studies aim at human health improvements, including ergonomics [31], stress control [32], and the prevention of construction workers' musculoskeletal disorders [33]. An example specifically developed to identify the exposure of the worker to hazardous vibrations is presented by [34]. Yet, these solutions are not yet currently implemented by construction companies. In a recent review about OHS in the industry 4.0 era, the authors of [35] argue that "emotion sensors need to be developed to monitor workers and ensure their safety continuously". The work of [4] reinforces the need to monitor physical demands of construction workers because it is highly variable depending on the tasks. Monitoring of physiological status can be conducted at work but also off-duty, as presented in [26], opening opportunities for OHS improvements and uncovering new risks for privacy [36]. If taken in isolation, biometric measurements are also challenging, because the correlation between work and fatigue varies with each person and, to be valid, fatigue based on heart rate requires context information [37].

3.3 Integration of Smart OHS in Construction

Nine literature reviews were included in our survey (4 done in 2018, 4 done in 2017, and 1 done in 2015). One of the studies highlights that the number of papers addressing innovative technologies is increasing since 2012, but most of them remain in the academic field [38]. According to these authors, researchers and practitioners should work together in "the effective path of innovative technology transition from construction safety research into construction safety practice". Two recent works expressly mention "industry 4.0" in the title: [35], who anticipates significant changes in OHS practices due to the closer connection between humans and machines, including risk management in real time; and [39] mentioning the importance of wearables for worker safety.

Wearable technologies were the focus of different literature reviews. The work of [3] presents a comprehensive list of technologies applicable to physiological and environmental monitoring, proximity detection, and location tracking. The authors conclude that it is necessary to "derive meaning from multiple sensors" [3]. The authors of [40] also suggest involving researchers and practitioners in wearables adoption for monitoring, augmenting, assisting, delivering, and tracking in construction. These authors have previously detailed the application of wearables in the workplace [41].

On one hand, user oriented OHS reviews can be found in [5], that collected data via smart watch and showed the correlation between psychological status and physical indicators. On the other hand, [42] reviews earthmoving equipment automation. Examples of technologies used to improve safety with these types of equipment include artificial vision, GPS, and RFID, but the authors state that "advanced sensing technologies (e.g. computer vision) for tracking and safety, are still in experimental stage and yet have to prove their efficiency in practice". IoT is also important to implement visualization techniques for safety management [43]. Yet, we could not find a proposal that integrates the user physical indicators with the environmental conditions. The combination of human, machines, and context data is determinant for compliance.

4 Hands-On Experiment: Integrated Smart OHS

The design and development of our prototypes involved two researchers and a student of informatics, responsible for the coding. This phase was conducted in close collaboration with the practitioners designated by the construction group, namely, two OHS assessors, the top managers of the construction equipment company and of the consulting company, and two construction technicians designated to assist in the field tests. There were two major steps. First, a preliminary research with sensors, wristbands, and mobile devices, described in Sect. 4.1. Second, the design of an integrated smart OHS information system, described in Sect. 4.2.

4.1 Testing Opportunities and Integrating Technologies

In this step we tested proposals found in the literature (e.g. RFID for proximity warning system). Figure 2 presents the example for the worker oriented OHS.

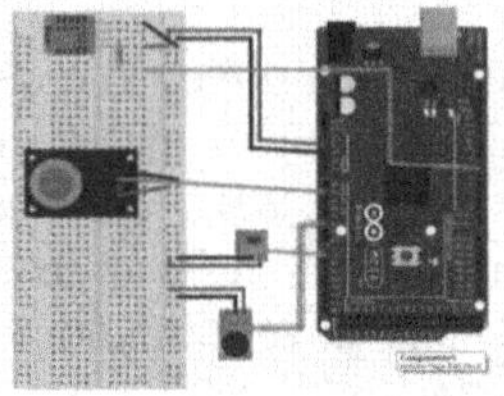

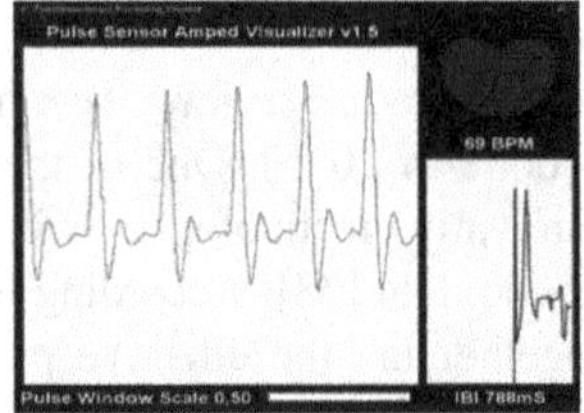

Fig. 2. Smart OHS – worker oriented preliminary (non-integrated) experiments.

Our purpose with the experiments presented in Fig. 2 was to test low cost equipment: Arduino UNO and RC522 RFID sensor presented on the left of Fig. 2; environmental sensors for temperature and humidity (DHT11), noise (SEN-12642 sound detector), and air quality (MQ135) in the middle; and a simple visualization tool to display data and compare to real values – rightmost image. It was important for three reasons. First, it facilitated the discussion with the practitioners about the potential of wearables and technologies emerging from industry 4.0. Second, it confirmed previous research pointing to the applicability of using low cost IoT material in real applications [12]. Third, it identified synergies in the systems described in the literature. For example, the RFID system for collision detection can also be used for site access (identification of workers in the site) and control if PPE is in use. Each of these examples have already been studied independently (e.g. [22] or [29]), missing an integrated solution.

These cases were selected on the basis of the (1) need to protect the worker from collisions and falls, (2) legal requirements of OHS (e.g. noise and air quality), and (3) potential to contrast biometric and environmental parameters to alert the user. We were also looking for synergies when using different technologies. For example, RFID to protect the worker from collisions and also (1) avoid unauthorized access to the site and (2) prevent equipment use by non-qualified staff (block the engine of the equipment). Other examples are the continuous monitoring of environmental conditions for OHS audits and the combination with biometric parameters (e.g. heart rate) and improvement of work conditions.

An integrated OHS system should be aware of the workers but also their context. Yet, according to the company managers, most of the prototypes found in the literature are difficult to deploy in practice: "it is already difficult to ensure the proper use of PPE; the wearable will also need some sensor to ensure that it is being used (…) the best way is to create a unique wearable such as wristband that integrates RFID, GPS, biometric monitoring, and other functionalities (…) which is available in the market but not yet adopted in OHS and not fully explored for compliance checking".

4.2 Proposing the Integrated Smart OHS Model in Construction

Table 1 presents the candidate technologies for integrated smart OHS that we found in the literature and discussed in our meetings.

Table 1. Summary of Industry 4.0 opportunities for integrated smart OHS.

	VR	AR	IoT	Cloud	M	ARO	BDA	BIM	BC
Site oriented	+	+	+	+	+/-	-	-	+	-
User oriented	+/-	+/-	+	+	+	-	-	+/-	-
Audit oriented	-	+	+/-	+/-	+/-	+/-	+	-	+

Legend: light grey – short-term priority for development; dark grey – medium-term; white: long-term project.

The list of technologies includes Virtual Reality (VR), Augmented Reality (AR), IoT, Cloud, Mobile (M), Autonomous Robots (ARO), Big Data and Analytics (BDA), Building Information Modeling (BIM), and Blockchain (BC). The list is not exhaustive, for example, we did not include 3D printing and additive manufacturing, which raise issues for safety while using them, but they are not tools for OHS. We also did not include fog computing [44] because it was considered too technical for the discussion with construction experts (although important for implementation purposes).

The possible applications vary according to each technology. For example, VR, AR, and BIM (dark grey background) are important to digitalize the construction information, which can assist in the users training and guidance while executing tasks. AR seems promising for inspection and auditing, contrasting the real with the virtual scenario (e.g. evaluate if specific safety equipment is on site). IoT (includes related technologies such as RFID, wireless, and location systems), cloud, and mobile can assist OHS efforts in all aspects of construction. However, when combined with big data and analytics and blockchain, the potential increases for inspection and auditing (light grey background in the table). One important conclusion is that full industry 4.0 potential requires a combination of technologies for specific purposes.

The interest in blockchain emerged during our discussions about the relation with external entities, such as government and insurance companies. The experts that we interviewed confirm the importance of wearables for improvement actions or standards audits (voluntary regulations). Yet, if lacking evidence of data quality and reliability, the use for insurance communications or legal compliance is limited. It is necessary to prove to third parties that the company implemented all the measures for prevention and that the OHS data was in fact collected on the specific site, user, or context. The proposal for our companies is presented in Fig. 3.

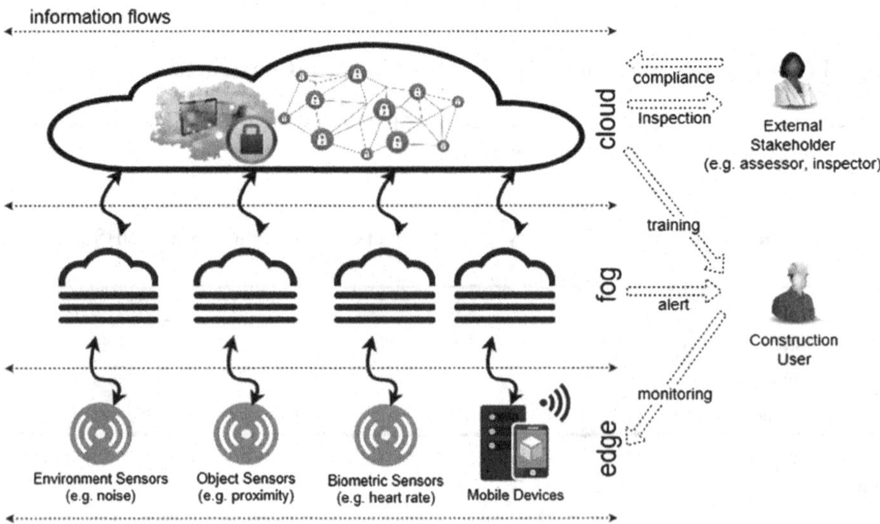

Fig. 3. Integrated smart OHS – conceptual model.

The proposed model is inspired in the architecture proposed by [44] for IoT, which considers a fog layer between the cloud and the edge devices. In this proposal we have integrated the platforms and sensors in the lower layer, considering four different elements in the cyber-physical IoT infrastructure (on the bottom): the *environment sensors* (nowadays unpractical to embed in the worker wearables) including light conditions, temperature and humidity, noise, and air quality (requires GPS tracking); the *objects sensors* including RFID antennas (collision detection or fall protection), personal protection (e.g. identify helmet presence), and site access; the *biometric sensors* including worker monitoring and alert system based on a smart wristband – biometric information, RFID, and GPS; and the *mobile devices* for communication with the OHS.

There are advantages in this design. First, minimizes the wearable burden while making it essential to enter the site. Second, concentrates the environmental parameters in a portable, low cost toolkit (<200€ each in our prototypes), making it affordable to use in different areas of the site. Third, ensure wearable use – if the wearable is not in use, the worker may not enter the site or use specific equipment. The arrows connecting the OHS stakeholders (on the right) represent the value obtained from data analysis in the OHS (training, alerts, and inspections), and the purpose of data collected via IoT (monitoring and compliance). On the top is represented the OHS cloud supported by a private blockchain and, bellow, the fog layer to support for IoT infrastructure [44].

5 Discussion

Few studies have addressed occupational health and safety in the industry 4.0, and those that do have identified challenges [35]. Most of the research in I4.0 technologies for safety in construction is diverse but also disperse. But integration is key, as stated by [35] "if the technologies driving Industry 4.0 develop in silos and the OHS initiatives of manufacturers are fragmentary, hazards will multiply and some of the gains made in accident prevention will be lost".

According to the company managers in our project consortium, a comprehensive model for wearable technologies in construction needs to support (1) training, (2) monitoring worker critical parameters, (3) user alerts (collision, fall, environment parameters), and (4) voluntary regulatory compliance (OHS standards audit and improvement evidences). Future research is needed to address requirements of insurance companies and government authorities for legal compliance checking.

Only the combination of technologies can turn industry 4.0 into a reality in OHS. According to the companies participating in our work, they were interested in investing only if (1) the solution is integrated, and (2) the data collected is not merely for monitoring and alert, but also valuable for improvement and compliance checking. According to our case companies "sometimes the problem is not in the existence of solutions for OHS; it is in convincing the users of the need to use it, in a daily basis". One of the OHS experts also told us that "we can monitor all the parameters in the world but for regulatory compliance, it is also crucial to prove that the data is reliable and not easily manipulated (…) that is essential to demonstrate our commitment to safety when dealing with insurance companies and assessors". She presents an example "if the environmental conditions are continuously monitored in 'friendly' locations – the term used to designate a location of the site that is far away from specific machinery – the data will be irrelevant for prevention and warning systems". Moreover, "if the collected data is not reliable" – she gave the examples of a worker that gives the wristband to another colleague or if the large amounts of historical data about warnings and environmental conditions can be changed by the organization – "then, the benefits will be limited".

The insights gathered in our literature review and the practitioners' feedback allowed us to identify critical elements of integrated smart OHS:

- Continuous monitoring is complementary to traditional sampling. We confirmed previous findings that used low cost sensors, however, at this moment, there are risks in trusting solely in this type of systems. For example, the error of measurements and obtaining data in wrong locations;
- Non-integrated developments may lead to duplicated investments and the proliferation of tools and may demotivate the workers from using a plethora of systems;
- Industry 4.0 opportunities are critical to improve health and safety, but the users must use it. Wearables must be friendly and useful, but the possibility of making them mandatory (e.g. if PPE is not identified, the worker is not allowed to enter the site; worker recognition for equipment start; site access), is interesting;
- Data quality should be a priority for integrated smart OHS. Trusting in single systems for health and safety is a risk too high. For example, there are risks of over confidence by the workers (e.g. collision detection) that may decrease the surveillance level. These aspects have not yet been addressed in the BIS literature.

6 Conclusions

We reviewed key literature for wearable implementation in OHS and proposed an integrated smart OHS system design for construction. Our findings emerged from a year-long design science research [13] project with two construction companies. Industry 4.0 opens the opportunity to adopt wearables for continuous monitoring and innovative regulatory compliance systems in OHS. Private blockchains [45] can be a viable solution to test in construction compliance and audit, when third party entities are involved, for example, insurance companies. To our knowledge, this is the first proposal taking advantage of wearables connected to site IoT-based solutions, and context information.

The system was designed in collaboration with practitioners, thus enabling testing and evaluation of current academic proposals. According to the construction group experts, independent single purpose solutions, such as collision detection, are interesting for specific stakeholders but it is difficult to equip workers with multiple wearables.

Continuous monitoring via wearables is complementary to the sampling techniques normally used for OHS, with the benefit of providing real-time alerts and valuable data for improvement and compliance checking. The prototypes that we have developed showed positive results for compliance with OHS standards, namely evidence of prevention efforts, but are not suitable for insurance and legal purposes. To that effect, additional requirements are mandatory, such as data source traceability (e.g. GPS location, timestamp, user); contextual data; and protection of the digital traces. Another important implication for practice is to ensure that the wearables are effectively used – one possibility is to ensure that the technology is friendly and necessary to use specific equipment or access the site, thus becoming a working tool.

Our research has limitations that need to be stated. First, is the selection of papers for our literature review. Industry 4.0 is a vibrant field of research and other studies could be included. Second, our research considers two different companies with important roles in OHS for construction, but belonging to the same group. Other companies may have different priorities for OHS investments. Third, despite including certified OHS auditors and senior consultants, the data collected with interviews with the experts did not involve external assessors and legal advisors.

Future research is necessary to test the part of the model that involves third party entities using a private blockchain. In addition, there are opportunities for comparative studies of different models, for example, mobile systems for diagnostics procedures [46] and solutions for ambient intelligence in OHS [47]. Our preliminary contacts with an integrated management system consulting company and two associations of construction in Portugal already provided results. A quantitative survey will be deployed with the support of the associations that represent hundreds of companies to understand their perspective about wearable investments for OHS and identify potential actions by their associations (e.g. ask government to enforce collision detection systems or continuous monitoring of critical parameters, with benefits in insurance policies).

We hope that our work may inspire other researchers to empirically evaluate the vast amount of academic proposals for industry 4.0 in traditional sectors of the economy. It is important to identify combinations of technologies to fully explore the potential in the fourth industrial revolution to improve business information systems.

References

1. Teizer, J.: Wearable, wireless identification sensing platform: self-monitoring alert and reporting technology for hazard avoidance and training (SmartHat). J. Inf. Technol. Constr. **20**, 295–312 (2015)
2. Le, Q.T., Pham, H.C., Pedro, A., Park, C.S.: Application of wearable devices for real time construction safety management. In: IPC Proceedings, South Korea, pp. 28–31 (2015)
3. Awolusi, I., Marks, E., Hallowell, M.: Wearable technology for personalized construction safety monitoring and trending: review of applicable devices. Autom. Constr. **85**, 96–106 (2018)
4. Hwang, S., Lee, S.H.: Wristband-type wearable health devices to measure construction workers' physical demands. Autom. Constr. **83**, 330–340 (2017)
5. Guo, H., Yu, Y., Xiang, T., Li, H., Zhang, D.: The availability of wearable-device-based physical data for the measurement of construction workers' psychological status on site: from the perspective of safety management. Autom. Constr. **82**, 207–217 (2017)
6. Choi, B., Hwang, S., Lee, S.H.: What drives construction workers' acceptance of wearable technologies in the workplace? Indoor localization and wearable health devices for occupational safety and health. Autom. Constr. **84**, 31–41 (2017)

7. Hwang, S., Seo, J.O., Jebelli, H., Lee, S.H.: Feasibility analysis of heart rate monitoring of construction workers using a photoplethysmography (PPG) sensor embedded in a wristband-type activity tracker. Autom. Constr. **71**, 372–381 (2016)
8. Seo, J., Han, S., Lee, S., Kim, H.: Computer vision techniques for construction safety and health monitoring. Adv. Eng. Informatics **29**, 239–251 (2015)
9. Kim, K., Kim, H., Kim, H.: Image-based construction hazard avoidance system using augmented reality in wearable device. Autom. Constr. **83**, 390–403 (2017)
10. Sacks, R., Perlman, A., Barak, R.: Construction safety training using immersive virtual reality. Constr. Manag. Econ. **31**, 1005–1017 (2013)
11. Dogan, O., Akcamete, A.: Detecting falls-from-height with wearable sensors and reducing consequences of occupational fall accidents leveraging IoT. In: Mutis, I., Hartmann, T. (eds.) Advances in Informatics and Computing in Civil and Construction Engineering, pp. 207–214. Springer, Cham (2019). https://doi.org/10.1007/978-3-030-00220-6_25
12. Thomas, G., et al.: Low-cost, distributed environmental monitors for factory worker health. Sensors **18**, 1411 (2018)
13. Hevner, A.R., March, S.T., Park, J.: Design science in information systems research. MIS Q. **28**, 75–105 (2004)
14. Simon, H.: The Sciences of the Artificial, 3rd edn. MIT Press, Cambridge (1996)
15. Lee, A., Thomas, M., Baskerville, R.: Going back to basics in design science: from the information technology artifact to the information systems artifact. Inf. Syst. J. **25**, 5–21 (2015)
16. Peffers, K., Tuunanen, T., Rothenberger, M.A., Chatterjee, S.: A design science research methodology for information systems research. J. Manag. Inf. Syst. **24**, 45–78 (2007)
17. Tranfield, D., Denyer, D., Smart, P.: Towards a methodology for developing evidence-informed management knowledge by means of systematic review. Br. J. Manag. **14**, 207–222 (2003)
18. Webster, J., Watson, R.T.: Analyzing the past to prepare the future. MIS Q. **26**, xiii–xxiii (2002)
19. Zhou, C., Ding, L.Y.: Safety barrier warning system for underground construction sites using Internet-of-Things technologies. Autom. Constr. **83**, 372–389 (2017)
20. He, V.: Application of sensor technology for warning unsafe conditions from moving objects above construction workers. In: ICEI Proceedings, pp. 69–74 (2018)
21. Lam, R., Junus, A., Cheng, W., Li, X., Lam, L.: IoT application in construction and civil engineering works. In: CSCI Proceedings, pp. 1320–1325 (2017)
22. Kanan, R., Elhassan, O., Bensalem, R.: An IoT-based autonomous system for workers' safety in construction sites with real-time alarming, monitoring, and positioning strategies. Autom. Constr. **88**, 73–86 (2018)
23. Park, J.W., Yang, X., Cho, Y.K., Seo, J.: Improving dynamic proximity sensing and processing for smart work-zone safety. Autom. Constr. **84**, 111–120 (2017)
24. Teizer, J., Allread, B.S., Fullerton, C.E., Hinze, J.: Autonomous pro-active real-time construction worker and equipment operator proximity safety alert system. Autom. Constr. **19**, 630–640 (2010)
25. Alberdi, A., Aztiria, A., Basarab, A.: Towards an automatic early stress recognition system for office environments based on multimodal measurements: a review. J. Biomed. Inform. **59**, 49–75 (2016)
26. Lee, W., Lin, K.Y., Seto, E., Migliaccio, G.C.: Wearable sensors for monitoring on-duty and off-duty worker physiological status and activities in construction. Autom. Constr. **83**, 341–353 (2017)
27. Liu, Y., Pharr, M., Salvatore, G.A.: Lab-on-skin: a review of flexible and stretchable electronics for wearable health monitoring. ACS Nano **11**, 9614–9635 (2017)

28. Bauk, S., Schmeink, A., Colomer, J.: An RFID model for improving workers' safety at the seaport in transitional environment. Transport **33**, 353–363 (2016)
29. Kritzler, M., Tenfält, A., Bäckman, M., Michahelles, F.: Wearable technology as a solution for workplace safety. In: MUM Proceedings, pp. 213–217 (2015)
30. Schall, M., Sesek, R., Cavuoto, L.: Barriers to the adoption of wearable sensors in the workplace: a survey of occupational safety and health professionals. Hum. Factors **60**, 351–362 (2018)
31. Nath, N.D., Akhavian, R., Behzadan, A.H.: Ergonomic analysis of construction worker's body postures using wearable mobile sensors. Appl. Ergon. **62**, 107–117 (2017)
32. Van Hoof, C.: Addressing the healthcare cost dilemma by managing health instead of managing illness - an opportunity for wireless wearable sensors. In: 5th International Workshop on Advances in Sensors and Interfaces Proceedings, p. 9 (2013)
33. Yan, X., Li, H., Li, A., Zhang, H.: Wearable IMU-based real-time motion warning system for construction workers' musculoskeletal disorders prevention. Autom. Constr. **74**, 2–11 (2017)
34. Kortuem, G., Kawsar, F., Sundramoorthy, V., Fitton, D.: Smart objects as building blocks for the Internet of Things. IEEE Internet Comput. **14**, 44–51 (2010)
35. Badri, A., Boudreau-Trudel, B., Saâdeddine, A.: Occupational health and safety in the industry 4.0 era : a cause for major concern? Saf. Sci. **109**, 403–411 (2018)
36. Mettler, T., Wulf, J.: Physiolytics at the workplace: affordances and constraints of wearables use from an employee's perspective. Inf. Syst. J. **29**, 245–273 (2019)
37. Bowen, J., Hinze, A., Griffiths, C.: Investigating real-time monitoring of fatigue indicators of New Zealand forestry workers. Accid. Anal. Prev. (2017)
38. Zhou, Z., Goh, Y.M., Li, Q.: Overview and analysis of safety management studies in the construction industry. Saf. Sci. **72**, 337–350 (2015)
39. Dallasega, P., Rauch, E., Linder, C.: Industry 4.0 as an enabler of proximity for construction supply chains: a systematic literature review. Comput. Ind. **99**, 205–225 (2018)
40. Khakurel, J., Melkas, H., Porras, J.: Tapping into the wearable device revolution in the work environment: a systematic review. Inf. Technol. People **31**, 791–818 (2018)
41. Khakurel, J., Pöysä, S., Porras, J.: The use of wearable devices in the workplace - a systematic literature review. In: Gaggi, O., Manzoni, P., Palazzi, C., Bujari, A., Marquez-Barja, J.M. (eds.) GOODTECHS 2016. LNICST, vol. 195, pp. 284–294. Springer, Cham (2017). https://doi.org/10.1007/978-3-319-61949-1_30
42. Azar, E.R., Kamat, V.R.: Earthmoving equipment automation: a review of technical advances and future outlook. J. Inf. Technol. Constr. **22**, 247–265 (2017)
43. Guo, H., Yu, Y., Skitmore, M.: Visualization technology-based construction safety management: a review. Autom. Constr. **73**, 135–144 (2017)
44. Bonomi, F., Milito, R., Zhu, J., Addepalli, S.: Fog computing and its role in the Internet of Things. In: MCC Proceedings, pp. 13–15 (2012)
45. Nærland, K., Müller-Bloch, C., Beck, R., Palmund, S.: Blockchain to rule the waves - nascent design principles for reducing risk and uncertainty in decentralized environments. In: ICIS 2017 Proceedings, p. 12 (2017)
46. Chamoso, P., De La Prieta, F., Eibenstein, A., Santos-Santos, D., Tizio, A., Vittorini, P.: A device supporting the self management of tinnitus. In: Rojas, I., Ortuño, F. (eds.) IWBBIO 2017. LNCS, vol. 10209, pp. 399–410. Springer, Cham (2017). https://doi.org/10.1007/978-3-319-56154-7_36
47. Tapia, D.I., Fraile, J.A., Rodríguez, S., Alonso, R.S., Corchado, J.M.: Integrating hardware agents into an enhanced multi-agent architecture for Ambient Intelligence systems. Inf. Sci. **222**, 47–65 (2013)

Author Index

Printed in the United States
By Bookmasters